APPLICATIONS OF NANO MATERIALS: ELECTRONICS, ENERGY AND ENVIRONMENT

APPLICATIONS OF NANO MATERIALS: ELECTRONICS, ENERGY AND ENVIRONMENT

Editors

V. Rajendran
R. Yuvakkumar
K. Thiyagarajah
K.E. Geckeler

BLOOMSBURY

London • New Delhi • New York • Sydney

First Published, 2013

BLOOMSBURY PUBLISHING INDIA PVT. LTD.
London New Delhi New York Sydney

ISBN: 978-93-82563-35-8

Published by Bloomsbury Publishing India Pvt. Ltd.
VISHRUT Building, DDA Complex, Building No. 3, Pocket C-6 & 7
Vasant Kunj, New Delhi 110 070

Laser Typeset by SHINU SIMON
A-5/56, Moti Nagar, New Delhi 110 015

Printed at Anvi Composers
1 DDA Market, Block A-1B, Pachim Vihar, New Delhi 110 063

प्रो. समीर के. ब्रह्मचारी
महानिदेशक, वै.औ.अ.प.
एवं सचिव, भारत सरकार
वैज्ञानिक और औद्योगिक अनुसंधान विभाग

Prof. Samir K. Brahmachari
Director General, CSIR
& Secretary, Government of India
Department of Scientific & Industrial Research

वैज्ञानिक तथा औद्योगिक अनुसंधान परिषद्
अनुसंधान भवन, 2, रफी मार्ग, नई दिल्ली-110001
COUNCIL OF SCIENTIFIC & INDUSTRIAL RESEARCH
Anusandhan Bhawan, 2, Rafi Marg, New Delhi-110001

Foreword

The past few decades have seen unprecedented advancements in the field of nanotechnology with uplifted developments in wide area of Material Science. To share the research activities worldwide, Centre for Nano Science and Technology (CNST), K. S. Rangasamy College of Technology, KSR Group of Institutions, Tamil Nadu, India is organising an International Symposium on Macro- and Supramolecular Architectures and Materials (MAM-12) with a special topic on Nano systems and Applications during November 21-25, 2012.

CNST has come up with all essential infrastructures. CNST is developing highly talented human resource to meet future challenges in the area of Nanotechnology. Dr. V. Rajendran who led this centre ahead with several funded research projects from DST-Nanomission, DRDO, BRNS and IGCAR etc, should be highlighted.

In this conference, 4 plenary talks including two noble laureates, 36 invited speakers and more than 350 delegates are interacting and sharing their research experiences. Conference books will bring the collections of all the latest developments and salient features of the research pursuit in the field of nanoscience and nanotechnology. A total of 70 contributed papers under the book entitled Applications of Nanomaterials: Electronics, Energy, & Environment will be beneficial to the readers and pave a way to come out with new ideas in developing green environment.

I take this opportunity to thank all the contributors as well as all our well wishers without whose support in bringing out these proceedings of MAM-12 would never turn to reality. I thank the Chairman Dr. K. S. Rangasamy, Principal Dr. K. Thyagarajah and Director & Organising chair Dr. V. Rajendran and his team to organise this mega event at right time in India.

I would like to wish the organising team to have good deliberations and more deliverables on the areas of interest.

New Delhi
October 18, 2012

[Samir K Bramachari]

Telephone Office : 91-11-23710472, 23717053, Fax : 91-11-23710618; Residence : 91-11-26842276
Email : dgcsir@csir.res.in & dg@csir.res.in

EDITORS

Dr. V. Rajendran FUSI, FASI, FInstP, is Director, Research & Development, K S R Group of Institutions and Centre for Nano Science and Technology, K.S Rangasamy College of Technology, Tamil Nadu, India. Under his able guidance, 11 scholars have completed and 14 scholars are pursuing their Ph.D. degrees. He has published more than 150 research papers in reputed international and national journals, 50 papers in conference proceedings, 28 refereed books, 4 R&D books and 11 patents. He has won many awards including the Prof. K. Arumugam National Award in 2011, Best faculty award in 2010, Raman-Chandra Sekhar silver medal (2010), Tamil Nadu Scientist Award, NDT Man of Year 2004, DAAD from Germany (2002), INSA, TNSCST Young Scientist, DAE/BRNS Visiting Scientist, Best paper award from MRSI, ASI, ASA and USI and Outstanding Organiser Award for the 7th National Symposium on Ultrasonics, 1996.

Dr. R. Yuvakkumar is an Assistant Professor, Centre for Nano Science and Technology (CNST), K.S. Rangasamy College of Technology. He completed his Ph.D. in April 2012 from Anna University. He has carried out extensive research work on synthesis, characterisation and application of silica, iron and molybdenum nanopowders. He has published ten research papers in peer reviewed international journals, one Indian patent and one paper in international conference proceedings and presented his research work in four international conferences.

Dr. K. Thyagarajah, an erudite academician, an able administrator and outstanding scholar has the distinction of receiving his doctoral degree in power electronics & AC motor drives from the world renowned Indian Institute of Science, Bangalore in 1993. He has 31 years of rich teaching, research and industrial experience. He has produced 6 Ph.Ds so far, and is currently guiding 6 Ph.D scholars. His areas of interest include higher performance motor drives, insulating materials and industrial automation. He has authored 24 research papers in international and national journals and attended many conferences. He is a member of the Board of Examinations of various universities in Tamil Nadu and neighboring states. Currently, he is the Principal of K.S. Rangasamy College of Technology, Tiruchengode.

Dr. K.E. Geckeler is the Founding Chair of the Department of Nanobio Materials and Electronics, World Class University (WCU), Gwangju, South Korea. He is also a Professor in the Department of Materials Science and Engineering and the Department of Medical System Engineering at the Gwangju Institute of Science and Technology (GIST), South Korea and serves as Vice Director of the Gruenberg Center for Magnetic Nanomaterials (GCMN). He received his Ph.D. and M.D. degrees from the University of Tuebingen, Germany (both degrees: "magna cum laude") and spent sabbatical leaves at Harvard University, University of Montana, Clemson University (USA) and at the University of Montpellier (France). He received a series of awards such as the "Fonds der Chemischen Industrie", the "Fritz-Ter-Meer Award", the "Science Price of the President of Korea" and the "Outstanding Research Award of GIST". He has to his credit initiation and organizing the biannual MAM Symposia series. He has published more than 300 research articles and short communications, 12 book chapters, over 10 books, and over 80 patents.

PREFACE

The International Symposium on Macro- and Supramolecular architectures and materials (MAM-12) with a special topic on Nano systems and Applications is organised by Centre for Nano Science and Technology (CNST) of K. S. Rangasamy College of Technology. This conference is jointly organised with World Class University, Gwangju Institute of Science and Technology (GIST), South Korea.

More than 525 abstracts have been received from India and abroad. Out of 400 full papers received, 260 have been selected and peer reviewed by the expert committee for the publications in conference books. All the received books are classified under four headings namely Nano Material Synthesis and Characterisation, Advanced Nanomaterial for industrial applications, Nano biomaterials and Applications of Nanomaterials: Electronics, Energy, & Environment.

A total of 260 accepted full papers, 60 of contributed papers for the inclusion in the book entitled Applications of Nano Materials: Electronics, Energy, & Environment after peer review. This book has the collection of contributed papers containing the role of nano materials in environmental applications like detecting toxic effects, solar cells and sensor applications. The most effort and care have been taken in getting the review of all the contributed papers. All the contributed authors are extended by our sincere thanks for their timely submission and cooperation in carrying out suggestions by the referees.

The CNST has well established the start–of–art of experimental facilities and looking for industrial collaboration and partners for the technology transfer in thrust areas of nanotechnology. We are sure that the outcome of MAM-12 open up new avenue between the researchers and industrialists for the development of nano products.

We owe our special thanks to members of technical committee for peer review of contributed papers. The various government funding agencies, private organizations and industries are thankful for their munificent support and sponsor for the successful conduct the conference. The various committee chairs and members are highly acknowledged for making this event a grant success.

The support extended by Mr. Suresh Gopal and Mrs. Jyoti Mehrotra of Bloomsbury Publishing India Pvt. Ltd in bringing out this book on time is highly appreciated.

V. Rajendran
R. Yuvakkumar
K. Thyagarajah
Kurt E. Geckeler

CONTENTS

INVESTIGATION ON THE DEVELOPMENT OF SILICENE BASED NANO THIN FILM DEVICES

V. GUNASEKARAN

Department of Nanoscience and Technology, School of Nanoscience and Nanotechnology, Karunya University, Coimbatore, Tamil Nadu
E-mail: gunasekaran@karunya.edu

ABSTRACT

Silicene, a two dimensional nano material similar to graphene, is an excellent material that has been attracted world research community. This single layer of atoms can form monolayer in which Si atoms tend to form honeycomb structures like hexagonal on silver substrates. Also it has the electronic properties like graphene. In this paper, we would report the synthesis of silicene, a monolayer of silicon, by chemical method. We have also characterized the synthesized material by using UV-Vis spectroscopy, Photoluminescence, SEM, and TEM. Also we patterned nano thin film silicene devices for electronic application like field effect transistors (FET).

INTRODUCTION

Silicene, an atomic layer thick layer of silicon, nowadays, has been attracted world scientific community [1–4] as it has similar electronic characteristics of graphene. Silicene, which was first reported by European researchers, is predicted to share some of the impressive properties of its carbon cousin, graphene. Importantly, however, silicene could be integrated with current silicon-based electronic devices, whereas graphene is incompatible. Silicene may turn out to be a better bet than graphene for smaller and cheaper electronic devices because it can be integrated more easily into silicon chip production lines. For the first time, silicon has been turned into a sheet just one atom thick. Silicene is the thinnest possible form of silicon, which is the unbeatable, dominant electronic material used in many devices around us. Silicene is thought to have similar electronic properties to graphene but ought to be more compatible with silicon-based electronic devices. Patrick Vogt et al reported the silicene can be produced by condensing silicon vapour onto a silver plate to form a single layer of atoms. They then measured the optical, chemical and electronic properties of the layer, showing it closely matched those predicted by theory.[5] An important challenge now will be to grow silicene on insulating substrates to learn more about its electrical properties and understand how they can be exploited to build future electronic devices. Ultimate miniaturisation of such devices should be achieved if silicene is made available for applications.

EXPERIMENTAL

The recent studies revealed that the synthesized silicene with a buckled, bumpy appearance, with silicon atoms at three different heights. With this buckled structure, researchers observed a

V. Rajendran, R. Yuvakkumar, K. Thyagarajah and K.E. Geckeler (eds.)
Applications of Nano Materials: Electronics, Energy and Environment, pp. 1–2 (2012).

bandgap-a no-go zone for electrons. Experimental studies are revealed that silicene can be grown on Ag substrate. Many methods including chemical methods are being under investigation to get the single layer of silicene. The detailed reports on this excellent material will be revealed elsewhere.

RESULTS AND DISCUSSION

Silicene was reported to grow on Ag (111) surface [6] Silicene grown on Ag (111) was demonstrated to be a Dirac-fermionic system, similar as grapheme.[7] Recently three groups have independently reported ordered phases on the same surface [8] Silicene has also been reported to grow on a ZrB_2 substrate.[9] Our experimental results will be reported elsewhere.

CONCLUSION

Silicene was reported to grow on Ag (111) surface [6] Silicene grown on Ag (111) was demonstrated to be a Dirac-fermionic system, similar as graphene [7]. The detailed results on our experimental results will be reported elsewhere. This could be integrated with current silicon-based electronic devices.

ACKNOWLEDGEMENTS

The author would like to extend his thanks to the Management, Vice-Chancellor, Registrar and Deans of Academic Affairs and Academic Research of Karunya University, Coimbatore, for providing research funding through Seed Money Research Grant (KSTG seed money grant No. KU/Reg/D(R)/OO/12/29 dated 21.01.2012) and DST Nanomission, New Delhi.

REFERENCES

[1] Guzman-Verri, G.G. and Lew Yan Voon, L.C., Electronic structure of silicon-based nanostructures, *Physical Review B* 76(7), 075131, 2007.

[2] Aufray, B., Kara, A., Vizzini, S., Oughaddou, H., Léandri, C., Ealet, B. and Le Lay, G., Graphene-like silicon nanoribbons on Ag(110): A possible formation of silicone, *Applied Physics Letters* 96, 183102, 2010.

[3] Jose, Deepthi and Datta, Ayan, Structures and Electronic Properties of Silicene clusters: A promising material for FET and hydrogen storage. *Phys. Chem. Chem. Phys.* 13: 7304, 2011.

[4] Research highlight (2012). "Graphene's silicon cousin". *Nature* 485: 9. DOI:10.1038/485009a

[5] Patrick Vogt *et al.*, Physical Review Letters, DOI: 10.1103/*PhysRevLett.*108.155501).

[6] Lalmi, B., Oughaddou, H., Enriquez, H., Kara, A., Vizzini, S., Ealet, B. and Aufray, B. Epitaxial growth of a silicene sheet. *Applied Physics Letters* 9, 223109, 2010.

[7] Patrick Vogt, Paola De Padova, Claudio Quaresima, Jose Avila, Emmanouil Frantzeskakis, Maria Carmen Asensio, Andrea Resta, Benedicte Ealet and Guy Le Lay., Silicene: Compelling experimental evidence for graphenelike two-dimensional silicon. *Physical Review Letters* 108: 155501, 2012.

[8] Chun-Liang Lin, Ryuichi Arafune, Kazuaki Kawahara, Noriyuki Tsukahara, Emi Minamitani, Yousoo Kim, Noriaki Takagi and Maki Kawai., Structure of silicene grown on Ag(111), *Appl. Phys. Expr.:* 045802., 2012.

[9] Antoine Fleurence, Rainer Friedlein, Taisuke Ozaki, Hiroyuki Kawai, Ying Wang and Yukiko Yamada-Takamura, Experimental evidence for epitaxial silicene on diboride thin films". *Physical Review Letters,* 108: 245501, 2012.

HIGHLY EFFICIENT, REUSABLE Z-SCHEME PHOTOCATALYST CdO–Ag–ZnO FOR THE DEGRADATION OF AZO DYES UNDER SUN LIGHT

SUBRAMANIAN BALACHANDRAN AND MEENAKSHISUNDARAM SWAMINATHAN

Photocatalysis Laboratory, Department of Chemistry, Annamalai University, Annamalainagar
E-mail: chemres50@gmail.com

ABSTRACT

In this article, we report the fabrication of a potential new class of Z-scheme photocatalyst, CdO–Ag–ZnO, by a simple co-precipitation method. This catalyst has been characterized by surface analytical techniques. HR-SEM and TEM images of the catalyst show ZnO particles have 3D pentagonal or hexagonal plate-like structure. This three-component nano junction system exhibited enhanced photocatalytic activity for the degradation of Acid Black1 (AB 1) and Acid Violet 7 (AV 7) under sun light far exceeding those of the single and two component systems. CdO–Ag–ZnO was found to be stable and reusable without appreciable loss of catalytic activity up to five runs. Z-scheme mechanism is proposed for dye degradation.

INTRODUCTION

Photocatalysis is a promising technology for the treatment of toxic contaminants, especially for the removal of organic compounds with solar energy.[1] ZnO is a good semiconductor photocatalyst due its unique optoelectronic properties, environmental stability and low cost.[2,3] CdO with band gap of 2.2eV is a promising photocatalyst for the optoelectronic applications.[4]

Recent research is focused on the construction of a two-step photoexcitation system, often called a Z-scheme photocatalysis that mimics natural photosynthesis. In this paper, we report the preparation of Z-scheme CdO–Ag–ZnO photocatalyst by facile co-precipitation process for the first time and its photocatalytic activity in the degradation of two azodyes Acid black 1 and Acid violet 7 under natural sun light.

V. Rajendran, R. Yuvakkumar, K. Thyagarajah and K.E. Geckeler (eds.)
Applications of Nano Materials: Electronics, Energy and Environment, pp. 3–6 (2012).

EXPERIMENTAL

Preparation of CdO–Ag–ZnO

11.90 g of $Zn(NO_3)_2.6H_2O$ (0.4 M) and 0.128 g of $AgNO_3$, 0.193g of Cd $(CH_3COO)_2.2H_2O$ were dissolved in 100 ml deionized distilled water under stirring to form a clear solution. Then the solution of oxalic acid in distilled water (0.6 M) was introduced into the above solution. The mixed precipitate of Zinc oxalate and cadmium oxalate with Ag, obtained was filtered, washed with distilled water and ethanol, and finally dried in air at 90°C for 12 h. The mixed precipitate was calcined at 450°C for 12 h in a muffle furnace to obtain CdO–Ag–ZnO. Characterization and solar photocatalysis experiments were carried out as described in our earlier paper[5].

RESULTS AND DISCUSSION

Catalyst Characterization

A comparison of X-ray diffractograms of prepared ZnO and CdO–Ag–ZnO shows the characteristic peaks for metallic silver(2θ = 38.2°) and CdO(2θ = 33.8°, 54.7°, 65.7° and 68.1°) in the catalyst. HR-SEM images of CdO–Ag–ZnO at three different magnifications are shown in Figure 1.

Fig. 1

ZnO particles have 3D-pentagonal or hexagonal plate-like structure. Cadmium oxide and silver clusters are formed on the clear smooth surface of ZnO and show aggregations. From TEM images, it is observed that CdO–Ag–ZnO particles posses predominantly three dimensional pentagonal and hexagonal structures. The selected area electron distribution pattern recorded from the silver shows concentric ring with intermittent bright dots, indicating that the samples are highly crystalline in nature. X-ray Photoelectron spectra show the binding energy peaks for the presence of Ag metal, Zn^{2+} (ZnO) and Cd^{2+} (CdO) in the catalyst. DRS of CdO–Ag–ZnO indicates an increased absorption both in UV and visible region of light. This reveals that CdO–Ag–ZnO can be used as an UV and Visible light active semiconductor photocatalytic material. Reduction of photoluminescence intensity at 416 nm by CdO–Ag–ZnO when compared to prepared ZnO indicates the suppression of recombination of the photogenerated electron-hole pair. The N_2 adsorption/desorption isotherms of the synthesized CdO–Ag–ZnO exhibited a hysteresis loop, typical of type II pattern representing the predominant nonporous structure according to the classification of IUPAC.[6] Average pore radius of CdO–Ag–ZnO, shown by pore size distribution curve is 250Å. The pore size distribution of the CdO–Ag–ZnO sample thus confirms the nonporous structure.

Photocatalytic Activity

Experiments were carried out under the same conditions to test the photodegradability of Acid black 1 (AB 1) by ZnO, CdO–ZnO, Ag–ZnO, and CdO–Ag–ZnO catalysts and the results are presented in Figure 2.

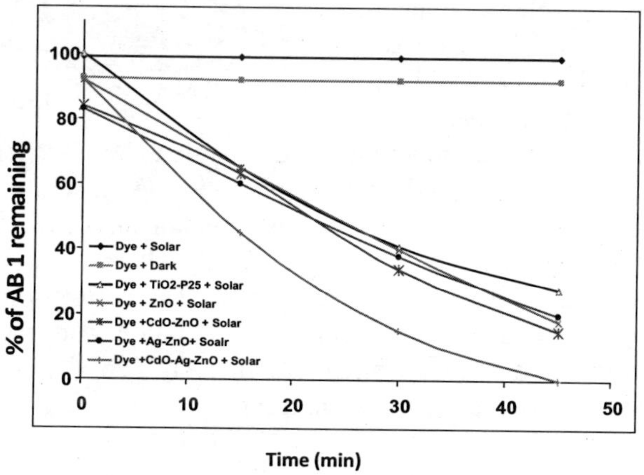

Fig. 2

AB 1 undergoes almost complete degradation in the presence of CdO–Ag–ZnO and natural sunlight light in 45 min. But, prepared ZnO, Cd–ZnO, Ag–ZnO, and TiO_2–P25 produced 82, 85, 80 and 72% degradations, respectively in 45 min. This shows that CdO–Ag–ZnO is more efficient in AB1 degradation than other photocatalysts. Similar trend was observed in AV7 degradation with these photocatalysts. UV spectral changes of AB 1 and AV 7 at different irradiation times with CdO–Ag–ZnO catalyst shows a gradual decrease in intensity with out the appearance of new absorption peaks. This reveals that the intermediates formed during degradation do not absorb at analytical wavelength. A study on the influence of solution pH reveals that the degradation rate increases with increase in pH up to 11 and then it decreases. The optimum pH is found to be 11 for both AB 1 and AV 7 dyes. Experiments on the effect of catalyst amount shows that 3 g/L for AB 1 and 2 g/L for AV 7 are found to be optimum for 5×10^{-4} M dye degradation. The reusability of CdO–Ag–ZnO was tested in the degradation of both dyes under optimum conditions for five runs. CdO–Ag–ZnO exhibited 95% efficiency even at fifth run. It was found that the crystal structure of CdO–Ag–ZnO photocatalyst after 5 cycles did not change during the reaction, indicating the stability of photocatalyst. Hence the catalyst can be reused for continuous treatment of wastewater.

Mechanism OF DEGRDATION

Based on the band energy levels of CdO and ZnO, a mechanism for the photodegradation using charge separation and photocatalytic reaction of CdO–Ag–ZnO photocatalyst is proposed in Figure 3.

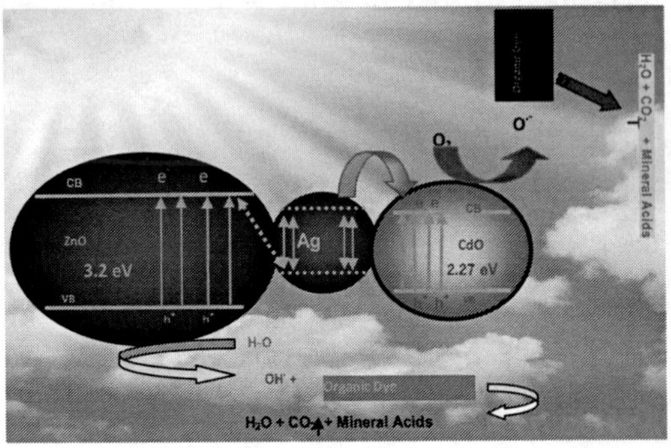

Fig. 3

Photogenerated electrons in the conduction band of ZnO easily flow through Silver metal into the conduction band of CdO. Simultaneously, the photo-generated holes in the valence band of CdO also easily flow into valence bond of ZnO. Holes remained on ZnO valence band and electrons concentrated on CdO conduction band, enabled the Z-scheme three component system with strong oxidation and reduction capabilities to generate hydroxyl (·OH) and superoxide radicals for the dye degradation. In this mechanism, a two step excitation of CdO and ZnO with Ag as a mediator, which is similar to Z-Scheme in photosynthesis of green plants might have occurred as proposed by Tada *et al.*[7] The realization of a biomimectic Z-scheme charge transfer mechanism permits higher degradation of dyes than can be achieved by a typical single and two component systems (ZnO, TiO_2–P25, CdO–ZnO and Ag–ZnO) in dye degradation under irradiation by natural sun light.

CONCLUSION

Characterization study revealed the morphology of the catalyst. A mechanism involving a two step excitation of CdO and ZnO with Ag as mediator, similar to Z-Scheme in photosynthesis is proposed for its enhanced photocatalytic activity in degradation of Acid Black and Acid Violet 7 under natural sunlight. Stability and reusability of CdO–Ag–ZnO makes the catalyst very useful as industrial green catalyst for effective treatment of dye effluents under natural sunlight.

REFERENCES

[1] Hoffman, M.R., Martin, S.T. and Choi, W., Bahnemann, D.W. *Chem.Rev.* 1995, 95, 69.
[2] Lizama, C., Freer, J., Baeza, J. and Mansilla, H.D., *Catal. Today* 2002, 76, 235.
[3] Akyol, A. Yatmaz, H.C. and Bayramoglu, M. *Appl. Phys. B* 2004, *54*, 19.
[4] Wu, Y.M., Xing, M.Y., Zhang, J.L. and Chen, F., *Appl. Catal. B: Environ.,* 2010, 97, 182.
[5] Velmurugan, R. and Swaminathan, M. *Sol. Energy Mater. Sol. Cells* 2011, 95, 942.
[6] Yu, J.G., Zhang, L.J., Cheng, B., Su, Y.R., *J. Phys. Chem. C* 2007, *111*, 10582.
[7] Tada, H., Mitsui, T., Kiyonaga, T., Akita, T. and Tanaka, K. *Nature Materials* 2006, 5, 782.

DROPLET MANIPULATION USING ZnO/SAPPHIRE BASED SAW DEVICES

S. KANDASAMY,[1,2] D.K. MAURYA,[3] R.T. TJEUNG,[4]
J.R. FRIEND[1,5] AND K. ALAMEH[3]

[1]*Melbourne Centre for Nanofabrication, 151 Wellington Rd, Clayton, VIC 3168, Australia*
[2]*School of Physics, University of Melbourne, Parkville, VIC 3010, Australia*
[3]*Electron Science Research Institute, Edith Cowan University, Joondalup, WA 6027, Australia*
[4]*Dept. of Mechanical and Aerospace Engineering, Monash University, Clayton, VIC 3168, Australia*
[5]*School of Electrical and Computer Eng., RMIT University, Melbourne, VIC 3004, Australia*
E-mail: d.maurya@ecu.edu.au

ABSTRACT

The hybrid ZnO/Sapphire material is selected to exploit the collective benefits of Zinc Oxide (ZnO) and sapphire for the development of a high performance Surface Acoustic Wave (SAW) device. The microfluidic applications of ZnO/Sapphire layered surface acoustic wave devices are discussed. In addition, acoustic streaming in droplets is investigated using a custom-made ZnO/Sapphire layered device. Experimental results show that a maximum acoustic streaming of ~600 μm/s is achieved when the device is operated at 158 MHz.

INTRODUCTION

There has recently been an increasing interest in developing high frequency surface acoustic wave (SAW) by depositing a piezoelectric thin film on a high acoustic velocity substrate.[1,2] SAW technology offers simpler and more compact devices without involving moving parts. The interaction of fluid with acoustic energy arising from SAW can be exploited for fluid manipulation at micro to nano scale.

ZnO is a multifunctional material for several applications because of its structural, optical and electrical properties.[3,4] Piezoelectric ZnO materials possess high electromechanical

V. Rajendran, R. Yuvakkumar, K. Thyagarajah and K.E. Geckeler (eds.)
Applications of Nano Materials: Electronics, Energy and Environment, pp. 7–10 (2012).

coupling coefficients, which make them attractive for realizing thin films SAW devices. In contrast, sapphire (Al_2O_3) substrates are widely used for their high acoustic velocity ((~10,000 m/s) and relatively low loss.[5] The high acoustic velocity of sapphire eliminates the need for sub-micron IDT patterning. The potential approach of developing a hybrid ZnO/Sapphire [6–8] SAW device is desirable as it enables the fabrication of microfluidic devices for various applications, such as biomedical and environmental sensing.

Various conventional deposition techniques, such as sol-gel, pulsed laser deposition (PLD), molecular beam epitaxy (MBE), metal organic chemical vapour deposition (MOCVD) and sputtering have been used to prepare ZnO thin films. Each deposition method has advantages, depending on the specific application that it is used for. Sputtering, however, is currently the most commonly used deposition technique for the preparation of polycrystalline ZnO, which is the most favored material for acoustic applications. Sputtering offers several advantages, such as deposition at low substrate temperatures and high sputtering rates, the fabrication of thin films with a uniform surface and excellent crystallinity. In this paper, we investigate acoustic streaming in droplets placed on a ZnO/Sapphire layered device. Experimental results show that a maximum acoustic streaming of ~600 μm/s can be attained when the device is driven by a 158 MHz sine wave.

EXPERIMENTAL

In this study, a ZnO thin film was deposited on a sapphire substrate using a radio-frequency magnetron sputtering system. A 99.99% pure, 3-inch diameter ZnO target was used to sputter ZnO onto a sapphire substrate. The sputtering power was 150W and the deposition process parameters are listed in Table 1. Post-annealing treatments were sequentially carried out at 500°C for 5 hrs in ambient air to eliminate the intrinsic stress induced by sputtering and also to provide sufficient energy to ensure that ZnO molecules are crystallized. The ZnO film thickness was measured using a surface profilometer, and X-ray diffraction was used to analyze the crystallographic structure of the ZnO film.

After the required film quality has been evaluated, SAW devices were fabricated on the ZnO thin films. Inter Digital Transducer (IDT) patterns were transferred on to the ZnO surface by conventional lift-off process. The IDT consists of an adhesion layer of 10 nm Cr and 200 nm of Au, which were thermally evaporated. The IDT was designed to have 60 numbers of finger-pair with 8 μm dimension for both the width and the spacing, which corresponds to a wavelength of 32 μm.

Table 1: ZnO Deposition Parameters

Target	ZnO
Process gas	60% Ar + 40% O_2
Substrate temperature	150°C
RF power	150 W
Process pressure	10 mTorr

RESULTS AND DISCUSSION

The XRD data plotted in Figure 1 show the high 2θ intensity at 34.4°, which represents a (002) diffraction peak, indicating that the ZnO deposited on sapphire is c-axis oriented. The deposited ZnO film was required to be c-axis oriented, due to its high electromechanical coupling coefficient. Several SAW devices were made with Cr/Au fingers for $\lambda = 32$ µm. All ZnO films had a thickness (h) of 1 µm ($h/\lambda = 0.03$).

The measured centre frequency and the calculated phase velocity were 161 MHz and 5152 m/s, respectively. The fabricated SAW devices were utilized to induce acoustic streaming in a sessile droplet. Figure 2 shows the successive images of the SAW streaming inside a 1µL water droplet seeded with 4.8 µm fluorescent particles. The IDT is located to the right of the droplet. The particles are concentrated in the two vortices within 4 seconds of IDT activations. The streaming speed with respect to increasing input power is shown in Figure 3.

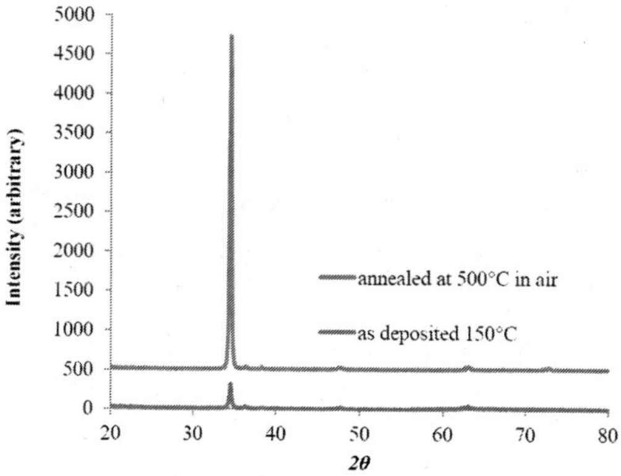

Fig. 1: XRD Spectra of Sputtered ZnO Film on Sapphire Substrate

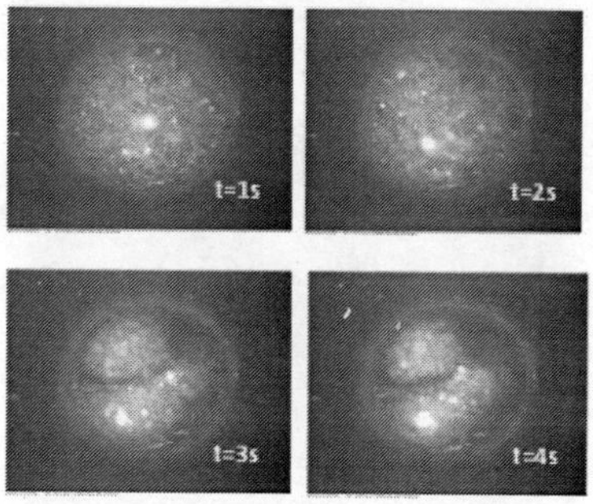

Fig. 2: Successive images of the SAW streaming inside a 1 µL water droplet seeded with 4.8 µm fluorescent particles. The IDT is located to the right of the droplet. The particles are concentrated in the two vortices within 4 seconds of IDT activations.

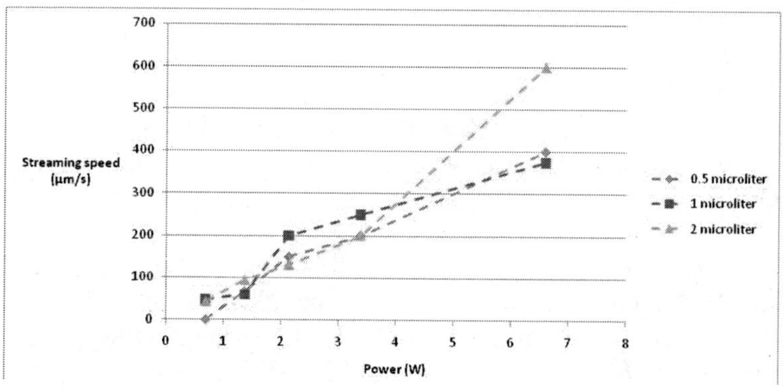

Fig. 3: Acoustic Streaming Speed of the Water Droplet Versus Input Power (RF frequency of 158 MHz)

Three different volumes of water droplets were used in the experiments, and a maximum streaming speed of 600 μm/s was observed.

Acoustic streaming was a steady fluid motion that resulted from the propagation of acoustic waves through a dissipative fluid medium. As the acoustic waves attenuated in the fluid medium, it generated a gradient in the momentum flux. This induced a Reynolds stress, causing the steady fluid velocity.

CONCLUSION

Acoustic streaming in micro droplets has been investigated using a ZnO/Sapphire based layered device. Acoustic streaming has been observed in the form of two vortices. Results have shown that micro particles can concentrate in the form of vortices within 4 seconds of IDTs activations. A maximum acoustic streaming of ~600 μm/s has been measured when the device was operated at 158 MHz. These results have application in biomedical and environmental sensing.

ACKNOWLEDGEMENT

This work was supported by DPMC, ARC, NHMRC, Edith Cowan University and the Department of Industry, Innovation, Science, Research and Tertiary Education, Australia.

REFERENCES

[1] Kim, Y.-J. and Kim, K.-W., Characteristics of Epitaxial ZnO Films on Sapphire Substrates Deposited using RF-Magnetron Sputtering, Jpn. J. Appl. Phy., vol. 36, pp. 2277–2280, 1997.

[2] Kadota, M. and Minakata, M., Piezoelectric Properties of ZnO Films on a Sapphire Substrate Deposited by an RF-Magnetron-Mode ECR Sputtering System, Jpn. J. Appl. Phy., vol. 37, pp. 2923–2926, 1998.

[3] Winey, J. M., Gupta Y. M. and Hare, D.E., r-axis sound speed and elastic properties of sapphire single crystals. J. App. Phy., Vol. 90, p. 3109, 2001.

[4] Emanetoglu, N.W., Gorla, C., Liu, Y., Liang, S. and Lu, Y., Epitaxial ZnO piezoelectric thin Film for SAW Filters, Material Science in Semiconductor Processing, vol. 2, pp. 247–252, 1999.

[5] Emanetoglu, N.W., Patounakis, G., Liang, S., Gorla, C., Wittstruck, R. and Lu, Y., Analysis of SAW Properties of Epitaxial ZnO Films Grown on R-Al2O3 Substrates, IEEE Transactions on Ultrasonics, Ferroelectrics, and Frequency Control, Vol. 48, pp. 1389–1394, 2001.

[6] Emanetoglu, N.W., Liang, S., Gorla, C. and Lu, Y., Epitaxial Growth and Characterization of High Quality ZnO Film for Surface Acoustic Wave Applications, IEEE Ultrasonics Symposium, 1997.

[7] Mitsuyu, T., Ono, S. and Wasa, K., Structures and SAW properties of rf sputtered single-crystal films of ZnO on sapphire, Journal of Applied Physics., Vol. 51, pp. 2464–2470, 1980.

[8] Wu, T-T. and Wang, W.-S., An Experimental Study on the ZnO/Sapphire layered surface acoustic wave device, *J. App. Phy.*, vol. 96, pp. 5249–5253, 2004.

MAGNETO-THERMAL BEHAVIOUR OF Ni-Mn-Ga SINGLE CRYSTAL

B. Rekha[1,2], S. Seenithurai[1], S. Vinodh Kumar[1], C. Mahalakshmi[1,3], A. Sajitha Banu[1,4] and M. Mahendran[1]

[1]Smart Materials Lab, Department of Physics Thiagarajar College of Engineering, Madurai
[1,2]Department of Physics, SACS MAVMM Engineering College, Madurai
[1,3]Department of Physics, V.V. VanniaPerumal College for Women, Virudhunagar
[1,4]Department of Physics, Sri GVG Visalakashi College for Women, Udumalpet
E-mail: bsrekha@ymail.com

ABSTRACT

The martensitic transformation and magnetic properties of Ni-Mn-Ga single crystals are investigated using Differential Scanning calorimetry and vibrating sample magnetometer. A structural transition from the cubic austenite to tetragonal martensite is observed. The measurement confirms that the Ni-Mn-Ga crystal attains the low temperature martensitic transformation. The heating curve shows a rapid decrease in the magnetization in the temperature range between 80 and 100°C, indicating the Ferro-Para magnetic transition at the Curie temperature T_C = 98°C.

INTRODUCTION

Ferromagnetic shape memory alloys (FMSAs) are a new class of smart materials which exhibit high magnetic field induced strains (MFIS) with faster response as compared to conventional shape memory alloys (SMAs) due to the rearrangement of martensite variants under a magnetic field.[1] The Ni-Mn-Ga Heusler alloys have been extensively studied as FMSA in recent years since the single crystals of these alloys shows MFIS up to 6–10% making them a potential material system for the device applications such as actuators and sensors.[2] Ni-Mn-Ga alloys exhibit

V. Rajendran, R. Yuvakkumar, K. Thyagarajah and K.E. Geckeler (eds.)
Applications of Nano Materials: Electronics, Energy and Environment, pp. 11–14 (2012).

thermo-elastic martensite transformation which involves transformation of ordered cubic $L2_1$ austenite phase to a low-symmetry non-modulated tetragonal or modulated (5M, 7M) martensite phase on cooling.[3] Based on TM and TC, Ni-Mn-Ga alloys are classified into three different groups (I) TM << TC, (II) TM ~ TC, and (III) TM > TC.[4] The alloys belonging to the first two groups exhibit shape memory effect (SME), whereas the third group, which shows high thermal stability of the thermo elastic martensite is one of the potential candidate for the high temperature shape memory alloys [5–7] . The TM for the alloys belonging to the groups I and II is found to be influenced by the quenching temperature, suggesting that the degree of $L2_1$ long range order of the austenite phase also effects the martensite transformation temperature.[8,9] The atomic order of the austenite phase is inherited by the martensite phase due to diffusion less characteristic of the martensite transformation. In the present work, we have studied the magnetic and thermal properties of the Ni-Mn-Ga alloy to confirm the martensite to austenite transformation.

EXPERIMENT

The Ni–Mn–Ga single crystals were annealed at different temperatures.[10] The transformation temperatures were measured in a Differential Scanning Calorimeter (PERKIN-ELMER) with heating and cooling rates of 5 K/min. The Vibrating Sample Magnetometer (VSM-883 A) has been carried out to study the temperature dependence of magnetic properties.

RESULTS AND DISCUSSION

The results DSC measurements were depicted in Figure 1 showing the phase transformation temperatures of Ni-Mn- Ga and corresponding peak temperatures are marked by arrows. The transformation temperatures are $M_S = 38°C$, $M_f = 32°C$, $A_s = 42°C$ and $A_f = 45°C$ respectively. The martensitic transformation temperature is calculated using $T_M = (T_{Af} + T_{Ms})/2$. In thermodynamic calculations, TM refers to the temperature at which the Gibbs energies of the martensitic and parent phases are equal.[11] The curves show the forward (cooling) and reverse (heating) martensitic transformations. It can be seen that exothermic and endothermic peaks in the cooling and heating curves, respectively, representing the appearance of phase transformation. Figure 1 reveals that, small kink observed near 98°C indicates the curie transition (TC) and the martensitic transformation (TM) at 41°C which is above room temperature.[12] In this alloy the TM << TC, confirming the SME. This clearly exhibits the actuation mechanism of single crystals for potential applications at room temperature. The sharp transformation peaks reveals the tetragonal structure of the single crystal.

THERMO-MAGNETIC PROPERTIES

Temperatures versus magnetization curves of the Ni-Mn- Ga were described in Figure 2. The measurements are carried out at a magnetic field of about 200 Oe and the temperature range is between 10 and 100°C. In the presence of applied field, the measurements were also made while sample was allowed to cool. Heating curve in Figure 2 shows a rapid increase in magnetization near 51°C, followed by a gradual decrease related to the martensitic transformation.[13] This sudden increase in the magnetization is the first-order transformation over a range of temperature, and is due to the elastic energy accumulation which increases the martensite phase energy.[14,15]

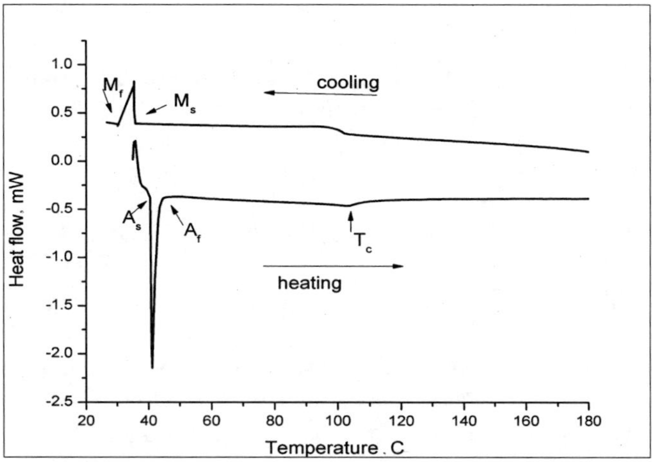

Fig. 1: Variation of Heat Flow with Temperature. The Graph Vividly Shows the Phase Transformations of Ni-Mn-Ga Alloy

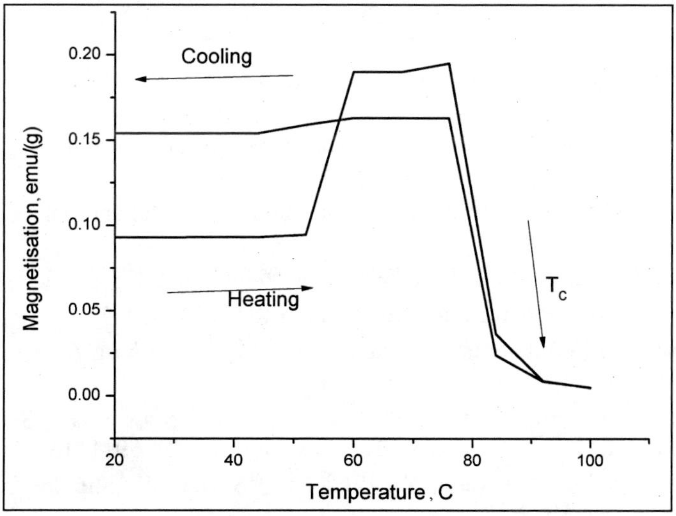

Fig. 2: Variation of Magnetization with Temperature. The Hysteresis Behavior of the Ni-Mn-Ga Alloy is Depicted Here

CONCLUSION

The phase transformation reveals that at low temperatures the martensitic phase is observed. This makes the alloys to be suitable for room temperature sensor and actuator mechanisms. The variation in the martensite transformation temperature attributes to the degree of long range atomic order of the alloy and which in turn affects the magnetic properties of the alloy. This alloy can be a suitable candidate material for actucator applications.

ACKNOWLEDGEMENTS

One of the authors (MM) would like to thank Prof. Dr. Robert C. O' Handley for introducing him to the subject and providing Ni–Mn–Ga single crystals.

REFERENCES

[1] Ullakko, K., Huang, J, K., Kanther, C., O'Handley, R.C. and Kokorin, V,V., *Appl. Phys. Lett.,* 69, 1966, 1996.

[2] Zhua, F.Q., Yanga, F.Y., Chiena, C.L., Ritchieb, L., Xiaob G. and Wuc, G.H., *J. Magn. Magn. Mater.* 288, 79–83, 2005.

[3] Brown, P.J., Crangle, J., Kanomata, T., Matsumoto, M., Neumann, K.U., Ouladdiaf B. and Ziebeck, K.R.A., *J. Phys. condens. Matter,* 14, 10159, 2000.

[4] Chernenko, V.A., Pons, J., Segui C. and Cesari, E., *Acta. Mater*, 50, 53, 2003.

[5] Soderberg, Y. Ge, O., Lanska, N., Sozinov, A., Ullakko, K. and Lindroos, V.K., *J. Phys. IV,* 112, 921, 2003.

[6] Jiang, Y. Ma, C., Xu, Y. Li, H., Wang C. and Liu X., *Acta Mater.,* 55, 1533, 2007.

[7] Pushpanathan, K., Vallal Peruman, K., Seenithurai, S., Kodi Pandyan, R. and Mahendran, M., *Mod. Phy. Lett. B*, 25, 1–13, 2011.

[8] Kreissl, M., Neumann, K.U., Stephens T. and Zieback, K.R.A., *J Phys. Condens. Matter.,* 158, 3831, 2003.

[9] Sanchez-Alarcos, V., Recarte, V., Perez-Landazabal J.I. and Cuello, G.J., *Acta. Mater.,*55, 3883, 2007.

[10] Pushpanathan, K., Chokkalingam, R., Senthurpandi R. and Mahendran, M., *Mater. Manuf. Proc.,* 26, 223–229, 2011.

[11] Straka, L., Heczko, O., Lanska, N., *IEEE Trans. Magn.*, 38, 2835, 2002.

[12] Manickam Mahendran, Smart Mater. Struct. 14, 1403–1409, 2005.

[13] Vallal Peruman, K., Chokkalingam R., and Mahendran, M., *Phase Transitions* 83, 509–517, 2010.

[14] Vasil'ev, A.N., Bozhko, A.D., Dikshtein, I.E., Shavrov, V.G., Buchelnikov, V.D., Matsumoto, M., Suzuki, S., Takagi, T., Tani, J., *Phys. Rev.* B, 59, 1113, 1999.

[15] Wedel, B., Suzuki, M., Murakami, Y., Wedel, C., Suzuki, T., Shindo, D. and Itagaki, K., *J.Alloys Compd.*, 137, 1999.

AN INVESTIGATION ON THE OPTICAL PROPERTIES OF NOVEL $SnWO_4$-REDUCED GRAPHENE OXIDE NANOCOMPOSITES

T. SAKTHIVEL AND V. GUNASEKARAN

Department of Nanosciences and Technology, School of Nanosciences and Nanotechnology, Karunya University, Karunya Nagar, Coimbatore, Tamil Nadu
E-mail: gunasekaran@karunya.edu

ABSTRACT

In this paper first time we report the optical property of tin tungstate reduced Graphene Oxide ($SnWO_4$-rGO) nanocomposite. $SnWO_4$-rGO was prepared by simple microwave method. Further the crystalline phase was analyzed by X-Ray Diffraction (XRD). The composites morphology was study by scanning electron microscopy (SEM). The grain size was analysed through by XRD. The composite structural property was analysed by Raman spectroscopy.

INTRODUCTION

Two dimensional crystals have been great interest in recent years there are graphene,[1] graphene is two dimensional nanomaterials, among this synthesized method chemical synthesis gives one promising application to making composites. After reduction treatment the rGO has some amount of function groups attach with the plane of the sheet. Thus the functional groups are easily attached with metal and metal oxide nanoparticles. GO is an electrical insulator due to the functional group of aromatic rings and partially oxidized GO can act as a semiconductor.[2]

As of our knowledge nobody consider work was reported of graphene-$SnWO_4$ nano composite. In this paper we will discuss about the first time synthesis of $SnWO_4$-rGO nanocomposite.

V. Rajendran, R. Yuvakkumar, K. Thyagarajah and K.E. Geckeler (eds.)
Applications of Nano Materials: Electronics, Energy and Environment, pp. 15–18 (2012).

SYNTHESIS OF GRAPHENE OXIDE

The graphene oxide was synthesized by modified Hummers method.[3]

Synthesis of Tin Tungstate—Reduced Graphene Oxied Nanocomposite

Briefly, a small amount of GO (0.5 g) was added in 100 ml of deionized water and the mixture was sonicated for 30 min. Then 3 g of tin chloride and 0.6 g sodium tungstate was added in that solution followed by addition of 0.5 ml of hydrazine hydrate. The above solution was heated by microwave oven for 20 s in two cycles. The change of color from brown to black is observed which confirms that the GO composite could be reduced to rGO composites. Then the solution was washed several times by distilled water. Finally the precipitate was collected and dried in oven around 50°C over night.

RESULTS AND DISCUSSION

The amount of rGO was varied in the composite material tune be adjusted the amount GO in used in the preparation of the composites.

Figure 1 shows the XRD (XRD Lab X Shimadzu) pattern of GO and $SnWO_4$-rGO composite respectively. Figure 1 (a) resemble to GO which lattice constant was 8.3 Å calculated from XRD pattern. Furthermore fully oxide the graphite 26.3° does not show in the XRD pattern. This was compare to higher then graphitic lattice constant. Figure 1 (b) shows the XRD pattern of $SnWO_4$-rGO. According to this Figure three maximum peaks were observed at the plane (210), (220), (410). The Bragg angle was resembled to file no 70–1497 (Joint Committee for Power Diffractions Standards. β-$SnWO_4$). The grain size was calculated by using Scherer formula which was to be around 14 nm.

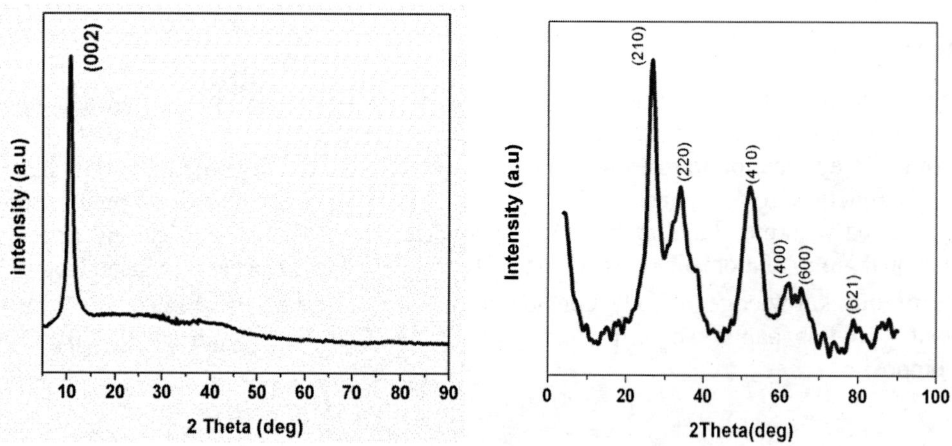

(a) GO (b) $SnWO_4$-rGO

Fig. 1: XRD Patterns of GO and $SnWO_4$-rGO

SEM

Figure 2 shows the SEM (JEOL JSE-6390) image of rGO and $SnWO_4$–rGO nano composites. Which was shows the morphology of $SnWO_4$–rGO nano composites. We can see the reduced graphene sheets, a curled morphology, consisting of a thin wrinkled paper-like structure (Figure 2(a)). Figure 2(b) shows the morphology of $SnWO_4$–rGO nano composites, the individual crystals are not clearly seen in these micrographs structure.

(a) rGO (b) $SnWO_4$-rGO

Fig. 2: SEM Images of rGO and $SnWO_4$-rGO

RAMAN STUDY

Raman spectroscopy is an importune characterization tool to analyze carbon based material, like CNT and Graphene. For using this Raman spectroscopy we could study about the number of layers, defects and doping levels.

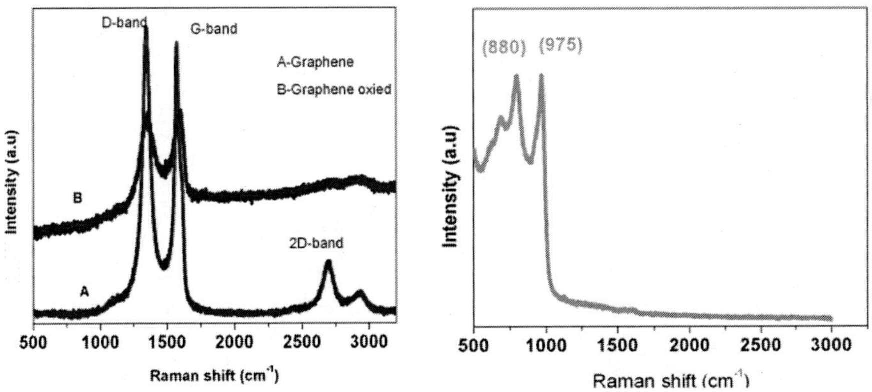

(a) GO and rGO (b) $SnWO_4$-rGO

Fig. 3: Raman Spectrum of GO, rGO and $SnWO_4$-rGO

In Figure 3 (a) shows the compares the Raman spectra of GO and rGO (HORIBA Scientific). Figure 3 (b) shows the Raman spectra of $SnWO_4$-rGO. Both samples were measured in 514 nm. The four most peaks can be observed namely G, D and second order 2D peaks for GO and rGO.

RAMAN SPECTURM OF $SNWO_4$-RGO NANOCOMPOSITE

In the Raman spectrum of the nanocomposite, sharp peak of $SnWO_4$ and low intensity peak of reduced graphene oxide observed. We can observe two sharp peaks at 890 cm^{-1} and 955 cm^{-1}. From this Raman we may conclude as prepared composite was contain β-.

CONCLUSION

In summary, we have synthesized $SnWO_4$-rGO nanocomposite by using simply chemical method. As prepared material were characterized by SEM and Raman. The Raman spectra were concluding as synthesized rGO was a single layer.

ACKNOWLEDGEMENTS

The author (T.S) is so thankful to the management of Karunya University for providing Silver Jubilee Fellowship (SJF) scholarship to carry out the research work.

REFERENCES

[1] Loh, K.P., Bao, Q., Eda, G. and Chhowalla, M., *Nature Chemistry* 2, 1015, 2010.

[2] Marcano, D.C., Kosynkin, D.V., Berlin, J.M., Sinitskii, A., Sun, Z., Slesarev, A., Alemany, L.B., W.

[3] Krishnamoorthy, K., Mohan, R. and Kim, S.J., Graphene oxide as a photocatalytic material, *Applied Physics Letters*, 98, 244101, 2011.

HETROPOLY ACID PILLARED TiO_2 NANOCOMPOSITE FOR ENVIRONMENTAL APPLICATION

R. NIVEA, R. KANNAN[1] AND V. GUNASEKARAN

School of Nanosciences and Technology, Karunya University, Coimbatore, Tamil Nadu
[1]Department of Chemistry, Sona College of Technology, Salem, Tamil Nadu
E-mail: gunasekaran@karunya.edu

ABSTRACT

TiO_2 is a well-known semiconductor with photocatalytic activities whose energy exceeds the bandgap of 3.3 eV in the anatase crystalline phase. However, the photocatalytic efficiency at the present stage is still very low, due to the fast recombination of photogenerated electron-hole pairs. Thus, the improvement of the photocatalytic efficiency is still a major challenge. In this work attempts were made to increase the photocatalytic efficiency by doping it with HPA .The prepared composite was characterized by means of powder X-ray diffraction and SEM. We have also examined the photodegradation of a dye (methylene blue) in the presence of HPAs namely $H3PMo_{12}O_{40}$-TiO_2 [HMA]. The results are discussed in detail.

INTRODUCTION

It is well-known that nano-TiO_2 plays vital role as a photo-catalyst in treatment of environmental pollutants owing to its outstanding features such as inexpensiveness, good photo-stability, high reactivity and environmentally benign.[1]

There are many reports [2–4] that investigated various types of dyes can be photo degraded effectively in TiO_2 dispersions subjected to visible radiation. Among various electron scavengers for enhancing charge separation of TiO_2, the Polyoxometalates (HPAs) seem mostly to be interesting because of their unique characteristics. HPAs can undergo stepwise multi electron redox reactions while their structure remains intact.[5] HPAs have been employed as electron

V. Rajendran, R. Yuvakkumar, K. Thyagarajah and K.E. Geckeler (eds.)
Applications of Nano Materials: Electronics, Energy and Environment, pp. 19–22 (2012).

scavengers to retard the electron-hole recombination, consequently show positive influence on enhancing the TiO_2 photocatalytic efficiency for degradation of pollutants.[6]

In this communication we report the synthesis of HPA-TiO_2 nanocomposite and its photocatalytic efficiency was studied towards Methylene blue dye.

EXPERIMENTAL

Materials

Titanium (IV) Isopropoxide (TTIP, 98%) was purchased from Merck, Hetropoly acids (Phosphomolybdic acid hydrate) purchased from Acros Organics.

Preparation of Photo-Catalytic Material

200 mg of HPA ($H_3PMo_{12}O_{40}$-TiO_2) and 5 ml of Titanium (IV) Isopropoxide 98% dissolved in 20 ml of ethanol, and the contents were sonicated for about 4hrs and kept at 100°C overnight and calcined at 400°C for 2 hrs. This sample was used for the photo catalytic studies.

Characterization of the Photo Catalysts

The photo catalysts thus synthesized were characterized using XRD, UV–vis diffuse reflectance. Powder X-ray diffraction studies were carried out using XRD instrument Shimadzu X-600 Japan, with Cu-Kα radiation, Scanning electron microscopy (SEM) images were taken on JEOL 6390, Japan. Photo-catalytic performance of sample was studied using Shimadzu UV –VIS double beam spectrophotometer (model No. 1800)

RESULTS AND DISCUSSION

Retention of the structural integrity of HPA in the TiO_2 was confirmed by powder X-ray studies. The XRD patterns of HPA-TiO_2 indicate the anatase phase. All peaks holds good for

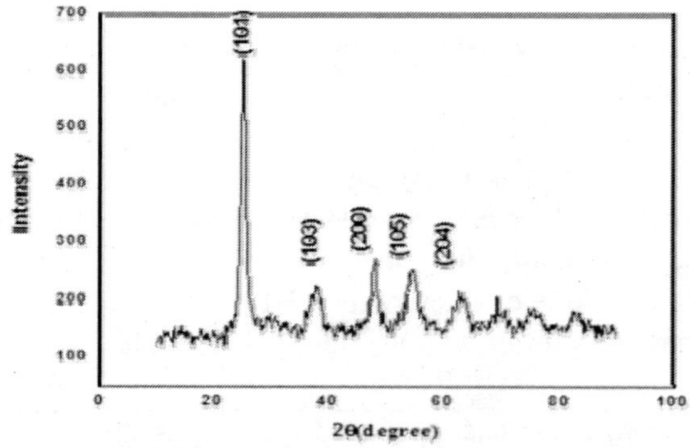

Fig. 1: XRD Pattern of HMA-TiO_2 Nanocomposite

JCPS number 21–1272. Figure 2 shows the SEM image of the prepared nanocomposites. This image confirms that the prepared sample is a nanocomposite.

Fig. 2: SEM Image of HMA-TiO_2 Nanocomposite

Photocatalytic Effect of TiO_2-HPA on MB

Figure 3 shows photocatalytic performance of prepared sample. All the reactions were carried out at room temperature. In order to check the catalytic efficiency, three different experiments were carried out as follows: one in which the dye without the catalyst, the second one in which the dye and TiO_2-HPA as catalyst and the third one in which the dye and the TiO_2 – HPA nanocomposite as catalyst was irradiated with visible light. When irradiated with UV for 30 minutes photocatalysts showed better performance.

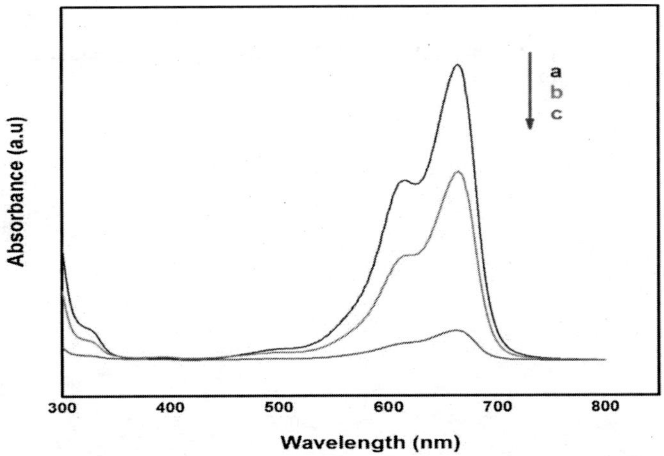

Fig. 3: UV-Vis Spectrum of (a) MB Dye, (b) Adsorption of MB on Nanocomposite and (c) Photocatalytic Decolorization of MB Dye catalyzed by $H_3PMo_{12}O_{40}$ -TiO_2 Nanocomposite for 30 Irradiation

CONCLUSION

The TiO_2-HPA was synthesized by simple chemical method. The resultant nanocomposite shows excellent photocatalytic activity towards the decolorization of MB. The present study reveals that the $H3PMo_{12}O_{40}$-TiO_2 nanocomposite has great potential for environmental application.

ACKNOWLEDGEMENTS

The authors thank the management, Vice-Chancellor, Registrar and Deans of Academic and Research of Karunya University, Coimbatore, and DST Nanomission for their kind financial support and also for their constant encouragement and support.

REFERENCES

[1] Yoon, M., Chang, J.A., Kim, Y., Choi, J.R., Heteropoly Acid-Incorporated TiO_2 Colloids as Novel Photocatalytic, *Systems J. Phys Chem B* 105, 2539–2545, 2011.

[2] Zhang, F., Zhao, J., Hidaka, H., Pelizzetti, E. and Serpone, N., TiO_2-assisted photodegradation of dye pollutants II. Adsorption and degradation kinetics of eosin in TiO_2 dispersions under visible light irradiationAppl. *Catal. B*, 15, 147–156, 1988.

[3] Ruya, Ozer, John, Investigation of the Photocatalytic Activity of TiO_2-Polyoxometalate Systems, *Environ. Sci. Technol.*, 35, 3242–3246, 2001.

[4] Chen, Chuncheng; Zhaom Wei; Lei, Pengxiang and Zhao, Jincai, Photosensitized degradation of dyes in polyoxometalate solutions versus TiO_2 dispersions under visible-light irradiation: mechanistic implications, *Chem. Eur. J.*, 10, 1956–65, 2004.

[5] Chen, Chuncheng; Zhao, Wei; Lei, Pengxiang, Zhao, Jincai and Serpone, Nick, Photosensitized Degradation of Dyes in Polyoxometalate Solutions Versus, TiO_2 Dispersions under Visible-Light Irradiation: Mechanistic Implications, *Chem. Eur. J.*, 10, 1956–1965, 2004.

[6] Changjun Yang, Lihong Tian, Liqun Ye, Enhancement of photocatalytic degradation activity of poly(vinyl chloride) - TiO_2 nanocomposite film with polyoxometalate, *J Applied Polymer Science,* 120, 2048–2053, 2011.

ON THE EFFECTIVE ELECTRON MASS IN QUANTUM WELLS OF NONLINEAR OPTICAL AND OPTOELECTRONIC MATERIALS

S.N. Banerjee, S. Bhattacharya[1] and K.P. Ghatak[2]

Additional Director, Integrated Test Range, Defence Research and Development Organisation, Chandipur, Orrisa
[1]Nanoscale Device Research Laboratory, Center for Electronics Design and Technology, Indian Institute Science, Bangalore
[2]Department of Electronics and communication Engineering, National Institute of Technology, Jirania, Tripura

ABSTRACT

An attempt is made to study the two dimensional (2D) effective electron mass (EEM) in quantum wells (QWs) of nonlinear optical materials on the basis of a newly formulated electron dispersion law within the framework of k.p. formalism. The results of QWs of III-V, ternary and quaternary materials form a special case of our generalized analysis. It has been found, taking QWs of $CdGeAs_2$, n-InGaAsP and n-HgCdTe, as examples, that the EEM in QWs increases with decreasing film thickness in various manners which are totally band structure dependent. The EEM in QWs of nonlinear optical materials is quantum number dependent. The appearance of the humps of the respective curves is due to the redistribution of the electrons among the quantized energy levels when the quantum numbers corresponding to the highest occupied level changes from one fixed value to the others. Although the EEM varies in various manners with all the variables as evident from all the curves, the rates of variations totally depend on the specific dispersion relation of the particular 2D structure. Under certain limiting conditions, all the results as derived in this paper get transformed into well known formulas of the EEM and the electron statistics and thus confirming the compatibility test.

V. Rajendran, R. Yuvakkumar, K. Thyagarajah and K.E. Geckeler (eds.)
Applications of Nano Materials: Electronics, Energy and Environment, pp. 23–28 (2012).

INTRODUCTION

The concept of the effective mass of the carriers in nano-materials which is connected with the motion of the carriers is vital in the realm of Nanoscience and Technology. Among the various definitions of the effective electron mass,[1] it is the effective momentum mass that should be regarded as the basic concept [1,2,3] since it is this mass which appears in the description of transport phenomena and all other properties of the conduction electrons in semiconductors having various band structures.[3] It is the effective momentum mass which enters in various transport coefficients and plays the most dominant role in explaining the experimental results of different scattering mechanisms through Boltzmann's transport equation.[4,5] Under degenerate conditions, only the electrons at the Fermi surface of n-type materials act in the conduction process and hence, the effective momentum mass of the electrons (EEM) corresponding to the Fermi level would be of interest in electron transport under such conditions. The nature of these variations has been investigated in the literature.[1–5] Some of the significant features, which have emerged from these studies, are:

1. The EEM increases monotonically with electron concentration.
2. The EEM increases with doping in heavily doped materials in the presence of band tails.
3. The nature of variations is significantly influenced by the energy band constants of various materials having different band structures.
4. The EEM oscillates with inverse quantizing magnetic field due to SdH effect. The EEM in Bismuth under magnetic quantization depends both on the Fermi energy and the magnetic quantum number due to the presence of band non-parabolicity only.

In this paper we shall study the EEM in QWs of nonlinear optical and optoelectronic materials.

THEORETICAL BACKGROUND

The EEM in QWs of Non-linear Optical Materials

The dispersion relation of the conduction electrons in bulk specimens of nonlinear optical materials can be written as,

$$\gamma(E) = f_1(E)k_s^2 + f_2(E)k_z^2 \qquad \ldots (1.1)$$

where the notations are defined in.[5]

For dimensional quantization along z-direction, the dispersion relation of the 2D electrons in this case can be written following (1.1) as,

$$\psi_1(E) = \psi_2(E)k_s^2 + \psi_3(E)(n_z\pi/d_z)^2 \qquad \ldots (1.2)$$

where $\psi_1(E) = \gamma(E)$, $\psi_2(E) = f_1(E)$, $\psi_3(E) = f_2(E)$, $n_z (=1,2,3,\ldots)$ and d_z are the size quantum number and the nano-thickness along the *z*-direction respectively.

The EEM at the Fermi level in the xy-plane can be written as,

$$m^*(E_{Fs},n_z)=\left(\frac{\hbar^2}{2}\right)\left[\psi_2(E_{Fs})\right]^{-2}\left[\psi_2(E_{Fs})\left\{\{\psi_1(E_{Fs})\}'-\{\psi_3(E_{Fs})\}'\left(\frac{n_z\pi}{d_z}\right)^2\right\}-\left\{\psi_1(E_{Fs})-\psi_3(E_{Fs})\left(\frac{n_z\pi}{d_z}\right)^2\right\}\{\psi_2(E_{Fs})\}'\right] \quad \ldots (1.3)$$

where, the primes denote the differentiation of the differentiable functions with respect to Fermi energy. Thus, we observe that the EEM is the function of size quantum number and the Fermi energy due to the combined influence of the crystal field splitting constant and the anisotropic spin-orbit splitting constants respectively.

The 2D carrier statistics in this case assumes the form,

$$n_{2D}=\frac{g_v}{2\pi}\sum_{n_z=1}^{n_{z_{max}}}\left[T_{51}(E_{Fs},n_z)+T_{52}(E_{Fs},n_z)\right] \quad \ldots (1.4)$$

where, $T_{51}(E_{Fs},n_z)\equiv\left[\frac{\psi_1(E_{Fs})-\psi_3(E_{Fs})(n_z\pi/d_z)^2}{\psi_2(E_{Fs})}\right]$ and $T_{52}(E_{Fs},n_z)\equiv\sum_{r=1}^{s}L(r)\left[T_{51}(E_{Fs},n_z)\right]$.

$L(r)=2(k_BT)^{2r}(1-2^{1-2r})\xi(2r)\frac{\partial^{2r}}{\partial E_F^{2r}}$, k_B is the Boltzmann constant, T is the temperature, r is the set of real positive integers whose upper limit is r_0 and $\xi(2r)$ is the Zeta function of order 2r.

The EEM from QWs of III-V and Optoelectronic Materials

Under the conditions, $\delta=0$, $\Delta_{\parallel}=\Delta_{\perp}=\Delta$ (isotropic spin orbit splitting constant) and $m_{\parallel}^*=m_{\perp}^*=m_c$ (isotropic effective electron mass at the edge of the conduction band), (1.1) gets simplified into the form,

$$\frac{\hbar^2k^2}{2m_c}=I_{11}(E),\ I_{11}(E)\equiv\frac{E\left(E+E_{g_0}\right)\left(E+E_{g_0}+\Delta\right)\left(E_{g_0}+\frac{2}{3}\Delta\right)}{E_{g_0}\left(E_{g_0}+\Delta\right)\left(E+E_{g_0}+\frac{2}{3}\Delta\right)} \quad \ldots (1.6)$$

which is known as the three band model of Kane [5] and is often used to study the electronic properties of III-V materials.

Thus, under the conditions $\delta=0$, $\Delta_{\parallel}=\Delta_{\perp}=\Delta$ and $m_{\parallel}^*=m_{\perp}^*=m_c$, (1.6) assumes the form

$$\frac{\hbar^2k_s^2}{2m_c}+\frac{\hbar^2}{2m_c}(n_z\pi/d_z)^2=I_{11}(E) \quad \ldots (1.7)$$

Using (1.7), the EEM in x-y plane for this case can be written as,

$$m^*(E_{Fs})=m_c\{I_{11}(E_{Fs})\}' \quad \ldots (1.8)$$

It is worth noting that the EEM in this case is a function of Fermi energy alone and is independent of size quantum number.

The 2D carrier concentration assumes the form,

$$n_{2D} = \frac{m_c g_v}{\pi\hbar^2} \sum_{n_z=1}^{n_{z\max}} \left[T_{53}\left(E_{Fs}, n_z\right) + T_{54}\left(E_{Fs}, n_z\right) \right] \quad \ldots (1.9)$$

where $T_{53}\left(E_{Fs}, n_z\right) \equiv \left[I_{11}\left(E_{Fs}\right) - \frac{\hbar^2}{2m_c}\left(\frac{n_z \pi}{d_z}\right)^2 \right]$ and $T_{54}\left(E_{Fs}, n_z\right) \equiv \sum_{r=1}^{s} L(r) T_{53}\left(E_{Fs}, n_z\right)$.

RESULT AND DISCUSSIONS

Using the appropriate equations and the values of the energy band constants as given in,[5] we have plotted in Figure 1 the EEM in QWs of n-$CdGeAs_2$ as a function of film thickness in accordance with the three and two band models of Kane together with the incorporation of the crystal field parameter. It appears that the effect of the crystal field splitting increases the EEM sharply below sub-20 nm. The EEM also seems to increase about 7% at 5 nm and converges to its bulk value beyond 20 nm at the same level of electron degeneracy.

Figures 2 and 3 exhibits the variation of the EEM with respect to the film thickness for the ternary and quaternary materials at same carrier degeneracy level. It appears that at an alloy composition x = 0.3, the EEM in both the cases tends to about 0.1 times the rest mass at film thickness of 5 nm. The effect of variation of EEM on the alloy composition for these two materials has been exhibited in Figure 4. The Figure 1 exhibits the plot EEM in QWs of

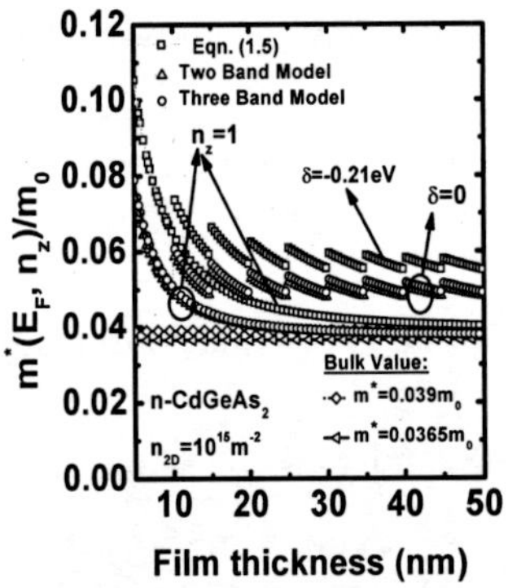

Fig. 1: Plot of the effective electron mass as function of QW film thickness for n-$CdGeAs_2$ in the presence of crystal field constant. The effective electron mass in accordance with three and two band models of Kane have also been plotted where, $m_{\perp}^{*} = 0.039 m_0$ and $m^{*} = \frac{1}{2}\left(m_{\parallel}^{*} + m_{\perp}^{*}\right) m_0 = 0.0365 m_0$ are the corresponding bulk values.

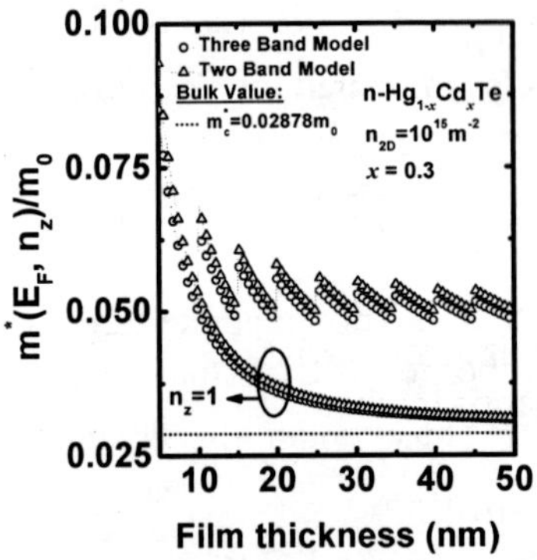

Fig. 2: Plot of the Effective Electron Mass as Function of QW Film Thickness for n-$Hg_{0.3}Cd_{0.7}Te$ for Three and Two Band Models of Kane.

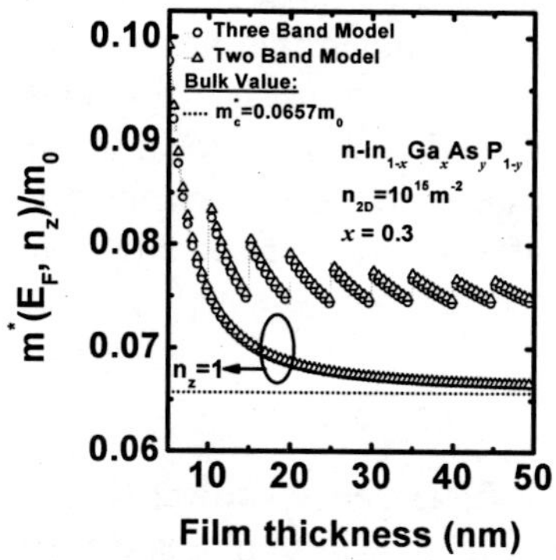

Fig. 3: Plot of the Effective Electron Mass as Function of QW Film Thickness for n-$In_{1-x}Ga_xAs_yP_{1-y}$ for all the Cases of Figure 2

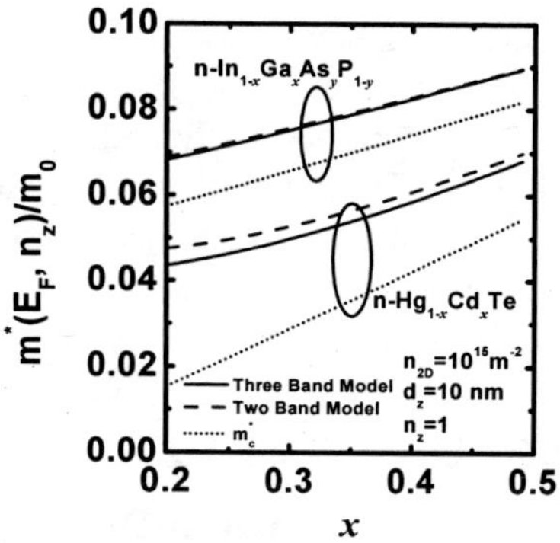

Fig. 4: Plot of the Effective Electron Mass at the Lowest Subband as Function of Alloy Composition in QW of n-$Hg_{1-x}Cd_xTe$ and n-$In_{1-x}Ga_xAs_yP_{1-y}$ for the Three and the Two Band Models of Kane

n-$CdGeAs_2$ as a function of film thickness in accordance with the three and two band models of Kane together with the incorporation of the crystal field parameter. It appears that the effect of the crystal field splitting increases the EEM sharply below sub-20 nm. The EEM also seems to increase about 7% at 5 nm and converges to its bulk value beyond 20 nm at the same level of electron degeneracy.

Figures 2 and 3 exhibits the variation of the EEM with respect to the film thickness for the ternary and quaternary materials at same carrier degeneracy level. It appears that at an alloy composition x = 0.3, the EEM in both the cases tends to about 0.1 times the rest mass at film thickness of 5 nm. The effect of variation of EEM on the alloy composition for these two materials has been exhibited in Figure 4. It appears that the effect of increasing the alloy composition increases the EEM for all the said two materials. For comparison, we have also plotted the variation of the bulk effective mass with the alloy composition. It appears that for the quaternary material, the difference in the two energy band model is not much.

The increment in EEM is rather linear in case of InGaAsP than that of HgCdTe. This also exhibits the variation of the electron mobility in these systems as the alloy composition changes. It appears that with increase in x, the mobility falls down assuming a constant relaxation rate.

The influence of quantum confinement is immediately apparent from the Figures since the EEM depends strongly on the thickness of the quantum-confined materials in contrast with the corresponding bulk specimens. The EEM changes with increasing carrier concentration suffering discontinuities with different numerical magnitudes. It appears from the aforementioned figures that the EEM exhibits spikes for particular values of film thickness which, in turn, depends on the particular band structure of the specific semiconductor. This effect of the discontinuity on the EEM will be less and less prominent with increasing film thickness. For bulk specimens of the same material, the EEM will be found to increase continuously with increasing electron degeneracy in a non-oscillatory manner.

With varying electron degeneracy, a change is reflected in the EEM through the redistribution of the electrons among the size-quantized levels. It may be noted that at the transition zone from one sub band to another, the height of the peaks between any two sub-bands decreases with the increasing in the degree of quantum confinement and is clearly shown in the respective figures. It should be noted that although the EEM changes in various manners with all the variables as evident from all the figures, the rates of variations are totally band-structure dependent.

ACKNOWLEDGEMENTS

S.N. Banerjee is grateful to Director, I.T.R. for his encouragement towards writing of the paper. K.P. Ghatak, and S. Bhattacharya are grateful to UGC for sanctioning the research project having the No-F 40–469/2011(SR) under which this paper has been completed.

REFERENCES

[1] Adachi, S., *J. Appl. Phys*. 58, R11 (1985).

[2] Dornhaus, R., and Nimtz, G., "Springer Tracts in Modern Physics," Vol. 78, (Springer, 1, 1976).

[3] Zawadzki, W., "Handbook of Semiconductor Physics", (Paul, W., ed.). Vol. 1, (North Holland, 719, 1982).

[4] Ghatak, K.P. and Mondal, M., Naturforschung, Z.F. 41A, 821 (1986).

[5] Bhattacharya, S. and Ghatak, K.P., Effective electron mass in low dimensional semiconductors, *Springer Series in Materials Science* 167, (2012).

ELECTROCHEMICAL STUDIES OF PLASTICIZED NANOCOMPOSITE POLYMER ELECTROLYTES FOR RECHARGEABLE MAGNESIUM ION BATTERIES

N. AMMAKUTTI SRIDEVI, K. KARUPPASAMY, C. VIJIL VANI, S. DAWN DHARMA ROY[1], PAITIP THIRAVETYAN[2] AND X. SAHAYA SHAJAN

Centre for Scientific and Applied Research (CSAR), School of Basic Engineering and Sciences (SOBES), PSN College of Engineering and Technology, Tirunelveli–627 152, Tamil Nadu, India
[1]Nesamony Memorial Christian College, Marthandam–629 165, Tamil Nadu, India
[2]School of Bioresources and Technology, King Mongkut's University of Technology, Thanburi, Bangkok 10150, Thailand
E-mail: shajan89@gmail.com

ABSTRACT

Flexible, transparent and self-supporting plasticized nanocomposite polymer electrolytes (PNCPE) consisting of polyethylene oxide (PEO) as host with magnesium triflate ($Mg(CF_3SO_3)_2$), succinonitrile (SN) and nanochitin were prepared by membrane hot press method. The effect of succinonitrile and nanochitin on electrochemical stability of PNCPE was investigated. The complexation behaviour of PNCPE was confirmed by means of FT-IR analysis. The prepared membranes were subjected to various electrochemical studies such as ionic conductivity, cyclic voltammetry. The ionic conductivity of PNCPE shows some novel results and attains a maximum conductivity of $10^{-3.1}$ S/cm at 60°C.

INTRODUCTION

The development of an environmentally favourable and nontoxic electrolyte system with superior safety is more and more vital. In view of negligible hazards and enhanced safety,[1] studies on rechargeable magnesium batteries are expected to be the best choice. The new

V. Rajendran, R. Yuvakkumar, K. Thyagarajah and K.E. Geckeler (eds.)
Applications of Nano Materials: Electronics, Energy and Environment, pp. 29–34 (2012).

methods of measuring ionic conductivity in the polymer salt complexes directed the interest of the researchers towards the study of solid polymer electrolytes. However a useful conductivity for solid polymer electrolyte which comprises of PEO-metal complex is obtained only above the crystalline to amorphous phase transition temperature (60°C) of PEO.[2] Various methods has been put forth to maintain the amorphous nature of the polymer matrix. The physical properties of such polymer matrix composites have been tailored using different fillers.[3,4] Chitin nanofiber, a natural polysaccharide polymer is used as nanofiller which prevents the tendency of polymer chain reorientation. A non-ionic plastic material, succinonitrile (SN) has been used as a plasticizer which has been reported by us as a versatile additive to the solid state ionic conductors that leads to ambient ionic conductivity.[5]

EXPERIMENTAL

Polyethylene oxide (PEO) and magnesium trifluoromethane sulfonate ($Mg(CF_3SO_3)_2$ or MgTf) were dried under vacuum condition for 3 days at 40 and 100°C respectively. The plasticizer succinonitrile (SN) was used without further purification. All the chemicals were purchased from Sigma Aldrich. Chitin nanofiber was prepared in our collaborator's laboratory and the preparation method was described elsewhere.[6] The nanocomposite solid polymer electrolytes were prepared by dissolving appropriate amounts of PEO, nanochitin fibre, SN and MgTf salt in acetonitrile. The compositions taken in the present study is given in table 1. The PNCPE was prepared by membrane hot-press method and details of preparation method were published by us elsewhere.[7]

RESULTS AND DISCUSSION

FTIR Studies

In Figure 1(a) the peaks at 1468 cm^{-1}, 1095 cm^{-1}, 952 cm^{-1}, 838 cm^{-1} of PEO were assigned to CH_2 scissoring, C-O-C stretching, CH_2 twisting, CH_2 wagging respectively.[8] In fig 1(b) the peak at 2252 cm^{-1} was attributed to cyano functional group C≡N of succinonitrile [9] and the peaks at 1231 cm^{-1} and 764 cm^{-1} were attributed to the CH_2 twisting and CH_2 rocking. The fig 1(c) showed the amide I band of chitin fibre is split into two peaks at 1649 cm^{-1} and 1625 cm^{-1}. The 1649 cm^{-1} peak was assigned to the stretching of C=O group hydrogen bonded to N-H of the neighbouring intra-sheet chains of polyamides and proteins. The peak at 1625 cm^{-1} was attributed to specific hydrogen bond of C=O in the hydroxyl methyl group of next chitin residue of the same chain.[8] Further in Figure 1(d) the peaks at 1231 cm^{-1}, 1031 cm^{-1} and 639 cm^{-1} was assigned to the symmetric stretching mode of CF_3, symmetric stretching mode of SO_3 and symmetric bending mode of SO_3 [10] respectively. In fig 1(e) upon adding succinonitrile with the PEO-MgTf salt complex the peaks at 1031 cm^{-1}, and 1231 cm^{-1} disappeared resulting in complex formation of the plasticizer with polymer–salt complex.

Also the C≡N peak at 2252 cm^{-1} shifted to 2234 cm^{-1}. Further the peak broke on addition of nanofiller chitin fibre to the PEO – MgTf – Succinonitrile system which was shown in fig 1(f). The C-O-C stretching peak of PEO at 1095cm^{-1} shift to 1097cm^{-1} resulted in decrease in bond length supported the total complexation of the prepared PNCPE (PEO-MgTf-succinonitrile-nanochitin fibre).

Table 1: Composition of Polymer, Salt, Filler, Plasticizer in wt %

Sample	*PEO*	*MgTf*	*CNF*	*SN*
N1	95	5	–	–
N2	90	5	5	–
N3	90	5	–	5
N4	85	5	8	2
N5	85	10	3	2
N6	85	8	5	2
N7	80	10	5	5
N8	80	8	10	2

Thermal Analysis

The DSC thermograms in Figure 2 depicted the existence of dominant endothermic peak for electrolytes N1, N5, N6, N4, N8 which corresponded to semi crystalline-amorphous phase transition and melting temperature (T_m) of PEO. The dispersal of nanochitin fibre (3 to 10 wt %) in the PEO-MgTf salt complex with plasticizer SN (2 wt %) substantially influenced T_m. The crystalline melting temperature T_m of pure PEO was 66.2°C as reported by Xing-Long Wu *et al.*[11] From our work it is revealed that with the introduction of nanochitin fibre into PEO-salt matrix, the T_m, decreased the ratio of crystalline PEO to 62.4°C which showed the enhancement of volume fraction of amorphous phase.

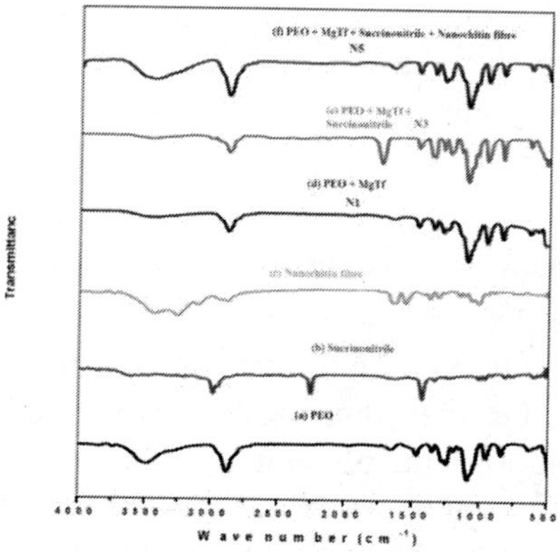

Fig. 1: a-f FTIR Spectra of Pure PEO, SN, CNF, PEO-MgTf(N1), PEO-MgTf-SN(N3), PEO-MgTf-SN-CNF(N5)

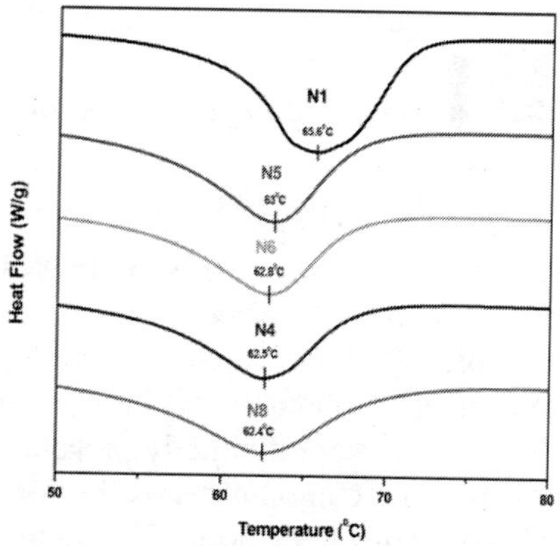

Fig. 2: DSC Thermograms for Samples N1, N5, N6, N4, N8

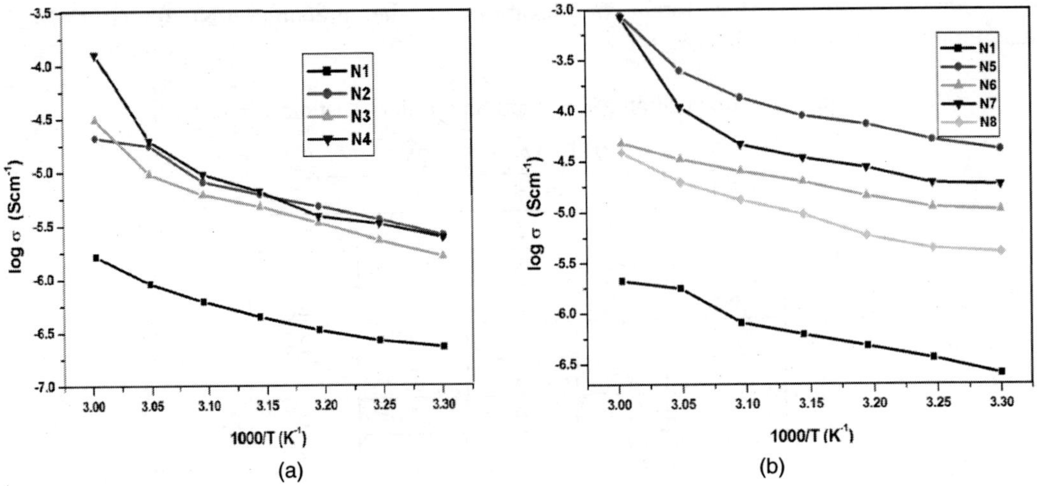

Fig. 3a&b: Temperature Dependence of Ionic Conductivity for Samples N1, N2, N3, N4 and N1, N5, N6, N7, N8

Ionic Conductivity

The overall mobility of the ion in polymer chains and in amorphous phase enhances the ionic conductivity of polymer electrolyte. The Figure 3a and 3b depicts the temperature dependence of ionic conductivity for samples N1, N2, N3, N4 and N1, N5, N6, N7, N8 respectively. The highest conductivity was observed for the sample N5 and N7 containing 10 wt% of the salt content in the order of $10^{-3.1}$ Scm^{-1} at 60°C. Beyond 10 wt% of the salt content, the viscosity of the electrolyte increased. The plasticizer SN does not supply ions to the electrolyte system but it dissociates more salts into ions. The filler found to increase the transport properties and provide the resistance to crystallization thereby increasing the ionic conductivity of electrolyte system. The competition between SN and nanochitin fibre particles with respect to salt concentration facilitated the migration of Mg^{2+} in the polymer chains.

Cyclic Voltammetry Studies

The cyclic voltammograms of samples N3 (without filler), N7 (PNCPE) sandwiched between symmetrical stainless steel (SS) electrodes were recorded at a scan rate of 30 mVs^{-1}. The figure 4 depicts that there was no current peaks on the voltammogram of N3 due to negligibly small current throughout the voltage range –3v to 3v. On the other hand, both the cathodic and anodic peaks were distinctly present on the voltammograms of sample N7. These results suggested that the cathodic deposition of Mg was not favoured on SS for electrolyte without filler. The ratio of the cathodic charge and the anodic charge was about 0.98, thus indicating that the quasireversible reaction.[12]

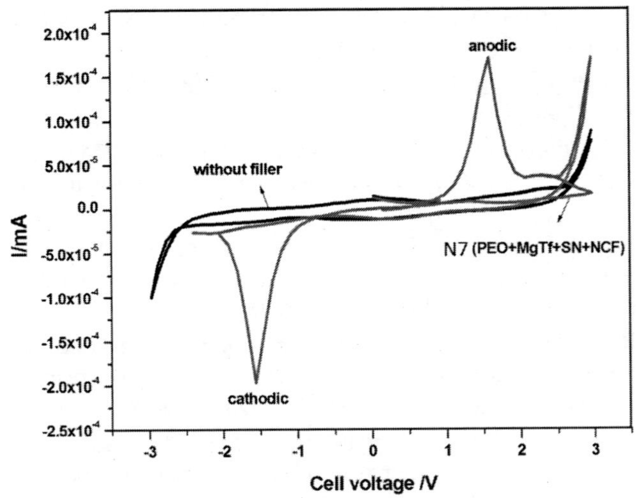

Fig. 4: Cyclic Voltammograms for Samples N3 and N7

CONCLUSION

Effect of incorporating plasticizer and nanofiller to PEO-MgTf complex were investigated and the results were summarized. The complexation of salt, plasticizer, nanofiller at the molecular level was confirmed using FTIR spectral analysis. The DSC results showed the enhancement of amorphous nature of the polymer matrix in the temperature range of 65–62°C. The ionic conductivity studies confirmed it showing the maximum conduction at 60°C. Maximum ionic conductivity of the order $10^{-3.1}$ Scm^{-1} was obtained for the sample with 10 wt % of Mg salt at 60°C. Cyclic voltammetry indicated the existence of an electrochemical equilibrium between the electrodes and Mg^{2+} ions in NCPE.

REFERENCES

[1] Sheha, E., Ionic conductivity and dielectric properties of plasticized $PVA_{0.7}(LiBr)$ $(H_2SO_4)_{2.7M}$ solid acid membrane and its performance in a magnesium battery, Solid State Ionics 180,1575–1579, 2009.

[2] Sundar, M., Selladurai, S., Effect of fillers on magnesium—poly(ethyleneoxide) solid polymer electrolyte, Ionics 12, 281–286, 2006.

[3] Thomas Hanemann, Dorothee Vinga Szabo., Polymer–nanoparticle composites: From synthesis to modern applications, Material 3, 3468–3517, 2010.

[4] Jaroslaw Syzdex, Michel Armand, Marek Marcinek, Aldona Zalewska, Grazyna Zukowska, *et al.*, Detailed study on the fillers modification and their influence on composite, poly (oxyethylene) based polymeric electrolytes, Electrochim. Acta 55, 1314–1322, 2010.

[5] Karuppasamy, K., Antony, .R, Thanikaikarasan, S., Balakumar, S., Sahaya Shajan, X., Combined effect of nanochitosan and succinonitrile on structural, mechanical, thermal and

electrochemical properties of plasticizd nanocomposite polymer electrolytes (PNCPE) for lithium batteries, Ionics (under review), 2012.

[6] Riyira Dolphen, Paitip Thiravetyan., Adsorption of Melanoidins by chitin nanofibers, *Chem. Eng. J.,* 166,890–895,2011.

[7] Ammakutti, N., Karuppasamy, K., Balakumar, S., Sahaya Shajan, X., Structural and ionic conductivity studies on nanochitosan incorporated polymer electrolytes for magnesium batteries, *Chem Sci. Trans*, 1 (2), 311–316,2012.

[8] Angulakshmi, N., Premkumar, T., Sabu Thomas, Manuel Stephen, A., Ionic conductivity and interfacial properties of nanochitin – incorporated polyethyleneoxide-$LiN(C_2F_5SO_2)_2$ polymer electrolytes Elecrochim Acta 55, 1401–1406,2010.

[9] Myung-Hyun Ryou, Dong Jin Lee, Je-NamLee, Hong Kyeong Lee, Myungwon Seo, *et al.*, Cross-Linkable polymer matrix for enhanced thermal stability of succinonitrile based polymer electrolyte in lithium rechargeable batteries, *J. Electrochem Sci. Tech.* 2, 198–203, 2011.

[10] Ramesh, S., Liew Chiam Wen., Investigation on the effects of addition of SiO_2 nanoparticles on ionic conductivity, FTIR, and Thermal properties of nanocomposite PMMA – $LiCF_3SO_3$-SiO_2 Ionics 16: 255–262, 2010.

[11] Xing-Long Wu, Sen Xin, Hyun-Ho Seo, Jaekook Kim, Yu-Guo Guo,Jong-Sook Lee., Enhanced Li^+ conductivity in PEO-LiBOB polymer electrolytes by using succinonitrile as a plasticizer, *Solid State Ionics,* 186, 1–6, 2011.

[12] Girish Kumar, Munichandraiah, N., Effect of plasticizer on magnesium-poly(ethyleneoxide) polymer electrolyte, *Journal of Electroanalytical Chemistry*, 495, 42–50.

DIELECTRIC PROPERTIES AND AC CONDUCTIVITY OF CERIUM PHOSPHATE AND ITS LITHIUM EXCHANGED PHASE

A.P. Apsara and B. Beena[1]

Department of Chemistry, FMN College, Kollam, Kerala
[1]Department of Chemistry, DB College, Sasthamcotta, Kerala
E-mail: apsaraap@gmail.com

ABSTRACT

The frequency dependence of dielectric behaviour and ac conductivity of synthesized Cerium phosphate and its lithium exchanged phase was examined. The dispersion in dielectric properties has been explained in the light of space charge polarization. Conduction in the materials is attributed to hopping of charge carriers. It has been observed that the lithium exchanged phase exhibits encouraging conductance behaviour.

INTRODUCTION

Lithium batteries that are safer, less expensive, non toxic are choices for portable electronic devices. Lithium ion conductors are also attractive[1]. TMA salts have been widely investigated for ion conduction behavior.[2–4] Due to the increased significance of dielectrics and eco-friendly safe cathode materials for lithium batteries, in the present endeavor, the frequency dependence of dielectric behaviour and ac conductivity of amorphous Cerium phosphate (CP) and its Lithium–exchanged phase (LiCP) have been examined.

EXPERIMENTAL

CP in the present study was prepared by adding aqueous solution of 0.1 M disodium hydrogen phosphate to half its volume of 0.5 M ceric sulphate slowly, with continuous stirring at pH~2.

V. Rajendran, R. Yuvakkumar, K. Thyagarajah and K.E. Geckeler (eds.)
Applications of Nano Materials: Electronics, Energy and Environment, pp. 35–40 (2012).

The gel obtained was stirred for another 1 hour and kept overnight at room temperature in contact with the mother liquor for the growth of fine particles. The granular precipitate thus obtained, was filtered, washed with conductivity water till free of chloride and sulphate ions and dried at 40°C. The material was then converted to the hydrogen form completely by repeated treatment with 1 M HCl solution.

LiCP was prepared by equilibrating 5 g CP with 500 ml of 0.2 M lithium acetate solution with continuous stirring at 50°C for 48 h. The solid was separated by filtration and washed with conductivity water for removal of adhering ions and dried at room temperature.

Elemental analysis of the prepared material was carried out by ICP-AES. X-Ray diffractograms were taken from Bruker D8 Advance difffractometer using Cu K_α radiation. FTIR was recorded on a Perkin-Elmer IR spectrometer. The dielectric measurement as a function of frequency in the range of 316 Hz to 316 MHz were carried out using LCR meter H10K1 3532–50 LLR HI TESTER.

RESULTS AND DISCUSSION

CP and LiCP were obtained as pale yellow solids. Elemental analysis carried out using ICP–AES showed the ratio of Ce: P in CP and LiCP to be 1: 2. The composition of LiCP indicates that, about 32% of Ce ion and 0.62% Li ion by mass are present in the sample. The X-ray diffractogram of LiCP indicates some crystallinity in the material as against CP which is purely amorphous. No characteristic changes are observed in the FTIR spectrum of lithium exchanged phase of CP. However, additional metal-oxygen stretches have been observed around ~521 cm^{-1} attributed to the formation of Li-O bond.

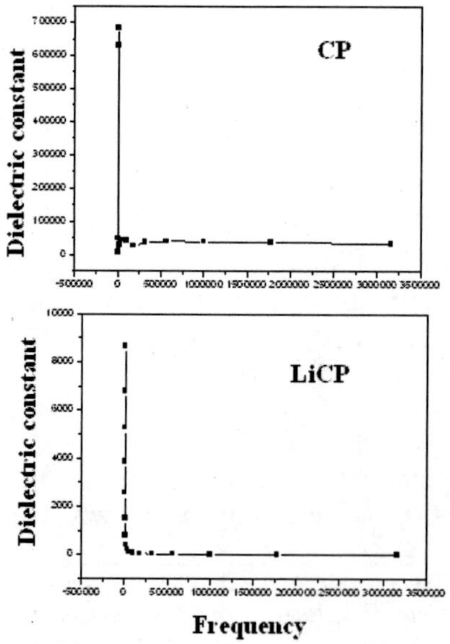

Fig. 1: Variation of Dielectric Constant with Frequency

In the present study, the frequency dependence of various parameters like dielectric constant, ac conductance, impedance and dielectric loss of CP and LiCP have been analyzed. The dielectric constant decreases rapidly with increase in frequency and reaches a constant value (Figure 1). The variation in dielectric constant may be explained on the basis of space-charge polarization.

According to Maxwell and Wagner two layer model,[5] space-charge polarization is because of the inhomogeneous dielectric structure of the material. The material may be formed of large well conducting grains separated by thin poorly conducting intermediate grain boundaries. This produces localized accumulation of charge under the influence of an electric field.[6]

Space-charge polarization caused by mobile charges which are present at the interfaces may be responsible for the high dielectric constant observed in the low frequency region for CP and LiCP. LiCP shows a gradual rise in conductance with frequency (Figure 2). Conductance of a solid depends on the number of charge carriers and also on the mobility of the species in the system.[7] In synthesizing lithium exchanged phases, it is known that all H^+ sites cannot be exchanged with lithium ions.

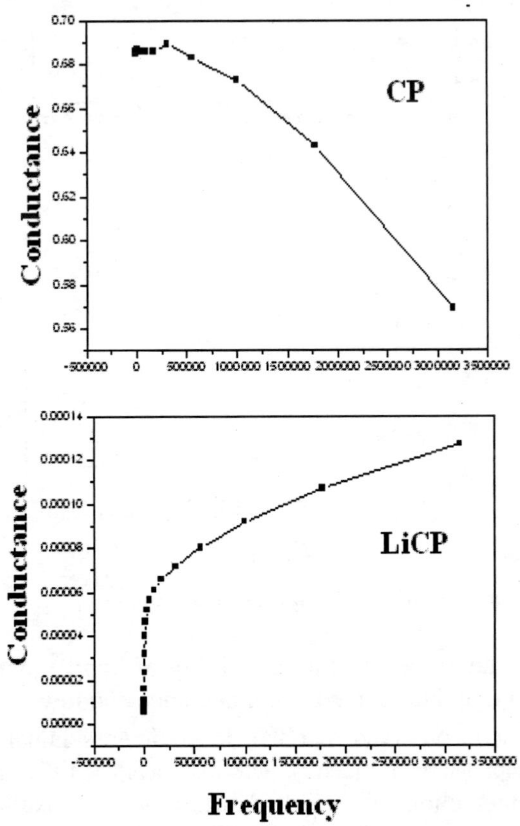

Fig. 2: Variation of Conductance with Frequency

Thus, the exchanged phases contain two types of ions available for conduction namely H^+ and Li^+. The H^+ ion is smaller in size than Li^+ and so the O-Li bond is weaker than O-H bond. Li-O framework bond allows faster lithium conduction as observed in LiCP. The variation of impedance with frequency shows the reverse trend of ac conductance (Figure 3). LiCP shows high impedance in the low frequency region, may be due to confined passage of Li^+ and H^+ ions within the grains.[8]

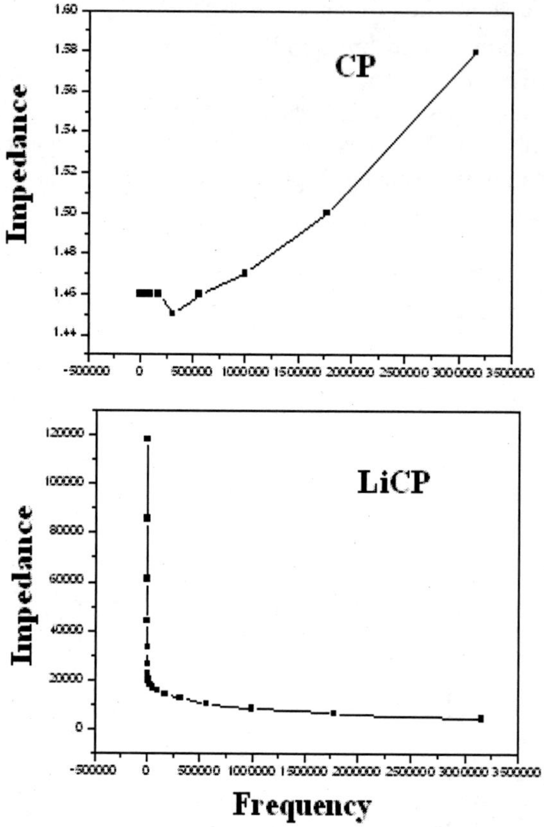

Fig. 3: Variation of Impedance with Frequency

As the applied frequency increases, the probability of jump attempts made by the charge carriers also increases, thereby increasing the conductance and lowering the impedance.

For the sample CP, it is observed that the dielectric loss initially remains constant and thereafter decrease with increase in frequency. The loss peak of LiCP appears broad suggesting the presence of dipolar interactions with wide distribution of relaxation times.[9] The low loss factor at higher frequencies observed for both CP and LiCP, shows the potential of these materials for high frequency application.

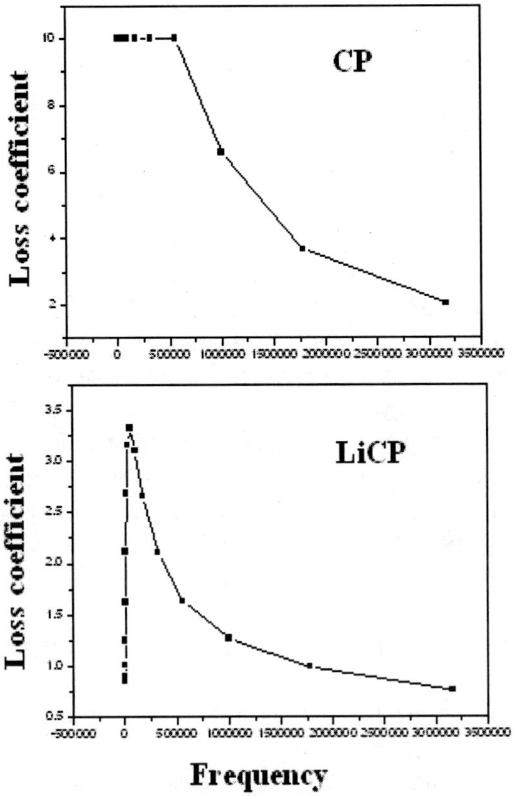

Fig. 4: Variation of Dielectric Loss with Frequency

The dielectric and ac conductivity properties of Mn ferrites are found to be similar to the properties of LiCP in the present study.[10] Ferrites are important commercially on account of their technological applications in many electronic devices such as high frequency transformer cores, phase shifter, switches, resonators etc. The material namely LiCP used in the present study, may also find application in such fields.

CONCLUSION

The dispersion in dielectric properties has been explained in the light of space charge polarization discussed in Maxwell-Wagner model. The dielectric loss coefficient for CP shows a normal behaviour with frequency but for LiCP the broad peak indicates the presence of dipolar interactions and wide distribution of relaxation times. LiCP shows high impedance in the low frequency region, may be due to confined passage of Li^+ and H^+ ions within the grains It has been observed that LiCP exhibits dielectric properties and encouraging conductance behavior similar to ferrites.

ACKNOWLEDGEMENT

This work was supported by UGC (FIP PlanXI).

REFERENCES

[1] Osaka, T. and Momma, T., Lithium metal/polymer battery, *J Power Sources*, 97–98, 2001, 765–767.

[2] Rakesh, Thakkar and Uma, Chudasama, Synthesis, Characterization and proton transport properties of a mixed material-Zirconium titanium phosphate, a tetravalent bimetallic acid salt, Electrochimica Acta, 54, 2720–2726, 2009.

[3] Casciola, M. and Fabiani, D., Ionic conduction and dielectric properties of anhydrous alkali metal salt forms of α-zirconium phosphate, Solid State Ionics, 11, 31–38, 1983.

[4] Bianchi, D. and Casciola, M., Frequency response of polycrystalline samples of α-$Zr(HPO_4)_2.H_2O$ with different relative densities, Solid State Ionics, 17, 7–12, 1985.

[5] Maxwell, J.C., Electricity and Magnetism, Vol. 1, Oxford University Press, Oxford, 1929.

[6] Chanda, M. (Ed.), Science of Engineering Materials, Vol. 3, The Machmillan Company of India Ltd., New Delhi, 1980.

[7] Bunde, Armin and Halvin, Schlomo (Eds.), Fractals and Disordered Systems, Springer, 2nd Edn., 1995.

[8] Maier, J., Physical chemistry of ionic materials. Ions and electrons in solids*, John Wiley and Sons, Ltd, Chichester,* 2004.

[9] Cole, K. and Cole, R., Dispersion and absorption in dielectrics*, J. Chem Phys,* 9, 341–351, 1941.

[10] Batoo, K.M., Study of dielectric and impedance properties of Mn ferrites, *Physica B: Condensed Matte*r, 406, 382–387, 2011.

SYNTHESIS AND FABRICATION OF HIGH STRETCHABLE CONDUCTING POLYPYRROLE FILMS

SEKKARAPATTI RAMASAMY MADESHWARAN AND JAE WHAN CHO

Department of Textile Engineering, Konkuk University, Seoul 143–701, Korea
E-mail: jwcho@konkuk.ac.kr

ABSTRACT

This paper presents highly stretchable conducting films composed of polypyrrole (PPy) and poly(styrene-b-(ethylene-co-butylene)-b-styrene) (SEBS). The PPy was synthesized by chemical oxidative polymerization method with ferric chloride as oxidant. The synthesized PPy at different content was incorporated into SEBS matrix, and then the PPy/SEBS films were prepared in the solvent casting. The synthesis of PPy was confirmed by FT-IR and NMR measurements. SEM measurement showed a good PPy dispersion in SEBS matrix. High electrical conductivity was obtained for the PPy/SEBS films and increased with increasing the PPy content. The synthesized SEBS/Ppy composite films were fabricated as actuator material using double side adhesive tape. The composite films showed good actuation with applied voltage in the presence of dodecyl benzenesulfonic acid as the electrolyte solution.

INTRODUCTION

Actuators and artificial muscle technology has received much attention due to their potential biomedical applications. The attractive properties of conducting polymers such as light weight, low driving voltages, large strain, and high strength make them as a potential candidate for use as actuator material. Among the many kinds of conducting polymers, polypyrrole (PPy) may be one of the most attractive materials due to their excellent electrical conductivity, good mechanical properties, and large actuation strain. However, the processability of polypyrrole may be difficult due to their solubility problem. On the other hand, the blends or composites

V. Rajendran, R. Yuvakkumar, K. Thyagarajah and K.E. Geckeler (eds.)
Applications of Nano Materials: Electronics, Energy and Environment, pp. 41–46 (2012).

of conducting polypyrrole with other polymers may be more attractive for fabrication of actuator materials. There have been some papers reported on PPy composites.[1–5] The electromechanical and dynamic mechanical properties of the SEBS/PPy blends have been prepared by solvent casting approach.[1] The synthesis of PPy/Sulfonated-SEBS by inverted emulsion polymerization method was also reported.[2]

In this study, the composite films of poly(styrene-b-(ethylene-co-butylene)-b-styrene (SEBS) and PPy are prepared and their actuation behaviour has been studied in the presence of electrolyte solution.

EXPERIMENTAL

Synthesis of Polypyrrole

An amount of 12.16 g (0.075 mol) of $FeCl_3$ was dissolved in 354 ml of the distilled water under nitrogen atmosphere at 0°C. After stirring for 40 minutes, 10.06 g (0.15 mol) of pyrrole monomer was slowly added to the reaction mixture. Then the mixture was stirred for 12 hours to give crude black polypyrrole. The resultant product was purified by washing several times with water, methanol, and acetone. The final product was dried in vacuum at 40°C for 12 hours.

Fabrication of SEBS/PPy Composites

Different weight ratios of polypyrrole (10, 20 and 30 wt %) were dispersed in SEBS by sonication and mechanical stirring. In typical experiment, PPy was bath sonicated for 10 minutes using toluene as solvent. Then the PPy solution was transferred in to a 20 ml vial containing SEBS solution (0.5 g in 4 ml toluene). The mixture solution was stirred until PPy was completely dispersed in SEBS matrix. Then the solution was cast in to petri dish and dried at oven at 50°C. The resultant composite films were characterized by SEM, TGA, conductivity and tensile test measurements. The composite films are coded as SEBS/PPy10, SEBS/PPy20, and SEBS/PPy30 where the numbers 10, 20 and 30 represent the wt% of PPy in the SEBS matrix.

Characterization of SEBS/PPy Composites

The synthesized polypyrrole was characterized by using Fourier transform infrared spectroscopy (FT-IR), thermogravimetric analysis (TGA), scanning electron microscopy (SEM), and conductivity measurements in four probe method. The conducting polymer actuator was fabricated by applying the SEBS/PPy composite films on both sides of the double sided tape. The wet actuation measurement was carried out using dodecyl benzene sulfonic acid sodium salt as the electrolyte solution. Different mole ratios of electrolyte solution were employed to study the effect of salt concentration on the actuation mechanism.

RESULTS AND DISCUSSION

Synthesis and Morphology of SEBS/PPy Composites

The FT-IR spectra of the synthesized polypyrrole were measured (Figure 1). The characteristic peaks appeared due to the ring fundamental vibration at 1524.9 and 1461 cm^{-1}, C-N stretching vibration at 1141.3 cm^{-1}, C-H in-plane vibration at 1320 and 1030.5 cm^{-1} and C-H out of plane vibration at 885.5 cm^{-1}. The results confirmed the polymerization of pyrrole.[6] Different loadings of PPy particles were dispersed in the SEBS matrix by solvent casting approach. Figure 2 shows the SEM images of the pure PPy and SEBS/PPy composite films. It was found that the PPy particles were present within the SEBS matrix as in the form of small spherical granules.

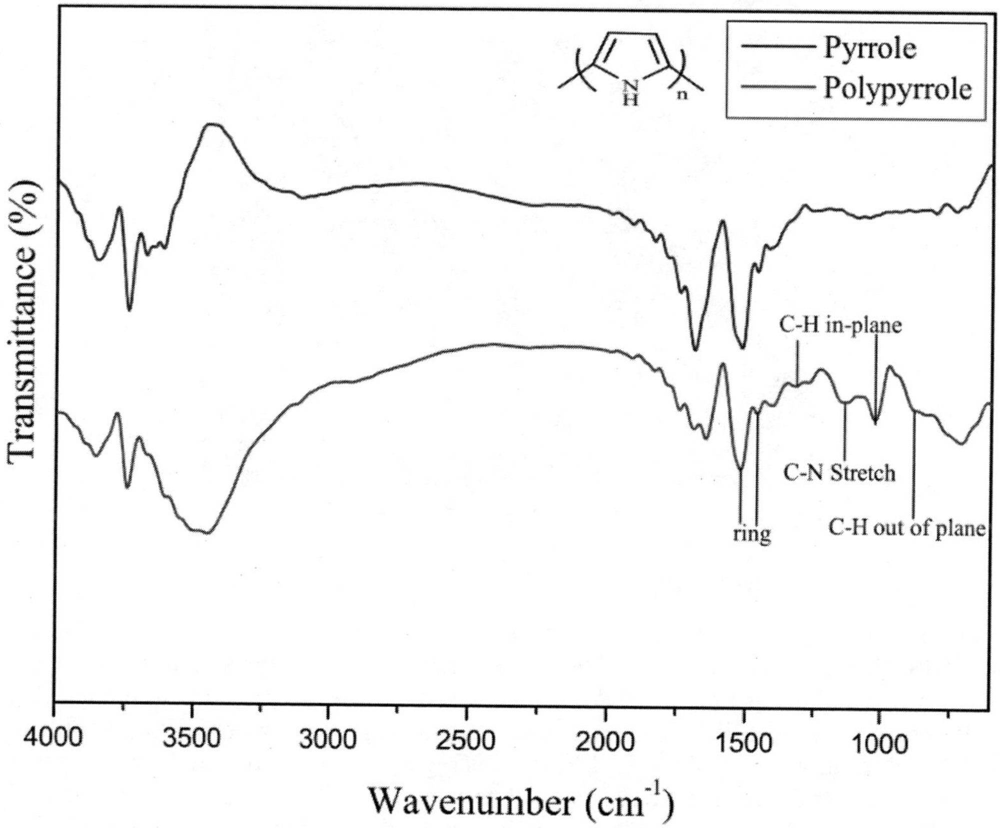

Fig. 1: FT-IR Spectra of Pyrrole and Polypyrrole

Fig. 2: SEM Images of Pure PPy and SEBS/Ppy30 Composite Films

Mechanical and Conducting Properties of SEBS/PPy Composites

The mechanical properties such as breaking stress, elongation-at-break, and modulus of the pure SEBS, SEBS/PPy10, SEBS/PPy20, and SEBS/PPy30 are summarized in Table 1. The breaking stress and elongation-at-break of the composite films decreased with the increase of the PPy content in the SEBS. However the stretchability of composite films is still good though the elongation-at-break for SEBS/PPy30 is highly decreased due to the high content of brittle PPy. The modulus of the composite films increased after addition of PPy, depending on the PPy content. Table 1 also represents the conductivity of the conductivity of the pure PPy, PPy10, PPy20 and PPy30. The conductivity of the composite films increased as the content of PPy particles was increased.[7] These results indicate the success of the high stretchable conducting films by incorporating the PPy in the SEBS matrix. The actuation of the SEBS/PPy composite films was measured in the presence of dodecyl benzenesulfonic acid sodium salt as the electrolyte solutions.[8] When the dodecyl benzenesulfonic acid sodium salt was used as the electrolyte solution, we could observe the good displacement of the actuator due to the large anion size which facilities the high volume change in PPy.[9] For the SEBS/PPy30 composites, the obtained displacement was 10.76 mm on the application of 7 V.

Table 1: Electrical Conductivity and Mechanical Properties of Samples

Sample	*Conductivity (S/cm)*	*Breaking stress (MPa)*	*Elongation-at-Break (%)*	*Modulus (MPa)*
PPy	9.3×10^{-3}	–	–	–
SEBS	–	32.2	1133.5	14.7
SEBS/PPy10	1.2×10^{-3}	25.7	984.8	32.4
SEBS/PPy20	3.6×10^{-3}	21.4	928.4	21.2
SEBS/PPy30	8.6×10^{-3}	10.7	640.7	20.3

CONCLUSIONS

The SEBS/PPy composites were successfully synthesized by simple solvent casting. They showed the increased conductivity with the increase of PPy content and the excellent stretchable property, indicating the good stretchable conducting properties of SEBS/PPy composites. The good electrical actuation was also demonstrated in the presence of dodecyl benzenesulfonic acid sodium salt electrolyte solution.

ACKNOWLEDGEMENTS

This research was supported by the grant from the Industrial technology development program of the Ministry of Knowledge Economy (MKE) of Korea (No KEIT 10033449–2009–11) and the Defense Acquisition Program Administration (DAPA) and the Agency for Defense Development (ADD).

REFERENCES

[1] Muller, D., Garcia, M., Salmoria, G.V., Pires, A.T.N., Paniago, R. and Barra, R.M.O., SEBS/PPy.DBSA Blends: Preparation and Evaluation of Electromechanical and Dynamic Mechanical Properties, *Journal of Applied Polymer Science*, 120, 351, 2011.

[2] Lee, Y.H., Ju, Y.W., Jung, H.R., Huh, Y. and Lee, W.J., Preparation of Polypyrrole/Sulfonated-SEBS Conducting Composites through an Inverted Emulsion Pathway, Journal of Industrial Engineering and Chemistry, 11, 550, 2005.

[3] Chen, S., Chen, W., Si, Z., Zhou, D. and Xue, G., Preparation of Size Controllable Polypyrrole Sub-microcapsules Using SEBS Copolymer as the Building Block, *Macromolecular Rapid Communications,* 27, 328, 2006.

[4] Ju, Y.W., Park, J.H., Jung, H.R. and Lee, W.J., Electrochemical Properties of Polypyrrole/ Sulfonated SEBS Composite Nanofibers Prepared by Electrospinning, Electrochimica Acta, 52, 4841, 2007.

[5] Laurer, J.H., Mulling, J.F., Khan, S.A., Spontak, R.J. and Bukovnik, R., Thermoplastic Elastomer Gels. I. Effects of Composition and Processing on Morphology and Gel Behaviour, *Journal of Polymer Science: Polymer Physics,* 36, 2379, 1998.

[6] Sahoo, N.G., Jung, Y.C., So, H.W., and Cho, J.W., Polypyrrole Coated Carbon Nanotubes: Synthesis, Characterization, and Enhanced Electrical Properties, Synthetic Metals, 157, 374, 2007.

[7] Rana, S. and Cho, J.W., Core-shell Morphology and Characterization of Carbon Nanotube Nanowires Click Coupled with Polypyrrole, *Nanotechnology,* 22, 275609, 2011.

[8] Kang, H.C. and Geckeler, K.E., Enhanced Electrical Conductivity of Polypyrrole Prepared by Chemical Oxidative Polymerization: Effect of the Preparation Technique and Polymer Additive, Polymer, 41, 6931, 2000.

[9] Park, S.J., Cho, M.S., Nam, J.D., Kim, I.H., Choi, H.R., Koo, J.C. and Lee, Y., The Linear Stretching Actuation Behavior of Polypyrrole Nanorod in AAO Template, Sensors and Actuators, 135, 592, 2009.

STRUCTURAL, SPECIFIC HEAT AND CONDUCTION MECHANISM IN $Gd_{0.7}Ca_{0.3}MnO_3$

P. NEGI, H.M. AGRAWAL AND R.C. SRIVASTAVA

Department of Physics, G.B. Pant University of Ag. and Technology, Pantnagar, Uttarakhand
E-mail: puneetphymat@gmail.com

ABSTRACT

The structural, specific heat and electronic transport properties of orthorhombic perovskite structured $Gd_{0.7}Ca_{0.3}MnO_3$, synthesized by modified sol-gel method has been reported. X-ray diffraction along with the Rietveld refinement confirms the pure phase of the sample and the crystallite size comes out to be ~42 nm. Scanning electron micrograph shows the grain distribution having average particle size ~71 nm. Energy dispersive spectroscopy confirms the stoichiometry of the composition. Sample is also characterized by Raman spectroscopy for structural and phase confirmation. The specific heat of the sample is measured in the temperature range of 2–300 K. By using the specific heat data the Debye temperature was also calculated. Resistivity measurements with and without magnetic field has been done. The activation energy in absence of magnetic field is 0.162 eV which reduces to 0.154 eV in presence of 8 T magnetic field. The observed electrical behavior is explained in terms of polaron hopping mechanism.

INTRODUCTION

$R_{1-x}A_xMnO_3$ type perovskite manganites (R = rare-earth and A = alkaline-earth elements) have received much attention due to their interesting electronic and magnetic properties. Recently, very large magnetoresistance has been found in some of these materials, which show the metal insulator transition.[1–3] It has been anticipated that a structural transition accompanies the ferromagnetic transition, and the relative importance of each in promoting the metal-insulator transition is of particular interest.[4] Emin and Holstein, suggested a model to explain the conduction mechanism above metal-insulator transition, and gave $\rho = BTe^{E_A/k_BT}$.[5] In the adiabatic approximation, the polaron has a high probability of jumping into the next site, and it was assumed that the polaron makes several jumps back and forth before the

V. Rajendran, R. Yuvakkumar, K. Thyagarajah and K.E. Geckeler (eds.)
Applications of Nano Materials: Electronics, Energy and Environment, pp. 47–52 (2012).

phonon wave recedes. Although the electron phonon interaction is one of the most relevant contributions in determining the conduction mechanism in these materials but lattice part of the specific heat also deserves proper attention.[6]

In the present work, we studied the structural, specific heat and electronic transport properties of $Gd_{0.7}Ca_{0.3}MnO_3$. The composition x = 0.3 was chosen because this doping concentration is supposed to maximize the double exchange effect as seen in the lanthanum compounds.[7]

EXPERIMENTAL

$Gd_{0.7}Ca_{0.3}MnO_3$ was synthesized by modified sol-gel method. Aqueous solutions of gadolinium, calcium and manganese salts were prepared separately in stoichiometric proportion by dissolving, $Gd(NO_3)_3$ $6H_2O$, $Ca(NO_3)_2$ xH_2O and $Mn(NO_3)_2.4H_2O$ in distilled water along with citric acid. The cation to citric acid molar ratio was kept as 1:2. The solution was stirred at 80–85°C until the viscous gel was formed. This gel was further dried in an oven at 100°C for 8 h and grounded into fine black powder. Finally, the black precursor was sintered at 800°C for 2 h to get the desired $Gd_{0.7}Ca_{0.3}MnO_3$ powder. The powder was pelletized by applying a pressure of 2 tone and then heated at 1200°C for 13 hours. The sample was characterized by X-ray diffractometer (XRD), Raman spectrometer, field emission scanning electron microscope and energy dispersive X-ray detector. The PPMS was used for the specific heat measurement. DC four-probe resistivity measurement was done to investigate the conduction mechanism of the sample with and without magnetic field.

RESULTS AND DISCUSSION

Structural Study

XRD Analysis

The structural investigation of the sample was done by Rietveld refinement of the XRD pattern and is shown in Figure 1. The Rietveld refinement analysis of XRD data was done by assuming orthorhombic structure with pnma space group. The pseudo-Voigt function was used to fit the

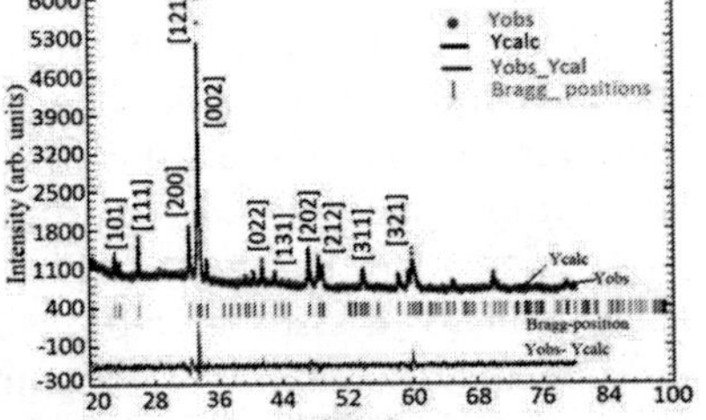

Fig. 1: Rietveld Refined XRD Pattern of $Gd_{0.7}Ca_{0.3}MnO_3$

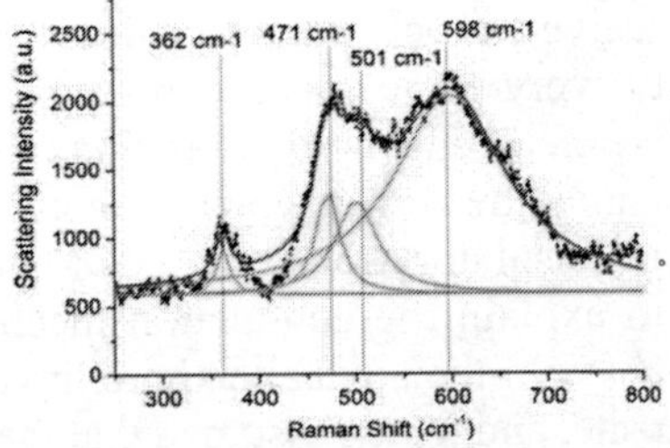

Fig. 2: Lorentz Fitting of Raman Spectra of $Gd_{0.7}Ca_{0.3}MnO_3$ (Showing modes associated with MnO_6 Octahedra)

peak profile. Rietveld refinement gives the value of lattice parameters as, a = 5.55 Å, b = 7.51 Å and c = 5.34 Å. Scherrer's formula gives the crystallite size of 42 ±1 nm. It is clear from the analysis of the XRD data that the sample has single phase, without any detectable impurity.

Raman Analysis

In order to confirm the phase, the Raman spectroscopy was also carried out. Figure 2 shows the Lorentz fitted Raman spectra of $Gd_{0.7}Ca_{0.3}MnO_3$ in the spectral range of 200–800 cm^{-1}. The spectrum shows four modes associated with MnO_6 octahedra as observed in $GdMnO_3$ [8] (Table 1). There is a small shifting in position of Raman modes in comparison to that of pure $GdMnO_3$, as shown in Table 1. This may be due to the doping effect of Ca in place of Gd^{+3} ions as the band position depends on the masses, positions of the atoms, the interatomic force and the bond length. Therefore any effect altering these features will produce a change in the frequency of the band.[9]

Table 1: Raman Modes of $GdMnO_3$ [8] and $Gd_{0.7}Ca_{0.3}MnO_3$

Raman Mode Associated with MnO_6 Octahedra	**$GdMnO_3$**	**$Gd_{0.7}Ca_{0.3}MnO_3$**	**Shift in Raman modes**
Stretching mode (Symmetric) (υ_s)	615 cm^{-1}	598 cm^{-1}	17 cm^{-1}
Bending mode (δ)	511 cm^{-1}	501 cm^{-1}	10 cm^{-1}
Stretching mode (Asymmetric) (υ_{as})	493 cm^{-1}	478 cm^{-1}	15 cm^{-1}
Tilting mode (τ)	378 cm^{-1}	362 cm^{-1}	16 cm^{-1}

SEM and EDS Analysis

The morphology of the $Gd_{0.7}Ca_{0.3}MnO_3$ was observed by SEM, as shown in Figure 3 (a). The average particle size is estimated to be ~71 nm, larger than its averaged crystallite size of 42 nm.

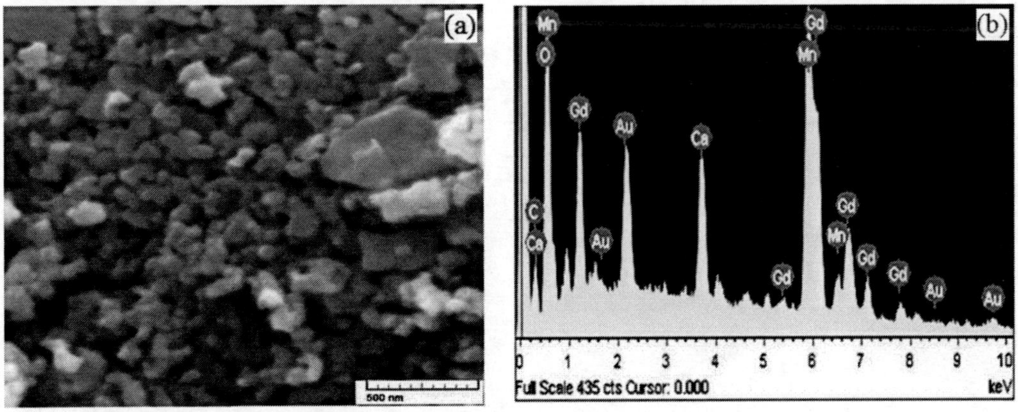

Fig. 3: (a) High-Magnification SEM and (b) EDS Images of the $Gd_{0.7}Ca_{0.3}MnO_3$

The EDS spectrum was recorded to determine the chemical composition of the $Gd_{0.7}Ca_{0.3}MnO_3$. Figure 3 (b) shows that main elements in our sample are Gd, Ca, Mn, and O, close to the nominal concentration. The C and Au peaks are due to the coating of the sample. The unmarked peaks in Figure 3(b) are of Cu and these are originated from the sample grid.

Specific Heat Study

Figure 4 shows the variation of specific heat (C_p) with temperature in the absence of magnetic field. In $GdMnO_3$ a sharp peak at 40 K was attributed to the antiferromagnetic ordering of Mn^{3+} ion designated as $T_N(Mn^{3+})$ and the broad hump at 5 K indicates the ordering of Gd^{+3} ion spin and is hereafter designated as $T_N(Gd^{3+})$.[10] But in $Gd_{0.7}Ca_{0.3}MnO_3$, no such type of transition was observed, however at 5 K there is a sudden decrease in the specific heat which indicates the possibility of ordering of the Gd^{+3} ion on application of magnetic field. At low temperature, the specific heat satisfies the well known law: $C_p = \gamma T+\beta T^3$. Inset of Figure 4 shows the linear fitting of C_p/T vs T^2 plot to evaluate the value of γ and β. In the absence of spin fluctuations, β does not depend on the temperature and the magnetic field and is related to the Debye temperature θ_D, $\theta_D = (12\pi^4NR/5\beta)^{1/3}$, where R is the universal gas constant and N the number of atoms per formula unit.[11–12] The calculated value of Debye temperature θ_D ~545 K, which is higher than that of $GdMnO_3$ (~474 K).[11] The higher value of Debye temperature indicates the presence of higher phonon frequency in $Gd_{0.7}Ca_{0.3}MnO_3$, which may be due to the doping of the Ca^{+2} ions.

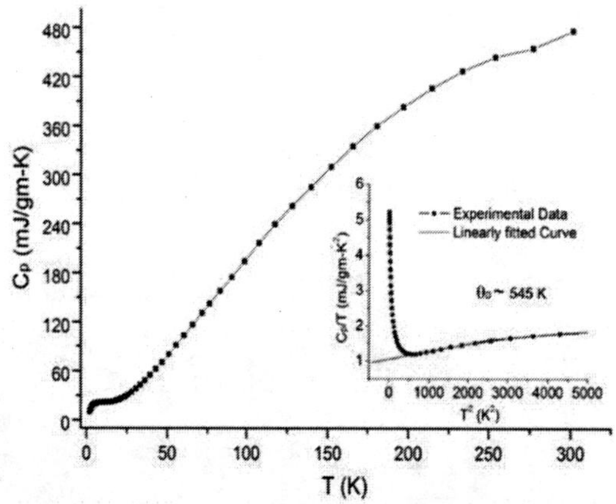

Fig. 4: Specific Heat vs. Temperature Plot of $Gd_{07}Ca_{0.3}MnO_3$. Inset Shows the Cp/T vs. T^2 Plot in the Absence of Magnetic Field

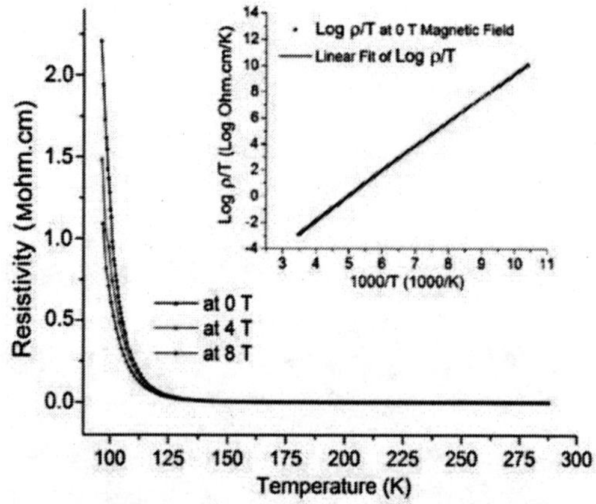

Fig. 5: Resistivity vs. Temperature Plot of $Gd_{0.7}Cs_{0.3}MnO_3$. Inset Shows the Log ρ/T vs. 1000/T Plot in the Absence of Magnetic Field

Resistivity Study

The conduction process in manganites for the insulating phase is attributed mainly to the presence of thermally activated polarons.[13] The polaron model for the temperature dependence of resistivity is given as,

$$\rho = BT\exp\left(\frac{E_A}{k_B T}\right) \qquad \dots (1)$$

where, B = resistivity coefficient (ohm.cm/ K), E_A = Activation energy (eV), k_B = Boltzmann constant. Figure 5 shows the resistivity vs. temperature graph at different magnetic fields. The resistivity decreases with increase in magnetic field and temperature. At temperature above 150 K the resistivity is almost constant and is independent of magnetic field. Inset of Figure 5 shows log ρ/T vs. 1000/T graph in the absence of magnetic field, which is linear. Similar linear graphs are obtained for 4 T and 8 T magnetic fields too. The linear fitting was done according to eq. (1). Fitting gives the value of polaron activation energy (E_A) and is collated in table 2. The activation energy in the absence of magnetic field is 0.162 eV which reduces to 0.154 eV in presence of 8 T magnetic field. The heating and cooling (resistivity vs. temperature) plot at 8 T magnetic field, gives the reversible nature of the resistivity in this temperature range.

Table 2: Polaron Activation Energy (E_A)

Magnetic Field (T)	*Polaron Activation Energy (eV)*
0	0.162
4	0.159
8	0.154

CONCLUSIONS

$Gd_{0.7}Ca_{0.3}MnO_3$ in pure orthorhombic perovskite phase was prepared by modified sol gel method. The average crystallite size is ~42 nm. The specific heat measurement gives the Debye temperature of the sample ~545 K, which is higher in comparison to $GdMnO_3$ and it signifies the higher phonon frequency in $Gd_{0.7}Ca_{0.3}MnO_3$. The resistivity decreases with increasing magnetic field and temperature and is consistent with small polaron hopping in the observed temperature range.

ACKNOWLEDGMENTS

H.M.A. is thankful to IUAC, New Delhi for funding the research project no. UFR-49309. Puneet Negi is also thankful to Dr. V. Ganesan, Dr. R. Rawat and Dr. Mukul Gupta for their keen interest and suggestions in specific heat, resistivity and XRD measurements, respectively at UGC-DAE-CSR, Indore.

REFERENCES

[1] Snyder, G.J., Hiskes, R., DiCarolis, S., Beasley, M.R. and Geballe, T.H., Intrinsic electrical transport and magnetic properties of $La_{0.67}Ca_{0.33}MnO_3$ and $La_{0.67}Sr_{0.33}MnO_3$ MOCVD thin films and bulk material, *Phys. Rev. B,* 53, 14434–14444, 1996.

[2] Gomes, I.T., Pereira, A.M., Ventura, J., Carpinteiro, F., Moreira, J.M., *et al.*, High-Field Magnetoresistance of $La_{0.67}Sr_{0.33}MnO_3$ Thin Films Deposited on $LiNbO_3$ Substrates, Low Temp. Phys., 159, 156–159, 2010.

[3] Segal, Y., Garrity, K.F., Vaz, C.A.F., Hoffman, J.D., Walker, F.J., *et al.*, Resonant phonon coupling across the $La_{1-x}Sr_xMnO_3/SrTiO_3$ interface, Phys. Rev. Lett., 107, 105501, 2011.

[4] Millis, A.J., Shraiman, B.I. and Mueller, R., Dynamic Jahn-Teller Effect and Colossal Magnetoresistance in $La_{1-x}Sr_xMnO_3$, *Phys. Rev. Lett.*, 77, 175–178, 1996.

[5] Emin, D. and Holstein T., Studies of small-polaron motion IV. Adiabatic theory of the Hall effect, *Ann. Phys.,* 53, 439–520, 1969.

[6] Choithrani, Renu, Gaur, N.K., Heat capacity of $EuMnO_3$ and $Eu_{0.7}A_{0.3}MnO_3$ (A = Ca, Sr) compounds, *J. Magn. Magn. Mater.* 320,612–616, 2008.

[7] Schiffer, P., Ramirez, A. P., Bao, W. and Cheong, S.W., Low Temperature Magnetoresistance and the Magnetic Phase Diagram of $La_{1-x}Ca_xMnO_3$ Phys. Rev. Lett., 75, 3336–3339, 1995.

[8] Ferreira, W.S., Agostinho Moreira, J., Almeida, A., Chaves, M. R., Araujo, J. P., Spin-phonon coupling and magnetoelectric properties: $EuMnO_3$ versus $GdMnO_3$, *Phys. Rev. B*, 79, 054303, 2009.

[9] Jawhari, T., Micro-Raman spectroscopy of the solid state: applications to semiconductors and thin films, Analusis, 28, 15–21, 2000.

[10] Pavan Kumar, N, Lalitha, G and Venugopal Reddy, P, Specific Heat and Magnetization study of $RMnO_3$ (R = Sm, Eu, Gd, Tb and Dy) multiferroics, Phys. Scr., 83, 045701, 2011.

[11] Choithrani, Renu, Gaur, N. K., Singh, R. K., Specific Heat and Transport Properties of $La_{1-x}Gd_xMnO_3$ at $15 \leq T \leq 300$ K, Solid State Commun., 147, 103–106, 2008.

[12] Viennois, Romain, Girard, Luc, Ravot, Didier, Paschen, Silke, Charar, Salam, Thermoelectric properties of $Ce(La)Fe_4Sb_{12}$ skutterudites under a magnetic field, *Phys. Rev. B,* 80, 155109, 2009.

[13] Worledge, D.C., Snyder, G.J., Hiskes, R., DiCarolis, S., Beasley, M.R. and Geballe, T.H., Anneal-tunable Curie temperature and transport of $La_{0.67}Ca_{0.33}MnO_3$, *J. Appl. Phys.*, 80, 5158, 1996.

EFFECT OF ANNEALING AND γ-IRRADIATION ON THE STRUCTURE AND OPTICAL TRANSITIONS OF NANOCRYSTALLINE NICKEL OXIDE

K. Maniammal, Vipin C. Bose, G. Madhu, L.R. Asitha and V. Biju

Department of Physics, University of Kerala, Kariavattom, Thiruvananthapuram, Kerala
E-mail: bijunano@gmail.com

ABSTRACT

Nanocrystalline nickeloxide (NiO) is a very interesting system due to the less well understood electronic structure dominated by the narrow '3d' band of Ni and contribution of defects associated with surface. The varied electrical and electronic properties can be advantageously used in diverse fields such as positive temperature coefficients, temperature sensors, smart windows, electrodes for supercapacitors, catalysts, switching devices, etc. γ-irradiation is evolving as an effective technique for material processing especially in the case of polymers and ceramics. Herein, we report a comparison of the effect of annealing and γ-irradiation on the structure and optical absorption of nanocrystalline NiO with an average crystallite size of 2–22 nm. Nanocrystalline NiO was synthesized through a two step route and decomposed at 250°C. The sample was exposed to different doses, for the as prepared samples using γ-irradiation from a ^{60}Co source. A comparison, on the changes in the average crystallite size, microstrain, lattice constants, was estimated from X-ray diffraction (XRD). It was found that unlike annealing γ-irradiation does not cause any significant change in crystallite size, while the optical transitions show changes due to both processes. A detailed investigation of the changes in intensities in the various optical transitions in the UV- Visible region of as-prepared and γ-irradiated is presented in the paper.

INTRODUCTION

Nickel oxide, NiO with a wide intrinsic band gap of ~3.6 eV [1,2] shows interesting optical, electrical and magnetic properties, which are advantageous in areas such as smart windows,

V. Rajendran, R. Yuvakkumar, K. Thyagarajah and K.E. Geckeler (eds.)
Applications of Nano Materials: Electronics, Energy and Environment, pp. 53–58 (2012).

thermistors, spin valves, etc. These properties are determined by the behaviour of electrons in the 2p band of oxygen and localized 3d band of Ni and show large variations depending on the structure, defect concentration, processing conditions, etc. Annealing in air, vacuum or in the presence of gases such as O_2, N, Ar etc., are effective methods for varying the physical properties of nanocrystalline transition metal oxides.[3] γ-irradiation by swift heavy ions or high energy radiation such as γ-rays is an emerging area of materials processing.[4] γ-irradiation would lead to the creation or annihilation of oxygen of oxygen vacancies in metal oxides and hence could be employed as an effective tool in tuning the physical properties of nanocrystalline transition metal oxides.[4–6] In this paper, we report a comparison of the effect of annealing and γ-irradiation on the structure and optical transitions of nanocrystalline NiO.

EXPERIMENTAL

Nanocrystalline NiO was prepared by the thermal decomposition of carbonate precursor synthesized through the controlled chemical precipitation route.[7] Nickel chloride ($NiCl_2$), Ammonium carbonate (NH_4CO_3) were used as starting materials and Ethylene dinitrilotetraacetic acid disodium salt dehydrate (EDTA) was used as the stabilizer.

All the chemicals used were of analytical grade and were used without further purification. The nanocrystalline NiO sample was annealed at different temperatures, in the range of 500–800°C. Further, the as prepared NiO samples was enclosed in a self lock polythene cover and were exposed to ^{60}Co (γ) radiation source at a dose rate of 8.47 kG/hour, in different time and gray. Sample codes were assigned for convenience and are listed in Table 1 together with the processing conditions.

X-ray diffraction pattern of the samples were recorded in the range using a Philips XPERT PRO diffractometer with Cu Kα as radiation (λ = 1.54056Å) source. The nanocrystalline samples were dispersed in distilled water using an ultrasonic bath and the UV-Visible absorption spectra of the dispersion in the range of 200–800 nm was recorded using Shimadzu UV-2550 Spectrophotometer.

RESULTS AND DISCUSSION

The XRD patterns of the nanocrystalline NiO samples are shown in Figures 1 and 2. These can be indexed with JCPDS-ICDD pattern no. 47–1049, corresponding to NiO (bunsite) with space group of Fm3m (225). No peak corresponding carbonate precursor or any other oxide of Ni such as Ni_2O_3 is present in the XRD pattern of the as-prepared sample N1 ensuring phase purity. Further, no phase change occurs on annealing or on γ-irradiation as detectable from the XRD analysis.

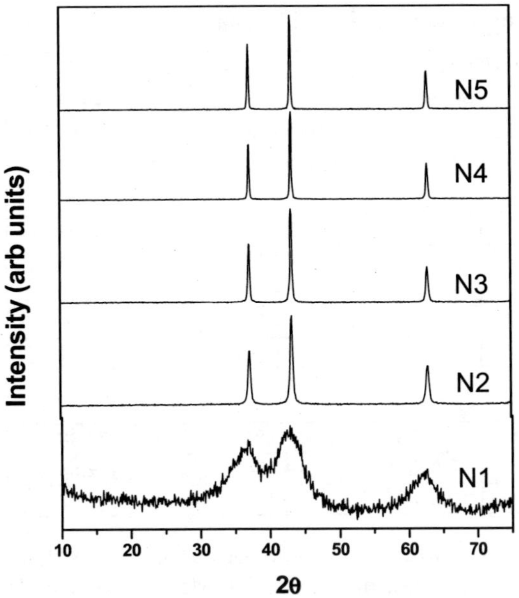

Fig. 1: The XRD Pattern for the Annealed Samples

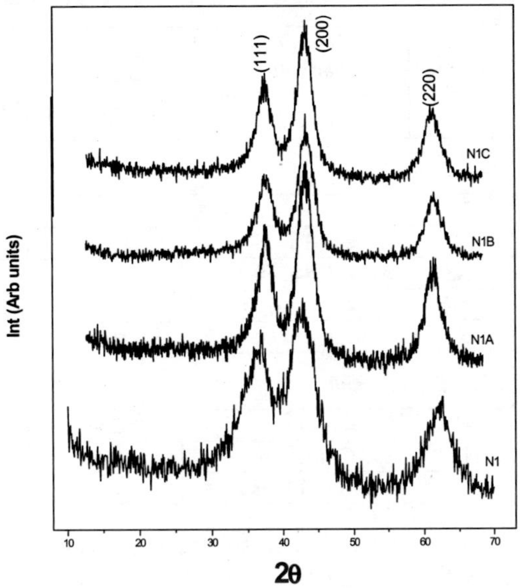

Fig. 2: The XRD Pattern for the N1 and γ-Irradiated Samples

The crystallite size of all the samples were calculated from the line broadening of the diffraction peaks corrected for instrumental broadening using the Scherrer formula, $D = K\lambda/\beta \cos\theta$ where, K is a geometrical shape factor, λ is the wavelength of light used in XRD; θ is the Bragg angle; β is full width at half maximum of diffraction peaks.[8] The results are included in Table 1. It can be noted that there is an increase in crystallite size with increase in the annealing temperature. Hall-Williamson analysis was carried out to separate the size and microstrain contributions to XRD line broadening and the results are listed in Table1. According to Hall-Williamsons model, both small crystallite size and microstrain contribute to the XRD line broadening. This leads to the relation, $\beta \cos\theta = k\lambda/D + 4\varepsilon \sin\theta$ where, λ is the wavelength of X-rays used; $\langle\varepsilon\rangle$ is the rms microstrain, λ the wavelength used, k the geometrical shape factor and D the crystallite size. The rms microstrain, $\langle\varepsilon\rangle$ is obtained from the slope of the straight line obtained by plotting ($\beta \cos\theta$) vs ($4 \sin\theta$) while the average crystallite size corrected for microstrain is obtained from the Y-intercept. Results of Hall-Williamson analysis are also included in Table 1.

For samples N1 and N5, the Hall-Williamson plots have a negative slope which is physically unacceptable as rms microstrain cannot be negative. The very small microstrain values in the present case for all the samples leads to close agreement between the crystallite sizes estimated from Scherrer equation and Hall-Williamson analysis. Thus from Hall-Williamson analysis, it can be concluded that in the case of nanocrystalline NiO samples of the present study, microstrain is minimal.

Table 1: Results of XRD Analysis of Nanocrystalline NiO Samples

Sample Code	*Annealing Temperature/γ-Irradiated in Gray and Time*	*Crystallite Size Schrerrer Equation (nm)*	*Hall-Williamson*		*Lattice Constant (Å) Bulk a = 4.1771*	*Cell Volume ($Å^3$) Bulk V = 72.87*
			Size (nm)	Strain		
Annealed samples						
N1	Decomposed at 250°C	1.79			4.1857	73.33
N2	Annealed at 500°C	15.50	15.94	0.00237	4.1728	72.65
N3	Annealed at 600°C	18.94	19.01	0.001795	4.1757	72.80
N4	Annealed at 700°C	21.62	22.24	0.0032	4.1754	72.79
N5	Annealed at 800°C	21.70			4.1777	72.91
γ-irradiated samples						
N1A	50 G in 21 sec	2.96	2.81	0.00035	4.1851	73.30
N1B	150 G in 1'4 sec	2.60	2.6	0.00035	4.1768	72.86
N1C	250 G in 1'46 sec	2.94	3.15	0.00885	4.1799	73.02

The precise lattice constant 'a' was estimated from the exptrapolation method of Nelson-Riley function.[9,10] The lattice constant for single crystalline NiO is reported to be JCPDS-ICDD pattern no. 47–1049.[11] It may be noted that the annealing and γ-irradiation process causes no change in lattice constant as inferred from the XRD analysis.

The as prepared sample which is black in colour changes into green on annealing indicating a decrease in the concentration of uncompensated Ni^{2+} vacancies. This is because the effect of one Ni^{2+} vacancy is compensated by the creation of one oxygen vacancy which results due to high temperature annealing.[12] A slight change of colour from black to green was observed on γ-irradiation also but the colour was not similar to the one obtained on annealing. The UV Visible spectra of the samples are show in Figures 3 and 4. The UV Visible spectra of the annealed and γ-irradiated samples in the range 2 to 6 eV were deconvoluted assuming Gaussian profile for each transition. The spectra for samples N1, N5 and N1B which are representative of the procedure are shown in Figures 5, 6 and 7 respectively. The peak energy values of transitions together with the assignments are given in Table 2. The transitions below 4 eV are the purely intraionic $3d^8$-$3d^8$*transitions. It was noted that the maxima of these peaks are slightly shifted towards higher energies with annealing and also on γ-irradiation. In the case of irradiated samples the transition involving $2p^6$–$2p^5$4S transition which appear in the 4–5 eV range also shoe a shift towards higher energy values on annealing and γ-irradiation. Further, the intensity of $3d^8$-$3d^8$*transition with peak in the range 3.5–3.7 eV is the one with maximum intensity in the case of annealed samples while in the case of γ-irradiated samples, the maximum intensity peak is the $2p^6$–$2p^5$4s transition which appear in the 4–5 eV range.[13,14] Thus the UV Visible analysis reveals that even though both annealing and γ-irradiation causes change in the concentration of defects as inferred from the change ion colour from black to green the two processes are distinctly different according to the electronic transition probability. Another important difference is that while annealing causes grain growth, γ-irradiation does not result in any appreciable change in grain size.

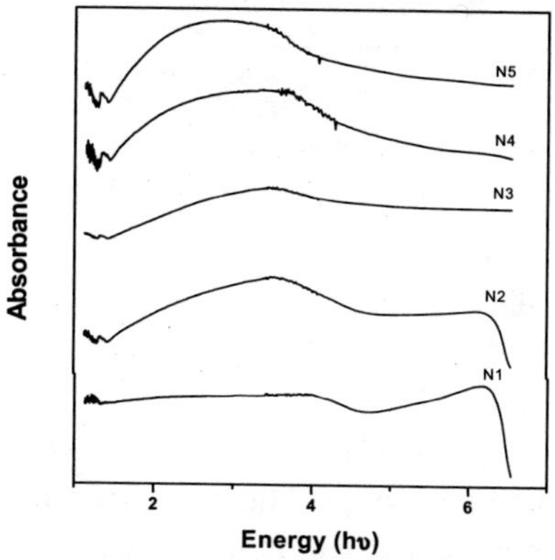

Fig. 3: The Optical Absorption Spectrum for the Annealed NiO Samples

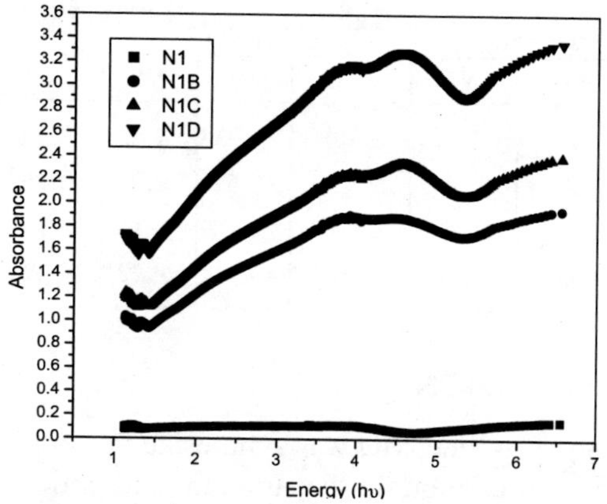

Fig. 4: The UV-visible Spectrum of N1 and γ-Irradiated Samples

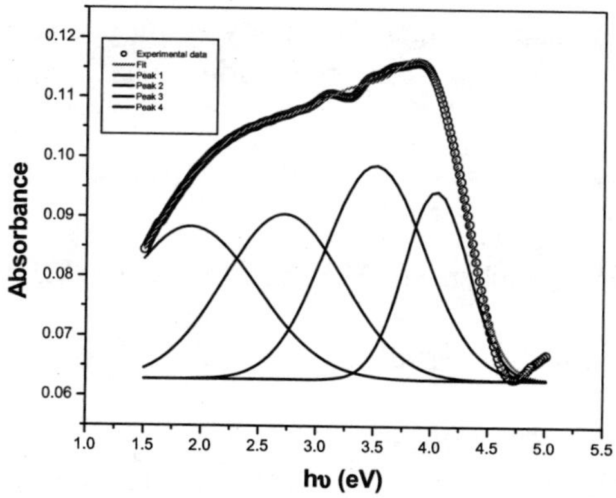

Fig. 5

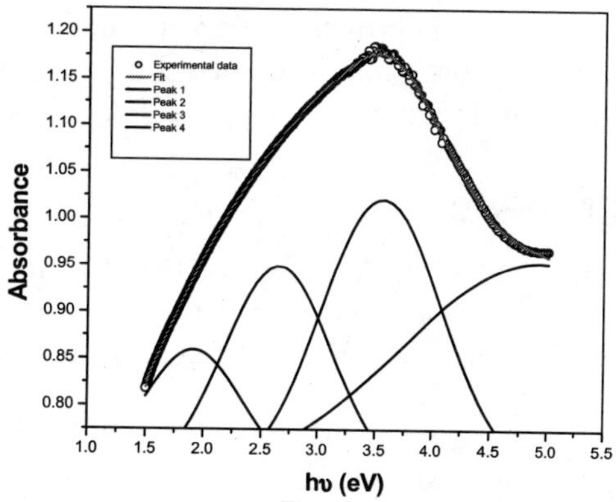

Fig. 6

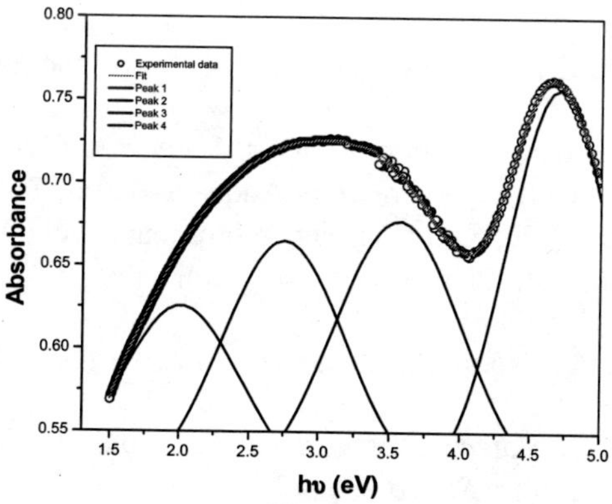

Fig. 7

Figs. 5–7: Shows the Optical Absorption Spectrum of Samples Like N1, N5 and N1B

Table 2: Results of UV-Visible Analysis for the NiO Samples

N1	*N2*	*N3*	*N4*	*N5*	*N1A*	*N1B*	*N1C*	*Assignments*
2.07	2.06	2.18	1.93	1.88	2.34	2.34	2.31	Purely intraionic $3d^8$-$3d^8$*transition
2.56	2.70	2.72	2.62	2.38	3.16	3.11	3.07	Purely intraionic $3d^8$-$3d^8$*transition
3.50	3.52	3.40	3.42	3.14	3.74	3.72	3.70	$3d^8$-$3d^8$* and onset of 3d-3d74s
4.03	4.89	4.54	4.21	4.22	4.62	4.70	4.65	3d-3d74s transitions

CONCLUSION

Nanocrystalline NiO was synthesized through a facile chemical route. The effect of annealing and γ- irradiation on the structure and optical transitions were studied. Both annealing and γ-irradiation causes a decrease in uncompensated Ni^{2+} vacancies due to creation of oxygen valences. However the UV visible spectra reveals that the maximum probable transition is annealed and γ-irradiated samples are different. Another notable distinction is that annealing introduces grain growth while γ-irradiation causes no such change. The study employed as an effective processing technique for processing of nanostructured transition metal oxides without grain growth.

REFERENCES

[1] Swagten, H.J.M., Strijkers, G.J., Bloemen, P.J.H., Willekens, M.M.H. and De, Jonge W.J.M. (1996). *Phys Rev B.* 53: 1039.

[2] Suman Nandy, Biswajit Saha, Manoj K. Mitra, K.K. Chattopadhyay, *J Mater Sci.* (2007) 42: 5766–5772.

[3] Suslick, K.S. and Bang, J.H., *Adv. Mater.* 2010, 22, 1039–1059

[4] Arshak, K. and Korostynska, O.S., Clifford, Sensors and Actuators A 110 (2004) 354–360.

[5] Anspoks, A., Kuzmin, A, Kalinko, A. and Timoshenko, J., Solid State Communications 150 (2010) 2270–2274.

[6] Nakamura, Y., Ogawa, H., Nakashima, T., Kishimoto, A. and Yanagida, H., *J. Am. Ceram. Soc.* 80, 1609, (1997).

[7] Biju, V. and Abdul, Khadar, *Journal of Material Science* 36 (2001) 5779–5787

[8] Cullity, B.D., Elements of X-ray diffraction, Addison-Wesley Publishing Company,INC.

[9] Klug, H.P and Alexander, L.E., X-ray Diffraction Procedures, John Wiley, New York (1974).

[10] Cullity, B.D. and Stock, S.R., Elements of X-ray diffraction, 3rd Edition (Prentice- Hall, Upper Saddle River, NJ, 2001).

[11] Martin, K., McCarthy, G., North Dakota State, Univ., Fargo, ND, USA, ICDD Rant-in-Aid, (1991).

[12] Biju, V., Khadar, M.A., *J Materials Research Bulletin,* 36 (2001), 21–33.

[13] David, A. and Julius, F., *Phys. Rev. B* 2, 3112 (1970).

[14] Kuzmin A., Mironova-Ulmane N., and Ronchin S., "Origin of visible photoluminescence in NiO and NicMg1−*c*O singlecrystals," *Proceedings of SPIE*, Vol. 5122, pp. 61–67, 2003.

EFFECT OF Cu AND Mn DOPING ON STRUCTURAL, ELECTRICAL AND MAGNETIC PROPERTIES OF Ni-Co NANOCRYSTALLINE FERRITE SYSTEM

P.S. JADHAV, K.K. PATANKAR[1], V.L. MATHE[2] AND VIJAYA PURI

Thick and Thin Film Device Lab, Department of Physics, Shivaji University, Kolhapur
[1]Physics Department, Rajaram College, Kolhapur
[2]Novel Material Research Lab, Department of Physics, University of Pune
E-mail: vijayapuri1@gmail.com

ABSTRACT

$Ni_{0.8}Co_{0.2-2x}Cu_xMn_xFe_2O_4$ [x = 0.01, 0.03, 0.05, 0.07, 0.09] ferrite system was successfully prepared via combustion route. The effect of Cu and Mn doping in Ni-Co ferrite was studied. All the ferrites show pure phased cubic spinel structure. SEM images show the formation of nanocrystalline ferrites. Resistivity studies show semiconducting behavior and also slight decrease in resistivity with Cu and Mn content. Magnetostriction enhances with Cu and Mn content.

INTRODUCTION

The great interest has been developed in the field of nanoferrites due to its interesting properties like high resistivity, low eddy current losses applicable in most of the useful areas like transformer core, magnetic recording, gas sensing. Due to their high resistivity and low eddy current losses these ferrites are used in radio frequency circuits; high quality filters, rod antennas, transformer cores, read/write heads for high-speed digital tape and operating devices.[1] Number of methods like hydrothermal route, co-precipitation, high energy ball milling have been developed for synthesis of complex nanoferrite system. The co-precipitation method is very sensitive to pH and most of the methods are expensive or energy consuming. The main reasons to select combustion technique for synthesis the mixed nanoferrite system is that it is very simple, cost

V. Rajendran, R. Yuvakkumar, K. Thyagarajah and K.E. Geckeler (eds.)
Applications of Nano Materials: Electronics, Energy and Environment, pp. 59–64 (2012).

effective, consumes low external energy.[2] The method is of great technical interest due to high homogeneity and good chemical and improved physical properties of finally synthesized ferrite powder. This particular composition was selected so as to get high resistivity and large magnetostriction. The nickel ferrite has high resistivity of the order of 10^{11} Ω-cm and the additional doping of the John-Tellar ions will enhance the magnetostriction of the ferrite system. This composition will be used as one of the candidate of magnetoelectric composite to achieve considerable ME coefficient. The highly resistive ferrite is one of the requirement to get high ME coefficient

EXPERIMENTAL

Material Synthesis

$Ni_{0.8}Co_{0.2-2x}Cu_x Mn_xFe_2O_4$ [x = 0.01, 0.03, 0.05, 0.07, 0.09] ferrite system was synthesized by adopting the combustion technique. The highly pure nickel nitrate, cobalt nitrate, cupric nitrate, manganese nitrate, ferric nitrate and glycine were used as starting materials. All the nitrates were mixed into double distilled water and kept for continuous stirring with addition of proper amount of glycine into the mixture. This formed the starting precursor solution. The pH of the solution was maintained to 7 using ammonia and kept for constant heating and stirring till the solution transformed into reddish brown gel. The whole assembly was then placed on hot plate and in less than 5 minutes the gel gets ignited and bursts throughout the whole volume. Finally porous black ash was formed as a final product. The powder was ground using agate mortar pre-sintered at 600°C for 5 hrs and final sintered at 850°C for 9 hrs and pelletized for further characterization.

Material Characterization

The structural analysis was carried out using powder X-ray diffractometer model Phillips (PW–3710) with Cu Kα radiation. The SEM images of the samples were obtained by JEOL-JEM-6360 scanning electron microscope. The resistivity measurement was carried out using the two probe method from room temperature to 600°C. The strain measurements were carried out using P3 strain recorder and analyzer at room temperature.

RESULTS AND DISCUSSION

X-ray Diffraction

Figure 1 shows the representative X-ray diffraction pattern for $Ni_{0.8}Co_{0.1}Cu_{0.05}Mn_{0.05}Fe_2O_4$ with most intense (311) peak. No peak other than cubic spinal structure was observed which reveals formation of highly pure ferrite. The crystallite size was measured using formula:

$$D = \frac{0.9\lambda}{\beta\cos\theta} \quad \ldots (1)$$

Where

λ is the wavelength of target = 1.5418 Å

β is F.W.H.M. and ρ is glancing angle.

The crystallite size was in the range of 50 nm (23 nm for particularly x = 0.05 composition).

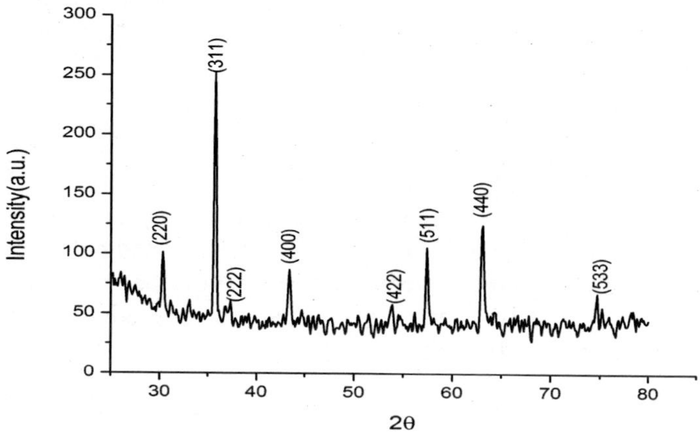

Fig. 1: XRD Pattern for $Ni_{0.8}Co_{0.1}Cu_{0.05}Mn_{0.05}Fe_2O_4$ Ferrite System

Scanning Electron Microscopy

Figure 2 shows the representative SEM image for x = 0.05 composition. The SEM micrograph shows agglomerated grains with slightly porous structure. The agglomerates formed due to high local reaction temperature during synthesis. The porous material formation is characteristic of combustion synthesis because the ingredients are mixed at the atomic level during synthesis. The average grain size was calculated using line intercept method. The average grain size was 144nm for this composition.

Fig. 2: SEM Image for $Ni_{0.8}Co_{0.1}Cu_{0.05}Mn_{0.05}Fe_2O_4$ Ferrite System

Resistivity

Figure 3 shows the variation of resistivity with increase in temperature. The DC electrical resistivity is an important property of low-temperature sintered ferrite for MLCI application.[3] The plot shows decrease in resistivity with increase in temperature. The plot shows three different regions The conduction mechanism in ferrite is considered due as hopping mechanism between Fe^{2+} and Fe^{3+} in B site. The following equilibrium may exist during sintering in ferrites.

$Ni^{2+} + Fe^{3+}$ $\quad$ $Ni^{3+} + Fe^{2+}$

$Co^{2+} + Fe^{3+}$ $\quad$ $Co^{3+} + Fe^{2+}$

$Cu^{+} + Fe^{3+}$ $\quad$ $Cu^{2+} + Fe^{2+}$

$Mn^{2+} + Fe^{3+}$ $\quad$ $Mn^{3+} + Fe^{2+}$

Where Ni^{2+}, Co^{2+}, Cu^{2+}, Mn^{2+} and Fe^{2+} ion prefer to occupy octahedral sites. Conduction occurs as a result of the electron transfer between Fe^{2+} and Fe^{3+} and hole transfer between Ni^{2+} and Ni^{3+}, Co^{2+} and Co^{3+}, Cu^{2+} and Cu^{3+}, $Mn^{2+}$and Mn^{3+} at octahedral sites. It is obvious from resistivity plot that it shows usual semiconducting behavior. The resistivity decreases with increase in Cu and Mn content except x = 0.03 composition.

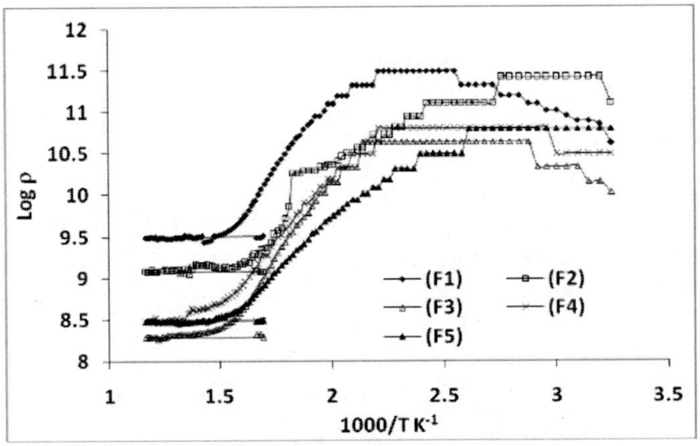

Fig. 3: Variation of Log ρ with Variation in Temperature

Strain with Applied Magnetic Field

Figure 4 shows variation of strain with applied magnetic field. The strain increases with increase in Cu and Mn content. The doping of Jahn-Tellar ions in pure nickel ferrite is responsible for the increase in strain with dopant content. It is known that doping of Jahn-Tellar ions in nickel ferrite enhances the magnetostriction. The strain values along X axis are comparable with that of Z but are observed to be half of that along Y axis. It clearly indicates that the particular composition of doped nickel ferrite is magnetically anisotropic in nature. The magnetic anisotropy due to ordering in one direction is strongly related with cation distribution.[4] The saturation magnetostriction is directly related to saturation magnetization, hence that accounts for increase in strain along with increase in magnetization. The strain measurement in these ferrites may prove useful for selecting the superior composition amongst the series and thereby choosing a competent ferrite in ME composites.

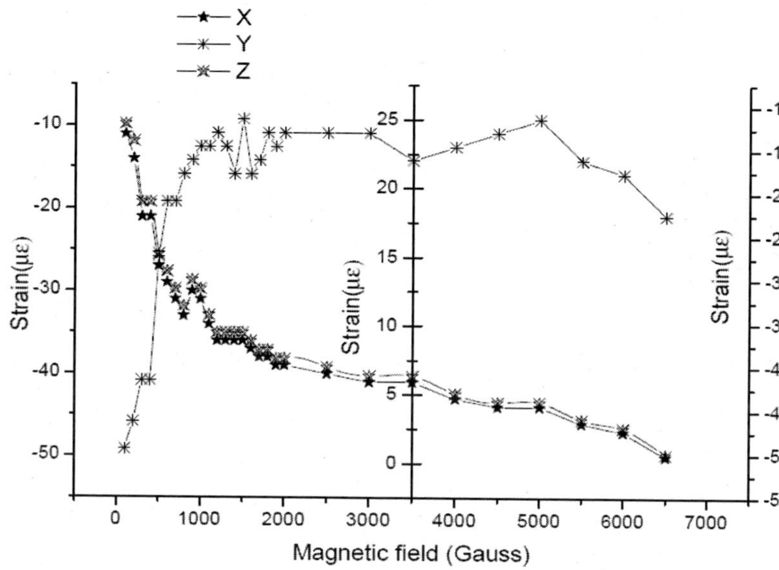

Fig. 4: Variation of Strain with Variation in Magnetic Field for x = 0.07

CONCLUSIONS

The combustion technique was found to be economical and cost effective as it reduces the decomposition temperature of nitrates. The single phase nanocrystalline ferrites were successfully synthesized by solution combustion method. The formation of nanocrystalline ferrites is very important for industrial application like transformer core, magnetic recording, MLCI etc. The incorporation of Co, Cu and Mn enhances the strain in the particularly Nickel ferrite hence which will found the potential candidate as a transducer, actuator and sensing device. .

ACKNOWLEDGEMENTS

One of the authors Vijaya Puri is thankful to UGC for awarding Research Scientist C.P.S. Jadhav and K.K. Patankar are thankful to UGC for financial assistance to carry out this research work.

REFERENCES

[1] Mangalaraja, R.V., Ananthakumar, S., Manohar, P. BS Gnanam , F.D., Jour. Magn. Magn. Mater. 253 (2002), 56–64.

[2] Ming-Ru Syue, Fu-Jin Wei, Chan-Shin Chou, Chao-Ming Fu, Thin Solid Films 519 (2011) 8303–8306.

[3] Zhenxing Yue, Ji Zhou, Longtu Li, Xiaohui Wang, Zhilun Gui Mater. Sci. Engin. B86 (2001) 64–69.

[4] Caltun, O.F., Rao, G.S.N., Rao, K.H., Rao, B. Parvatheeswara, Kim, CheolGi, Kim, Chong-Oh, Dumitru1, I., Lupu4, N. and Chiriac, H., *Sens. Lett.*, Vol. 5 (2007), 1–3.

MICROWAVE PROPERTIES OF STRONTIUM CALCIUM MANGANITE IN THE 8–12 GHz FREQUENCY SPECTRUM

R.P. PAWAR, S.N. MATHAD AND VIJAYA PURI

Thick and Thin Film Device Lab, Department of Physics, Shivaji University, Kolhapur
E-mail: vijayapuri1@gmail.com

ABSTRACT

Cost effective co precipitation method was used for synthesizing strontium calcium manganite using nitrate precursors. The structural properties were studied by X-ray diffraction and scanning electron microscopy. The high frequency (8–12 GHz) absorbance and reflectance of strontium calcium manganite were investigated and was found to be dependent on synthesis parameters. The absorption measured was more than 60% under certain condition. The complex permittivity of these materials showed decreasing behavior in 8 to 12 GHz but at 9 GHz imaginary part of complex permeability shows peak like behavior.

INTRODUCTION

Due to saturation in the sub-gigahertz range, microwaves in the higher gigahertz range are being increasingly utilized in wireless communication, radar, local area network, etc. Along with the development of higher gigahertz electronics and the trend towards miniature circuitry, electromagnetic interference (EMI) is also becoming a serious problem and a matter of crucial concern in higher gigahertz range.[1] The reduction of electromagnetic backscattering using the microwave absorbing materials therefore has important implications in the field of electromagnetic compatibility (EMC).[2] The substantial importance of microwave devices in the area of commercial and military communication, surveillance and navigation technologies etc. have increased in recent years. The requirement for new functional microwave modulation devices pushes research into the exploration of new materials. Clearly, for optimum device

V. Rajendran, R. Yuvakkumar, K. Thyagarajah and K.E. Geckeler (eds.)
Applications of Nano Materials: Electronics, Energy and Environment, pp. 65–70 (2012).

design, the choice of materials and therefore knowledge of their permittivity in the microwave region is vital. Microwave dielectric ceramics with a high dielectric constant have become very important for the miniaturization of microwave device such as filters, duplexers, voltage-controlled oscillators and down converters.[4] In this paper we report the effect of calcium content on the microwave properties of $Sr_{(1-x)}Ca_xMnO_3$ ($0.0 \leq x \leq 1.0$) ceramic. Effect of calcium content in the strontium manganite was also studied by structural characterization using X-ray diffraction spectroscopy (XRD) and Scanning electron microscope (SEM).

EXPERIMENTAL

Material Synthesis

The $Sr_{(1-x)}Ca_xMnO_3$ (x = 0.0, 0.2, 0.4, 0.6, 0.8, 1.0) was prepared by chemical co precipitation method employing strontium nitrate, calcium nitrate, manganese nitrate and oxalic acid as starting materials. 0.5 M solution of strontium, manganese and calcium nitrates were prepared then added drop wise to 2 M hot (75°C) oxalic acid solution with constant stirring. After cooling to room temperature the precipitate was filtered, washed with distilled water and dried at room temperature, pre-sintered at 3500°C for 3 hr and mixed with PVA (polyvinyl alcohol) binder. The powder was pressed into small disk of 1.8cm diameter and ~0.2 cm thickness under a pressure of 10 ton for 5–7 min. The pellets of samples $0.0 \leq x \leq 1.0$ were sintered at temperature 1000°C in a muffle furnace for 8 h followed by cooling

Results and Discussion

X-ray Diffraction

The perovskite structure of $Sr_{(1-x)}Ca_xMnO_3$ ($0.0 \leq x \leq 1.0$)ceramic was confirmed from X-ray spectrum as shown in Figure 1. A small amount of Ca doping changes the crystal symmetry from Monoclinic to Orthorhombic.

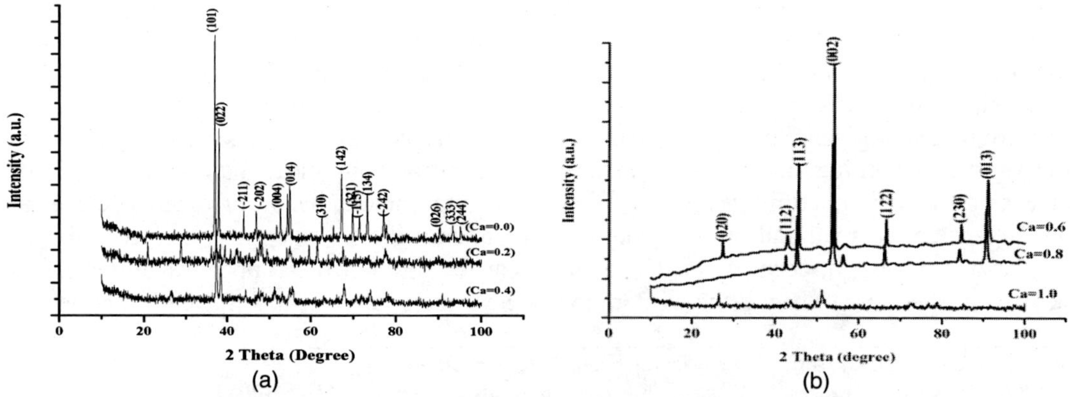

Fig. 1: X-ray Diffraction Images (a) Monoclinic Structure (b) Orthorhombic

Scanning Electron Microscopy

Figure 2 shows the SEM (JSM-6360 JEOL Japan) morphology of $Sr_{(1-x)}Ca_xMnO_3$ ($0.0 \le x \le 1.0$). For both monoclinic and orthorhombic structure it is observed that as calcium content increase density increases and porosity decreases.

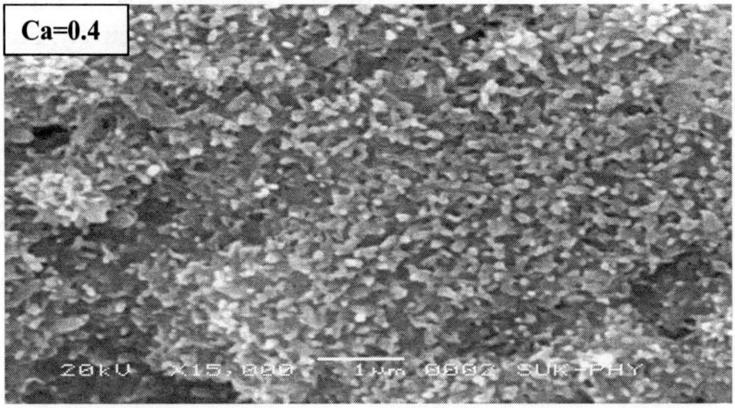

Fig. 2: Scanning Electron Microscopic Images of $Sr_{(1-x)}Ca_xMnO_3$ ($0.0 \le x \le 1.0$)

Microwave Characterization-Permittivity and Permeability by VSWR Method

From the position of the voltage standing wave minima and using the Smith chart [5] the real and imaginary permittivity was obtained.

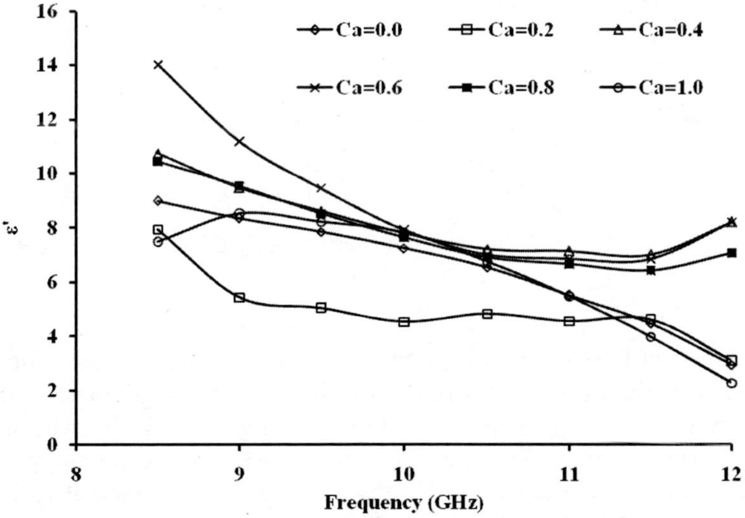

Fig. 3: Real Part of Complex Permittivity of $Sr_{(1-x)}Ca_xMnO_3$ ($0.0 \le x \le 1.0$)

The Smith chart was used to find the phase change due to the manganite material in the path of microwaves.Figure 3 shows the frequency dependence real part (ε') of complex permittivity of Sr1–xCaxMnO3 (0.0 ≤ x ≤ 1.0) at microwave frequency (8 GHz to 12 GHz), which was measured by VSWR method. It was observed that no significant variations were observed with respect to compositions but frequency dependent variations were observed. The real part (ε') of complex permittivity of $Sr_{(1-x)}Ca_xMnO_3$ decreased with increasing frequency from 14 to 7.8 and 9 to 2.2 at 8.5 GHz and 12.0 GHz respectively. Figure 4 shows the energy reduction due to magnetism indicated by magnetic loss that decreased generally with increasing frequency. Magnetic loss is caused by the time lag of the magnetization vector M behind the magnetic field vector H. The change of the magnetization vector is generally brought about by rotation of the magnetization or the domain wall displacement. These motions lag behind the change of the magnetic field and contribute to μ''. The smaller the particle size, the weaker the spins coupling at the surface of the particle, which makes the magnetic relaxation behavior more complex, and will give rise to a magnetic loss mechanism.[6]

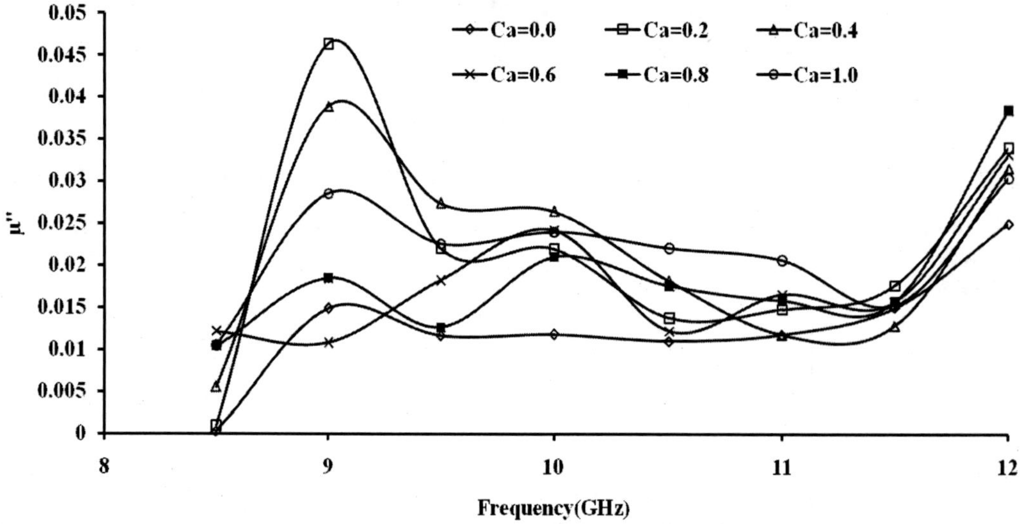

Fig. 4: Imaginary Part of Complex Permeability of $Sr_{(1-x)}Ca_xMnO_3$ (0.0 ≤ x ≤ 1.0)

Microwave Absorption

Generally, the excellent EM-wave absorption of materials is known from efficient balancing between the relative permittivity and permeability in materials. Figure shows the microwave absorption of strontium calcium manganite. From the figure it is observed that initially at frequency of 8.2 GHz absorption increases from 50% to 90% as calcium concentration goes on increasing. Then again at frequency of 9 GHz absorption decreases, again at 10 GHz absorption increases up to 80% and again decreases at 11 GHz to 10%. Thus as frequency increases strontium calcium manganite shows wavelike absorbing nature

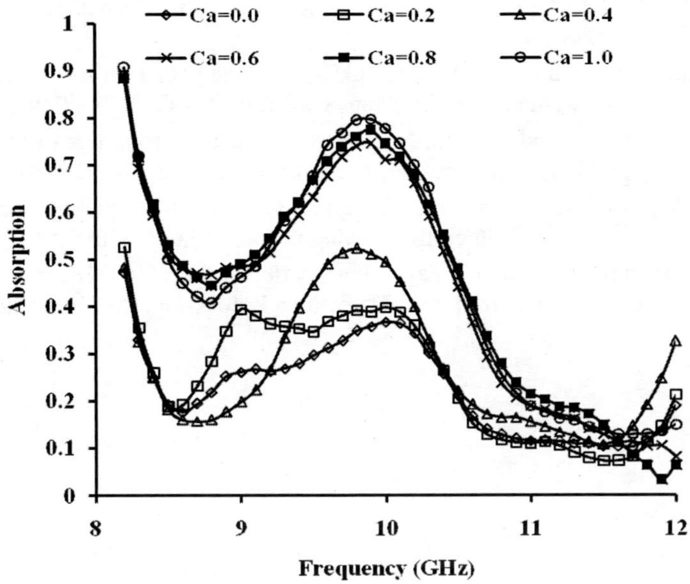

Fig. 5: Microwave Absorption of $Sr_{(1-x)}Ca_xMnO_3$ ($0.0 \leq x \leq 1.0$)

CONCLUSIONS

The perovskite $Sr_{(1-x)}Ca_xMnO_3$ was successfully synthesized by oxalic precursor method. These manganites can be a potential candidate as microwave absorber which is eco friendly and cost effective. The composition dependent microwave absorption was observed and it increased with Ca concentration from 47% to 68% at 9.7 GHz. The real part of complex permittivity shows composition dependent behavior. The investigation of complex dielectric permittivity and permeability indicates that the absorption can be ascribed to the dielectric and magnetic losses. The unique properties of doped manganese oxides provide an important foundation for developing the materials with strong microwave absorption in wide band and materials can have wide application prospects.

ACKNOWLEDGEMENTS

One of the authors Vijaya Puri gratefully acknowledges the UGC India for Award of Research Scientist 'C'. R.P. Pawar acknowledges DST PURSE.

REFERENCES

[1] Yusoff, A.N., Abdullah, M.H., Ahmad, S.H., Jusoh, S.F., Mansor, A.A. and Hamid, S.A., Electromagnetic and absorption properties of some microwave absorbers, *J. Appl. Phys.* 92, 876–882, (2002).

[2] Chung, D.D.L., Electromagnetic interference shielding effectiveness of carbon materials, Carbon 39, 279–285 (2001).

[3] Awai, I., Zhang, Y., Ishida, T. and Suzuki, T., Principles of time domain calculus of microwave resonator parameters IEICE Trans Electron E90–C, 2198–2204 (2007).

[4] Huange, C.L., Tsai, J.T. and Li, J.L., Microwave dielectric properties of (1– x)CaO-xBaO-Li_2O-(1–y) Sm_2O_3-yNd_2O_3-TiO_2 ceramics system J Mate Sci 35, 4901–4905 (2000).

[5] Kulkarni, D.C., Lonkar, U.B. and Vijaya, Puri, High-frequency permeability and permittivity of $Ni_xZn_{(1-x)}Fe_2O_4$ thick film J. Magn.Magn. Mater. 320 1844–1848 (2008).

[6] Pawar, R.P., Jadhav, R.N. and Vijaya, Puri, Effect of Ca on Microwave Properties of Strontium Manganite Synthesized by Microwave Processing, J. Elec.Mater.Letts.8, 321–324, (2012).

ROLE OF Mn-DOPING ON THE STRUCTURAL, MORPHOLOGICAL AND OPTICAL PROPERTIES OF ZnO NANORODS

N. Rajamanickam, S. Rajashabala and K. Ramachandran

School of Physics, Madurai Kamaraj University, Madurai
E-mail: nrajamku@gmail.com

ABSTRACT

In this study, the effect of doping concentration on the morphological, structural and optical properties of $Zn_{1-x}Mn_xO$ nanorods synthesized by solvothermal technique has been investigated. The morphological studies showed nanorod-like morphology and the preferential growth direction along the c-axis. The doping of Mn^{2+} ions has significant influence on the optical properties of ZnO nanorods. Room-temperature photoluminescence spectra of Mn-doped ZnO nanorods exhibit a weak emission at 381 nm and a strong emission around 552 nm and the spectra also indicate that the concentration of the defects increased by doping. The dynamic process of the UV absorption for Mn-doped ZnO nanorods was discussed in detail.

INTRODUCTION

One-dimensional (1D) nanostructures such as wires, rods, and tubes have gained extensive research interest during recent years due to their potential use as inter connects, nanoscale electronic, optoelectronic, and sensing devices.[1] Zinc oxide, as a wide direct band gap (3.37 eV) semiconductor with a large exciton binding energy of 60 meV at room temperature has attracted considerable attention due to the applications in spintronic devices, ultraviolet (UV) luminescence devices, light emitting diodes (LEDs), gas sensors, solar cells and transparent thin film transistors.[2] There has been much attention on modifying ZnO by doping with transition metals [3] such as Ag, Ni, Cu, Co, Cr, Ti, Al, Mn, etc. Doping of metals into ZnO can change the band gap, the crystallite size, and other properties. ZnO doped with Mn, the

V. Rajendran, R. Yuvakkumar, K. Thyagarajah and K.E. Geckeler (eds.)
Applications of Nano Materials: Electronics, Energy and Environment, pp. 71–76 (2012).

most promising diluted magnetic semiconductors (DMS) has been extensively studied. Unfortunately, the value of the energy band gap and optimum amount of Mn as dopant for spintronic applications has not been reported clearly yet. Also nowadays, significant effort has been devoted to develop various 1D semiconductor nanostructures. Hence in this work, Mn-doped ZnO nanorods were synthesized in order to study the structural, morphological, and optical properties exclusively.

EXPERIMENTAL

$Zn_{1-x}Mn_xO$ (x = 0.000, 0.005, 0.010, 0.030, and 0.050) was synthesized here by solvothermal method. The $Zn_{1-x}Mn_xO$ samples are designated as MnX, where X is the dopant concentration in at.%. The crystallite structure of the samples was characterized by XRD (PANalytical X'Pert pro). The morphologies of as-prepared samples were observed by SEM (JEOL JSM-6390). Energy dispersive X-ray spectroscopy (EDS, Oxford Instruments) was used for determining the elements present in the nanostructure. FT-IR spectrum of as-synthesized sample was recorded by FT-IR (Shiraz) spectrometer in the range of 4000–400 cm^{-1}. Optical spectra in the wavelength range of UV and visible light were recorded by Shimadzu-2450 at room temperature (RT). Photoluminescence (PL) spectra of the samples were obtained at RT using Shimadzu RF-5000 spectrofluorophotometer in the wavelength range of 350–650 nm using Xe-laser lamp as the excitation source at excitation wavelength (λ_{ex} = 300 nm).

RESULTS AND DISCUSSION

XRD Analysis

The XRD patterns of undoped and Mn-doped ZnO are shown in Figure 1, which revealed that all the samples possess hexagonal wurtzite structure. No additional reflections are observed

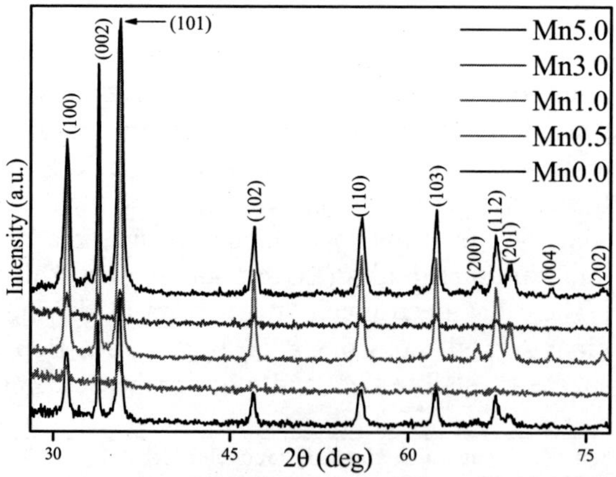

Fig. 1: X-ray Diffraction Patterns of Undoped and Mn-doped ZnO Nanorods

indicating that, there are no additional crystalline phase present in the samples. The strong and sharp diffraction peaks demonstrate that the products are well crystalline. The absence of the diffraction peaks of Mn or oxides of Mn in the XRD implies the substitution of Mn at the Zn site in the ZnO lattice. The crystallite size of undoped and Mn-doped ZnO samples are estimated from the full width at half maximum (FWHM) of the XRD peaks using Scherrer formula and are tabulated in Table 1. The XRD patterns show slight shift in the diffraction peaks of ZnO for Mn incorporation. This shift is also reflected in the lattice parameters, which decrease with the addition of Mn when compared to that of undoped ZnO. Low concentration (0.5 at.%) of Mn doping doesn't affect the crystal lattice of ZnO. But with the increase in the Mn doping concentration, rapid compression of the lattice is observed. Substitution of 0.74 Å sized Zn^{2+} ions with 0.80 Å sized Mn^{2+} ions is expected to expand the lattice but here it compresses the interatomic spacing significantly, due to local coordination.

Table 1: XRD Results of Undoped and Mn-doped ZnO Nanorods

Sample	*Crystallite Size D (nm)*	*Lattice Parameters*			*Intensity Ratio of Peaks*	
		a (Å)	*c (Å)*	*c/a*	*(100)/(002)*	*(101)/(002)*
Mn0.0	23	3.2838(2)	5.2606(4)	1.6020	0.717	1.208
Mn0.5	31	3.2836(2)	5.2613(4)	1.6023	1.022	1.149
Mn1.0	25	3.2816(2)	5.2590(4)	1.6026	0.898	1.486
Mn3.0	25	3.2825(2)	5.2579(4)	1.6018	1.027	1.115
Mn5.0	18	3.2815(2)	5.2582(4)	1.6024	0.680	1.190
JCPDS: 89–0511	–	3.249	5.205	1.602	1.371	2.390

The intensity peak ratios of (100)/(002) and (101)/(002) planes are different from that of the standard ZnO (JCPDS) peak ratios, which indicates that ZnO has preferential growth direction. From the *c*/*a* ratio, it is inferred that ZnO has *c*-axial growth. The 1D growth of the nanorods is further confirmed by SEM analysis.

Morphological Analysis

Figure 2(a and b) shows the SEM images of undoped and Mn (5 at. %) doped ZnO nanorods, respectively. It appears from the images that the diameter of the rods is larger than the estimated crystallite size. This is because, the SEM gives images of the rods formed due to combination or aggregation of particles whereas the Scherrer formula gives the average crystallite size of the particles. From the SEM images, the average diameter of undoped and Mn (5 at.%) doped ZnO nanorods are estimated as ~36 nm and ~22 nm respectively. The diameter and the length of the ZnO nanorods decrease with Mn doping. Figure 3 shows the EDS spectra of undoped and 5 at.% Mn-doped ZnO nanorods. Figure 3a indicates that the undoped sample is composed of

Zn and O elements and Figure 3b indicates that the Mn-doped sample is composed of Zn, O and Mn elements and there is no additional elements. Quantitative analysis confirms that the concentration of Mn in 5 at.% Mn-doped sample is 3.3 at.%, which shows that the amount of Mn incorporated into ZnO lattice is slightly lower than the nominal amount of Mn added during the synthesis.

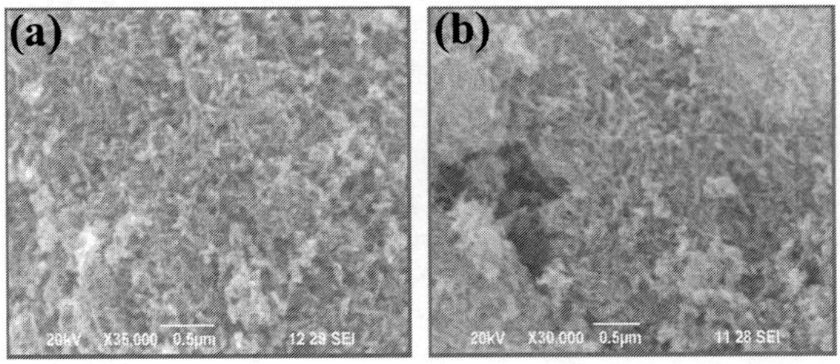

Fig. 2: SEM Image of (a) Undoped and (b) 5 at.% Mn-doped ZnO Nanorods

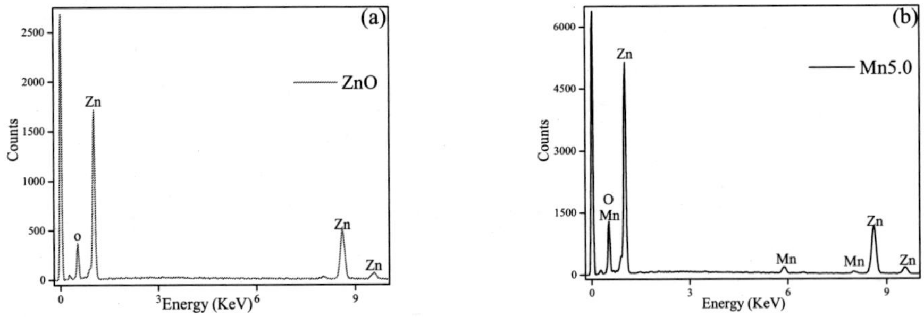

Fig. 3: EDS Image of (a) Undoped and (b) 5 at.% Mn-Doped ZnO Nanorods

UV-vis Spectral Analysis

Analysis of optical absorption spectra is one of the most productive tools for determining the optical band gap of the synthesized samples. Figure 4a shows the UV-Vis absorption spectra of undoped and Mn-doped ZnO nanorods, which exhibit strong absorption in the UV region and high transparency in the visible region. The absorption peaks occurred at 368 nm, which are slightly blue shifted from bulk ZnO (368.5 nm) due to the weak quantum confinement effect. Figure 4b shows the Tauc's plot of undoped and Mn-doped ZnO nanorods. The band gap of ZnO nanorods is around 3.25 eV and the decrease in the band gap when compared to bulk (3.37 eV) is attributed to the high carrier concentration.[5] The band gap energy of MnO system is about 3.6–4.2 eV.[6] Therefore, the band gap of Mn-doped ZnO nanorods should be higher than that of undoped ZnO nanorod. However, the band gap of Mn-doped ZnO nanorods is lower than that of undoped ZnO. Also it is seen that as the concentration of Mn increases, the band gap decreases first and then increases. The low band gap of Mn-doped samples may be due to the s–d and p–d exchange interactions between electrons in conduction and valence bands or due to the localized Mn^{2+} spins contributing to band gap bowing [7] or due to the narrowing effects caused by the high carrier concentration.[5] The inset of Figure 4b shows the effect of concentration of Mn^{2+} ions on the band gap energy of ZnO nanorods.

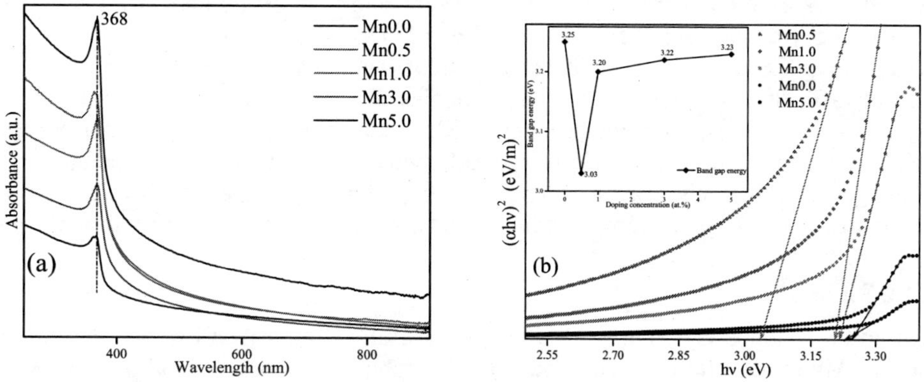

Fig. 4: (a) UV-Vis Spectra and (b) Tauc's Plot of Undoped and Mn-doped ZnO Nanorods. The Inset in (b) Shows the Variation of Band Gap Energy with Mn Concentration

FT-IR Analysis

The composition, quality and the molecular structure of the synthesized samples are analyzed by FT-IR spectroscopy. Figure 5 shows the IR spectra of undoped and Mn-doped ZnO nanorods. The peak at 440 cm^{-1} is ascribed to Zn-O stretching vibration. The two peaks at 3428 and 1407 cm^{-1} are due to O-H stretching vibrations and the peak at 1589 cm^{-1} is due to the H-O-H bending vibration, all of which are associated with the small amount of H_2O molecules existing in the nanorods. The two acetate peaks at 2924 and 2854 cm^{-1} are attributed to C-H stretching vibration. The band observed at about 2368 cm^{-1} corresponds to the O = C = O vibration, which is due to the existence of CO_2 molecule in air.

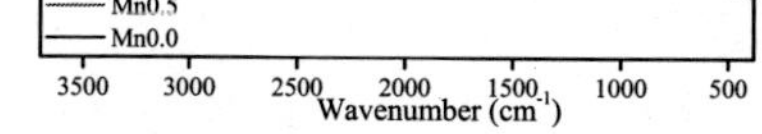

Fig. 5: FTIR Spectra of Undoped and Mn-doped ZnO Nanorods

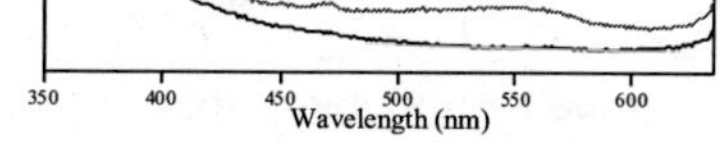

Fig. 6: PL Spectra of Mn (0.0, 1.0 and 5.0 at.%) Doped ZnO Nanorods

Photoluminescence Analysis

PL is an effective method to investigate the presence of defects in semiconductors. Figure 6 displays the characteristic PL spectra of undoped and Mn-doped ZnO samples at RT. The PL spectra reveals sharp UV emission at ~ 3.42 eV (362 nm) and ~ 3.25 eV (381 nm) and broad green emission at ~ 2.246 eV (552 nm). The UV emission is originated from the excitonic recombination corresponding to the near band-gap emission of ZnO, while the emission band in the visible range is due to the recombination of photogenerated holes with singly ionized charge states in intrinsic defects such as oxygen vacancies, Zn interstitials, or impurities.[2,8] The intensity of the visible emission increases with Mn addition and is very high for Mn (5 at.%) doped ZnO nanorods compared with undoped ZnO, which is mainly due to the defects originated from doping. These defects play an important role in the appearance of room temperature ferromagnetism (RTFM) in this sample, which is extensively needed for spintronic applications.

CONCLUSIONS

Undoped and Mn-doped ZnO nanorods were successfully synthesized by solvothermal method with various Mn concentrations. The structural, morphological, and optical properties of these nanorods were investigated. The rods showed preferential orientation along (002) crystal plane. It was seen from the optical investigations that Mn-doped ZnO nanorods have low band gaps compared to undoped one. The PL spectra of the nanorods exhibit strong UV emission and broad visible emission at room temperature. The intense visible emission in Mn (5 at.%) doped ZnO nanorods is due to the high concentration of defects in this sample, which may induce RTFM needed for spintronic applications. Thus Mn (5 at.%) doped ZnO nanorods showing intense visible emission may find potential application in spintronic devices.

REFERENCES

[1] Xia, Y.N., Yang, P.D., Sun, Y., Wu, Y., Mayers, B., *et al.*, One Dimensional Nanostructures: Synthesis, Characterization and Applications, Adv. Mater., 15, 353–389, 2003.

[2] Özgür, Ü., Alivov, Y.I., Liu, C., Teke, A., Reshchikov, *et al.*, A Comprehensive Review of ZnO Materials and Devices, J. Appl. Phys., 98, 41301–103, 2005.

[3] Abdollahi, Y., Abdullah, A.H., Zainal, Z., Yusof, N.A., Synthesis and Characterization of Manganese doped ZnO nanoparticles, Int. J. Basic Appl. Sci., 11, 62–69, 2011.

[4] Wu, C., Shen, L., Zhang, Y.C., Huang, Q., Solvothermal synthesis of Cr-doped ZnO nanowires with visible light-driven photocatalytic activity, Mater. Lett., 65, 1794–1796, 2011.

[5] Akyuz, I., Kose, S., Atay, F., Bilgin, V., The optical, structural and morphological properties of ultrasonically sprayed ZnO:Mn films, Semicond. Sci. Tech., 21, 1620–1626, 2006.

[6] Kasinathan, D., Kunes, J., Koepernik, K., Diaconu, C.V., Martin, R.L., *et al.*, Mott Transition of MnO under pressure: A comparison of correlated band theories, Phys. Rev. B, 74, 195110–12, 2006.

[7] Chen, L.C., Tien, C.H., Fu, C.S., Magneto-optical characteristics of Mn-doped ZnO films deposited by ultrasonic spray pyrolysis, Mater. Sci. Semicond. Processing, 15, 80–85, 2012.

[8] Kenanakis, G., Androulidaki, M., Koudoumas, E., Savvakis, C., Katsarakis, N., Photoluminescence of ZnO nanostructures grown by the aqueous chemical growth technique, Superlattices Micros., 42, 473–478, 2007.

PHOTO-SWITCHING AND NONLINEAR OPTICAL BEHAVIOR OF BIPHENYL BASED MAIN CHAIN LIQUID CRYSTALLINE POLYMERS

D. Suresh, M. Vijay Srinivasan, P. Kannan, A.N. Dhinaa[1] and P.K. Palanisamy[2]

Department of Chemistry, Anna University, Chennai
[1]K.S. Rangasamy College of Arts and Science, Tiruchengode
[2]Department of Physics, Anna University, Chennai
E-mail: pakannan@annauniv.edu

ABSTRACT

We report a series of biphenyl based main chain liquid crystalline polymers (BP-AZ, BP-TA and BP-AE); their mesomorphic and optical behaviors. The mesomorphic property of polymers were observed through polarized optical microscopy (POM) and confirmed by differential scanning colorimetry (DSC). Thermal stability of polymers was examined by thermogravimetric analysis and the polymers are stable upto 260°C. The BP-AZ polymer shows trans to cis photoisomerization occur at 275s whereas reversible processes take place at 12 h. The BP-AZ polymer exhibits negative nonlinear refractive index and good optical limiting behavior at 3.5 wt % of polymer in THF as solvent.

INTRODUCTION

Liquid crystalline polymers (LCPs) find technological application, owing to their unusual anisotropic, optical, electrical and mechanical properties.[1–4] Due to the unusual properties, the polymers have been used widespread commercial applications such as high-strength fibers, thermoplastics, displays and optical storage devices achieved using LCPs. Generally, LCPs have thermally stable and easy processable; the thermal stability and liquid crystalline properties of polymers were influenced by different mesogen (biphenyl moiety, terephthalic

V. Rajendran, R. Yuvakkumar, K. Thyagarajah and K.E. Geckeler (eds.)
Applications of Nano Materials: Electronics, Energy and Environment, pp. 77–82 (2012).

derivative, polyether's and azo benzene have been reported in main-chain as well as side-chain LCPs.[5–6]

In the present investigation, we have focus design and synthesized azobenzene containing biphenyl based main-chain LCPs (BP-AZ) and study their optical property. Since, most of the research work have been focused on azobenzene containing polymers and they have attracted much attention due to their potential utility such as, holography,[7] optical switching,[8] NLO,[9] biomedical application.[10] To find out thermal stability and liquid crystalline properties of polymer, the polymers (BP-TA and BP-AE) were synthesized, here biphenyl moiety (BP) is kept constant azo linkage is changed in to polyether, polyesters. The mesophase behaviors of polymers were studied by POM along with DSC analysis.

EXPERIMENTAL

Materials

Terephthalic acid, thionyl chloride, p-hydroxybenzaldehyde, 4-nitro benzoic acid, sodium hydroxide, α-D-glucose, glacial acetic acid, potassium dichromate, 4,4′-dihydroxybiphenyl, potassium iodide, 1,6-bromohexanol, 1,8-dibromoctane, and anhydrous potassium carbonate were obtained from Merck chemicals.

Synthesis

Three different biphenyl based main-chain liquid crystalline polymers have been synthesized (BP-AZ, BP-TA and BP-AE) by condensation polymerization technique using respective monomers. Synthetic rote of precursors, monomers and polymers were shown in the Figure 1.

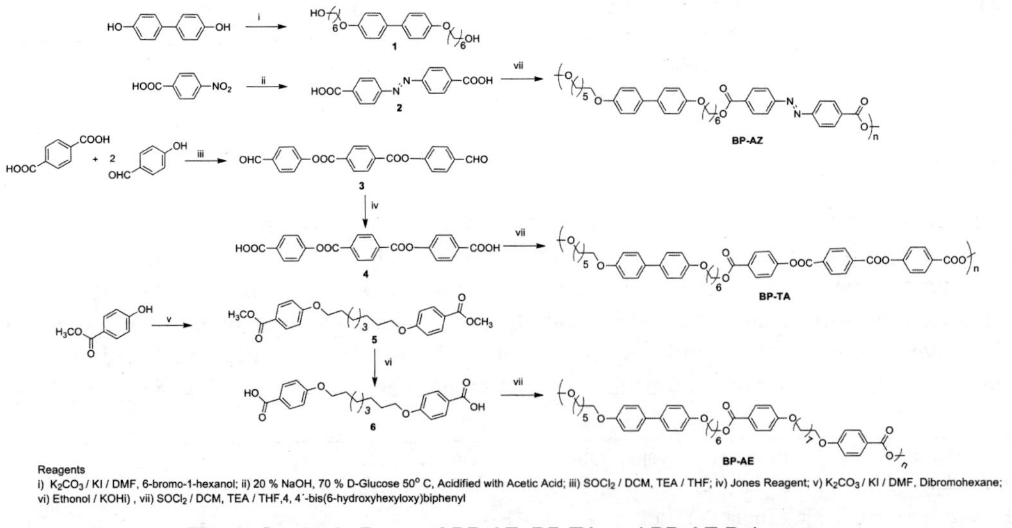

Reagents
i) K_2CO_3 / KI / DMF, 6-bromo-1-hexanol; ii) 20 % NaOH, 70 % D-Glucose 50° C, Acidified with Acetic Acid; iii) $SOCl_2$ / DCM, TEA / THF; iv) Jones Reagent; v) K_2CO_3 / KI / DMF, Dibromohexane; vi) Ethonol / KOHi) , vii) $SOCl_2$ / DCM, TEA / THF,4, 4´-bis(6-hydroxyhexyloxy)biphenyl

Fig. 1: Synthetic Route of BP-AZ, BP-TA and BP-AE Polymers

RESULTS AND DISCUSSION

POM and DSC Studies of Biphenyl Based Main Chain Liquid Crystalline Polymers

Three polymers were synthesized from their respective monomers (varied by linking group such as azo, ester and ether where biphenyl moiety is kept constant as core unit. These linking groups play an important role in the formation of mesophase. The mesomorphic behaviors of BP-AZ, BP-TA and BP-AE are shown in Figure 2. The BP-AE is shows grainy like texture, whereas BP-AZ exhibited a schlieren nematic texture and BP-TA displayed as nematic phase. The observed nematic phase of BP-AZ and BP-TA polymers are evidenced from its threaded and marble textures respectively,[11–12] whereas the grainy texture of BP-AE is might be due to presence of ether linkage. These observed LC textures are confirmed by DSC analysis (Figure 3a), which indicates three endothermic peaks. The phase transition temperatures of polymers are shown in Table 1.

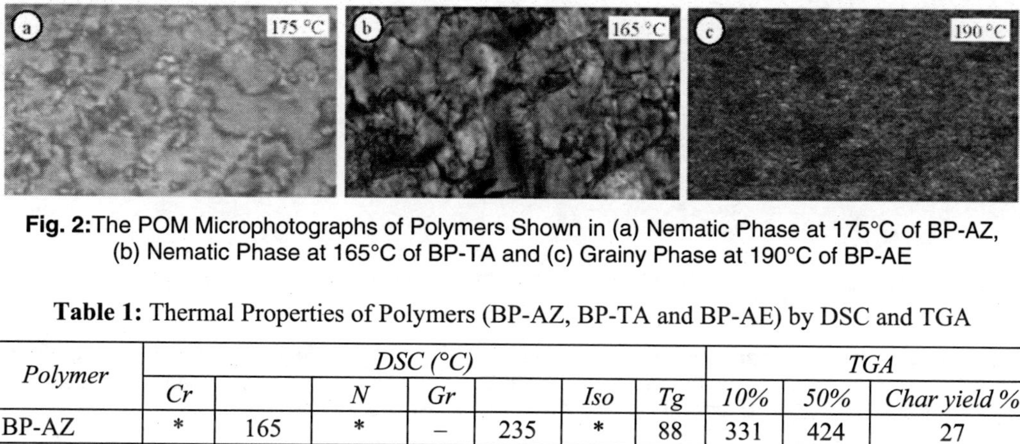

Fig. 2: The POM Microphotographs of Polymers Shown in (a) Nematic Phase at 175°C of BP-AZ, (b) Nematic Phase at 165°C of BP-TA and (c) Grainy Phase at 190°C of BP-AE

Table 1: Thermal Properties of Polymers (BP-AZ, BP-TA and BP-AE) by DSC and TGA

Polymer	*DSC (°C)*							*TGA*		
	Cr		*N*	*Gr*		*Iso*	*Tg*	*10%*	*50%*	*Char yield %*
BP-AZ	*	165	*	–	235	*	88	331	424	27
BP-TA	*	151	*	–	250	*	97	305	418	25
BP-AE	*	148	–	*	202	*	101	360	430	20

Cr = Crystal, N = Nematic, Gr = Grainy, Iso = isotropic of liquids, char yield measured at 600°C

The DSC analysis revealed that BP-AZ and BP-TA shows higher T_m and T_i compare with BP-AE polymer. Which is attributed due to presence of azo and terephthalate groups in the BP-AZ and BP-TA. Whereas the presence of longer alkyl chains in BP-AE expected to introduce greater flexibility to the polymer and consequently bring down T_m and T_i. The thermal stability of polymers is given in table 1 and corresponding of TGA traces are shown in Figure 3b. The polymers are stable up to 258°C, and thermal stability of the polymers increases in the following order BP-AZ < BP-TA < BP-AE. The high char yield due to more number of aromatic rings present in BP-AZ, BP-TA, whereas the presence of longer alkyl chains in BP-AE polymer possess greater flexibility and the polymer consequently bring down the char yield.

Photo-Switching Properties

The photo-switching properties of BP-AZ have been characterized using UV-Visible spectrophotometer by before and after illumination of UV light with 365 nm in chloroform. Absorption spectrum of 4, 4′-Azobenzenedicarboxylic acid (monomer) and BP-AZ (polymer) shows two peaks around 330 nm and 430 nm which is corresponds to π–π* and n–π* transition respectively,[13] in this BP-AZ polymer exhibit one more peak at 275 nm which is due to aliphatic chain. Figure 3c and 3d shows photo-isomerization behaviors of monomer and its polymer (BP-AZ) is irradiated at 365 nm UV light.

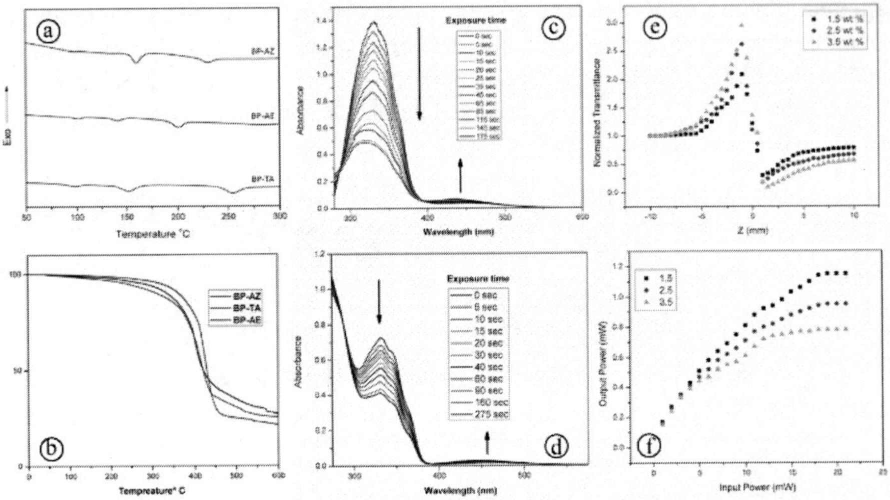

Fig. 3: a) DSC, b) TGA Thermograms of Polymers, c) Absorption Spectra of Monomer, d) Absorption Spectra of BP-AZ Polymer at UV Irradiation with 365 nm, e) Closed Aperture Z-scan Signature and f) Optical Limiting Approach for BP-AZ Polymer

The BP-AZ shows *trans* to *cis* photo-isomerization (275 s), which slower compare its monomer (175s). This slower isomerization is suggested that lack of free mobility of polymeric chain. Reversible photo-switching behavior of BP-AZ polymer slowly occurs almost 12 h whereas white light illumination polymer rapidly reversible at 360 s. The photo-switching processes of BP-AZ polymer precedes faster as compared with reported PMCLCP and AZO-PA polymers.[14–15] Generally, Photo-isomerization depends on the nature and size of terminal substituents in azobenzene.[2,12] The faster isomerization processes in azobenzene containing polymers is a key factor for application in optical data storage device.

Nonlinear Refractive Index

The nonlinear optical property of polymeric material makes it possible for the use of photonic applications. Recently, Z-scan technique has been widely used to study the nonlinear optical properties of different materials.[2,16–17] In the present investigation, nonlinear refractive index n_2, of the BP-AZ polymer has been investigated with single beam Z-scan technique. The Z-scan sign of the BP-AZ polymer was shown in Figure 3e. A closed aperture Z-scan curve of 0.3, 0.5 and 0.7 wt % of polymers were dissolved in chloroform and measured the nonlinear refractive index at the incident intensity of I_0 = 7.824 kW/cm^2. This normalized transmittance curve of BP-AZ was characterized by a peak followed by a valley. Based on this result, it confirms the negative nonlinear refractive index of the BP-AZ polymer.

Optical Limiting Behavior

The optical limiting behavior of BP-AZ polymer is investigated using with (SHG) 532 nm Nd:YAG laser, the 1mm quartz cuvette containing BP-AZ polymer solution placed before the lens (focal length of 3.5 cm), where the defocusing takes place. Three different concentrations of polymer solutions such as 1.5, 2.5 and 3.5 wt % of the BP-AZ polymer have been used for optical limiting studies. Figure 3f illustrate that the optical limiting behavior of BP-AZ polymer. Based on the investigation, it can be concluded that the transmittance increases up to certain level of input power with respect to concentration, then transmitted power become stable with increase in the input power. The results are exposed that the 3.5 wt % of the BP-AZ polymer has good optical limiting behavior than 2.5 and 1.5 wt % of BP-AZ polymers, because the absorption of the sample is directly proportional to their concentration, obviously the sample possess the respective optical liming behavior due to their absorption property.

CONCLUSION

The present work comprises of synthesis, characterization and optical studies of biphenyl based liquid crystalline polymers BP-AZ, BP-TA and BP-AE. The mesomorphic properties of BP-AZ, BP-TA polymers exhibit nematic phase whereas BP-AE polymer shows grainy phase, the observed mesophase is confirmed by DSC thermogram. The photo-switchable properties of BP-AZ polymer shows *E* to *Z* isomerization takes place around 275s whereas reversible processes take place at 360s. The nonlinear optical studies of BP-AZ polymer possess the negative nonlinear refractive index and also the optical limiting behavior.

ACKNOWLEDGEMENT

The authors sincerely acknowledged to Council of Scientific and Industrial Research (CSIR), New Delhi, India [CSIR scheme no. 01(2095)/07/EMR-II] for the financial support.

REFERENCES

[1] Yamada, M., Kondo, M., Mamiya, J.I., Yu, Y., Kinoshita M, Barrett C J, Ikeda T, Photomobile Polymer Materials: Towards Light-Driven Plastic Motors, Angew Chem., 120, 5064–5066, 2008.

[2] VijaySrinivasan, M., Kannan, P., Photo-switching and nonlinear optical behaviors of center linked bent-core azobenzene liquid crystalline polymers, J Mater Sci., 46, 5029–5043, 2011.

[3] Balamurugan, R., Kannan, P., 1, 3, 4-Oxadiazole epoxy resin-based liquid crystalline thermosets and their cure kinetics, J Mater Sci., 45,1321–1327, 2010.

[4] Weissflog, W., Nadasi H., Dunemann, U., Pelzl, G., Diele, S., Eremin, A., Kresse, H., Influence of lateral substituents on the mesophase behaviour of banana-shaped mesogens, J Mater Chem., 11, 2748–2758, 2001.

[5] Fodor-Csorba, K., Szabo, E., Vajda, A., Jakli, A., Bata, L., Synthesis and SmC* properties of chiral liquid crystalline terephthalates, Liq Cryst., 24, 861–869, 1998.

[6] Bao-Quan, C., Kameyama, A., and Nishikubo, T., New Combined Liquid Crystalline Polymers from Polyaddition of Biphenol Diglycidyl Ether and Trimeric Esters, Macromolecules, 32, 6485–6492, 1999.

[7] Saravanan, C., Kannan, P., Dual-mode optical switching property of copolymers containing pendant nitro and cyano substituted azobenzenes and fulgimide units, Polym Degrad Stabil., 94, 1001–1012, 2009.

[8] Branger, C., Lequan, M., Lequan, R.M., Large, M., Kajzar, F., Polyurethanes containing boron chromophores as sidechains for nonlinear optics, Chem Phys Lett., 272, 265–270, 1997.

[9] Rida, S.M., Ashour, F.A., El-Hawash, S.A. M., Elsemary, M.M., Badr, M.H., Shalaby, M.A., Synthesis of some novel benzoxazole derivatives as anticancer, anti-HIV–1 and antimicrobial agents, Eur J Med Chem., 40, 949–959, 2005.

[10] Rosalyn, P.D.S., Kannan, P., Vinitha, G., Ramalingam, A., Effect of halogen substitution on azo-naphthol dye based poly(alkyloxymethacrylate)s and their Z-scan measurements, J Mater Sci: Mater-El., 20, 835–843, 2009.

[11] Ilis, M., Bucosa, M., Dumitras, F., Circu, V., Mesomorphic behaviour of N-benzoyl-N′-aryl thioureas liquid crystalline compounds, J Mol Struct., 987, 1–6, 2011.

[12] Jayrang, S.D., Meera, R.M., Pratik, R.P., Liquid crystalline behaviour of mixtures of structurally dissimilar mesogens in binary systems, Proc Indian Acad Sci (Chem Sci)., 114, 213–221, 2002.

[13] Abellana, G., Garciab, H., Carlos, J.G.G, Ribera, A., Photochemical behavior in azobenzene having acidic groups, Preparation of magnetic photoresponsive gels J Photochem and Photobio A: Chemistry., 217, 157–163, 2011.

[14] Jayalatha, D., Balamurugan, R., Kannan, P., Novel photo□crosslinkable liquid crystalline polymers containing vanillylidene cycloalkanones and azobenzene units, Liquid Cryst., 35, 275–285, 2008.

[15] Yao, J., You, Y., Lei, Y., Dong, L., Xiong, C., Sun, Z., Main-chain azo polyaramides with high thermal stability and liquid crystal properties, J Polym Res., 16, 455–460, 2009.

[16] Sheik-Bahae, M., Said, A.A., Tai-Hue,W., Hagan, D.J., VanStryland, E.W., Sensitive measurement of optical nonlinearities using a single beam, IEEE J Quantum Electron., 26, 760–769, 1990.

[17] Wang, Y.H, Gu,B., Xu, G.D., Zhu, Y.Y., Nonlinear optical properties of neodymium-doped bismuth titanate thin films using Z-scan technique, Appl Phy Lett., 84, 1686–1688, 2004.

EFFECT OF ZnO NANOPARTICLES ON STRUCTURAL, MECHANICAL, OPTICAL AND SURFACE PROPERTIES OF HPMC POLYMER FILMS

B. Lakshmeesha Rao, Mahadeviah[1], S. Asha, R. Somashekar[2] and Sangappa

Department of Studies in Physics, Mangalore University, Mangalagangothri, Mysore
[1]Department of Polymer Science, Sir MV PG Center, University of Mysore, Mandya, Mysore
[2]Department of Studies in Physics, University of Mysore, Manasagangothri, Mysore
E-mail: syhalabhavi@yahoo.co.in

ABSTRACT

The Zinc Oxide (ZnO) nano-particles were synthesized by electrochemical method. Then prepared nano-particles were dispersed in hydroxypropyl methylcellulose (HPMC) solution. The HPMC films were prepared by solution casting method and ZnO concentration was varied from 0.01 to 0.04%. The films were examined for structural, optical, mechanical and surface properties verifying how the addition of ZnO nano-particles affected the properties using X-ray diffraction (XRD), Ultra-Violet-Visible spectrophotometer (UV-Vis), Universal Testing Machine (UTM) and Atomic force microscopy (AFM). In general, addition of ZnO nano-particles negatively affected the performance of the HPMC films. The microstructuralline parameters like <N>, g decreases with increasing concentration of ZnO nano-particles. A decrease of 30% in tensile strength, 37% in Young's modulus, and 28% decrease in the elongation at break of the films was observed, when the ZnO concentration is 0.04%. Average surface roughness (R_a) of the film increases with ZnO concentration and 69% in 0.04% ZnO concentration.

V. Rajendran, R. Yuvakkumar, K. Thyagarajah and K.E. Geckeler (eds.)
Applications of Nano Materials: Electronics, Energy and Environment, pp. 83–88 (2012).

INTRODUCTION

The addition of nano sized inorganic particles into polymer matrices, the new composites material will exhibit enhanced properties like physicochemical stability, high catalytic activity, effective antibacterial and bactericide function, intensive UV and infrared absorption, etc., which greatly differ from that of conventional materials. Researchers have studied the addition of nanoparticles to polymers to provide additional barrier or functional properties for food packaging purposes which include metal and metal oxide nanoparticles, nanofibres and nanotubes.[1] Silica nanoparticles were inserted into polypropylene or starch polymer matrices to improve mechanical or barrier properties of composites and biodegradable films.[2] Starch nanocrystals have been found to improve the mechanical properties of polymer films and inhibit there elongation.[3] Whiskers have also been incorporated into hydroxypropyl methylcellulose (HPMC) films and have been shown to improve mechanical and barrier properties.[4]

Hydroxypropyl methylcellulose is a biopolymer which has become a successful alternative material for two-piece capsule and is actually on the market. HPMC is also being adopted as a film coating or a sustained-release tablet material in the pharmaceutical field.[5] Researchers showed that Zinc Oxide (ZnO) nanoparticles have biodegradability and biocompatibility are potential useful for bio-applications.[6] In this paper Zinc Oxide (ZnO) nanoparticles were prepared and incorporated in hydroxypropyl methylcellulose films. The aim of the study was to investigate the effect of nano-ZnO particles on the HPMC films physical, optical, mechanical and surface properties. Hence in this paper we have carried out X-ray profile analysis to determine microstructuralline parameters ($<N>$, g in % and D_s), optical band gap (E_g), tensile strength, elongation (%) and Young's modulus and surface (R_a) roughness properties of ZnO incorporated HPMC polymer films.

EXPERIMENTAL

The metal wires (cathode electrode is platinum and Zinc metal rod is an anode) are used as electrodes. The experiment was run for 3–4 hrs with continuous stirring (40mA: 12V), the anodic dissolution of Zinc gives Zn (II) ions which are electrochemically reacted with aqueous $NaHCO_3$ (0.5%) to form Zn (II) oxide. The solid obtained was washed with deionised water till complete removal of unreacted $NaHCO_3$. The wet powder was then dried at a temperature 250˚C–400˚C for dehydration and removal of hydroxides to get ZnO. The synthesis takes place at the electrode -electrode interface or closed to the electrode within the electrical double layer.[7] The HPMC polymer films were prepared by solution casting method and concentration of ZnO varied from 0.01 to 0.04%.

The X-ray diffractograms (XRD) of the ZnO dispersed polymer samples were recorded using Rigaku Miniflex-II X-ray diffractometer with Ni filtered, CuKα radiation of wavelength $\lambda = 1.5406$ Å, with a graphite monochromator. The scattered beam was focused on a detector. The samples were scanned in the 2θ range 10–50° with a scanning speed and step size of 1°/min and 0.02°. Ultraviolet-visible (UV-Vis) absorption spectra of were recorded using a SHIMADZU UV–1800, UV-Vis spectrophotometer in the wavelength range 190–800 nm, at room temperature. The mechanical properties of polymer samples were measured with Universal Testing Machine (Lloyd Instruments, UK, 5kN load cell). Each sample was tested at the specifications of 50 mm of gauge length, with the crosshead speed of 20mm/min at room temperature. A model LRX Plus was used to determine the maximum tensile strength (TS), maximum percentage elongation at break (%), and elastic modulus (or Young's modulus). The surface morphology of polymer samples was carried out using Nanosurf easyscan2, Atomic Force Microscope (AFM-Switzerland) in contact mode with a scanner size of 12 μm × 12 μm.

RESULTS AND DISCUSSION

Wide angle X-ray diffractograms for pure and ZnO incorporated HPMC films are given in Figure 1. Wide angle X-ray scattering patterns of these films show that broad peak corresponding to the Bragg's angle 2θ = 22.2°. The microstructural parameters such as <*N*>, (*g*) and D_s are determined by matching the simulated intensity profile, using equations [8] with the experimental data for all the samples studied here and the obtained results are tabulated in Table 1. For the sake of completeness, we have reproduced in Figure 2 the simulated and experimental intensity profiles for some of the ZnO incorporated HPMC films. From Table 1, the following results emerge. They are: 1. Reduction of crystal size with the incorporation of ZnO has been observed. This indicated that the addition of nanoparticles increases the disorderliness in the polymer network of HPMC. This is due to the weakening of van der Waal's forces between the polymer molecules. 2. This decrease in crystal size is in consonance with the observed physical property like tensile strength that has also decreased. The reason for such changes can be attributed to indiscriminate dispersed nanoparticles in the matrix of HPMC, which generally tends to lower the short range interaction between the layers of HPMC, leading to increase in disorder in the lattice. These changes result in more broadening of

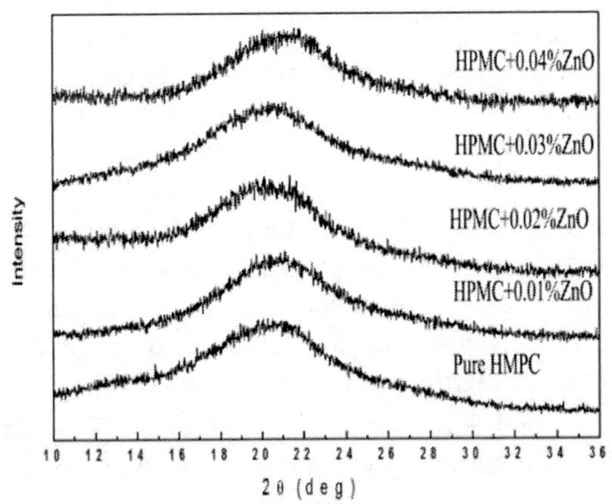

Fig. 1: XRD Scans of Pure and ZnO Dispersed Samples

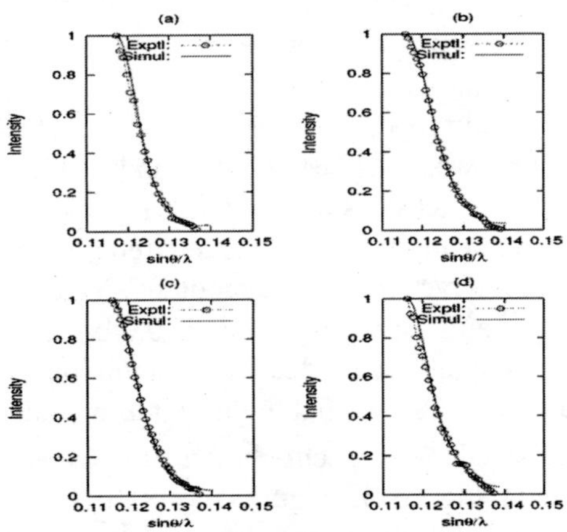

Fig. 2: Experimental and Simulated Intensity Profiles of Samples

Table 1: Microstructural Parameters of ZnO Incorporated Polymer Samples Computed by Exponential Distribution Function

Samples	d_{hkl} *(Å)*	*<N>*	*g in %*	*Delta*	D_s *(Å)*
Pure	4.307	4.41±0.21	0.4±0.02	0.049	18.99
0.01%	4.241	3.63±0.16	0.3±0.01	0.045	15.39
0.02%	4.213	3.17±0.11	0.3±0.01	0.049	13.36
0.03%	4.253	2.55±0.15	0.3±0.02	0.053	10.85
0.04%	4.212	2.23±0.09	0.2±0.01	0.042	09.39

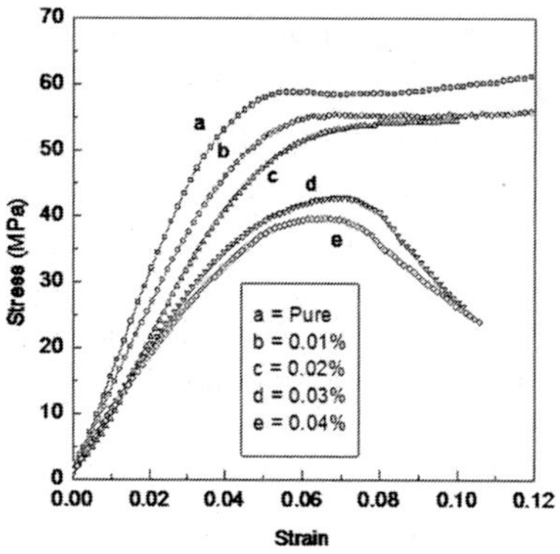

Fig. 3: Stress vs Strain Curves

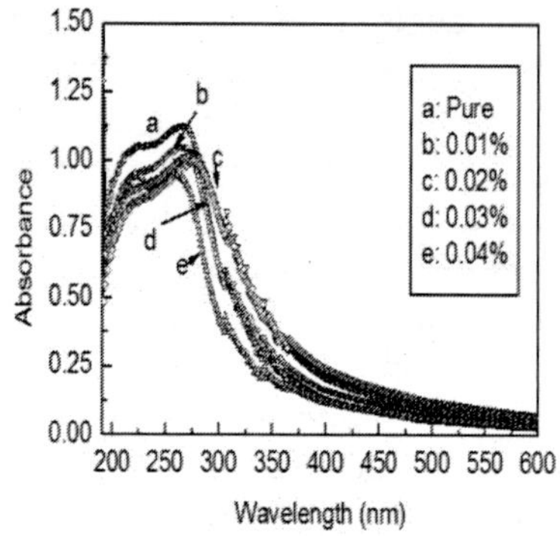

Fig. 4: UV-Vis Scans

X-ray Bragg reflections and hence, decrease in microstructural parameters. On a macroscopic scale, these affect the physical parameters like tensile strength and percentage of elongation.

Tensile properties were calculated from the plot of stress (tensile force/initial cross-sectional area) versus strain (extension as a fraction of the original length) given in Figure 3. The mechanical properties were analyzed as a function of ZnO concentrations. The tensile strength in pure film was 59.12±4.18 MPa and in 0.01% ZnO sample it was 55.43± MPa. The extent of reduction in tensile strength is varying and we observed 30.46% reduction in 0.04% concentration. The decreasing trend of tensile strength was due to non-alignment of nanoparticles and strong surface interaction. Elastic modulus (Young's modulus) goes on decreasing as the concentration changes and 37% reduction in 0.04% ZnO concentration. We observed that percentage of elongation at break decreases with increasing ZnO concentration. Contrary to the above results, we observe that the maximum load (N) increases with increasing concentration.

The UV-Vis spectrophotometric scans of virgin and ZnO incorporated polymer films are given in Figure 4. Figure shows the absorption decreases very sharply and come to the minimum absorption at λ_g. The values of λ_g for virgin and ZnO with different concentrations are estimated as 400, 421, 391, 487 and 517 nm respectively. The optical band gap energy for the virgin as well as for ZnO incorporated samples has been calculated from the respective values of λ_g using the formula $E_g = hc/\lambda_g$. A shift in λ_g towards higher wavelength indicates the decrease in E_g of the ZnO dispersed samples. The microstructural modification produced in polymer matrix due ZnO dispersion would cause a change in the molecular structure of the polymer network. As a result a modification within the optical band gap was expected.

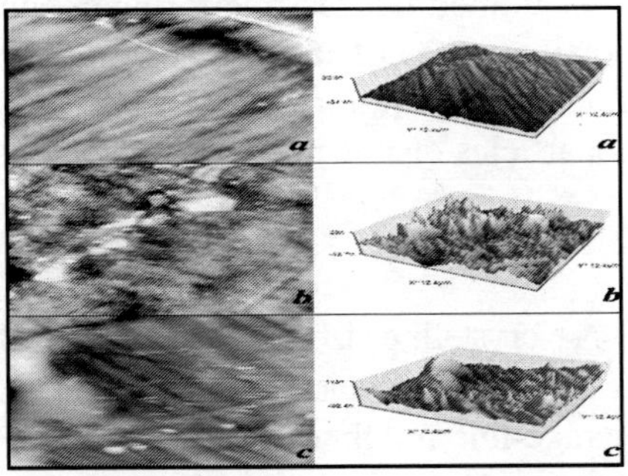

Fig. 5: AFM Topography Images of (a) pure, (b) 0.02% ZnO, (c) 0.04% ZnO

Table 2: Physical Scales Describing Roughness of the Virgin and ZnO Dispersed Samples

Samples	*R_a(nm)*	*R_q(nm)*
Pure	06.78±0.83	09.40±1.94
0.01%	07.33±1.72	10.16±4.58
0.02%	08.33±3.29	10.27±3.57
0.03%	19.39±3.69	23.57±3.89
0.04%	21.85±3.69	28.65±2.82
Samples	S_a(nm)	S_q(nm)
Pure	08.77±4.75	10.84±05.91
0.01%	07.62±1.81	11.19±04.74
0.02%	06.42±2.14	18.60±03.13
0.03%	17.14±8.62	21.88±10.78
0.04%	20.31±9.21	26.18±10.99

Figure 5 shows the topology and 3D AFM images of virgin and ZnO incorporated polymer samples. The surface topology parameters which are given in Table 2, presents some important physical scales, which are morphologically characterize the sample. The line roughness parameters (R_a and R_q) and surface roughness parameters (S_a and S_q) are the physical scales which describing the roughness degree of the samples. From the Table 2 it can be seen that all parameter increases as ZnO concentration increases. The roughness of HPMC films increases with ZnO concentration indicates non-alignment of nanoparticles in the polymer matrix. This fact is also evidenced by X-ray results as well as mechanical properties of the films.

CONCLUSIONS

We observe that there are changes, both microscopically and macroscopically, in HPMC films when ZnO nanoparticles are incorporated. These are quantified in terms of micro crystalline parameters and also physical parameters like tensile strength, percentage elongation, etc. We have also examined the reasons for such changes in terms of weakening of intermolecular van der Waal's forces. The UV-VIS study shows the optical band gap decreases with increasing ZnO concentration. AFM study shows the surface roughness increases with increasing nanoparticles dispersion.

NOMENCLATURE

$\mathbf{d_{hkl}}$ Inter planar spacing, (Å). **<N>** crystallite size. **g** Lattice strain. $\mathbf{D_s}$ Surface weighted crystallite size (Å), $\mathbf{E_g}$ Optical energy gap, (eV). **h** Planck's constant, (J s). **c** Speed of light, (m/s). $\boldsymbol{\lambda_g}$ Absorption edge (nm). $\mathbf{R_a}$ Average-line roughness (nm). $\mathbf{R_q}$ Root mean square-line roughness, (nm). $\mathbf{S_a}$ Average-surface roughness (nm). $\mathbf{S_q}$ Root mean square-surface roughness (nm).

REFERENCES

[1] Liu, Y-L., Chen, W.H. and Chang, Y., Preparation and properties of chitosan/carbon nanotube nanocomposites using poly(styrene sulfonic acid)-modified CNTs, Carbohydrate Polymers, 76(2), 232–238, 2009.

[2] Tang, S., Zou, P., Xiong, H. and Tang, H., Effect of nano-SiO_2 on the performance of starch/polyvinyl alcohol blend films, Carbohydrate Polymers, 72, 521–526, 2008.

[3] Chen, Y., Cao, X., Chang, P.R. and Huneault, M.A., Comparative study on the films of poly(vinyl alcohol)/pea starch nanocrystals and poly(vinyl alcohol)/native pea starch, Carbohydrate Polymers, 73, 8–17, 2008.

[4] De, Moura, M.R., Aouada, F.A., Avena-Bustillos, R.J., McHugh, T.H., Krochta, J.M. and Mattoso, L.H.C., Improved barrier and mechanical properties of novel hydroxypropyl methylcellulose edible films with chitosan/tripolyphosphate nanoparticlses, Journal of Food Engineering, 92, 448–453, 2009.

[5] Lehtola, V.M., Hainamaki, J.T., Nikupaavo, P. and Yliruusi, J.K., The Mechanical and Adhesion Properties of Aqueous Based Hydroxypropyl Methylcellulose Coating Systems Containing Polydextrose and Titanium Dioxide, Drug Dev. Ind. Pharm., 21, 675–685, 1995.

[6] Zhou, N.Xu. and Wang, Z.L., Dissolving behaviour and stability of ZnO wires in biofluids; A study on biodegradability and biocompatibility of ZnO nanostructures, Adv. Mater., 18(18), 2432–2435, 2006.

[7] Nuffenegger, R.M., Craft, B., Shaaban, M., Gorer, S., Erley, G. and Penner, R.M., A Hybrid electrochemical/chemical synthesis of Zinc Oxide nano particles and optically intrinsic thin films, Chem. mater., 10, 1120–1129, 1998.

[8] Sangappa, Demappa, T., Mahadevaiah, Ganesha, S., Divakara, S., Manjunath, Pattabi and Somashekar, R., Physical and thermal properties of 8 MeV electron beam irradiated HPMC polymer films, Nucl. Inst. and Meth. in Phy. (B), 266(18), 3975–3980, 2008.

CO-OXIDATION AGENT EFFECTS ON THE ELECTRICAL AND THERMAL PROPERTIES OF POLYANILINE NANOFIBERS

P. MANIVEL[1], S. RAMAKRISHNAN[1], M. DHAKSHNAMOORTHY[2], A. BALAMURUGAN, NIKHIL K. KOTHURKAR[1] AND C. VISWANATHAN

Department of Nanoscience and Technology, Bharathiar University, Coimbatore, Tamil Nadu
[1]Department of Chemical Engineering and Material Science, Amrita Vishwa Vidyapeetham, Coimbatore, Tamil Nadu
[2]Polymer Nanotechnology Centre, B.S.Abdur Rahman University, Chennai, Tamil Nadu
E-mail: viswanathan@buc.edu.in

ABSTRACT

Polyaniline (PANI) nanofibers were chemically prepared through in-situ polymerization in an aqueous medium using aniline, HCl and ammonium peroxydisulfate (APS) with additional use of sodium hypochlorite (NaOCl) as a co-oxidant agent. Variations in the experimental conditions (aniline concentration, oxidation agent, and reaction temperature) on the morphology of PANI nanofibers were investigated by scanning electron microscope (SEM). The chemical structures of PANI nanofibers were characterized by Fourier transform infrared spectra (FT-IR), Raman spectra, X-ray diffraction pattern (XRD), Electrical conductivity analyses and Thermal analysis (TG-DTA). Small diameter PANI nanofibers with uniform morphology and high thermal stability were successfully obtained in a wide range of aniline concentration (0.1 to 0.3 M) with co-oxidant agent and reaction temperature at 0–5°C. The electrical conductivity of PANI 0.6 S/cm, increased to ~0.8 S/cm after the addition of 2% NaOCl were also studied. This method provides large scale nanofibers along with the more potential applications in the synthesis of other conductive polymers.

V. Rajendran, R. Yuvakkumar, K. Thyagarajah and K.E. Geckeler (eds.)
Applications of Nano Materials: Electronics, Energy and Environment, pp. 89–94 (2012).

INTRODUCTION

Since the discovery of the conjugated polymers it was made to conduct electricity through doping, a tremendous amount of research has been carried out in the field of conducting polymers.[1] Polyaniline (PANI) has been one of the extensively studied intrinsically conducting polymers in the last two decades due to its easy of synthesis, low cost monomer, good environmental stability, exciting electrical, thermal, optical and doping-dedoping properties [.–4] High surface area PANI nanofibers have attracted even more interest because of their properties that make them possibly suitable for applications as sensors, batteries, capacitors, molecular electronic devices, and separation membranes.[2, 5]

Usually PANI can be synthesized from an aniline monomer by either chemical oxidative polymerization or electrochemical oxidative polymerization techniques.[6] Typically, chemical approach of PANI is based on an oxidative polymerization of aniline using an oxidation in the presence of a strong acid dopant. The chemical oxidative polymerization can be categorized into two categories, i.e., the template synthesis and template-free synthesis. The common template-free methods include rapid mixing polymerization,[7] sonochemical synthesis,[8] seeding polymerization,[9] and microwave-assisted synthesis.[10] Recently, Rahy *et al.* synthesized nanofibers by use of aniline concentration, co-oxidation agent and different dopant acids, and Rahy and Yang synthesized PANI nanofiber by using both potassium biiodate and sodium hyphochlorite oxidant shows high electrical conductivity.

In this paper, PANI was synthesized by use of co-oxidation agent and comprising the addition of sodium hyphochlorite of after APS drop wise addition. Always, the electrical and thermal properties were investigated of prepared nanofiber with different mole ratio of aniline and temperature at 0–5°C.

EXPERIMENTAL

The polyaniline nanofiber was syntheised by chemical polymerization.[11] Aniline (1.9 mL, 0.2 M) was dissolved in 100 mL of hydrochloric acid while 0.25 M of ammonium peroxydisulfate (APS) was dissolved in 100 mL of 1 M hydrochloric acid. Fifty milliliters of APS solution was added drop wise to the aniline mixture and stirred. This gives aniline to APS molar ratio of 1:1.25. The reaction vessel was maintained either at ice bath temperature (0 to 5°C) or room temperature. Once the reaction starts, 2.5 mL of an aqueous sodium hypochlorite (2% by wt) solution was quickly added drop wise with vigorous stirring. The reaction was then kept at 12 hrs under stirring. The dark green precipitated polymer was filtered, washed with 1 M HCl, double distilled water, and then acetone, followed by drying in vacuum oven 45°C overnight.

The components of material were analyzed by infrared spectroscopy (Thermo Nicolet is10 FTIR-ATR). The crystal nature of the samples were characterized by X-ray diffraction (XRD, X-pert Pro' PANlytical) using CuKα radiation. Morphology of the PANI fiber is determined by scanning electron microscope (ICON Quanta Make II Environmental SEM). Raman spectra were recorded at ambient temperature on a Horiba Jobin-LabRam-HR UV spectrometer with an Argon laser at an excitation wavelength of 514 nm. Electrical conductivity measurement was performed on pellets of the composite powder samples using a standard four probe technique using a Keithley 2182 A instrument.

RESULTS AND DISCUSSION

The molecular structure of the obtained PANI nanofiber was studied in the FTIR spectra. Figure 1 shows the FTIR spectra of PANI nanofiber. The spectrum of PANI shows vibration peaks at 1567, 1465, 1293, 1102 and 801 cm^{-1}, respectively. These peaks correspond to the characteristic peakof PANI, as described in literature.[12] The strong absorption peaks corresponding to the stretching of quinonoid assigned at 1567 cm^{-1}, benzenoid rings assigned at 1465 cm^{-1}, C-N stretching assigned at 1293 cm^{-1}, C = N stretching of a secondary aromatic amine assigned at 1102 cm^{-1} and aromatic CH out-of-plane bending is assigned at 801 cm^{-1}. However it is noted that these peaks are slightly shifted with respect to their normal position as seen for pure PANI samples.

The XRD technique was adopted to determine the crystallinity, orientation of conducting polymers and metallic conductive state.[13] Figure 2 shows the XRD pattern of co-oxidation agent used PANI nanofiber.

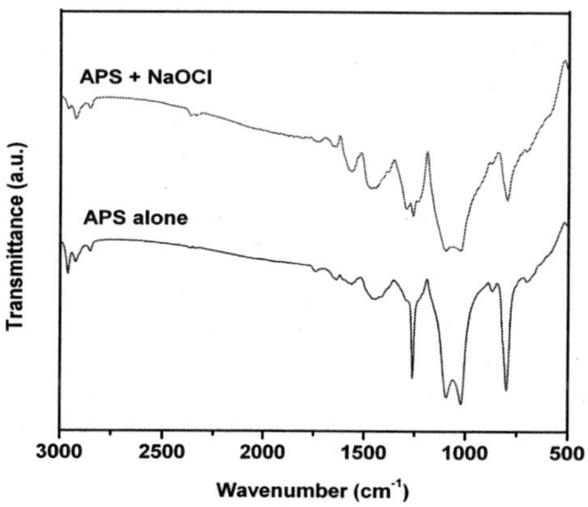

Fig. 1: FTIR Spectrum of PANI: (a) APS, (b) APS+NaOCl

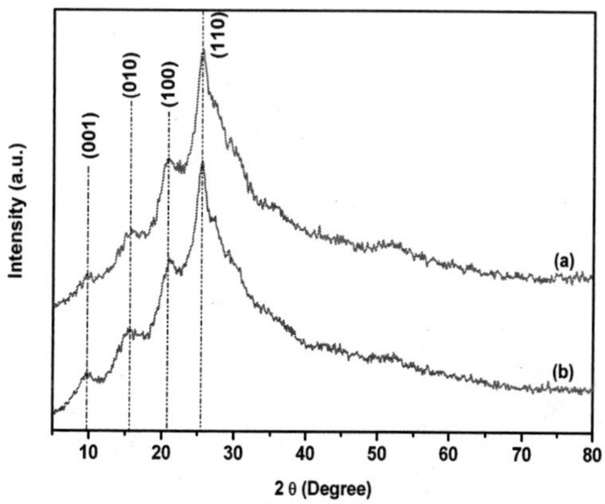

Fig. 2: XRD Pattern of PANI: (a) APS, (b) APS+NaOCl

The entire reflection of well defined peaks, which were obtained from XRD, confirms the formation of metallic state of PANI nanofiber (Figure 2(a) APS oxidant alone (b) APS and NaOCl co-oxidant). The crystallinity structure was revealed with 2θ values of 10°, 15°, 20° and 25° corresponding to the (001), (010), (100), and (110) crystalline planes, respectively.[14] It can be seen that single oxidant PANI consisting of four sharp peaks superimposed on a broad scattering background indicative of crystalline regions dispersed in an amorphous medium. However, for the more metallic samples (Co-oxidant PANI nanofiber) the broad background decreases, and the relative intensity of the four peaks (at 10°, 15°, 20° and 25°) changes: the peaks at 15° and 20° decrease and become nearly indistinguishable from the background, whereas the peak at 25° increases and sharpens.[15] Also, it can be clearly seen that the small intense peak at 2θ ~10° is attributed to doped state of PANI nanofiber.[16] From the result we can draw conclusion that our PANI nanofiber is crystalline materials.

The bond structure of the PANI EB nanofiber was studied by Raman spectroscopy. The Raman spectrum (obtained with 514 nm excitation) of the original PANI nanofiber (figure 3) synthesized at 0–5° for 12 hrs. The band at about 1606 cm^{-1} is due to C-C stretching vibrations of aromatic rings and the band at 1589 cm^{-1} corresponding to C = C stretching vibration of the quinonoid ring. The band at 1351 cm^{-1} is C-H bending vibration of benzene ring. The Raman active vibrational mode of perfectly ordered graphite is situated at 1589 cm^{-1} (band assigned for "G" for graphite). It originates from the ordered hexagonal rings consisting of conducting sp^2-bonded carbon. While increasing disorder, this band broadens and appears at 1351 cm^{-1} (band assigned for "D" for disordered form). The "D" peak in carbon-based materials has been observed for disordered graphite, such as cluster of hexagonal rings. This band has a strong resonant character, depending on the intensity and position on wavenumber.[17]

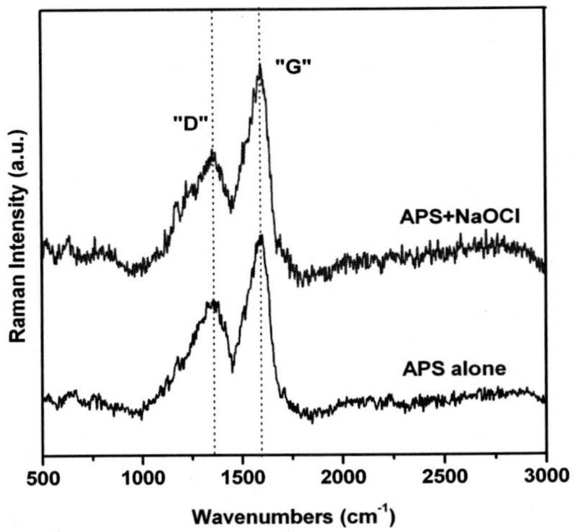

Fig. 3: RAMAN Specrum for PANI by Used of APS Oxidant and APS + NaOCl as co-Oxidant

To visually learn dimensional self-construction of the as-synthesized resultant sample, the surface morphologies and configuration of the prepared PANI with co-oxidation (NaOCl) and without co-oxidation agent were shown in Figure 4. It can be seen from figure 4(a) that the PANI indicates the presence of irregular mixture of short fibers and granular particles. However, we found that when NaOCl was used for polymerization process, it produced some long nanofiber (figure 4(b)) having uniform size with small amount of granular particles. The APS alone oxidant use produced PANI exhibit a few very short fiber mixed with more amount of granular particles having average diameter (60–100 nm). When co-used APS oxidant and

NaOCl (2%) as co-oxidant, we synthesized thinner diameter ~50 and below 50 nm nanofibers having long clear fibrillar morphology compared with single APS oxidant. we observed when we co-used APS oxidant with NaOCl oxidant and the hypochlorite use enhances selective fibrillar directional growth of polyaniline.[11]

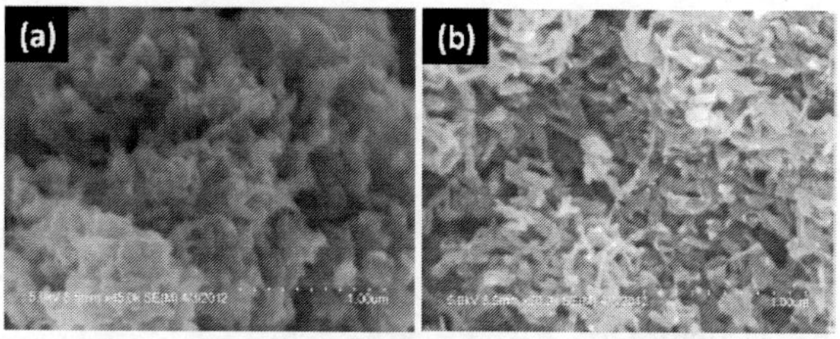

Fig. 4: SEM of PANI Obtained in 0–5°C. (a) APS Oxidant, (b) APS + NaOCl as Co-oxidant

The electrical conductivity could be affected by the morphology of π-conjugated polymers from respects of density of electrons and arrangement of PANI molecular chains.[18] Figure 5 shows the conductivity of PANI nanofiber. Clearly, with the increase of the pellet temperature, the conductivity of single oxidant used PANI and co-oxidant used PANI nanofiber was increased linearly, and the highest conductivity as 0.6 S/cm and 0.8 S/cm exhibited by the pellet tested at 120°C. The morphology of PANI has a significant effect on the conductivity of PANI.[2] The present nanofibrous morphology of PANI allows electrons to move throughout the length of nanofibers and further extend to the branch regions, leading to the increased delocalization of the charge, leading to an improvement in electrical conductivity. The addition use of NaOCl is likely to influence for the produce fiber structure of PANI molecular orientation. Therefore, the PANI nanofiber morphology shows the highly desirable electrical conductivity compared to single oxidant agent used of PANI.

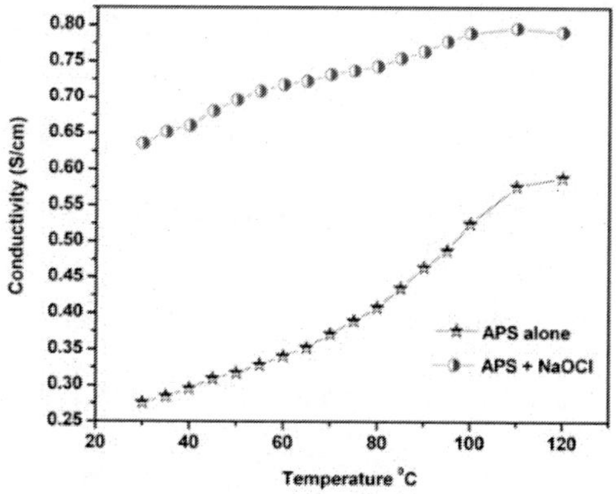

Fig. 5: Electrical Conductivity of PANI by Used of APS Oxidant and APS + NaOCl

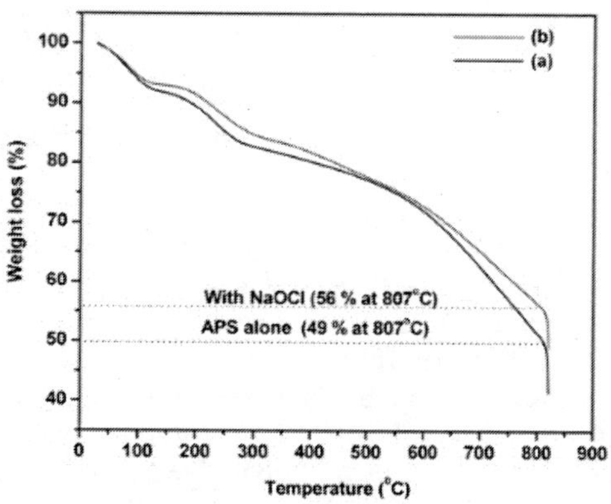

Fig. 6: TG Curves of (a) APS Oxidant Used PANI (b) APS+NaOCl Co-oxidant Used PANI

As shown in Figure 6, TG curves for PANI exhibited two weight losses. The initial mass loss step before 250°C is attributed to the evaporation of residual water and solvent in both samples. After 250°C, both curves showed a major moss loss due to the decomposition of PANI molecular chains.[19] It can be seen that the co-oxidant used of PANI obtained a total weight loss of 56 % at 807°C and without co-oxidant used of PANI obtained a total weight loss of 49% at 807°C, which is caused by the decomposition of PANI. However, it could be observed that the co-oxidant used of PANI shows slightly (difference at 7%) improved thermal stability compared with without co-oxidant used of PANI nanofiber.

CONCLUSIONS

PANI nanofiber was successfully prepared by a simple *in-situ* chemical polymerization method. The functional groups, structure of PANI were identified by FTIR, XRD and RAMAN respectively. The SEM image shows most of the thinner diameter ~50 and below 50 nm nanofibers having long clear fibrillar morphology compared with APS oxidant alone. We found that the co-oxidant used of PANI nanofibers having good electrical conductivity as high as 0.8 S/cm at 120°C. The co-oxidant used of PANI shows a little improved thermal stability compared to the producing PANI using APS alone.

REFERENCES

[1] Huang, J., Virji, S., Weiller, B.H. and Kaner, R.B., *Chem. Eur. J.*, 10 (2004), 1314.
[2] Lu, X.W., Wu, W., Chen, J.F. and Zhang, P.Y., Zhao, Y.B., *Ind. Eng. Chem. Res.*, 50 (2011) 5589.
[3] Li, Yu, Wang, Y., Wu, D. and Jing, X., *J. Appli Poly Sci.*, 113 (2009), 868.
[4] Rahy, A. and Yang, D.J., *Materials Letters*, 62 (2008), 4311.
[5] Zhang, L., Zujovic, Z.D., Peng, H., Bowmaker, G.A. and Kilmartin, P.A., *J. Travas-Sejdic, Macromolecules,* 41 (2008), 8877.
[6] Shirakawa , H., *Synth Met.*, 125 (2002) 3.
[7] Huang, J.X. and Kaner, R.B., *Angew. Chem., Int. Ed.* 43 (2004) 5817.
[8] Jing, X.L., Wang, Y.Y., Wu, D., She, L., Guo, Y., *J. Polym. Sci., Part A: Polym. Chem.,* 44 (2006) 1014.
[9] Zhang, X.Y., Goux, W.J., Manohar, S.K., *J. Am. Chem. Soc.*, 126 (2004) 4502.
[10] Nikolaidis, M.R., Stanisavljev, D.R., Easteal, A. J., Zujovic, Z.D., *J. Phys. Chem.*, C 114 (2010) 18790.
[11] Rahy, A., Sakrout, M., Manohar, S., Cho, S.J., Ferraris, J., Yang, D.J., Chem. Mater. 20 (2008) 4808.
[12] Jing, X., Wang, Y., Wu, D., Qiang, Ultrason. Sonochem. 14 (2007) 75.
[13] Li , T. *et al.*, Phys Rev B. 47, 1840.
[14] Pouget, J.P., Jdzefowiczt, M. E., Epstein, A.J., Tang, X., MacDiarmid, A.G., Macromolecules 24 (1991) 779–789.
[15] Lee, K., Cho, S., Park, S.H., Heeger, A.J., Lee, CW. and Lee, S.H., Nature. 441(2006) 65.
[16] MacDiarmid, A.G., *et al.* Macromolecules, 24 (2006) 779.
[17] Trchova, M. *et al.*, Polym. Degradation and Stability. 91 (2006) 114.
[18] Zhou, D., Li, Y., Wang, J., Xu, P., Han, X., *Materials Letters,* 65 (2011) 3601.
[19] Arenas, M.C., Andablo , E. and Castano, V.M., *J. Nanosci. Nanotechnol.,* 10 (2010) 549.

CHEMOSYNTHESIS OF ZnO NANORODS AND ITS PHOTOELECTROCHEMICAL PROPERTIES

S.B. PAWAR, S.A. PAWAR[1], P.S. PATIL[1] AND P.N. BHOSALE

Materials Research Laboratory, Department of Chemistry, Shivaji University, Kolhapur, Maharashtra
[1]Thin Film Materials Laboratory, Department of Physics, Shivaji University, Kolhapur, Maharashtra
E-mail: p_n_bhosale@rediffmail.com; pawar_sarita@yahoo.com

ABSTRACT

Suitable morphology for fast electron transportation is a crucial requirement for photo electrochemical (PEC) solar cells. Vertically aligned ZnO nanorods were grown on the glass and FTO coated glass substrates by a cost effective and low temperature aqueous chemical method. The grown nanostructures have been characterized by X-ray Diffraction (XRD), Scanning Electron Microscopy (SEM), Transmission Electron Microscopy (TEM), Selected Area Electron Diffraction (SAED) pattern, Fourier Transform Infrared (FT-IR) spectroscopy and optical absorption techniques. XRD pattern confirm the aligned ZnO nanorods formation takes place with the predominant orientation along (0 0 2) plane with hexagonal wurtzite structure. SEM images illustrates that the nanorods have a hexagonal morphology with average diameter of about 220 nm and length 3μm. TEM images shows that the diameter of nanorod tip is about 15 nm. The SAED pattern shows that the nanorods possess monocrystalline nature and grow along the (0 0 2) direction. Contact angle (CA) measurement shows the hydrophobic nature of ZnO nanorods. The PEC solar cell configuration of Glass/FTO /ZnO/0.5 M Na_2SO_4/Pt/FTO/Glass has been used to record the current voltage (I–V) characteristics of the films. The short circuit current (I_{sc}), open circuit voltage (V_{oc}), fill factor (FF) and efficiency have been estimated.

INTRODUCTION

ZnO is an important electronic and photonic material due to its wide direct band gap of 3.37 eV and large exciton binding energy of 60 meV. It crystallizes as wurtzite structure and exhibits

V. Rajendran, R. Yuvakkumar, K. Thyagarajah and K.E. Geckeler (eds.)
Applications of Nano Materials: Electronics, Energy and Environment, pp. 95–100 (2012).

n-type conductivity. ZnO is one of the semiconductors having good chemical and thermal stability. Because of its properties such as resistivity control over the range 10^{-3} to 10^{5} Ω.cm, transparency in visible range, high electrochemical stability, abundance in nature and absence of toxicity, it has great potential to apply as the functional materials for UV detectors, dye-sensitized photovoltaic cells, piezoelectronic devices, hydrogen storage devices, chemical sensors and biosensors.[1–5]

Films with well-aligned ZnO Nanorods (NRs) exhibit large surface-to-volume ratio allows for distinct structural and chemical behavior. The one unhindered dimension can direct the conduction of quantum particles such as electrons, photons and phonons. This control over various forms of energy transport recommends NRs as ideal materials from which to manufacture advanced solid-state devices. As a result, NRs provide a convenient platform through which researchers may study confined transport phenomena.

Aqueous chemical method is promising due to simple manipulation, low production cost, large-scale deposition, unique ability to tune the orientation and easy upward scaling. The as deposited ZnO NRs thin film was analyzed for their structural, optical and surface morphological study. Finally, the PEC performance including I-V characteristics and photo-conversion efficiency of the prepared sample was studied.

EXPERIMENTAL

In a typical experiment, efforts were taken to coat uniform deposition of ZnO seed layer on ultrasonically cleaned glass substrates. The seed solution was prepared in an 250 ml absolute ethanol with 0.05 M zinc acetate, Zn $(CH_3COO)_2{\cdot}2H_2O$ and 0.05 M diethanolamine (DEA), $HN(CH_2CH_2OH)_2$. The cleaned glass substrate was dip coated for 10 sec. in seed solution, removed it gently and then kept at room temperature over night for drying. The dried films were annealed at 400°C for 10 min. in air to yield a layer of ZnO seed on the substrate surface. This coating process was repeated twice to ensure the uniform coverage of ZnO seed on entire substrate. The seeded substrates were placed vertically in 200 ml solution of zinc acetate and hexamethylene tetramine (HMTA), $(CH_2)_6N_4$ with equimolar concentration of 0.05 M. The prepared solution was refluxed at 95±3°C for time period of 5 hrs. Finally, the refluxed film was washed with distilled water to eliminate loosely adhered material and then air dried at room temperature for further characterizations.

CHARACTERIZATIONS OF ZNO THIN FILM

The room temperature optical absorption measurement was performed in the wavelength range over 350–1100 nm by using a UV-visible spectrophotometer (UV-1800, Shimadzu, Japan). The X-ray diffraction spectrum of the film was recorded using X-ray diffractometer (Bruker AXS Model D8 Advance X-ray Diffractometer) with Cu Kα target having wavelength 1.542 Å. The surface morphology of the film was examined by using the Scanning Electron Microscopy Model JEOL-JSM-6360, Japan, operated at 20 kV. Transmission Electron Microscopy for ZnO thin film was performed using a TECNAI G2 20U TEM equipped with a field emission gun operating at 20 kV to 120 kV accelerating voltage.

The Fourier Transform Infrared spectrum of sample was collected using a spectrum-100 Perkin Elmer FT-IR spectrophotometer in the spectral range 450–4000 cm^{-1}. The wetting behavior of the sample was checked by means of contact angle (CA) measurement using Rame-hart model 500-F1, USA. The I-V characteristics were measured using semiconductor characterization system SCS-4200 Keithley, Germany using two electrode configurations.

RESULTS AND DISCUSSION

UV-Visible Absorption Study

The optical absorption spectrum recorded for the ZnO film in the range 350–1100 nm. Figure 1 shows variation in the optical absorption with wavelength.

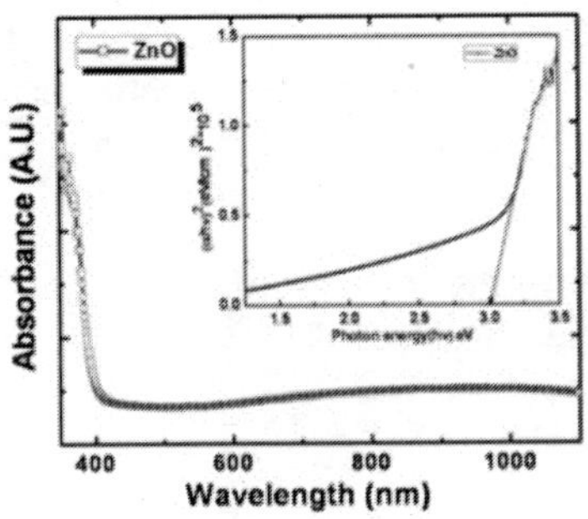

Fig. 1: Optical Absorption Spectrum of ZnO thin film. Inset shows Tauc Plots to Estimate the Direct band Gap of ZnO thin Film

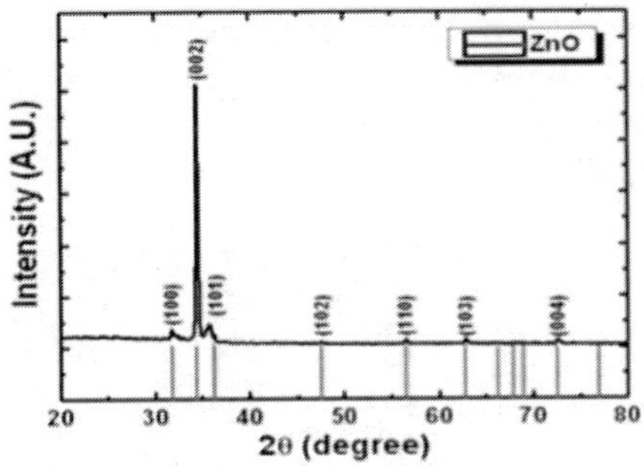

Fig. 2: X-ray Diffraction Pattern of the ZnO Film. The Standard JCPDS Data Card no. 76–0704 for ZnO is shown as Vertical Line

Inset of Figure 1 shows the variation of $(\alpha h\nu)^2$ versus hν for the ZnO thin film. The nature of the plot indicates the existence of direct optical transition. The extrapolation of straight line portions to zero absorption coefficient ($\alpha = 0$), leads to the estimation of band gap energy value. It is noted that the band gap energy is 3.0 eV. The change in Eg from bulk value is attributed to the crystallite size of grown structure.

X -Ray Diffraction (XRD) Study

XRD is a well known technique for the structural identification and determination of the crystallite size. XRD pattern of the ZnO nanorod arrays (ZNRs) is presented in Figure 2. It indicates that all ZNRs are highly oriented along (0 0 2) plane, corresponds to the wurtzite structure of single crystalline quality.

The comparison of observed XRD pattern with the standard JCPDS data (card no. 76–0704) confirms the formation of ZnO having hexagonal crystal structure with diffraction peaks (1 0 0), (0 0 2), (1 0 1), (1 0 2), (1 1 0), (1 0 3) and (0 0 4) planes at 31.71, 34.23, 35.68, 47.44, 56.54, 62.62 and 72.48 2θ degree respectively. The observed 'd' values are compared with JCPDS data and were found to be in good agreement with the standard 'd' values. The average crystallite size is deduced using the well known Scherer's formula. The calculated crystallite size is found to be 39 nm. The lattice parameter 'a' and 'c' for hexagonal ZnO was calculated, the value of a = 3.248 Å and c = 5.223 Å is in good agreement with the JCPDS data value, a = 3.253 Å and c = 5.213 Å respectively.

Surface Morphology Study

Figure 3 shows the SEM images of ZNRs. Highly uniform and densely packed arrays of ZnO nanorods formed on the entire substrate from the growth solution. From Figure 3 (a), the diameter of the ZNRs was found to be 220 nm. Cross sectional image of the ZNRs shown in Figure 3 (b) clearly depict that, ZNRs are vertically well aligned with compact structure having length 3.2 μm. From the cross section image of ZNRs the aspect ratio was calculated which found to be 9.6. The length of nanorods and the spacing between them are essential factors for effective light harvesting. The length of the nanorods provides effective surface area for light absorption and the spacing between nanorods helps in light scattering and subsequent trapping.

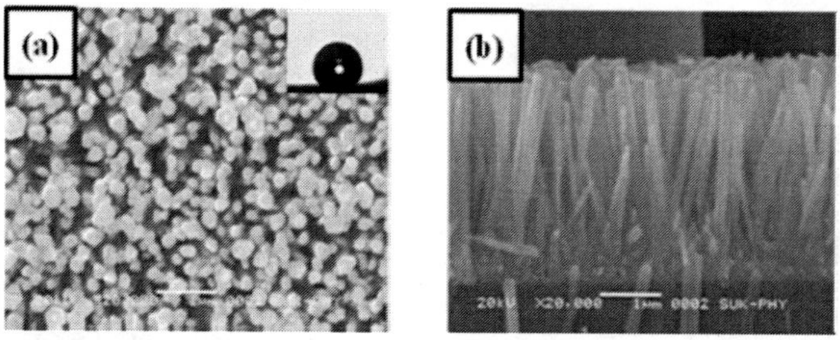

Fig. 3: SEM Image for Top View of ZnO thin Film is shown in (a); Inset Shows Water Contact Angle Measurement and Cross-section of ZnO thin Film Shown in (b)

The CA of a water drop is influenced by the roughness of the surface. Inset Figure 3(a) shows water CA measurement of ZnO thin film and are found to be 141^0 ($\pm 1^0$). A ZnO thin film shows hydrophobic nature this is unfavorable to the better access of electrolyte into the film structure which increases the PEC performance.

Transmission Electron Microscopy (TEM)

TEM images for ZnO sample are shown in Figure 4 (a) and (b). From Figure 4(a) the aspect ratio was calculated are found to be 16.7. The diameter of nanoneedle tip is about 14.28 nm and bottom 16.28 nm. The SAED pattern shown in inset of Figure 4(b) illustrates that the nanoneedles are monocrystalline in nature with wurtzite hexagonal structure and grow along the (0 0 2) direction.

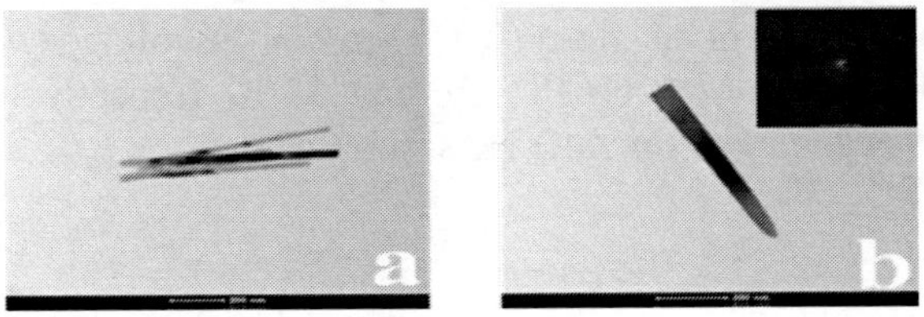

Fig. 4: TEM Images of the ZnO thin Film for Various Magnifications. Inset of (b) Shows SAED Pattern of ZnO thin Film

FT-IR Spectroscopy Study

The FT-IR spectrum of the sample, pelletized with KBr, was recorded over 450 to 4000 cm^{-1} which shown in Figure 5.

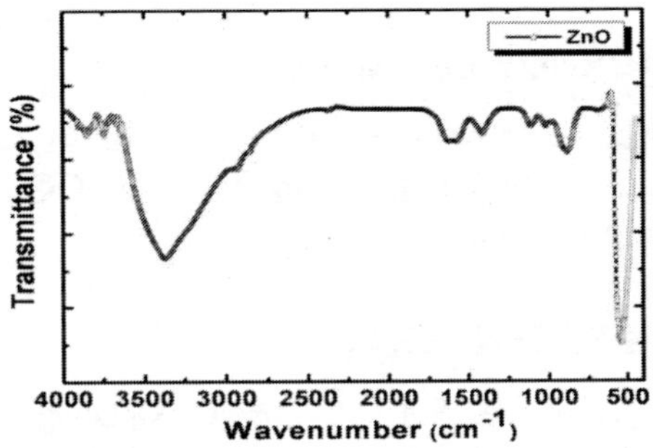

Fig. 5: FT-IR Spectrum of ZnO Thin Film

The spectra consist of seven vibrational bands. The transmittance spectrum of as-deposited ZnO film shows main absorption bands centered at 538 and 3366 cm^{-1} and several small features are located at 886, 1010, 1389, 1614, 2834 and 2926 cm^{-1}. The absorption bands centered at 1389 and 538 cm^{-1} can be associated to Zn-O bond vibration, comparing with standard IR of ZnO powder. It should be noted that for as-deposited ZnO film a strong peak at 3366 cm^{-1} is observed which is very sensitive to stretching vibrations of -OH group.[6] The other absorption bands at 2834, 1614 and 886 cm^{-1} can be related to N-H, N-O bond.[7]

Photoelectrochemical (PEC) Characterization

The PEC characterization of ZnO thin film sample was performed in 0.5 M Na_2SO_4 as an electrolyte in a two-electrode arrangement of following configuration:

Glass/FTO/ZnO/0.5 M Na_2SO_4/Pt/FTO/Glass

In configuration of PEC cell, ZnO thin film deposited on FTO acts as a working electrode (active area ~ 1.0 cm^2), and platinized FTO electrode, which acts as a counter electrode. The I-V characteristic was measured by a SCS-4200 unit in the dark and under illumination using ultra-violet (UV) lamp (28 mW/cm^2). Figure 6 shows the I-V characteristic of ZnO thin film.

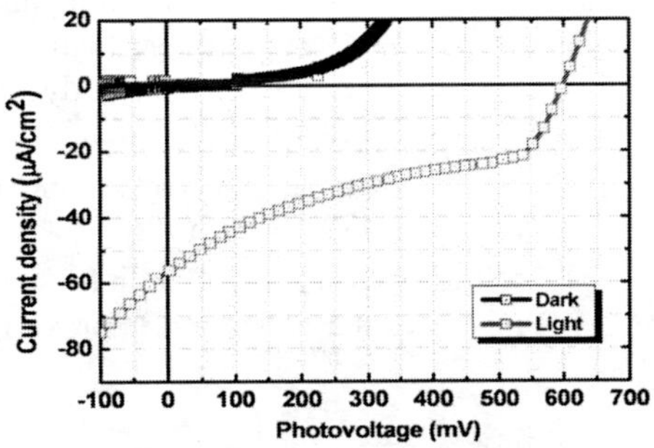

Fig. 6: The I-V Characteristic of ZnO Nanorods thin Film

The short circuit photocurrent densities (I_{sc}), open circuit voltage (V_{oc}) and fill factor (FF) obtained with ZNRs was 56 μA/cm^2, 618 mV and 33.19 respectively. The 0.03 % photo conversion efficiency was achieved for ZnO NRs sample.

CONCLUSIONS

ZnO thin films were successfully deposited by aqueous chemical method at room temperature from an aqueous zinc acetate solution. The XRD analysis shows the prepared ZnO sample have hexagonal crystal structures with 39 nm crystallite size. The SEM images show the formation of nanorods of ZnO samples. The existence of Zn-O peaks revealed from FT-IR spectra. The maximum efficiency is found to be 0.03 % for ZnO sample.

ACKNOWLEDGEMENT

Authors are very much thankful to Department of Science and Technology (DST) and University Grants Commission (UGC), New Delhi for providing financial assistance and infrastructural facilities.

REFERENCES

[1] Wei A., Sun X. W., Xu C. X., Dong Z. L., Yang Y., Tan S. T., Huang W., Growth mechanism of tubular ZnO formed in aqueous solution, *Nanotechnology,* 17, 1740–1744, 2006.

[2] Yu H. D., Zhang Z. P., Han M. Y., Hao X. T., Zhu F. R., A General Low-Temperature Route for Large-Scale Fabrication of Highly Oriented ZnO Nanorod/Nanotube Arrays, *J. Am. Chem. Soc.*, 127, 2378–2379, 2005.

[3] Liu B., Zeng H. C., Hydrothermal Synthesis of ZnO Nanorods in the Diameter Regime of 50 nm, *J. Am. Chem. Soc.*, 125, 4430–4431, 2003.

[4] Wang Z., Qian X. F., Yin J., Zhu Z. K., Large-Scale Fabrication of Tower-like, Flower-like, and Tube-like ZnO Arrays by a Simple Chemical Solution Route, Langmuir, 20, 3441–3448, 2004.

[5] Lupan O., Guérin V. M., Tiginyanu I. M., Ursaki V. V., Chow L., Heinrich H., Pauporté T., Well-aligned arrays of vertically oriented ZnO nanowires electrodeposited on ITO-coated glass and their integration in dye sensitized solar cells, *J. Photochem. Photobiol. A: Chem.,* 211, 65–73, 2010.

[6] Izaki M., Omi T., Transparent zinc oxide films prepared by electrochemical reaction, *Appl. Phys. Lett.,* 68, 2439–2440, 1996.

[7] Stolt L., Hedstrom J., Ruckh M., Kessler J., Velthaus K. O., Schock H. W., ZnO/CdS/$CuInSe_2$ thin-film solar cells with improved performance, *Appl. Phys. Lett.*, 62, 597–599, 1993.

INFLUENCE OF ALKYL THIOLS ON THE NANOSTRUCTURE OF P3HT/PCBM AND P3OT/PCBM BLENDS

R. Ramani and S. Alam

Polymer Science Division, DMSRDE, DRDO, G.T. Road, Kanpur
E-mail: sarfarazkazmi@yahoo.com

ABSTRACT

In this work, we have investigated the nanostructure of the blends of two poly(3-alkyl thiophenes) (P3ATs) (of tail lengths C_6 and C_8) with [6, 6]-phenyl-C61-butyricacid methyl ester (PCBM), using alkyl thiols of two different tail lengths (C_8 and C_{12}) as processing additives. In the blends, alkyl thiols act as selective solvents for PCBM. The investigation reveals that addition of dodecane thiol weakens the interaction of PCBM on P3AT chains in their blends as well as promotes their partial phase segregation during the film forming process. The PCBM phase separation is evident from SAXS measurements, DSC endotherms as well as from SEM. The dodecane thiol addition results to increase in periodicity of the self-organized lamellar structure of P3AT chains in the blends, as found from SAXS measurements. The UV-Visible measurements reveal that the absorption maximum together with the prominent shoulder peak of P3AT in their blends with PCBM show an appreciable red-shift with dodecane thiol addition and this effect is also evident in their photoluminescence emission spectra. This shift to lower energies is attributed to increased conjugation length, which is in agreement with the increase in lamellar periodicity obtained from SAXS results.

INTRODUCTION

High charge-separation efficiency combined with the reduced fabrication costs, and ease of processing make polymer solar cells a better option for future photovoltaic applications. Researchers are making continuous efforts to control the donor/acceptor morphology in Bulk

V. Rajendran, R. Yuvakkumar, K. Thyagarajah and K.E. Geckeler (eds.)
Applications of Nano Materials: Electronics, Energy and Environment, pp. 101–106 (2012).

Hetero Junction (BHJ) materials, which is required for achieving high power conversion efficiency.[1] The BHJ polymer solar cells are based on phase separated blends of conducting polymers and fullerene derivatives. Swift charge transfer from conducting polymers to fullerenes guarantees that the quantum efficiency for the charge transfer approaches unity with electrons on the fullerene network and holes on the polymer network. Thereafter, the electrons and holes migrate towards the respective electrodes of the photovoltaic device. Despite high charge separation efficiency, a significant fraction of carriers recombine at the donor/acceptor interface before extraction from the device owing, in part, to the inherently formed random interpenetrating network morphology in the film upon spin-casting. Carrier recombination before reaching the electrodes and low mobility limit the overall photon harvesting efficiency.[1]

The most efficient BHJ devices so far have used the high mobility conducting polymer, poly(3-hexyl thiophene) (P3HT), as the electron-donor, and the soluble fullerene derivative, [6,6]-phenyl C_{61}-butyric acid methyl ester (PCBM), as the acceptor.[1,2] But, the low T_g of P3HT (< 40°C) results to unstable morphology and thus limits its application, where the use of solar cell is desired at temperatures of 60–80°C.[2] Since, poly(3-octyl thiophene) (P3OT) has a T_g of ca. 100°C, research work on P3OT/PCBM based BHJ also considered significant, though its power conversion efficiency is less than P3HT/PCBM based BHJ.[3]

Thermal annealing, use of "unfriendly" solvents and controlled solvent evaporation upon film casting have proved critical for optimizing the charge separation and transport within the BHJ morphology.[4] The use of dithiooctane and dihalooctanes as processing additives with different conducting polymers/PCBM combinations have shown that their phase separation and polymer aggregation could be modified.[5] This prompted us to undertake a thorough investigation using monoalkyl thiols as processing additives for BHJs based on P3HT and P3OT with PCBM matrix. For this, we have used octane thiol and dodecane thiol as processing additives with P3HT/PCBM and P3OT/PCBM blends and investigated their influence on the phase separation and polymer segregation in these blends.

EXPERIMENTAL

Sample Preparation

The polymers P3HT and P3OT were obtained from American Dye Source Inc., Canada. The chemicals octane thiol, dodecane thiol and chlorobenzene were from Sigma-Aldrich. The P3AT/PCBM blends were prepared as follows: To the freshly prepared solution of P3AT in chlorobenzene (10 mg/ml), the solution of PCBM in chlorobenzene (8 mg/ml) was added slowly with constant stirring so as to get the blend in 50:50 ratio and the mixing continued for ca. 8 h. To this 2 vol% of dodecane/octane thiol is added and then allowed to mix again overnight. The solvent was evaporated by keeping them in fume-hood for 48 h and then in vacuum oven. For comparison, virgin samples of P3HT, P3OT and PCBM were also prepared in the same manner. These samples are designated as given in Table 1.

Table 1: Description of the Samples

Sample Code	*Description*
P3HT	P3HT prepared from solvent
P3OT	P3OT prepared from solvent
PCBM	PCBM prepared from solvent
P3HT-PCBM	P3HT/PCBM blend prepared from solvent
P3OT-PCBM	P3OT/PCBM blend prepared from solvent
P3HT-PCBM-OctT	P3HT/PCBM blend with octane thiol (2vol%) prepared from solvent
P3OT-PCBM-OctT	P3OT/PCBM blend with octane thiol (2vol%) prepared from solvent
P3HT-PCBM-DodT	P3HT/PCBM blend with dodecane thiol (2vol%) prepared from solvent
P3OT-PCBM-DodT	P3OT/PCBM blend with dodecane thiol (2vol%) prepared from solvent

Characterization

The SAXS measurements were done using Bruker SAXS system that consists of a 2D area detector (Bruker AXS) and a Bruker Microstar rotating anode X-ray source with Montel Optics (Cu K_α radiation, λ = 1.54Å). The magnitude of the scattering vector is given by $q = (4\pi/\lambda)\mathrm{Sin}\theta$, where '$\theta$' is the half-scattering angle. The DSC measurements were carried out using the DSC system of TA instruments (Model Q-200) in the nitrogen atmosphere (flow rate = 50 mL/min). The surface morphology of the blends was examined using a Carl Zeiss EVO 50 Scanning Electron Microscope. The UV-Vis measurements were carried out using a JASCO double-beam UV-Vis spectrophotometer. The photoluminescence measurements were done using Varian Cary Eclipse photoluminescence spectrophotometer.

RESULTS AND DISCUSSION

SAXS Results

To understand the detailed picture of the arrangement of self-organized structure of the P3AT chains in these samples, SAXS measurements were performed and the resulting scattering profiles are shown in Figure 1(A & B). The virgin P3HT exhibits a primary scattering peak at ca. q = 0.326 $Å^{-1}$, due to lamellar structure with a periodicity of ca. 19.3Å while the lamellar periodicity of P3OT is ca. 24.1Å. The PCBM shows two distinct peaks at ca. q = 0.426 $Å^{-1}$ and 0.455 $Å^{-1}$ (see Fig 1A). The addition of PCBM has only negligible change in the lamellar periodicity of P3HT (19.6 Å) and P3OT (24.5Å) but the intensities have reduced significantly. This indicates that the PCBM has reduced the crystalline structure of P3ATs as it is known that PCBM interacts in the amorphous domains of P3AT. Introduction of dodecane thiol has changed the scattering profile P3OT-PCBM (Figure 1B) and the lamellar periodicity is increased to 25.7Å. This reveals that the polymer alkyl tails are stretched to a greater extent. Additionally, two other weak peaks are evident at '*q*' values distinct from that of pure PCBM. In case of P3HT-PCBM, though no such significant change in lamellar structure is observed, atleast one of the weak peaks is evident. These additional peaks together with the main scattering peak do not fit to any of the generally known self-assembled structures. No such structural changes were observed with the addition of octyl thiol, in both the polymer blends.

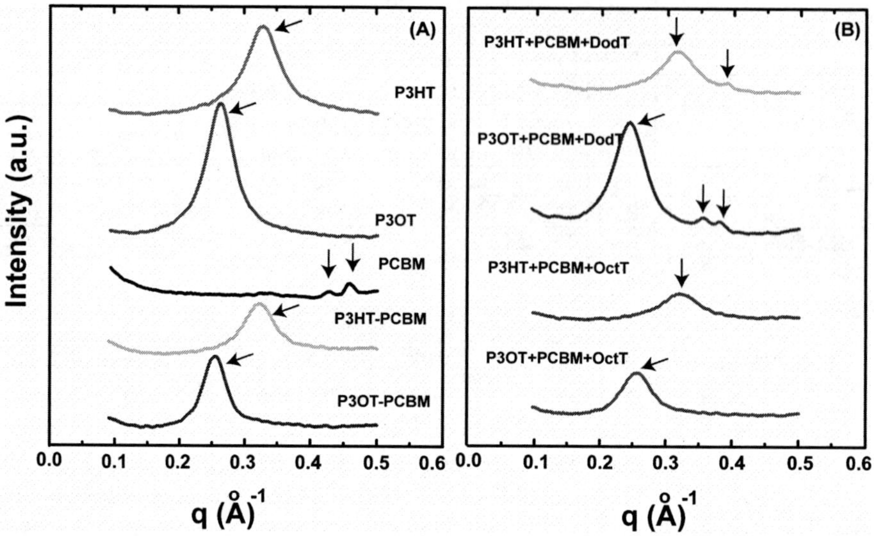

Fig. 1: SAXS Profiles of the Samples. The Profiles are Vertically Shifted for Clarity

DSC Results

In order to understand the change in crystalline nature of the polymer upon blending with PCBM as well as upon alkyl thiol addition, DSC measurements were made and are shown in Figure 2. Glass transition temperature of P3HT and P3OT appeared weak and observed at ca. 23 °C and 98°C, respectively. The crystalline fraction of P3HT is thermally more stable than P3OT, as can be identified from their melting temperatures (T_m) observed at ca. 228°C and 186°C, respectively. The PCBM crystals exhibit two melting peaks at 266°C and 289°C. Introduction of PCBM into the polymer altered the crystalline nature of the polymer drastically and the T_m of P3HT and P3OT are down shifted to 213°C and 178°C, respectively. Even the associated crystalline melting enthalpies have reduced by more than 60% upon blending with PCBM. This reveals that the crystalline part of the polymer has weakened and modified to amorphous part, so as to interact with the PCBM matrix.

The introduction of dodecane thiol modifies the crystalline nature of the polymer in the P3AT/PCBM blend. The first striking result that could be observed is the emergence of a melting endotherm at ca. 286°C close to that of virgin PCBM. On comparing the SAXS profiles and DSC melting endotherms of P3HT-PCBM-DodT and P3OT-PCBM-DodT samples, one could speculate that the phase separated PCBM has attained a different crystalline structure, a fact that will be cleared from SEM images that will be shown next. Similar to SAXS results, DSC plots showed no significant changes in the blends when octane thiol is used as the processing additive.

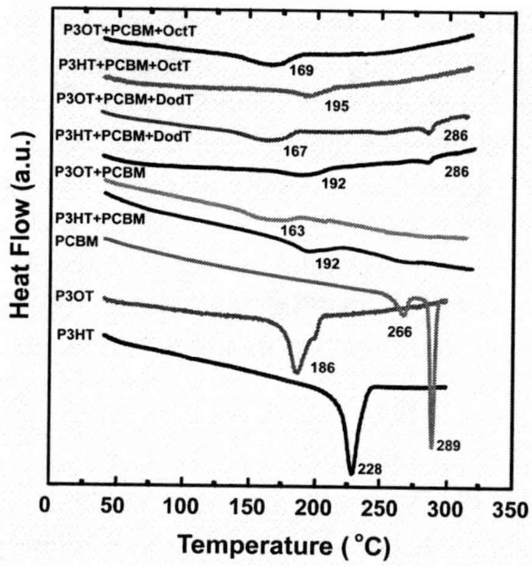

Fig. 2: DSC Plots of the Samples. The Plots are Vertically Shifted for Clarity

SEM Results

The SEM images of the samples P3OT, P3OT-PCBM and P3OT-PCBM-DodT and pure PCBM are presented in Figure 3 (A-D). For the sake of brevity, the images of P3HT series, which are similar, are not shown. The PCBM forms a well compatible blend with P3OT (Figure 3A&B). Introduction of dodecane thiol has modified the surface morphology and the phase separated PCBM is quite evident on the blend matrix (Figure 3C). For comparison, the image of pure PCBM is also shown (Figure 3D). The SEM images confirm that phase separated PCBM attained a different crystalline structure.

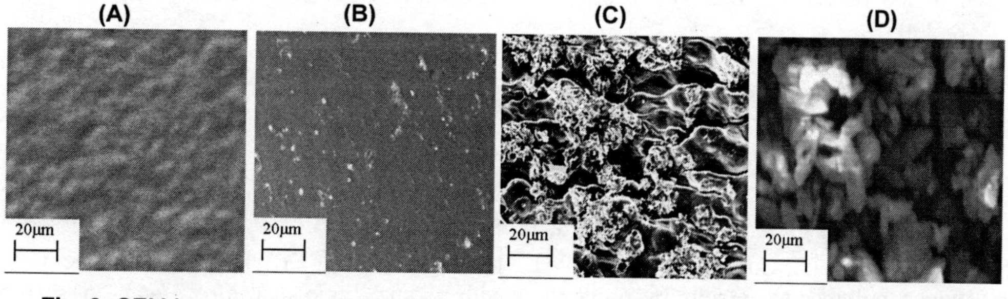

Fig. 3: SEM Images of (A) P3OT (B) P3OT-PCBM (C) P3OT-PCBM-DodT and (D) PCBM

UV-Visible and Photoluminescence Results

The P3HT and P3OT films show absorption maximum at 521 and 526 nm (due to π–π^* transition of the conjugated chains) with a vibronic absorption in the red region at 602 and 606 nm respectively, indicating good ordering of the polymer chains. The PCBM exhibits absorption maximum at ca. 269 nm and 337 nm. Both P3HT and P3OT upon blending with

PCBM, exhibit a blue shift with their absorption at 508 and 518 nm, respectively. This shift to high energy region refers to less ordering of the polymer matrix as PCBM hinders the chain alignment and goes in agreement with the SAXS and DSC results. Addition of dodecane thiol red shifts the absorption maximum close to their initial value. The red shift indicates enhanced conjugation length associated with more ordering and shift of the absorption to lower energies. Addition of octane thiol did not show any significant shift in the absorption maxima. The solid state photoluminescence spectra of P3HT and P3OT (excited at 520 nm) showed emission maximum at ca. 653 nm. Similar to UV-visible results, initially, the spectrum exhibited a blue shift and with the introduction of dodecane thiol, a red-shift is observed.

CONCLUSIONS

Blending of P3HT /P3OT with PCBM using dodecane thiol as the processing additive results to change in phase morphology together with partial phase separation of PCBM. Better phase separated morphologies are observed in P3OT/PCBM than with P3HT/PCBM. The results reveal that the device performance would be better when P3OT/PCBM BHJ is prepared with dodecane thiol as the processing additive, that will have better stable morphology when used in relatively high temperatures.

ACKNOWLEDGEMENTS

The authors thank the Director, DMSRDE, Kanpur for permitting to present this work.

REFERENCES

[1] Li, G., Shrotriya, V., Huang, J., Yao, Y, Moriarty, T. *et al.*, High efficiency solution processable polymer photovoltaic cells by self-organization of polymer blends, *Nature Materials,* 4, 864–868, 2005.

[2] Zhao, J., Swinnen, A., Assche, G.V., Manca, J., Vanderzande, D., *et al.*, Phase diagram of P3HT/PCBM blends and its implication for the stability of Morphology, *J. Phys. Chem. B.*, 113, 1587–1591, 2009.

[3] Kaur, M., Gopal, A., Davis, R.M. and Heflin, J.R., Concentration gradient P3OT/PCBM photovoltaic devices fabricated by thermal inter-diffusion of separately spin-cast organic layers, *Sol. Energy Mater. Sol. cells*, 93, 1779–1784, 2009.

[4] Peet, J., Kim, J.Y., Coates, N.E., Ma, W.L., Moses, D., *et al.*, Efficiency enhancement in low bandgap polymer solar cells by processing with alkane dithiols, *Nature Mater.*, 6, 497–500, 2007.

[5] Lee, K.J., Ma, W.L., Brabec, C.J., Yuen, J., Moon, J.S., *et al.*, Processing additives for improved efficiency from bulk heterojunction solar cells, *J. Am. Chem. Soc.,* 130, 3619–3623, 2008.

OPTICAL AND AC CONDUCTIVITY STUDIES OF CO DOPED TiO_2 NANOTUBES

P. Pournami Vijayan, Marykutty Thomas[1], Lakshmi Nair[2] and K.C. George

Department of Physics, S.B. College, Changanacherry, Kerala
[1]Department of Physics, B.C.M. College, Kottayam, Kerala
[2]Department of physics, M.G. College, Thiruvananthapuram
E-mail: poorni_asok@rediffmail.com

ABSTRACT

Cobalt doped TiO_2 nanotubes are prepared by the combination of sol-gel process with hydrothermal treatment. The morphology and crystalline structure of TiO_2 nanotubes are characterized by transmission electron microscopy (TEM), X-ray diffraction respectively. Co doping induces a structural transformation from anatase to rutile. The temperature dependence of the ac electrical conductivity is investigated in the temperature range 303–413 K. The transition from n type to p type conductivity is observed. The conductivity is maximum for the undoped samples. PL spectrum shows the presence of surface oxygen vacancies and self trapped excitons in Co doped TiO_2 nanotubes and undoped samples. Co doped samples shows a peak in PL spectrum at 526 nm, which is not in the undoped samples.

INTRODUCTION

TiO_2 is a wideband gap (3.32ev) n type semiconductor has a variety of promising applications in optical and electrical devices.[1] One dimensional TiO_2 nanostructures exhibit a high surface to volume ratio, improved charge transport afforded by their dimensional anisotropy and an increased number of delocalized carriers. In nanotubular geometrics, the electronic properties of TiO_2 are more important, because they determine how efficiently electrons can be transferred along the one dimensional path. These features make these structures promising for applications in photo catalytic and solar energy conversion.[2, 3]

V. Rajendran, R. Yuvakkumar, K. Thyagarajah and K.E. Geckeler (eds.)
Applications of Nano Materials: Electronics, Energy and Environment, pp. 107–112 (2012).

TiO_2 is a non conducting material, whose electrical conductivity can be enhanced by calcinations, doping etc. In recent years much attention has been paid to enhance the performance of TiO_2 by doping. Doping can alter the optical and electrical properties of the TiO_2 nanotubes. Dopants like Cr, Ge, Pb, Co, Fe could lead to the appearance of a metallic behavior of conductivity.[4–8] Bally *et al.* reported electrical conduction in TiO_2 thin films.[8] The ac conductivity studies give the evidence of surface defects present in the semiconductors. Electrical properties of the TiO_2 semiconductors are influenced by oxygen vacancies and extrinsic impurities. To the best of our knowledge the ac conductivity studies of Co doped TiO_2 nanotubes have not been exploited yet. The main objective of the present work is to investigate the ac electrical properties of the Co doped TiO_2 nanotubes for different Fe concentrations.

EXPERIMENTAL

Synthesis of Co doped TiO_2 nanotubes were performed by solgel method followed by hydrothermal method.[9] Tetrabutyl titanate and cobalt nitrate nitrate are precursors. Ethanol solution of tetrabutyl titanate was added dropwise to the mixture of 20 mL ethanol and .045 mol/l cobalt nitrate solution. Then the mixture was hydrolyzed at room temperature for 2h under vigorous stirring, a transparent sol was obtained. Then gel was obtained by aging the sol for 12h at room temperature. The resulting gel was dried at 80°C in oven for 12h and then calcined at 500°C in muffle for 3h. Then the Co doped TiO_2 powder was obtained. 1.2 g Co doped TiO_2 powder was mixed with 58 mL 10 mol/l NaOH aqeous solution in a Teflon lined bottle and maintained at 150°C for 12H. Obtained materials were washed with .1 mol/l HNO_3 solution and distilled water respectively and then dried at 80°C. Then Co doped TiO_2 nanotubes were obtained. Calculated cobalt concentration was .18%. The undoped sample named as Sa a and Co doped sample as Sa b.

The morphologies and composition of samples were observed on transmission electron microscope (TEM) using a GEOL 1011 MODEL. For TEM observation TiO_2 as prepared samples were ultrasonically dispersed in ethanol solution and the TEM micrographs were taken. The phase identification of the samples and the crystallite size determination were conducted with powder X-ray (XRD) diffractrometer with Cukα (λ = 1.54056 Å). The Photoluminescence spectrum was recorded using an LS 45, PerkinElmer fluorescence spectrometer and xenon lamp was used as the excitation source operating at 320 nm. The obtained sample was then crushed into fine powder and pelletized at a pressure of 4 tones for two minutes using hydraulic press. The pellet had a diameter of 10 mm and a thickness of 1 mm. The ac conducting measurements were carried out using Hioki 3532–50 Hi-tester LCR meter in the frequency range of 100 Hz–1 MHz and at various temperatures between 303 and 423 K.

RESULTS AND DISCUSSION

The surface morphology of the TiO_2 nanostructures formed is shown in Figure 1. As prepared sample shows nanotubes with an outer and inner diameter of 79 nm and 38 nm respectively (Figure 1a and b).

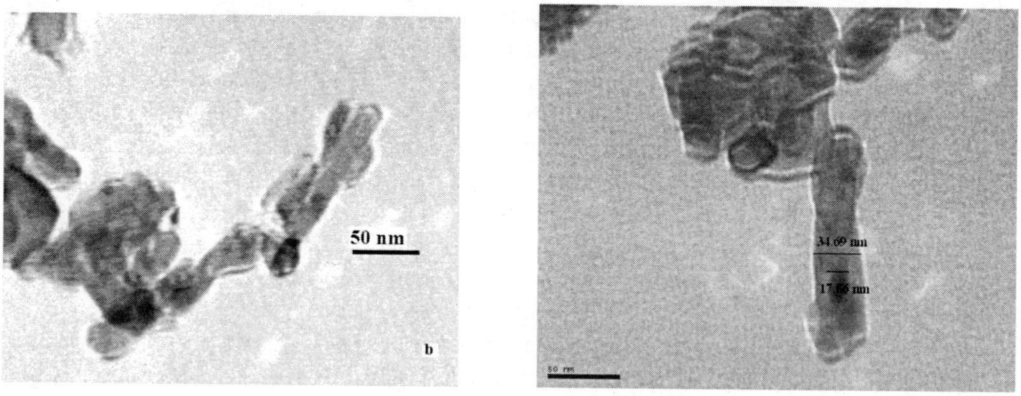

Fig. 1a&b: TEM Images of TiO_2 Nanotubes

STRUCTURAL ANALYSIS

The XRD pattern of the undoped and Co doped TiO_2 nanotubes is as shown in Figure 3. The undoped samples show only an amorphous structure. Crystallization can be seen for the Co doped samples.

Edax data reveals the existence of cobalt in the Co doped TiO_2 samples. The calculated weight percentage of Co is .18%.

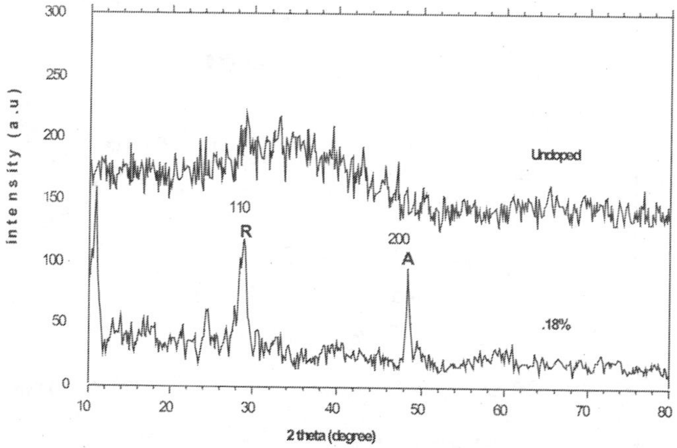

Fig. 2: The XRD Pattern of the Undoped and Co Doped TiO_2 Nanotubes

OPTICAL PROPERTIES

The PL spectrum gives information on the properties of the energy levels lying within the bandgap. PL spectrum gives information on the properties of the energy levels lying within the band gap. The room temperature PL spectra of the as prepared TiO_2 nanotubes and Co doped TiO_2 nanotubes under excitation with the 300 nm line are displayed in Figure The spectrum mainly consists of four emission bands at 388, 420, 440–460 and 486 nm. Co doped samples shows a peak in PL spectrum at 526 nm, which is not in the undoped samples. The peak at 388nm corresponds to the direct allowed transition X2-X1. The emission at 420 nm can be attributed to the indirect transition X1-Γ1.[10] Several shallow levels can be seen which is due to the defect related emission. Broad blue emission ranging from 440–470 nm is due to the self trapped excitons localized on TiO6 octahedra (11). The emission peak due to self trapped excitons for the Co doped samples is strong as compared to the undoped samples. The blue green emission at 486nm can be assigned to oxygen vacancies.[11,12] The more the oxygen vacancies the stronger the blue green emission intensity is. Fewer oxygen vacancies in rutile than in the anatase structure is reported.[8] .18 % Co doped TiO_2 nanotube has lower oxygen vacancies compared to the undoped one, due to the formation of rutile at Co doping. The process of PL emission identified with oxygen vacancies might occur with photo generated conduction band electrons trapped by ionized oxygen vacancies levels in titanium nanotube and subsequently recombine with holes in the valance band radiatively.

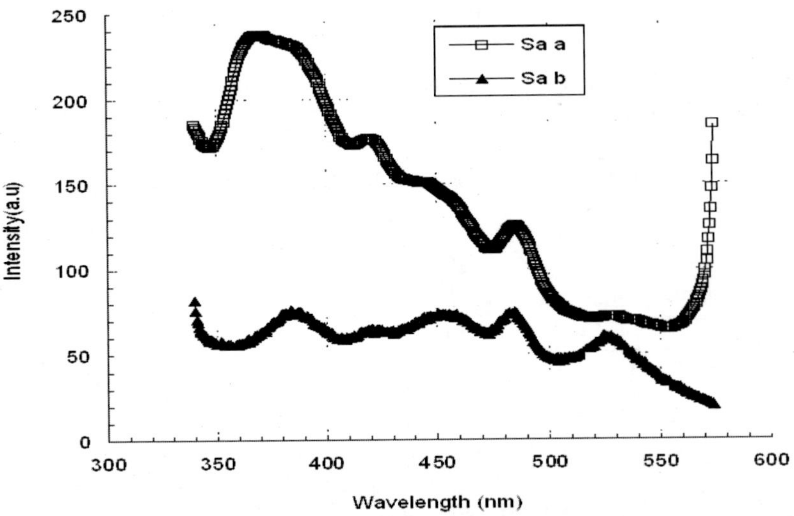

Fig. 3: PL Spectra of TiO_2 (a) as Prepared Nanotube (b) Co Doped Nanotube

AC CONDUCTIVITY

Plots of log σ_{ac} with inverse temperature (Arrhenius plot) at six different fixed frequencies of the as prepared TiO_2 nanotube for the temperature 303–413K is shown in Figure 6. A low temperature region is observed where slope is small and has strong frequency dependence. (303–353 K). Above 353 K a strong temperature dependent region is observed and all curves converge at high temperatures.

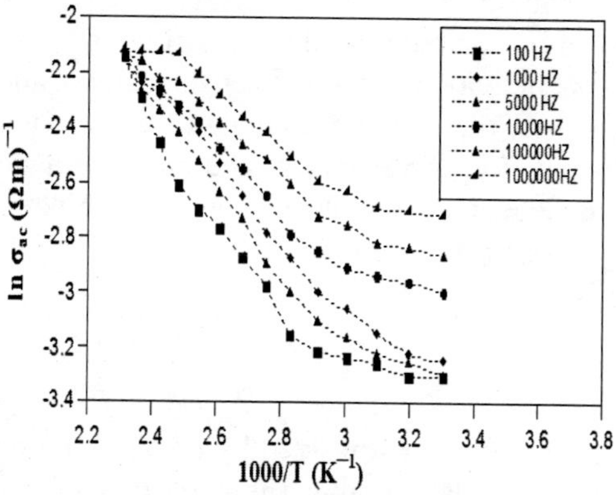

Fig. 4: Temperature Dependence of AC Conductivity at Various Frequencies for as Prepared Nanotube

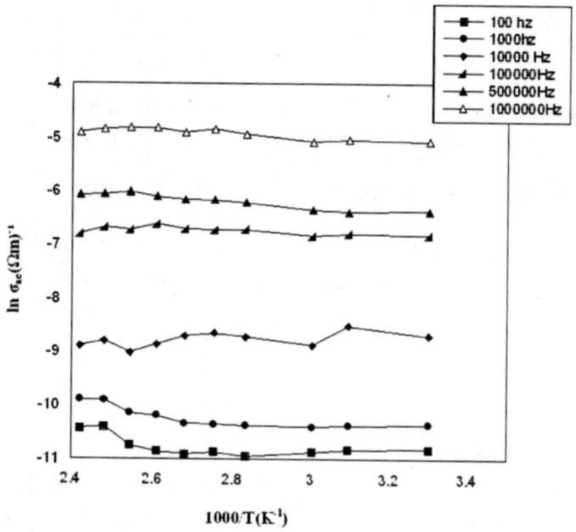

Fig. 5: Temperature Dependence of AC Conductivity of Co Doped Samples at Various Frequencies

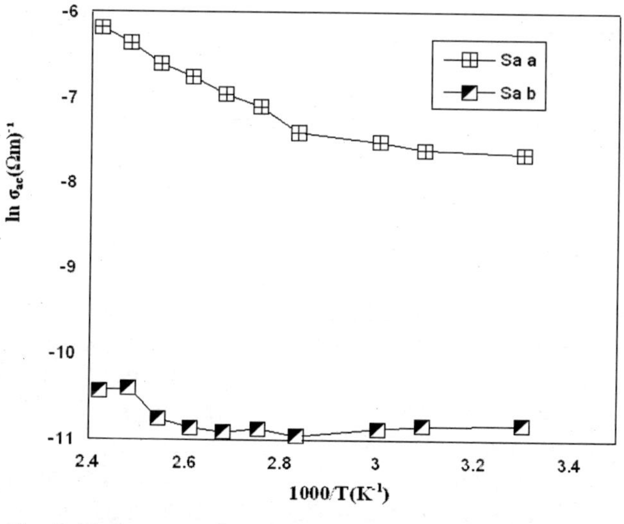

Fig. 6: Temperature Dependence of AC Conductivity of TiO_2 Nanotubes and Co Doped Samples at 100 Hz

To study the effect of Cobalt doping on the ac conductivity studies of TiO_2 nanotube log σ_{ac} with inverse temperature are plotted for as prepared TiO_2 nanotubes and for Co doped samples at 100 Hz. The ac conductivity of the Co doped samples is smaller than the undoped TiO_2 nanotube. A negative temperature coefficient of resistance is observed for Co doped samples. This p type conductivity was also observed for TiO_2 doped with impuries like Cr, Ge, Fe and Pb. Metallic behavior (p type conductivity) can be related to the Co doping and the concentration of oxygen vacancies. The p type conductivity is observed in Fe doped TiO_2 films [8]

In TiO_2, oxygen vacancies are known to be the most common defects. The p type conductivity observed in Co doped TiO_2 nanotube can be attributed to the formation of oxygen vacancies. As the Co concentration increases the ac conductivity decreases. Each oxygen vacancy gives two electrons in the titania conduction band. All the acceptors created by the cobalt atoms are compensated by the electrons given up by the oxygen vacancies. The rutile formation lowers the oxygen vacancies. Co impurities act as acceptors, which decreases the electron concentration produced by oxygen vacancies. Therefore p type conductivity is observed for cobalt doped samples.

CONCLUSION

Co doped TiO_2 nanotubes with Co concentration of .18% are synthesized by a solgel method followed by a hydrothermal method. An ac conductivity data analysis is carried out in the temperature range 303–413 K for the TiO_2 nanotubes and Co doped TiO_2 nanotubes P type conductivity is observed for Co doped TiO_2 nanotubes and is attributed to the formation of rutile phase and oxygen vacancies. The ac conductivity is at its maximum for undoped samples.

ACKNOWLEDGEMENT

Authors are grateful to Central instrumentation facility (CIF), St. Berchmans College and Rajiv Gandhi Centre for Science and Technology, Thiruvananthapuram.

REFERENCES

[1] Tang, H., Levy, F., Berger, H. 1995 Urbach tail of anatase TiO_2 Phys. Rev. B 52 7771

[2] Yuxiang Yu, Dongsheng Xu (2007) Single–crystalline TiO2 nanorods: Highly active and easily recycled photocatalysts. Applied catalysis B: Environmental 73 166–171

[3] Daibin Kuang, Jeremie Brillet, Peter Chen, Masakazu Takata, Satoshi Uchida, Hidetoshi Miura, Kohichi Sumioka, Shaik. M. Zakeeruddin, and Michael Gratzel(2008) Application of Highly Ordered TiO2 Nanotube Arrays in Flexible Dye-Sensitized Solar Cells. ACS Nano 2 1113–1116

[4] Matsumoto, Y., Murukami, M., Shono, T., Hasegawa, T., Fukumara, T., Kawasaki, M., Ahmet, P., Chikyov, T., Koshihara, S. and Koinuma, H., 2001 Science 291 854.

[5] Fromknecht, R., Wiss, T., Khubeis, I. and Meyer, O., 2000, Surf.Coat.Technol. 364 128–129

[6] dasilva, R.C., Alves, E. and Cruz, M.M. 2002 Nucl.Instrum.MethodsPhys.Res. b 191 158

[7] Song, H.-Q., Wang, Y., Yan, S.S., Mei, L.M., Zhang, Z., 920070 J.Magn. Mater. 53 312

[8] Bally, A.R., Korobeinikova, E.N., Schimid, P.E., Levy, F. Bussy J. 1998 Phys.D 31 1149

[9] Lixue Deng,Shurong Wang Daying Liu Baolin Zhu Welping Huang Shihua Wu Shoumin Zhang 2009 Catal.Lett 129 513–518

[10] Daude, N., Gout, C., Jounanin, C., 1977 Physical Review B 15 6

[11] Tang, H., Berger, H., Schmid, P.E., levy, F. 1993 Solid State Commun 87 847

[12] Siripala, W. and Tomkieviez, M. (1982). Interactions between photoinduced and dark charge transfer across n-TiO2-aqueous electrolyte interface. *J Electrochem Soc.* 129:1240.

PREPARATION AND CHARACTERIZATION OF NANOPOROUS ACTIVATED CARBON AND ITS APPLICATION AS A HETEROGENEOUS FLUIDIZED BED CATALYST FOR THE TREATMENT OF MUNICIPAL WASTEWATER

S. KARTHIKEYAN, R. BOOPATHY, P. MAHARAJA, J. SHARMILA, P. KUMARA SASHIDARA AND G. SEKARAN

Environmental Technology Division, Council of Scientific and Industrial Research (CSIR)—Central Leather Research Institute (CLRI), Adyar, Chennai
E-mail: ganesansekaran@gmail.com

ABSTRACT

In the present investigation an attempt was made to treat the municipal waste water in Fluidized Immobilized Catalytic Carbon Oxidation (FICCO) reactor. In this study rice husk carbon (nanoporuos Activated Carbon, NAC_{800}) is used as the catalyst, in which the contaminants adsorb the catalyst. The main objective of the present study is to assess the efficiency of reducing the organic pollutant in the waste water through fluidized immobilized carbon catalytic oxidation cell reactor with minimum sludge production. The effective removal of organic pollutants through various concentrations of starch, protein, surfactant, catalyst loading, and time was found to be most successful in reducing the COD and BOD removal efficiencies of the FICCO reactor.

INTRODUCTION

Increasing discharge and improper management of municipal waste water have created a great concern among the environment. Municipal waste water consists of toxic compounds or compounds which may be potentially mutagenic of carcinogenic. There is a need for the protection of the public health in a manner commensurate with environmental, economic,

V. Rajendran, R. Yuvakkumar, K. Thyagarajah and K.E. Geckeler (eds.)
Applications of Nano Materials: Electronics, Energy and Environment, pp. 113–118 (2012).

political and social concerns (Metcalf and eddy). The removal efficiencies of suspended solids (SS) and chemical oxygen demand (COD) are around 50% and 30% respectively in a typical primary treatment plant.

Hence application of biological methods such as fluidized bed reactors has generated a lot of interest in the recent past in the field of waste water treatment processes due to its high performance efficiency compared to the conventional suspended growth and fixed-film waste water treatment processes. The major advantage of fluidized bed reactor over other biodegradation systems is a higher biomass concentration, and a higher mass transfer, resulting in a higher rate of biodegradation.

Therefore the main aim of the present study is to evaluate the effectiveness of reducing the organic pollutant in the waste water through fluidized immobilized carbon catalytic oxidation cell reactor with minimum sludge production.

EXPERIMENTAL

The activated carbon samples from rice husk were prepared in two sequential steps: pre-carbonization at 400°C and chemical activation (Kennedy *et al*., 2004). The carbon samples have been characterized for surface area, pore volume and pore size, X-ray diffraction technique (XRD) and a SEM.

RESULTS AND DISCUSSION

Figure 1 presents nitrogen adsorption/desorption isotherms at 77°K for NAC_{700}, NAC_{800} and NAC_{900} respectively. The presence of nanoporous is justified by the hysteresis effect and the slope of the plateau increased to yield type IV isotherms with a significant increase in the nitrogen uptake through the entire pressure range. XRD pattern of MAC_{800} (Figure 2) contains a mixture of amorphous and crystalline phases of silica. The Low angle XRD spectrum of MAC_{800}, which was a characteristic of nanoporous material, is shown in the Figure 3. It shows a very intense diffraction peak (100) and three obvious peaks (110), (200) and (210), which suggest the preservation of the hexagonal structure (regularity) with a thicker wall. The thermo gravimetric analysis of NAC_{800} samples were carried out in the presence of nitrogen as shown in Figure 4. NAC_{800} exhibits three-step weight loss in the presence of nitrogen atmosphere. The SEM image of NAC_{800} sample, shown in Figure 5, depicts that the activated carbon matrix is highly porous in nature with major distribution of nanoporous, which is the characteristic feature of activation using phosphoric acid as the activant.

Design and Selection of FICCO Reactor

An acrylic sheet reactor of height 12 cm and width 9 cm was fabricated. such reactors were fabricated [Fig 6] with different dimensions of upper hopper and differing in the provision for aeration. In addition to these the raw water and FICCO treated water were analyzed for characteristics which include pH, sulphates, chloride, COD, BOD, total solids, dissolved solids and suspended solids level [Table 1]. Figure 7 shows the percentage removal of COD and BOD with and without adding catalyst varying with time. Oxidation removed BOD by 46% and 84% with and without carbon respectively. Figure 8 shows the effect of catalyst mass loading on COD during the catalytic oxidation of domestic waste water. The percent reductions of COD increased with the increase in catalyst loading from 1 to 12 g/620 ml. With further increase in catalyst mass loading from 12 to 15 g/620 ml, there has been no increase in percent COD removal. The performance of the FICCO reactor was studied with the concentration of anionic surfactants. The average influent concentration of surfactants was 23.8 mg/L, of which the FICCO reactor removed 80%. The COD and BOD removal efficiencies reached 82.7% and 81.5% respectively which is shown in the graph (Figure 9), leading to average effluent concentrations of 4.8 mg/L. When the reactor was operated at sufficient aeration condition with the addition of starch, the removal efficiencies of COD and BOD reached 88.1% and 87.7% (Figure 10) respectively. The performance of the FICCO reactor was studied with the concentration of adding oil and protein to the waste water in addition to surfactant and starch as shown in the Figure 11and Figure 12.

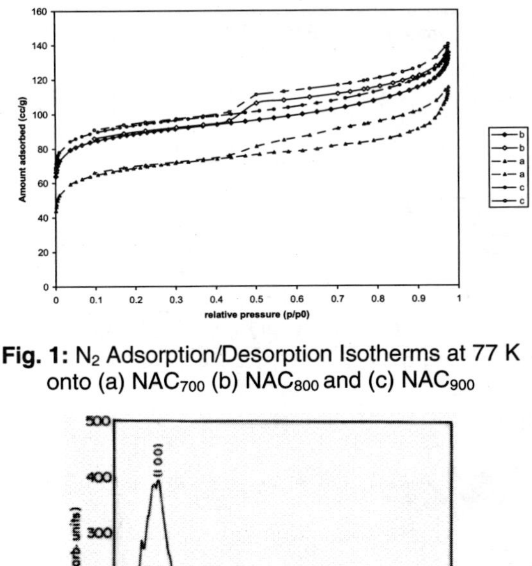

Fig. 1: N_2 Adsorption/Desorption Isotherms at 77 K onto (a) NAC_{700} (b) NAC_{800} and (c) NAC_{900}

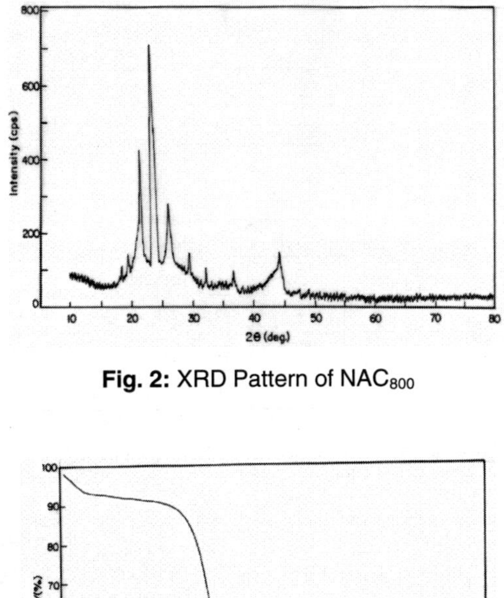

Fig. 2: XRD Pattern of NAC_{800}

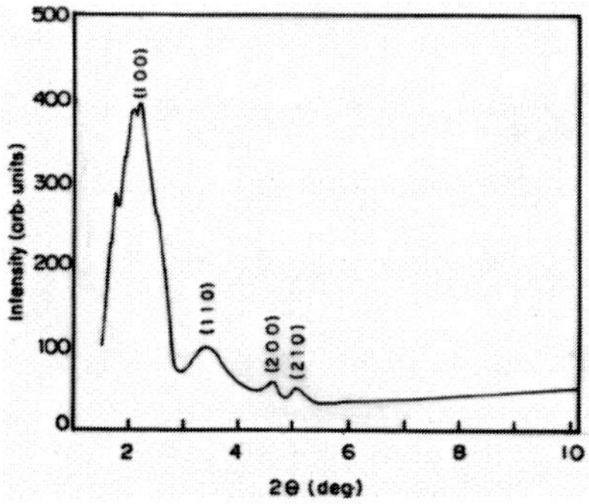

Fig. 3: Low Angle XRD Spectrum of NAC_{800}

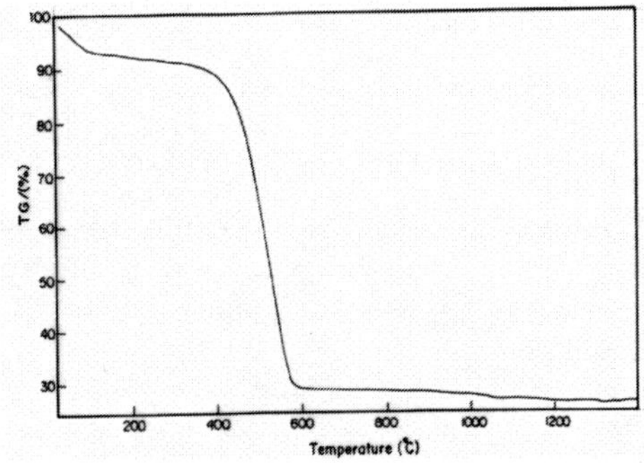

Fig. 4: TGA Spectrum of NAC

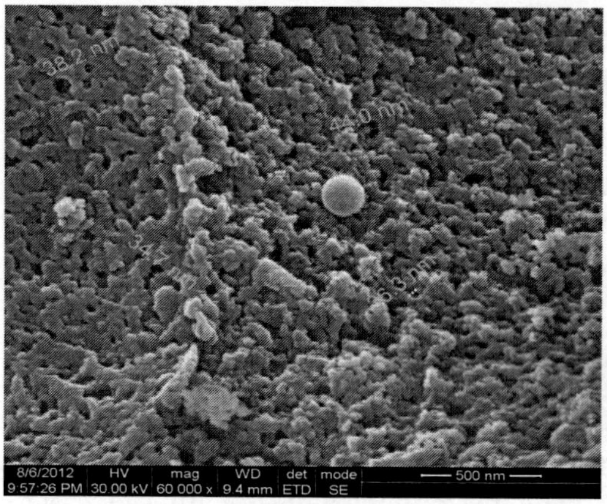

Fig. 5: SEM

Table 1: Physicochemical Parameters for Raw Domestic Waste Water and FICCO Treated Waste Water

S. No.	*Parameters*	*Raw Waste Water*	*FICCO Treated*
1.	pH	7.6 ± 0.6	8.2 ± 0.2
2.	BOD	171.1 ± 51.7	79.3 ± 5.2
3.	COD	514.7 ± 154.6	238.1 ± 16.5
4.	Total suspended solids	187.1 ± 60	82.6 ± 9.6
5.	Nitrogen	50.2 ± 21.6	31.1 ± 5.8
6.	Sulphate	36.1 ± 10.3	16.3 ± 5.9
7.	Chlorides	40 ± 6	13.2 ± 3.7
8.	Surfactant	11.1 ± 3.1	4.47 ± 0.4
9.	Oil and Grease	60 ± 8.9	16 ± 1.8

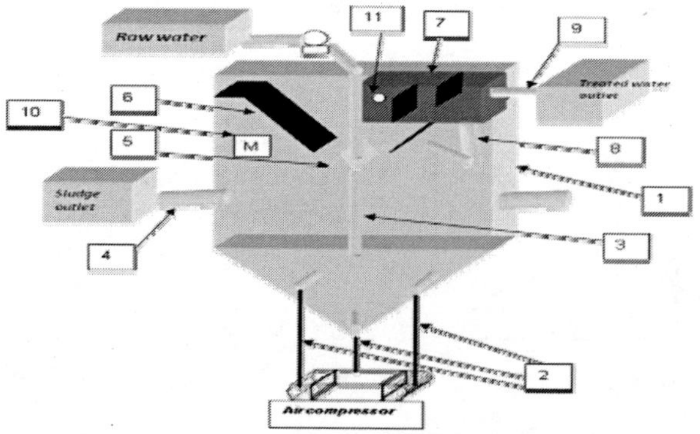

Fig. 6: Schematic Flow Diagram for Fluidized Bed Oxidation for the Treatment of Municipal Wastewater

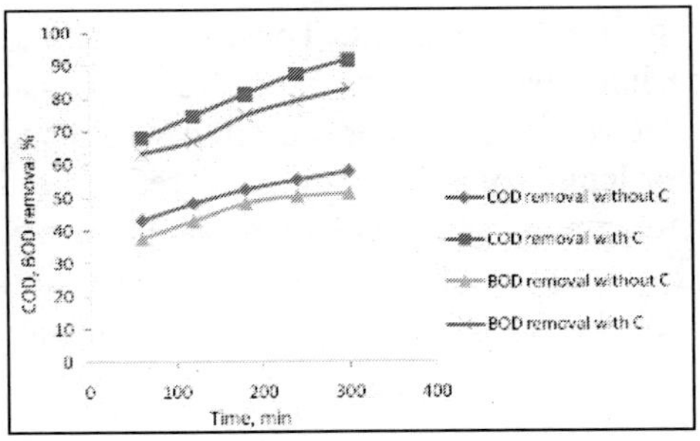

Fig. 7: Percentage Removal of COD and BOD with Function of Tzime

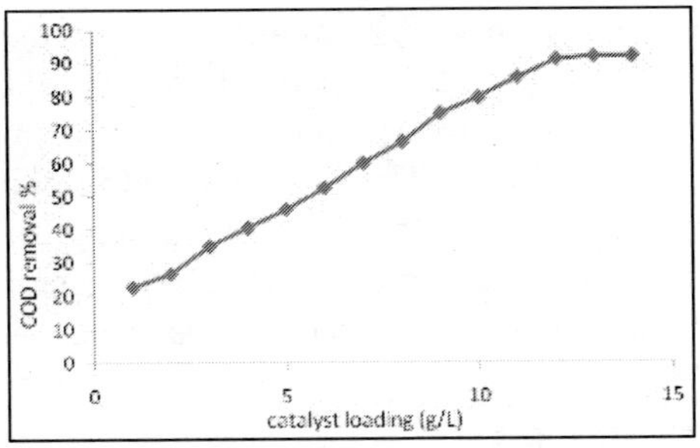

Fig. 8: Percentage Removal of COD as a Function of Catalyst Loading

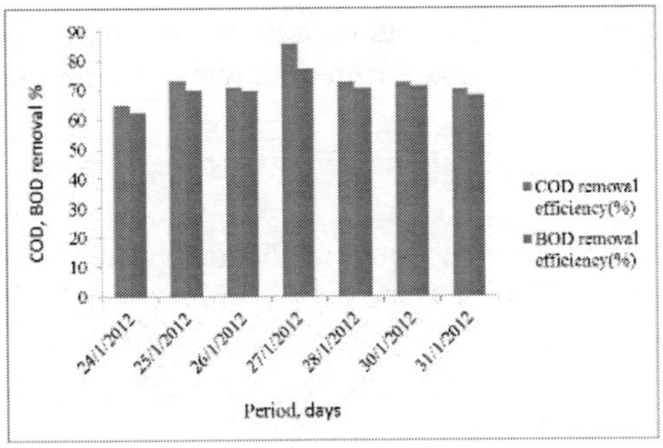

Fig. 9: Effect of Surfactant Loading on the COD Removal

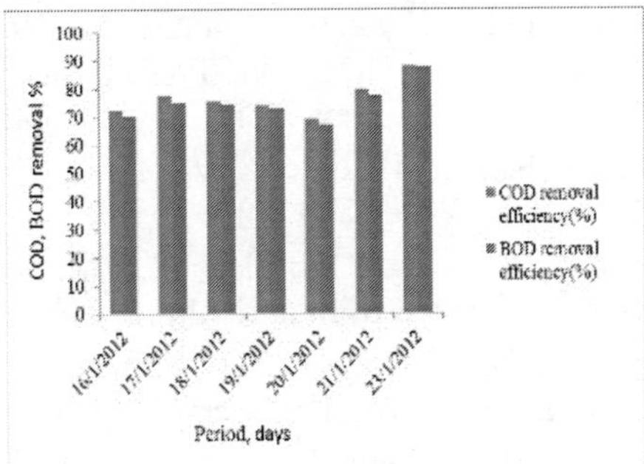

Fig. 10: Effect of Starch Loading on the COD Removal

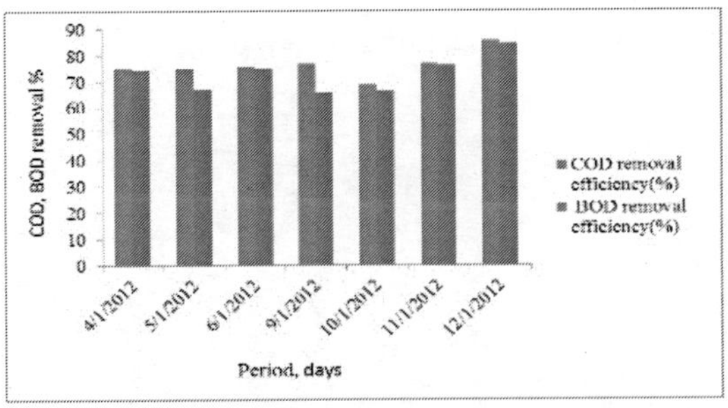

Fig. 11: Effect of Oil Loading on the COD Removal

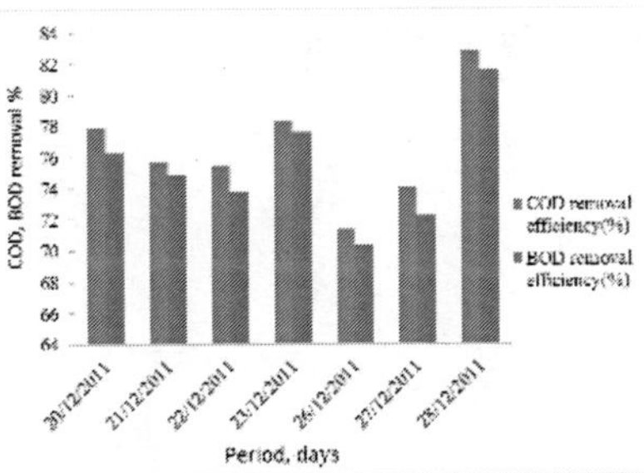

Fig. 12: Effect of Protein Loading on the COD Removal

CONCLUSION

The rice husk based carbon was used as catalyst for the oxidation of organics in wastewater discharged from domestic origin. The rice husk based was efficient enough to reduce COD, BOD by 91.6% and 84% respectively. The consistent removal of pollution parameters was accompanied with the removal of pathogenic and antibiotic resistant bacteria. The sludge production is only 60% of the conventional sewage treatment plant. The investment cost towards Sewage treatment plant by employing rice husk based carbon would be drastically reduced and thereby the operational cost towards electrical energy consumption will also be very much reduced. The reactor C with upper hopper length of 3.5cm is considered to be the desired optimum condition for running the reactor with high COD and BOD removal efficiencies. The outlet of the fluidized reactor was then transferred to the algal pond for further treatment through algal process to make the water suitable for release into the outer environment.

REFERENCES

[1] Ahmadna, A.L., Chong, M.F., Bhatia, S. and Ismail, S. (2006). Drinking water reclamation from palm oil mill effluent (POME) using membrane technology. Desalination, 191: 35–44.

[2] Eckenfelder, W. and Argaman, A. (1991). Principles of biological and physical/chemical nitrogen removal[M]. In: Phosphorus and nitrogen removal from municipal wastewater (Sedlak R., ed.). New York: Lewis Publishers.

[3] Faust, S.D. and Hunter, J.V. (1967). Principles and applications of water chemistry. New York: John Wiley and Sons.

[4] Girgis, B.S. and Daifullah, A.A.M., Removal of some substituted phenols by activated carbon obtained from agricultural waste, *Water Res*., 32 (1998), pp. 1169–1177.

[5] Hasler, B., Lundhede, T., Martinsen, L., Neye, S. and Schoi, J.S. (2005). "Valuation of groundwater protection versus water treatment in Denmark by Choice Experiments and Contingent Valuation", NERI Technical Report No. 543.

[6] Kennedy, John. L., Mohan das, K. and Sekaran, G. (2004). Integrated biological and catalytic oxidation of organics/inorganics in tannery wastewater by rice husk based mesoporous activated carbon-*Bacillus sp,* Carbon, 42: 2399–2407.

[7] Turovskiy, Izrail S. and Mathai, P.K., Wastewater Sludge Processing, 2006, p. 42.

SURFACE PLASMON RESONANCE BASED SENSOR FOR SIMULTANEOUS MONITORING OF WATER SALINITY AND TEMPERATURE

FENG XIAO, GUANGYUAN LI[1], KAMAL ALAMEH AND ANSHI XU[1]

Electron Science Research Institute, Edith Cowan University, Australia
[1]Department of Electronics, Peking University, Beijing 100871, China
E-mail: k.alameh@ecu.edu.au

ABSTRACT

A novel surface Plasmon resonance (SPR) based water salinity sensor enabling simultaneous measurement of water salinity and temperature is proposed and theoretically investigated. Two wavelengths are used for the measurement of the SPR angle, and the calculation of both the water refractive index and the temperature. To the best of our knowledge, this is the first report on simultaneous measurement of refractive index and temperature based on SPR. This approach requires no calibration process for water refractive index measurement at different temperatures.

INTRODUCTION

Salinity and temperature are among the most fundamental parameters for seawater characterization, and therefore, real-time monitoring of these parameters is crucial to desalination, marine ecosystem protection, and climate modeling.[1] Traditionally, water salinity has been monitored through the measurement of the electrical conductivity of the water, because the presence of chlorine ions in water increases its conductivity. Typical salinity concentration accuracy of 30 ppm can be attained with commercially available water salinity sensors. Conductivity has become the standard salinometric technique since 1960s and is now popularly used in oceanographic. However, this technique cannot account for species that may increase the seawater salinity but do not measurably conduct electricity. Recently, another technique for salinity measurement, based on refractive index (RI) measurement, has been

V. Rajendran, R. Yuvakkumar, K. Thyagarajah and K.E. Geckeler (eds.)
Applications of Nano Materials: Electronics, Energy and Environment, pp. 119–124 (2012).

proposed.[2] Several RI measurement methods and technologies have been proposed for water salinity measurement. For example, an ultrasonic technique based on measurement of the travel time of light was proposed for measuring the salinity in a solar pond. Refractometers can achieve sub ppm accuracy, but are inherently bench-top devices that require strict temperature control in their surrounding environment, which therefore limits their applications to where sample-capture is acceptable. Even though statistical treatment of salinity data has shown that RI correlates to density 80% better than conductivity, refractometry-based seawater salinity monitors have experienced less development than conductivity-base monitors. Until recently, due to fast development of photonic technologies, RI-based optical salinity sensors have been attracting intensive interest for deployment in specific application areas such as chemical oceanography and climatology.

Many research groups are currently engaged in the development of optical sensors for monitoring water salinity, with accuracies ranging from 1000 down to mid-100 ppm. Zhao *et al.* have proposed a remote monitoring technology of salinity based on the detection of beam deviation due to the refractive index changes of salt water.[3] This sensor offers a measurement resolution of 12 ppm within a measurement range from distilled water to 50,000 ppm salt water. Their design requires a saline standard and a prism in the sensor head, limiting the potential to miniaturize the sensor. Surface Plasmon Resonance (SPR) is another technique that enables the measuresment of RI changes, and has recently been successfully used in environmental sensors [4,5] In SPR sensors, light striking a thin metallic film is attenuated when surface plasmons are. excited. The light attenuation can be used to measure the RI of the water. Typically, no prisms or moving parts are needed in a fiber-optic based sensor tip. Therefore such a sensor allows a super-small sample size to be analyzed at low cost. Several groups have used SPR based fiber-optic sensors to measure salinity. Grunwald and Holst have attained a sensitivity of 2×10^{-4} RIU based on the use of a multimode fiber with a tapered tip [6]; Liu *et al.* have achieved an RI sensitivity of 8.9×10^{-5} RIU using multimode fiber [7]; and Esteban *et al.* have reported a sensitivity of 100 ppm using a side-polished single mode fiber [8]. However, none of these sensors can determine whether the RI change is caused by salinity change or temperature change, necessitating the calibration of the sensors for each RI measurement.

In this paper, we propose a novel SPR-based water salinity sensor that enables simultaneous measurement of water salinity and temperature, thus no calibration process is needed during the RI measurement. By measuring the SPR angles at two different wavelengths, both water refractive index and temperature can be measured. To the best of our knowledge, this is the first report on simultaneous measurement of water refractive index and temperature based on SPR.

NOVEL MEASUREMENT APPROACH

A fiber-optic based SPR geometry as shown in Figure 1 is used in the analysis. A small section of the cladding is removed from a step index multimode fiber. The uncladed portion of the fiber core having refractive index $n_p = 1.5$ is coated with a silver thin film of thickness $d =$ 42 nm and dielectric permittivityϵ_m. Light is launched from one end of the fiber, and detected, after going through the sensing section, using a photodetector, whereby the SPR angles are determined to evaluate the water salinity and temperature.

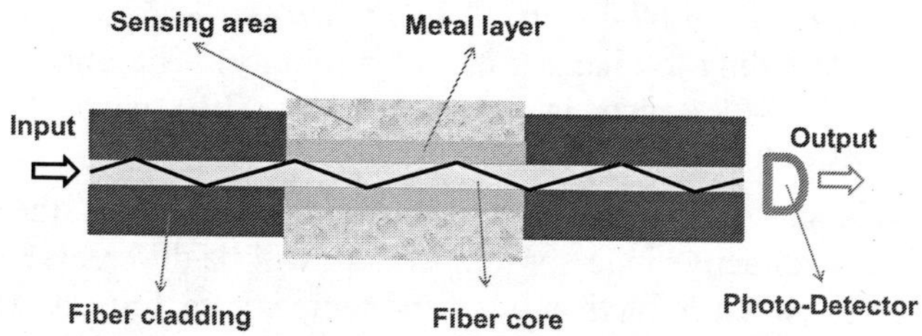

Fig. 1: Fiber-Optic Sensor Structure for Monitoring Water Senility and Temperature

When the sensing area is immersed in a water sample, the laser launched into the optical fiber produces an evanescent field that excites surface plasmons at the silver-water sensing interface. The coupling of evanescent field with the generated surface plasmon polaritons strongly depends on the incidence angle, wavelength, temperature, fiber parameters, probe geometry, and the silver layer properties. The angular response of the SPR sensor is evaluated by monitoring the photocurrent at the fiber end for different light incidence angles. Water salinity and temperature are determined by measuring the shift in the SPR angle.

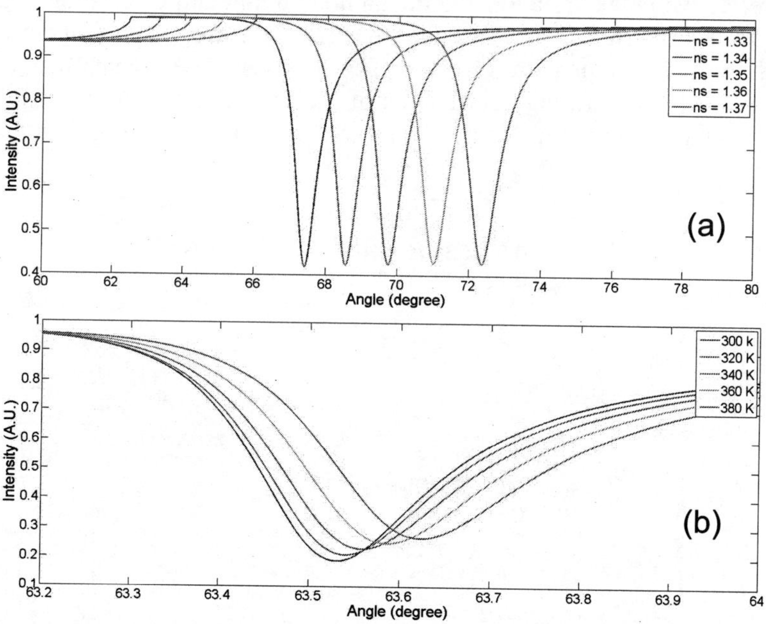

Fig. 2: SPR Angle Inerrogation for (a) Different Refractive Indexes of Sensing Water; (b) Different Water Temperatures, d = 42 nm, $n_p = 1.5$, $\lambda = 632.8$ nm

Numerical calculations were carried out to investigate the proposed SPR-based sensing approach. Figure 2(a) shows the detected current versus incidence angle for various water samples of different refractive indices, under a constant temperature of 300K. As can been seen, the resonance angle changes from 67.43° to 72.37°, when the water refractive index changes from 1.33 to 1.37 (corresponding to changes in water salinity). The impact of temperature on the angular response of the SPR sensor is shown in Figure 2(b), where the temperature was increased from 300K to 380 K.

If a different wavelength of the light is used for the measurements, the dependence of the SPR angle on either water refractive index or temperature will be different. Figures 3(a) and 3(b) show the SPR angle versus water refractive index and temperature, respectively, for wavelengths $\lambda = 632.8$ nm and $\lambda = 1550$ nm. Due to the fact that the SPR angle depends on both temperature, water salinity and wavelength, both changes in the ambient temperature and the water salinity can be resolved through the measurement of the SPR angular response at two different wavelengths. In the event of simultaneous variations in both temperature and salinity, the total SPR angle shifts of two selected wavelengths can be expressed in a matrix form as:

$$\begin{pmatrix} \Delta A_1 \\ \Delta A_2 \end{pmatrix} = K \begin{pmatrix} \Delta n \\ \Delta T \end{pmatrix} = \begin{pmatrix} k_{n1} & k_{T1} \\ k_{n2} & k_{T2} \end{pmatrix} \begin{pmatrix} \Delta n \\ \Delta T \end{pmatrix} \quad \text{... (1)}$$

where ΔA_1 and ΔA_2 are SPR angle shifts for the wavelengths λ_1, λ_2, respectively, and Δn and ΔT are the water refractive index variation and the ambient temperature variation, respectively. K is a sensitivity matrix which takes into account the cross-sensitivity between the water salinity variation and the ambient temperature variation. The sensitivity coefficients k_{ni} and k_{Ti} (i = 1, 2) are obtained from the linear fitting coefficients of the SPR angle shifts versus the refractive index and temperature shown in Figure 3. From Eq. (1) the water-salinity and ambient

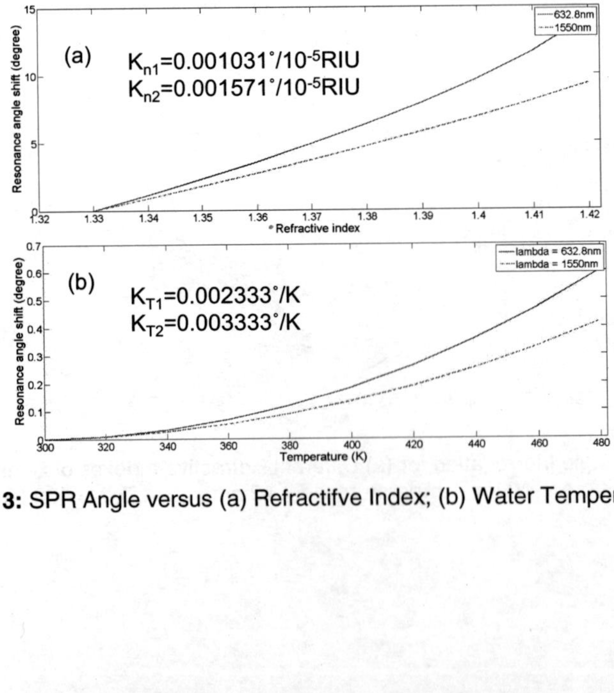

Fig. 3: SPR Angle versus (a) Refractifve Index; (b) Water Temperature

temperature variations can be derived by calculating the inverse matrix of K^{-1}. Note that the matrix K is always inversible because the SPR angle has different responses to both the water refractive index and its temperature variations.

Using the fitting coefficients for the SPR angle response pair obtained from Figure 3, the sensitivity matrix K can be determined. Note that the SPR angle shift is typically much less sensitive to temperature variation than to water refractive index variation. The unit of degree/10^{-5}RIU is chosen for the dependence of the SPR angle on the water refractive index. Therefore, the water salinity and ambient temperature variations can be calculated from the response of SPR angle shifts at the two chosen wavelengths:

$$\begin{pmatrix} \Delta n \\ \Delta T \end{pmatrix} = \begin{pmatrix} 14567 & -6866 \\ -10196 & 4506 \end{pmatrix} \begin{pmatrix} \Delta A_1 \\ \Delta A_2 \end{pmatrix} \qquad \ldots (2)$$

Once the SPR angle shifts for the two wavelengths are measured, the water-salinity shift and the ambient temperature shift can be obtained in units of 10^{-5} RIU and °K, respectively.

CONCLUSION

We have proposed and demonstrated the concept of a new SPR sensor scheme that enables simultaneous measurement of the refractive index and the ambient temperature using conventional signal interrogation. It has been shown that angular interrogation at different wavelengths enables both the refractive index and the temperature to be measured. This approach opens the way for developing calibration-free SPR sensors for water quality monitoring.

ACKNOWLEDGEMENT

The authors acknowledge the financial support of the National Centre of Excellence in Desalination Australia which is funded by the Australian Government through the Water for the Future initiative.

REFERENCES

[1] Gentleman , D.J. and Booksh, K.S., “Determining salinity using a multimode fiber optic surface plasmon resonance dip-probe,” *TALANTA,* vol. 68, pp. 504–515, Jan 15 2006.

[2] Rusby, J.S.M., “Measurements of the refractive index of seawater relative to Copenhagen Standard Sea Water,” *Deep-Sea Res.,* vol. 14, pp. 427–439, 1967.

[3] Yong, Z., Xinyuan, Z., Tingting, Z., Bo, Y. and Shuo, Z., “Optical Salinity Sensor System Based on Fiber-Optic Array,” *Sensors Journal, IEEE,* vol. 9, pp. 1148–1153, 2009.

[4] Esteban, Oacute, R. Alonso, Navarrete, M.C. and Gonzàlez-Cano, A., “Surface Plasmon Excitation in Fiber-Optics Sensors: A Novel Theoretical Approach,” *J. Lightwave Technol.,* vol. 20, p. 448, 2002.

[5] Obando, L.A., Gentleman, D.J., Holloway , J.R. and Booksh, K.S., "Manufacture of robust surface plasmon resonance fiber optic based dip-probes," *Sensors and Actuators B-Chemical,* vol. 100, pp. 439–449, May 15 2004.

[6] Grunwald , B. and Holst, G., "Fibre optic refractive index microsensor based on white-light SPR excitation," *Sensors and Actuators a-Physical,* vol. 113, pp. 174–180, Jul 5 2004.

[7] Liu, D., Huang , Y.I. and Huang, D., "Theoretical and experimental study on the optimum working wavelength for a fiber optic sensor based on the SPR used in monitoring the water quality," *Proc. SPIE,* vol. 3897, pp. 180–186, 1999.

[8] Esteban, O., Cruz-Navarrete, M., Gonzalez-Cano, A. and Bernabeu, E., "Measurement of the degree of salinity of water with a fiber-optic sensor," *Applied Optics,* vol. 38, pp. 5267–5271, Sep 1 1999.

GARNET NANOCOMPOSITE FOR USE IN MAGNETIC PHOTONIC CRYSTALS PREPARED BY COMPOSITION ADJUSTMENT IN THE $Bi_3Fe_5O_{12}$–$Bi_2Dy_1Fe_4Ga_1O_{12}$ MATERIAL SYSTEM

M. NUR-E-ALAM[1], MIKHAIL VASILIEV[1], V. A. KOTOV[2] AND KAMAL ALAMEH[1,3]

[1]*Electron Science Research Institute, Edith Cowan University, Joondalup, WA, 6027, Australia*
[2]*Institute of Radio Engineering and Electronics, RAS, 11 Mohovaya St, Moscow, 125009, Russia*
[3]*Department of Nanobio Materials and Electronics, Gwangju Institute of Science and Technology, World-Class University (WCU), Gwangju Institute of Science and Technology (GIST), Gwangju 500-712, South Korea*
E-mail: m.nur-e-alam@ecu.edu.au

ABSTRACT

RF co-sputtered nanocomposite ($Bi_3Fe_5O_{12}$-$Bi_2Dy_1Fe_4Ga_1O_{12}$) thin films are prepared using composition adjustment approach, implemented by varying the RF powers at each sputtering target. A new MO nanocomposite material is obtained by means of applying optimized annealing crystallization process to these as-deposited garnet layers. The optimally annealed garnet nanocomposites exhibit a significant red shift of the main MO transitions and show strong Faraday rotation of about 8.8°/μm at 532 nm simultaneously with having a high MO quality and magnetization direction perpendicular to the film plane.

INTRODUCTION

Bi-substituted iron garnets in thin film form are the best semi-transparent magneto-optic materials for photonics applications in the visible and near-infrared spectral regions. These materials can possess attractive magnetic properties and high specific Faraday rotation, if the deposited layers contain a high volumetric fraction of the garnet phase with good surface quality and

V. Rajendran, R. Yuvakkumar, K. Thyagarajah and K.E. Geckeler (eds.)
Applications of Nano Materials: Electronics, Energy and Environment, pp. 125–130 (2012).

microstructure. Significant numbers of studies have been conducted to understand the relationship between the MO properties (Faraday rotation) and the Bi-content in each material system. A strong dependence of Faraday rotation on the level of bismuth substitution in thin film materials was seen in the garnet films prepared by liquid-phase epitaxy (LPE). The extrapolated value of the specific FR of the fully bismuth-substituted $Bi_3Fe_5O_{12}$ was evaluated (based on these investigations) to be about 6 deg/micron at the wavelength of 633 nm. Another comparative study has been reported on the epitaxial $Bi_3Fe_5O_{12}$ films prepared by pulsed laser deposition (PLD) and reactive ion beam sputtering (RIBS) techniques where the Faraday rotation measured in PLD-BIG films at 635 nm was about 7.8°/μm, and 6.9°/μm in RIBS-BIG films.[1–3] Recently, electron cyclotron resonance (ECR) sputtered $Bi_3Fe_5O_{12}$ films have been shown to demonstrate the specific Faraday rotation of about 8.4°/μm at 633 nm.[3, 4] LPE films having a record-high Bi-content demonstrated a linear growth of specific Faraday rotation with increasing bismuth content which was obtained by the approach of applying compositional changes from $Lu_3Fe_5O_{12}$ to $Bi_{2.4}Lu_{0.6}Fe_5O_{12}$.[2] In order to investigate the source of such disagreements between the measured properties of highly Bi-substituted garnets and theory predictions and also to try and increase the number of bismuth atoms per unit formula in our sputtered films, we synthesized co-sputtered nano-composites of type ($Bi_3Fe_5O_{12}$-$Bi_2Dy_1Fe_4Ga_1O_{12}$). Another reason we decided to co-sputter these two compositions from two different targets was the fact that so far we could not successfully crystallize any RF-sputtered garnet-type materials from targets containing five iron atoms per stoichiometric formula unit (without any Ga or Al dilution), despite multiple trials. Therefore this co-sputtering approach was our selected way of synthesizing a garnet composition as close as possible to $Bi_3Fe_5O_{12}$ which still contained some Ga dilution and was expected to have out-of-plane magnetization due to some Dy substitution. The flexibility of co-sputtering approach also allows many final stoichiometry variations within nanocomposites.

In this paper, we report on the successful synthesis of a new type of high performance nanocomposite MO material of potential usefulness in magnetic photonics crystals (MPC) which exhibits a significant red-shift of the main MO transitions (thus improving its Faraday rotation), the existence of which we found conclusively after a series of sensitive magnetic circular dichroism (MCD) measurements.

NANOCOMPOSITE THIN FILMS GROWTH AND CHARACTERIZATION

Garnet nanocomposite films (about 1 μm thick) were sputtered onto the paramagnetic $Gd_3Ga_5O_{12}$ (GGG) substrates from two separate oxide-base-mixed targets. The sputtering conditions and process parameters used to deposit the nano-composite layers are detailed in Table 1.

The results of sputtering provided the garnet-type layers in amorphous phase onto the substrates and these amorphous layers were later crystallized by using an oven annealing process which we optimized for this material type by many annealing trials followed by MO characterization. Some of the already-annealed films were subsequently re-annealed at a slightly higher temperature to see if that changed any of MO properties. In particular, we initially annealed some films for 1h at 520°C and then re-annealed them for an additional 1 hour at 530°C. Later multiple crystallized nanocomposite films were characterized using magnetic circular dichroism (MCD) spectroscopy. MCD measurements are remarkably sensitive and allow detection of the very small spectral changes in the locations of the MO transitions fundamentally responsible for the observed features in the FR and ellipticity spectra of garnet materials. We show that it was possible to confirm that two slightly different annealing temperatures led to forming two different garnet compositions (the one with larger Bi-substitution crystallized at a colder temperature) showing an apparent red-shift in their MCD spectra with respect to each other as well as to other known garnets.

Table 1: Sputtering Conditions and Process Parameters Used to Synthesize the Magneto-Optic ($Bi_3Fe_5O_{12}$+ $Bi_2Dy_1Fe_4Ga_1O_{12}$) Nanocomposite Layers onto GGG Substrates

Sputtering target stoichiometries	$Bi_3Fe_5O_{12}$ (materials' purity 99.99%) and $Bi_2Dy_1Fe_4Ga_1O_{12}$ (99.99%)
Target size	3′′ (diameter) with the material layer thickness of 0.125′′, bonded to a 0.125″ Cu backing plate
Background Pressure	P(base) ~ 1–2x10^{-6} Torr
Process gas and pressure	Argon, P(Ar) = 1mTorr
RF power density at targets	4.35 W/cm^2 (200 W) for $Bi_3Fe_5O_{12}$ and 1.63 W/cm^2(75 W) for $Bi_2Dy_1Fe_4Ga_1O_{12}$
Substrate-target distance	18–20 cm
Substrate temperature during deposition	250°C

RESULTS AND DISCUSSION

The co-deposited annealed films exhibited rather high specific Faraday rotation of about 8.8°/μm at 532 nm (compared to around 6.9°/μm measured in typical $Bi_2Dy_1Fe_4Ga_1O_{12}$ films which usually required annealing at temperatures greater than 650°C) simultaneously with good surface quality. The MCD measurements performed over the spectral region 300–600 nm revealed a significant red-shift of the main MO transitions in all of our co-sputtered samples, in comparison with the spectral locations of these transitions measured in LPE garnet films with much smaller Bi-substitutions. The spectral dependencies of MCD signals were obtained using a sensitive dichrometer which measured the absorption difference between the left- and right-circularly polarized (LCP and RCP) light waves propagating through samples in the presence of strong magnetic field parallel to the light beam. The MCD signals from all samples were recorded in arbitrary units and plotted as a function of wavelength as shown in Figure 1. Our samples annealed at 520°C exhibited the first MCD peak position at 515 nm, and the samples annealed at 530°C displayed a two times broader MCD peak at 495 nm with a small shoulder at 515 nm, as distinct from the first MCD peak position at 442 nm observed in LPE films of composition $Bi_{0.45}Tm_{2.55}Fe_{3.8}Ga_{1.2}O_{12}$.[2] The long-wavelength negative peak of the FR moved from its position at 475 nm observed in $Bi_{0.45}Tm_{2.55}Fe_{3.8}Ga_{1.2}O_{12}$ up to 535 nm and even to 545 nm for nano-composite $Bi_3Fe_5O_{12}$- $Bi_2Dy_1Fe_4Ga_1O_{12}$ films annealed at 530°C and 520°C, respectively.

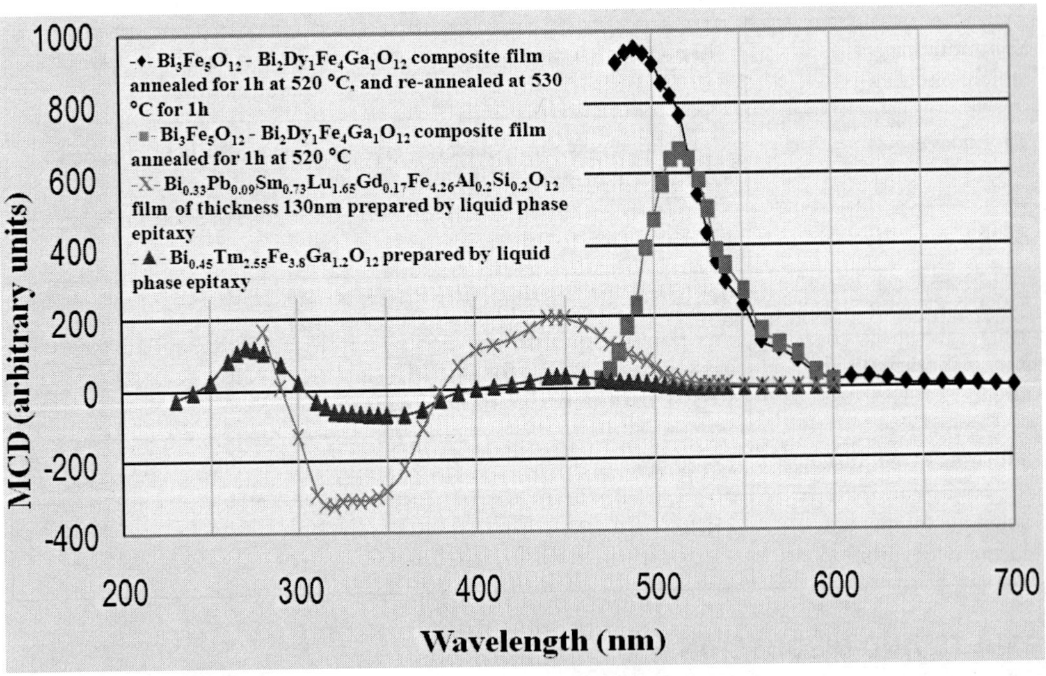

Fig. 1: Spectral dependencies of MCD signal (in arbitrary units) measured in different garnet samples. (-♦-) RF co-sputtered ($Bi_3Fe_5O_{12}$-$Bi_2Dy_1Fe_4Ga_1O_{12}$) composite film of thickness about 1 μm, annealed for 1 hour at 520°C at first stage and re-annealed for 1 hour at 530°C; (-■-) co-sputtered composite film of the same stoichiometry when still in its amorphous phase and the same thickness, but only annealed for 1 hour at 520°C; (-X-) LPE film - $Bi_{0.33}Pb_{0.09}Sm_{0.73}Lu_{1.65}Gd_{0.17}Fe_{4.26}Al_{0.2}Si_{0.2}O_{12}$ of thickness 130 nm, and (-▲-) $Bi_{0.45}Tm_{2.55}Fe_{3.8}Ga_{1.2}O_{12}$ film prepared by LPE. Sensitivity scale of the dichrometer was same during all measurements.

These results indicate that a slight difference in the annealing temperature led to a measureable difference in the material phase content and that two different garnet materials with non-identical Bi-substitutions may have formed at each of these temperatures since their MCD signatures were red-shifted with respect to one another. As is relatively well known (and confirmed by our experience), there exists a minimum annealing temperature suitable for effectively crystallizing each garnet type, with less Bi-rich compositions requiring higher temperatures. It was also expected that less Bi-rich (but still highly substituted) garnets would exhibit smaller red-shifts in their MCD signatures with respect to the reference MCD peaks measured in LPE garnets having very limited Bi-substitutions. The ways of synthesizing garnets with very high Bi-substitutions using RF-sputtering and co-sputtering are still a subject of ongoing research within our group.

CONCLUSION

RF magnetron co-sputtered ($Bi_3Fe_5O_{12}$-$Bi_2Dy_1Fe_4Ga_1O_{12}$) thin films resulted in obtaining new and interesting MO nanocomposite materials after annealing near 520°C, which exhibited simultaneously a high Faraday rotation, good MO quality, magnetization direction perpendicular to the film plane, and notable red-shift features in their MCD spectra. The results indicate that we synthesized a new garnet material type with rather high Bi-substitution well in excess of two formula units, which will be useful for applications in magnetic photonic crystals (MPC).

ACKNOWLEDGEMENT

We acknowledge the support of Edith Cowan University and the World-Class University Program funded by the Ministry of Education, Science, and Technology through the National Research Foundation of Korea (R31–10026).

REFERENCES

[1] Adachi, N., Denysenkov, V. P., Khartsev, S. I., Grishin A. M., and Okuda, T., Epitaxial $Bi_3Fe_5O_{12}$ (001) films grown by pulsed laser deposition and reactive ion beam sputtering techniques, J. Appl. Phys., 88, 1, 2734–2739, 2000.

[2] Zvezdin, A. K., and Kotov, V. A., Modern Magnetooptics and Magnetooptical Materials, Institute of Physics Publishing. Bristol and Philadelphia. (IOP Publishing Ltd.) 1997.

[3] Hansen, P., and Krumme, J. P., Magnetic and magneto-optical properties of garnet films, Thin Solid Films, 114, 69–107, 1984.

[4] Adachi, N., Isarai, K., Ota, T., Ishiyama, K., Ota, H, .and Arai, K., Synthesis of completely substituted $Bi_3Fe_5O_{12}$ garnet films by metal organic deposition technique", http://www.crl.nitech.ac.jp/ar/2010/0104_crl_ar2010_paper.pdf

MANGANITE NANORODS SUPPORTED PALLADIUM: A FACILE ELECTROCATALYST FOR GLYCEROL FUEL CELL

R. KANNAN[1,2], P. RAVICHANDIRAN[2], A. OBADIAH[2], K. KARUNAKARAN[1] AND S. VASANTHKUMAR[2]

[1]*Department of Chemistry, Sona College of Technology, Salem*
[2]*Department of Chemistry, School of Science and Humanities and Nanosciences and Technology, School of Nanosciences and Technology, Coimbatore*
E-mail: kumar2359@yahoo.com

ABSTRACT

Manganite (MnOOH, MN) nanorods were synthesized by a hydrothermal method and subsequently used as a supportive material for Palladium (Pd) towards glycerol electrooxidation in alkaline medium. The low quantity (5 wt%) palladium is coated on the Manganite nanorods by an insitu reduction method. The electrooxidation of glycerol at MN/Pd electrode exhibits, peak current vertexes at –0.2 V, and a negative shift in potential is observed when compared with C/Pd electrode. By varying the alkali (KOH) and glycerol concentrations the electrocatalytic behavior was studied. The possibilities for the removal of poisonous intermediates are also discussed. The effect of support material on the electrooxidation reaction is explained with the mechanism.

INTRODUCTION

Supported metal nanoparticles are widely employed in catalysis. Recent advances in controlling the shape and the size of nanoparticles have opened the possibility to optimise the particle geometry for enhanced catalytic activity and enhanced surface properties for suitable applications.[1]

V. Rajendran, R. Yuvakkumar, K. Thyagarajah and K.E. Geckeler (eds.)
Applications of Nano Materials: Electronics, Energy and Environment, pp. 131–136 (2012).

The poisoning of the active site of the catalyst is the major problem in DAFC. In order to reduce this poisoning, various methods have been adopted. Among these, placing a metal on a suitable support has been shown to have enhanced activity.[2] Researches in this direction have just begun. Recently, metal oxides supported Pd has been used as novel heterogeneous catalyst for various applications.[1–3]

Metal oxides provide best platform for the catalyst, which have provide the maximum utility of the catalytic activity was achieved. In this work, Pd nanoparticles supported on the MN catalyst was tested for the electrooxidation of glycerol in alkaline medium. The effect of glycerol, supporting electrolyte (KOH), were studied.

EXPERIMENTAL

Palladium acetate, glycerol (Sigma–Aldrich), $KMnO_4$, H_2O_2 (30% W/W), acetic acid, sodium acetate, potassium hydroxide, and ethanol were received from Merck, India. All the chemicals were of analytical grade and were used without further purification.

The Manganite was prepared by a reported procedure.[3] The Pd modified MN was synthesized by the insitu method. In brief, suitable quantity of 0.5 mm PdCl2 solution was taken in 5 ml ethanol, so as to achieve 10wt% of Pd and 100 mg of MN was added and stirred. After one hour formaldehyde was added to the mixture drop wise and a reductive reaction was performed at room temperature. After six hours, the resulting product was filtered washed with distilled water, and then dried at 60°C for 6 h.

MN/Pd nanocatalyst was ultrasonically dispersed in 0.5% of Nafion in ethanol solution. An aliquot of the Pd/Manganite nanocatalyst/Nafion ink was dropped on the surface of the freshly polished GC electrode and allowed to dry. The MN/Pd nanocatalyst modified GC electrode was the working electrode, a Pt wire served as the counter electrode and a standard calomel electrode was the reference electrode. For comparison, Pd supported on graphite powder and their electrodes were also prepared under the same preparative conditions. All the measurements were performed at room temperature.

The prepared Manganite and MN/Pd was characterized by powder XRD (Shimadzu Lab X–6000), EDS (INSTA FET, Oxford instruments), HRTEM (JEOL, JSM 2100) and the electrochemical studies on it was carried out by CHI660C electrochemical workstation (CH Instruments) and a conventional three electrodes cell equipped with computer controlled software.

RESULTS AND DISCUSSION

The crystal phase of the prepared MN and MN/Pd were analyzed by powder XRD and is shown in Figure 1.a The XRD peaks at $2\theta = 26$ (111), 33.9 (020) can be assigned to MN and this matches well with the JCPDS # 41–1379. The smoothened profile of MN/Pd nanocomposite exhibits a sharp intense peak for Pd (100) at $2\theta = 37.9°$ (JCPDS # 72–0710). In the MN/Pd, the Pd nanoparticles covered the Manganite nanorods, and therefore the characteristic XRD peaks for MN were masked. The surface morphology of the MN and MN/Pd were characterized by TEM (Figure 1 (b,c)). The TEM images show the presence of two isolated MN (Figure 1 b) and a Pd coated Manganite as shown in figure 1 c. The TEM images reveal that the prepared MN was in the form of nanorods (size is 10–15 nm). The Pd was uniformly coated on the MN and the particles size of the Pd was <10nm. EDS analysis was carried out to get local information about the MN/Pd catalyst. EDS was carried out by the microprobe analysis of various portions of the catalyst as a thin film. Figure 1d illustrates the EDS spectrum of MN/Pd catalyst. The average atomic ratio of Pd on the composite is about 5% on MN. Some oxygen peaks were observed, which is due to the MN.

In order to explain the electrocatalytic activity of MN/Pd nanocomposite catalyst towards the oxidation of alcohol, the voltammetric behavior of MN/Pd and C/Pd electrode under identical condition were studied As shown in Figure 2 A (a&b), the Cyclic Voltammograme (CV) of (a) C/Pd and (b) MN/Pd in 1M KOH has been recorded at the scan rate of 50 mV/s in the potential range of –1.1 to +0.2 V. A strong anodic peak around –0.9 to –0.7 V is due to the oxidation of surface adsorbed/absorbed hydrogen on the surface of the catalyst on both MN/Pd and C/Pd electrodes.

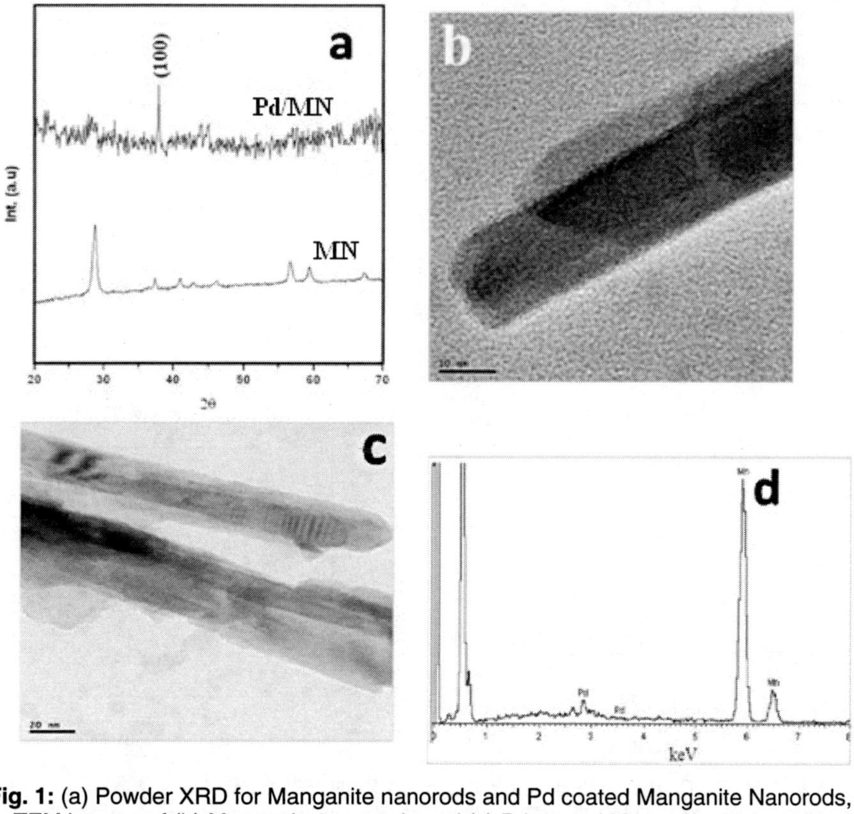

Fig. 1: (a) Powder XRD for Manganite nanorods and Pd coated Manganite Nanorods, TEM images of (b) Manganite nanorods and (c) Pd coated Manganite Nanorods; (d) EDS of Pd coated Manganite nanorods

A flat anodic (E = –0.18 V) and strong cathodic (E = –0.45) peaks are formed for the formation and reduction of PdO, respectively. This indicates that the Pd on both the support shows almost similar activity, except that in MN/Pd there is a improved current response compared to C/Pd. Xu *et al.* have also shown the recording of similar voltammogram on Pd/C in 1 M KOH.[2]

Figure 2 B demonstrates the glycerol oxidation reaction (GOR) on Pd/Manganite (5% Pd) electrode. The CV of the MN/Pd electrode in 1M KOH solution containing 1M Gly shows that, the hydrogen adsorption/absorption region is significantly suppressed in the presence of Gly in solution. This suppression of the peak can be attributed to the dissociative adsorption of Gly in the low potential region. The EG electrooxidation reaction starts at –0.7 V and two well defined anodic oxidation peaks were observed. A current peak (O^f) centered at – 0.2 V is observed during the forward scan (anodic condition) and in the reverse scan, another peak current peak (O^b) is found centered about –0.4 V (cathodic condition). In the forward scan (O^f), the oxidation current peak is produced due to the oxidation of freshly chemisorbed species from glycerol adsorption. The secondary oxidation peak (O^b) is primarily associated with oxidation of intermediate product formed during the forward scan.

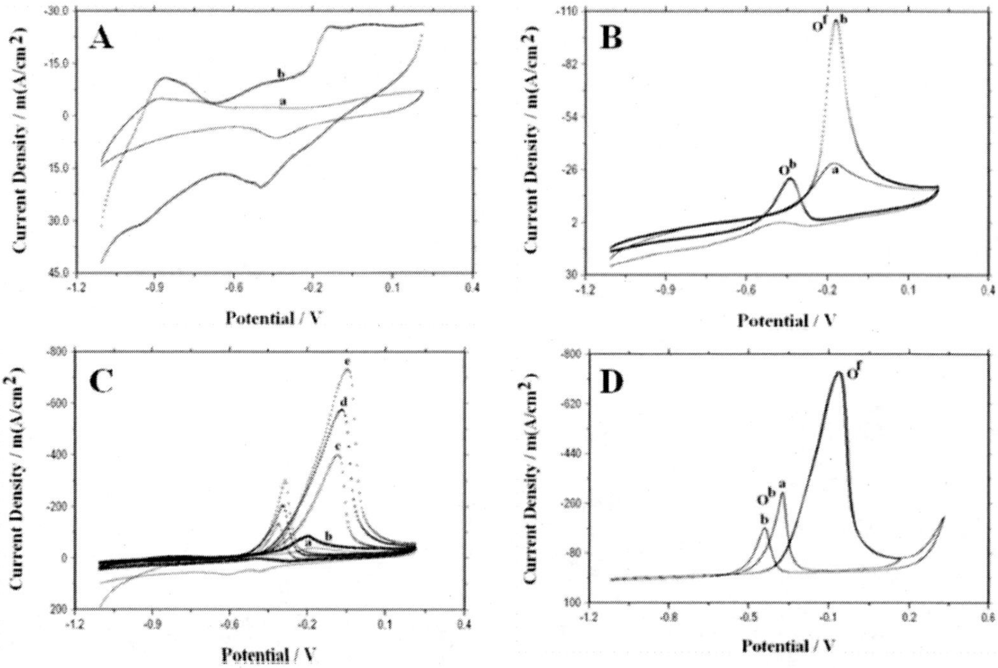

Fig. 2: (A) CVs of (a) Pd/C, (b) MN/Pd electrode in 1M KOH; (B) CVs of electrooxidation of glycerol at MN/Pd electrode in 1M KOH/1M Gly. (C) CVs of electrooxidation of glycerol at MN/Pd electrode in (a)0.01M, (b) 0.1M, (c) 0.5M, (d) 1 M and (e) 2M Gly/1 M KOH D) CVs for the electrooxidation of glycerol at MN/Pd electrode at the scanning potentials of –1.1 V to (a)+0;2 and (b) +0.4 V at 50 mV/s.

The CVs of Pd/MN electrode in Gly solutions of various concentration ranging from 0.01 to 2 M, is shown Figure 2C which shows, that the electrooxidation current increases monotonically as the concentration of Gly increases. The peak current first increases as the Gly concentration increases, but when the Gly concentration exceeds 2 M, the peak current starts to decrease. This is due to the hindrance of ionic mobility in the solution by the Gly ion to approach the electrode. The effect of supporting electrolyte plays a vital role in the GOR.

In the Gly electrooxidation, the complete removal of $C_2/(CO)_{ads}$ intermediates may not be possible, because it is strongly bounded with Pd active sites. To remove these intermediates high energy oxygen atom is required. The availability of such active oxygen may not be possible in the potential window of –1.1 to +0.2 V, Just increase the potential region upto +0.4 V [Figure 2D], the secoundary oxidation peak current density was decreased upto 50%. This is due to active oxygen supplied by the MN material through the reversible redox process [Reaction 1&2].

$$Pd/MnOOH + OH^- \longrightarrow Pd/MnO_2 + H_2O + e^- \quad (1)$$

$$Pd/MnO_2 + Pd(CO)_{ads} + OH^- \longrightarrow Pd + Pd/MnOOH + CO_2 + H_2O \quad (2)$$

CONCLUSIONS

The activity of glycerol electrooxidation reaction on palladium coated Manganite nanocatalyst can summarized as follows:

1. The Manganite nanorods act as excellent support material.
2. Small quantity of Pd (5%) was used and it exhibits excellent activity for Glycerol oxidation reaction.
3. By increasing the potential window upto +0.4 V, the secondary oxidation peak (O^b) was suppressed by about 50%, which is through the active active oxygen supplied by the material through the reversible redox process. Further research to assess the long-term stability of Pd/Manganite nanocatalyst is currently in progress.

ACKNOWLEDGEMENT

The authors thank the DST, New Delhi for the Lab facilities provided through DST Nanomission Project. The authors express their gratitude to the authorities and the management of the Karunya University, Coimbatore. RK&KK thankfully acknowledged the Management, the Secretary and the Principal of Sona College of Technology, Salem, India

REFERENCES

Campelo, J. M., Luna, D., Luque, R., Marinas, J. M., Romero, A. A., Sustainable Preparation of Supported Metal Nanoparticles and Their Applications in Catalysis, Chemsuschem, 2, 18–45, 2009.

Xu, C., Tian, Z., Shen, P., Jiang S. P., Oxide (CeO_2, NiO, Co_3O_4 and Mn_3O_4)-promoted Pd/C electrocatalysts for alcohol electrooxidation in alkaline media Electrochim Acta 53,2610–2618, 2008.

Kannan, R., Karunakaran, K., Vasanthkumar, S., PdNi-coated manganite nanorods as catalyst for electrooxidation of methanol in alkaline medium, Appl. Nanosci. 2, 149–155, 2012.

Serov A., Kwak C., Recent achievements in direct ethylene glycol fuel cells (DEGFC) Appl Catal B:Environ 97, 1–12, 2010.

HIGH CONTRAST TANDEM ORGANIC LIGHT EMITTING DEVICES EMPLOYING SEMITRANSPARENT INTERMEDIATE NANO METAL LAYERS AND A PHASE SHIFTING LAYER

BAO-FU DING AND KAMAL ALAMEH

Electron Science Research Institute, Edith Cowan University,
270 Joondalup Dr, Joondalup, WA 6027, Australia
E-mail: b.ding@ecu.edu.au; k.alameh@ecu.edu.au

ABSTRACT

A high contrast-ratio organic light emitting device (OLED) is proposed and experimentally demonstrated. The OLED is implemented by stacking two organic phase tuning layers between composite metal layers and optimizing their thicknesses. Such a tandem device can increase the current efficiency by 120%, and reduce the operating voltage by 1.1 V, in comparison to conventional high contrast OLEDs. Measured reflection spectra validate the high-contrast capability of the OLED, and demonstrate experimentally an average reflectivity of 6% under ambient light illumination. This is the lowest reflectivity reported to date for OLEDs employing organic phase tuning layers.

INTRODUCTION

Organic semiconductors offer several advantages, namely, variety of materials, simple fabrication processes, cost-effectiveness and transparency, which make organic optoelectronic devices attractive for many applications.[1–5] Organic light emitting diodes (OLEDs) have particularly been used in flat panel displays (FPDs) due to their wide viewing angle, ultra thin thickness requirements, low power consumption and ability to emit light without the need for external backlight sources, in addition to the possibility of growing them on flexible substrates.

In a conventional single-cell OLED, the reflective metal layer of the cathode enables the back emission from organic layer to be reflected forward resulting in a high light coupling

V. Rajendran, R. Yuvakkumar, K. Thyagarajah and K.E. Geckeler (eds.)
Applications of Nano Materials: Electronics, Energy and Environment, pp. 137–142 (2012).

efficiency. However, such OLEDs have the drawback of low contrast ratio due to the reflection of ambient light by the highly reflective cathode, which degrades the performance of OLEDs especially in outdoor applications where strong ambient light might be present.[2] Recently, a black composite layer has been introduced to OLED structures in order to increase the contrast ratio. This composite layer consists of a thin semi-transparent metal layer, a phase-tuning (PT) layer made of organic materials and a thick reflective metal layer,[6–8] and its low reflection is produced by the cancellation (destructive interference) of the two light waves reflected off the front thin metal layer and the rear thick metal layer which induces a π phase difference with respect to the front layer.[8] Due to their simple thermal evaporation, organic materials are considered as the best candidates for the realization of PT layers, such as tris8-hydroxyquinoline aluminum (Alq_3) (Refs. 8,9) and copper phthalocyanineand (Ref. 10). To obtain a π phase difference, the thickness of the organic PT layer must be around 80 nm, and this is the same order of the emissive layer thickness. However, with a high carrier injection barrier between the PT layer and the intermediate metal layer, the operating voltage more than doubles, whereas, the current efficiency is reduced by 50%, because the black cathode absorbs half of the generated light from the emissive layer. The twofold increase in the operating voltage and the 50% reduction in the current efficiency reduce the overall power efficiency by 75%, thus making them impractical for emerging applications.

In this paper, we propose the use of an alternative black-layer cathode employing a composite semitransparent LiF/Al/Au in conjunction with stacked PT NPB/Alq_3 layers to achieve much better current efficiency and even lower reflectivity of ambient light in comparison with conventional OLEDs employing LiF/Al as intermediate layers and Alq_3 as PT layers.[8–10]

SYSTEM DESCRIPTION AND EXPERIMENTAL RESULTS

Figure 1(a) shows the structures of a conventional single-cell OLED (Device 1), conventional high-contrast OLED (Device 2) and the proposed high contrast tandem OLED (Device 3), which were developed using thermal sublimation of organic materials in an ultra-high vacuum environment onto a transparent glass substrate coated with indium tin oxide (ITO), similar to the process used in Refs.[5–11] Prior to the gas treatment, the routine procedure of cleaning the substrate included ultrasonication in detergent for 30 minutes, spraying with de-ionized water for 2 minutes, ultrasonication in de ionized water for 20 minutes and drying by rotating at the spinning speed of 2000 rpm in a spin coater for 40 seconds. UV-ozone treatment was then made in a chamber with a high-purity oxygen flow. Immediately after the treatment, the samples were transferred into a growth chamber with a base pressure of 5×10^{-6} Pa for subsequent depositions of various layers. Alq_3 and NPB were chosen, respectively, as the electron transporting and emitting layer and the hole transporting layer, and Al on the top was chosen as the cathode. To increase the electron injection efficiency, a thin layer of LiF buffer was inserted between the Al and Alq_3 layers, as usually done for conventional OLEDs. The top Al layer was also in the shape of narrow strips, crossing the bottom ITO strips to form active device area of approximately 4×4 mm^2.

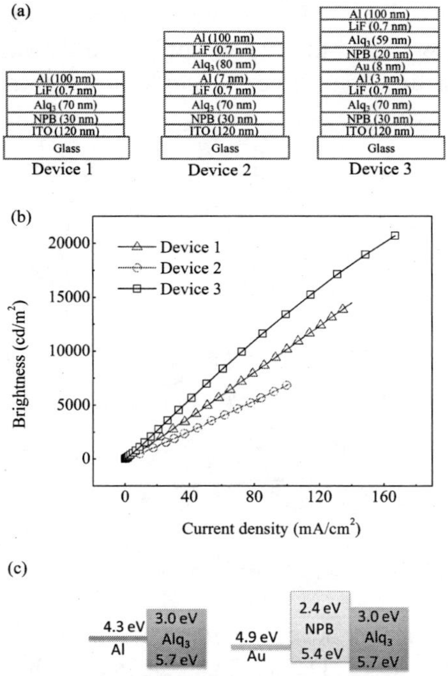

Fig. 2: (a) Architectures and (b) B-J Characteristics of Three Devices, Namely Device 1: Single-cell OLED, Device 2: High Contrast OLED with the PT Layer and Device 3: Proposed High Contrast Tandem OLED

RESULTS AND DISCUSSION

Figure 1(b) shows the measured brightness-current density characteristics for the three devices. It is obvious from Figure 1(b) that the brightness of Device 2 is approximately half of that of Device 1 at a given current density. For instance, at 40 mA/cm^2, the electroluminescence (EL) of Device 2 is 2500 cd/m^2, compared with 4050 cd/m^2 luminescence for Device 1 at the same current density. Such reduction in luminance for Device 2 is due to the additional metal-organic-metal (MOM) black cathode introduced on top of the bottom emissive cell.[8] Almost half of the photons emitted by the bottom emissive layer are reflected by the bottom and top surfaces of the black cathode. Therefore, even if the PT layer thickness is optimized to suppress the reflected light through destructive optical interference, the current efficiency of this OLED structure is limited since half of the generated light is lost by the black cathode. Therefore, the theoretical limit of current efficiency in such black-cathode- based OLEDs is only half of that of conventional counterparts. However, for the proposed Device 3, the EL at the same current density dramatically increases to 5500 cd/m^2, and in comparison with Device 2, the brightness increases by 120%. Interestingly, the EL of Device 3 is even higher than that of Device 1. Our measured results show that besides the bottom cell, the top MOM structure of Device 3 contributes to photon emission as well. In the MOM structures, the interfaces between intermediate layers and the PT layers in Device 2 and Device 3 are Al/Alq_3 and Au/NPB respectively. Generally, the hole-injection barrier ϕ_B in this interface is the energy difference between the work function of the composite intermediate layer and the highest occupied molecular orbit (HOMO), which is analogous to the top of the valence band in an inorganic semiconductor.

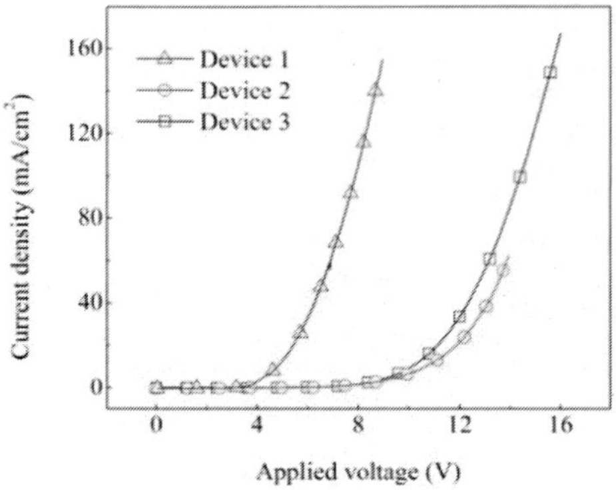

Fig. 3: *J-V* Characteristics of Devices 1, 2 and 3

Figure 1(c) illustrates the energy level diagrams of the interfaces between the composite intermediate layer and the PT layer for Device 2 and Device 3. The work function of Al and the HOMO of Alq_3 are respectively 4.3 eV and 5.7 eV,[12] the ϕ_B is 1.4 eV, which is high enough to prevent hole injection from the composite intermediate layer to the PT layer, restricting the exciton formation and photon emission in the PT layer. As for Device 3, the work function of Au and HOMO of NPB are 4.9 eV and 5.4 eV, respectively.[12,13] Thus ϕ_B is reduced from 1.4 eV for Device 2 to 0.5 eV for Device 3, and this barrier is low enough for hole injection from Au to NPB. The injected holes pass through the hole-transporting materials of NPB and are subsequently captured by the electrons in the Alq_3 layer to form excitons, resulting in the light emission from the PT layer. Thus, Device 3 has two emissive cells in tandem, namely, the lower cell and the upper MOM cell. We term Device 3 a "high contrast tandem OLED". Noticeably, the current efficiency of Devices 3 is even higher than that of Device 1. This implies that the current efficiency of the upper MOM cell is higher than that of the bottom cell. This originates from much more efficient hole injection from Au to NPB in the upper cell than that from ITO to NPB in the lower cell due to the lower ϕ_B at the interface of Au/NPB.[12]

Figure 2 shows the J-V characteristics for the three devices. At a current density of 40 mA/cm^2, the operating voltage of Device 2 is 13.2 V, almost 7 V higher than that of Device 1. This indicates that the extra 80 nm thick Alq_3 higher than that of Device 1, and that the extra 80 nm thick Alq_3 layer introduces a considerably high resistance, resulting in a remarkably higher operating voltage. This is consistent with previous reported results.[14] For Device 3, the operating voltage is 1.1 V lower than that of Device 2. Since ϕ_B in the MOM structures for Device 2 and Device 3 is higher than 0.3 eV, the current flow through the MOM is dominated by the injection limited current (J_{ILC}).[15] From the J_{ILC} expression of Eq. (1) in Ref.,[15] at a given current density, the operating voltage increases monotonically with the ϕ_B Therefore, the lower ϕ_B at the Au/NPB interface leads to the lower operating voltage in Device 3.

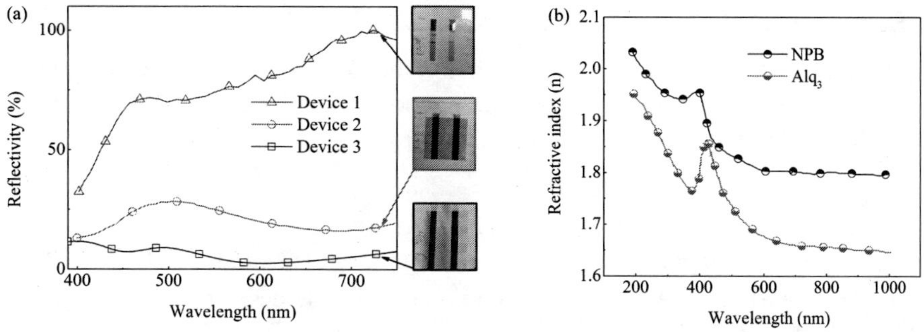

Fig. 4: (a) The Refractive Indices of NPB and Alq_3. (b) The Reflection Spectra of Three devices as Schematically Shown in Figure 1a

Figure 3(a) shows the optical reflectance spectra for Devices 1–3 measured at 5° off the surface normal. The average reflectance of the OLED is 80% for Device 1, 20% for Device 2, mainly due to the addition of the MOM structure. To attain maximum destructive interference with the stacked NPB/Alq_3 PT layers in Device 3, the phase difference between the two light waves reflected off the upper and lower cells should be π. This corresponds to half of the design wavelength (½ λ). If the refractive indices of NPB and Alq_3 are respectively $n_1(\lambda)$ and $n_2(\lambda)$, the optimized thicknesses, t_1 and t_2, of NPB and Alq_3, respectively, for obtaining destructive interference should satisfy:

$$2[n_1(\lambda)\cdot t_1 + n_2(\lambda)\cdot t_2] = \frac{\lambda}{2} \quad \dots (2)$$

The spectral range of the ambient visible light extends from 400 nm to 750 nm, however, the elimination of the light around 550 nm is the main concern since 550 nm is the most sensitive wavelength to the human eyes. As shown in Figure 3(b), at 550 nm, $n_1(\lambda)$ and $n_2(\lambda)$ are 1.7 and 1.8, respectively. With the NPB thickness being fixed to 20 nm, the optimal Alq_3 thickness is 59 nm, according to Eq. (2). The measured reflectance from the proposed Device 3 was only 6.0% over the range of 400 to 750 nm. As we know, this is the lowest reflectance among all high contrast OLEDs employing an organic PT layer. While in the human-eye-sensitive range of 500–600 nm, the reflection value is further reduced to 4%, which nearly approaches the minimum limitation value since the reflectance of air/glass is around 4%.

CONCLUSIONS

We have proposed and demonstrated the concept of a high contrast-ratio organic light emitting device (OLED) realized through the stacking of two organic phase tuning layers between composite metal layers. Experimental results have demonstrated an increase in current efficiency by 120%, a reduction in operating voltage by 1.1V and an ambient reflection as low as 6%, which are attributed to the photon emission from the PT layer, the reduced carrier injection barrier, between the intermediate layer and the PT layer, and a proper optical design.

ACKNOWLEDGEMENTS

This research is supported by Edith Cowan University, the Department of Industry, Innovation, Science, Research and Tertiary Education, Australia, and National Natural Science Foundation of China (NSFC] and Ministry of Science and Technology of China.

REFERENCES

[1] Tang, C.W., Two-layer organic photovoltaic cell. Appl. Phys. Lett. 48, 183–185, 1986.

[2] Tang, C.W. and Vanslyke, S.A., Organic electroluminescent diodes. Appl. Phys. Lett. 51, 913–915, 1987.

[3] Friend, R.H., Gymer, R.W., Holmes, A.B., Burroughes, J.H., Marks, R.N. *et al.*, Electroluminescence in conjugated polymers. Nature 397, 121–128, 1999.

[4] Dimitrakopoulos, C.D. and Malenfant, P.R.L., Organic Thin Film Transistors for Large Area Electronics. Adv. Mater. 14, 99–117, 2002.

[5] Ding, B.F., Zhan, Y.Q., Sun, Z.Y., Ding, X.M., Hou, X.Y. *et al.*, Electroluminescence and magnetoresistance of the organic light-emitting diode with a La[sub 0.7]Sr[sub 0.3]MnO[sub 3] anode. Appl. Phys. Lett. 93, 183307, 2008.

[6] Hung, L.S. and Madathil, J., Reduction of ambient light reflection in organic light-emitting diodes. Adv. Mater. 13, 1787–1790, 2001.

[7] Krasnov, A.N., High-contrast organic light-emitting diodes on flexible substrates. Appl. Phys. Lett. 80, 3853–3855, 2002.

[8] Feng, X.D., Khangura, R. and Lu, Z.H., Metal--organic--metal cathode for high-contrast organic light-emitting diodes. Appl. Phys. Lett. 85, 497–499, 2004.

[9] Zhou, Y.C., Ma, L.L., Zhou, J., Gao, X.D., Wu, H.R. *et al.*, High contrast organic light-emitting devices with improved electrical characteristics. Appl. Phys. Lett. 88, 2006.

[10] Lee, J.H., Liao, C.C., Hu, P.J. and Chang, Y., High contrast ratio organic light-emitting devices based on CuPC as electron transport material. Synth. Met. 144, 279–283, 2004.

[11] Ding, B.F., Yao, Y., Sun, Z.Y., Wu, C.Q., Gao, X.D. *et al.*, Magnetic field effects on the electroluminescence of organic light emitting devices: A tool to indicate the carrier mobility. Appl. Phys. Lett. 97, 163302, 2010.

[12] Ishii, H., Sugiyama, K., Ito, E. and Seki, K., Energy level alignment and interfacial electronic structures at organic metal and organic organic interfaces. Adv. Mater. 11, 605–625, 1999.

[13] Cao, J., Jiang, X.Y. and Zhang, Z.L., MoOx modified Ag anode for top-emitting organic light-emitting devices. Appl. Phys. Lett. 89, 2006.

[14] Burrows, P.E., Bulovic, V., Forrest, S.R., Sapochak, L.S., McCarty, D.M. *et al.*, Reliability and degradation of organic light emitting devices. Appl. Phys. Lett. 65, 2922–2924, 1994.

[15] Lee, S.G. and Hattori, R., Current density equations representing the transition between the injection□ and bulk□limited currents for organic semiconductors. Journal of Information Display 10, 143–148, 2009.

THEORETICAL ANALYSIS OF GaSb/GaAs QWIP

MD. AREF BILLAHA AND MUKUL K. DAS

Department of Electronics Engineering, Indian School of Mines, Dhanbad
E-mail: rf.billaha@gmail.com

ABSTRACT

In this paper, the intersubband transition in GaSb/GaAs quantum well infrared photodetector (QWIP) considering large valence band discontinuity is investigated. A broad detection window of 2–5 μm with peak responsivity of 53.8 μA/W at λ ~ 2.7 μm is obtained at 5V bias voltage. The quantum well structure is designed using a computational tool developed to solve the Schrödinger equation with the help of transfer matrix method.

INTRODUCTION

The area quantum well infrared photodetector (QWIP) design is very extensive. Recent advances in material science and semiconductor growth technology have allowed the use of bandgap engineering to cover, ideally, all interested infrared (IR) bands. The mid-wavelength (2–5 μm) and long-wavelength (8–12 μm) QWIPs have already been reported in literature.[1,2] Majority of the published works mainly focused on *n*-type QWIP where normal incidence is forbidden. The attractive feature of *p*-QWIPs comes from allowed normal incidence and hence elimination of a grating coupler.[3,4] However, choice of material is very important for the design of such QWIPs. The GaSb/GaAs heterostructure have received much attention because of their unusual type-II band alignment that can cause interesting electrical and optical properties. The state-of-art epitaxy technology has permitted growth of high quality GaSb quantum wells (QWs) and self-assembled quantum dots embedded in the GaAs matrix.[5,6]

There have been few studies computed by several researchers in recent past to figure out the electronic structure of the GaAs/GaSb quantum well.[7–15] Chidley *et al.*[7] have grown pseudomorphic GaAs/GaSb/GaAs heterostuctures. They have observed an optical transition at

V. Rajendran, R. Yuvakkumar, K. Thyagarajah and K.E. Geckeler (eds.)
Applications of Nano Materials: Electronics, Energy and Environment, pp. 143–148 (2012).

around 1.3 eV in both single and double well structures using photoluminescence and photoconductivity techniques. Ledentsov *et al.*[9] made a theoretical calculation of PL peak energy for the GaSb/GaAs structure but found a large discrepancy. Lo *et al.*[10] attributed the discrepancy to the choice of an incorrect valence band offset (VBO) and estimated the unstrained VBO for the GaSb/GaAs alignment to be about 0.4 eV from their PL measurement and theoretical calculation. Qteish and Needs [13] used self-consistent pseudopotential techniques to determine the band offset. They found a strained VBO of 1.1 eV, corresponding to a high unstrained VBO of 0.88 eV.

To be able to design such devices, it is necessary to use theoretical approaches with enough accuracy to estimate the confined quantum wells (QW) energy levels and its respective wave functions (the eigenvalues and the eigenfunctions). This work presents an approach to calculate the confined energy levels and its wave functions in valence bands of GaSb/GaAs QWIP using Transfer Matrix Method (TMM).[16] This is a versatile method in obtaining the transmission coefficient, the energy states and the corresponding wave functions of a QW structure with any potential profile. In this technique, the structure can be divided in N slices describing parts of the device, resulting in a matrix multiplication describing the propagation of a plane wave across the whole multilayer structure. Intersubband transitions have been analyzed considering strong hole confinements and large valence band discontinuity.

In this paper we describe a model for GaSb/GaAs QWIP which analyzes Eigen energies, the band profile, absorption coefficient and responsivity under different bias voltage.

THEORETICAL MODEL

We consider GaSb/GaAs quantum well structure which is formed by thin GaSb layers of 1.5 nm, is separated by 50 nm GaAs (nominally undoped, the acceptor concentration is 10^{15} cm^{-3}) barrier layer. Top and bottom contact layers are 150 nm and 500 nm p-GaAs (2×10^{18} cm^{-3}) respectively. The standard processes can be adapted to fabricate the mesa device with area of 100×100 μm^2. The key issue in the modelling of quantum well photodetector is to be able to compute quantized energy levels inside the wells as well as their respective wave functions. The wave functions in the heterostructure have computed using the effective mass and envelope function approximations for the one-dimensional potential profile along the growth direction. To compute the Eigen energies in the well, we have to solve the well known time-independent one dimensional Schrödinger equation for each square QW region of the multilayered QW structure is given by,

$$-\frac{\hbar}{2m_j^*}\frac{\partial^2\Psi(x)}{\partial x_j^2}+V_j(x)\Psi_j(x)=E\Psi_j(x) \quad \ldots (1)$$

where m_j^* is the effective mass in the region j.

The probability interpretation of the wave function requires the normalization condition for the bound states potential and the solution to Eq. (1) has the mathematically form:

$$\Psi_j(x) = A_j e^{ik_j x} + B_j e^{-ik_j x} \quad \text{... (2)}$$

where

$$k_j = \sqrt{2m_j^*(E - V_j) / \hbar} \quad \text{... (3)}$$

The potential profile of the square QW under an electric field can be represented as,

$$V(x, F) = V(x, 0) - eFx \quad \text{... (4)}$$

where $V(x,0)$ is the potential profile of the QW, F is the applied electric field and x is the associated spatial coordinate.

The modified Schrödinger equation under applied electric field can be written as:

$$-\frac{\hbar}{2m_j^*}\frac{\partial^2\Psi(x)}{\partial x_j^2} + (V_j - eFx)\Psi_j(x) = E\Psi_j(x) \quad \text{... (5)}$$

Since the potential energy V_j is a constant within the jth region of the QW, the solution of Eq. 5 can be written as a linear combination of Airy function as [7]:

$$\Psi(x) = A_j A_i(\rho_j(x)) + B_j B_i(\rho_j(x)) \quad \text{... (6)}$$

where A_i and B_i are the Airy functions of first and second kind respectively

and $$\rho_j(x) = \frac{-Fx - \eta_j}{\left(\frac{\hbar^2 F^2}{2m_j^* e}\right)^{1/3}} \quad \text{... (7)}$$

where, $\eta_j = (E - V_j) / e$ ... (8)

Proper boundary conditions have been used to solve Eq. (1) and Eq. (5) and hence to calculate Eigen energy of the QWIP using transfer matrix method. In this particular work, single quantum Well (single GaSb layer) is considered.

NUMERICAL RESULTS

The GaSb/GaAs QWIP has large valence band discontinuity approximately 0.7177 eV at 77K which can be used to formulate wide range of photodetection window. Energy Eigen values for different well widths and biases have been calculated and the plot of Eigen energies with well widths for different biases is shown in Figure 1. It is clear from figure that Eigen energy values decrease with increase in width, but it increases with increase in bias voltage, for a fixed well width. However, variation of spacing between two subband energies is not linear and hence variation in detection wavelengths is also non linear. We choose well width of 15

Å, which corresponds to the critical thickness,[14] the confinement energies of the first heavy and light-hole states are given 0.22 eV and 0.43 eV, respectively. But our numerical result shows confinement energies are to be about 0.1718 eV and 0.4329 eV, respectively. Absorption spectrum for different bias is plotted and shown clearly in Figure 2 for better understanding. It is seen from the Figure 2 that the peak absorption coefficient decreases nonlinearly with increase in bias and wavelengths corresponding to the peak absorption also shifted. These are due to the variation of relative shifts between two subband energy levels in the well with bias. An opposite nature of variation in responsivity versus wavelength graph is observed which is shown in Figure 3. This can be explained as follows. Though absorption coefficient decreases with increase in bias considering heavy hole in GaSb/GaAs QWIP, due to the high electric field, photocurrent density increases and effect of electric field variation on the photocurrent dominates over the effect of absorption coefficient. Figure 4 represents the calculated responsivity variation as a function of applied bias for different widths of the well. Responsivity initially increases with bias and saturates at high biases. At particular bias responsivity increases with well width.

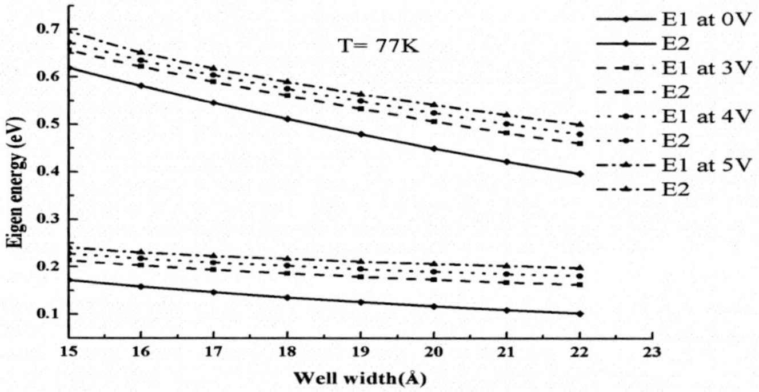

Fig. 1: Calculated Eigen Energies under Different Positive Bias: 0 V, 3 V, 4 V and 5 V, where E1 and E2 are Ground State and Excited State Energy

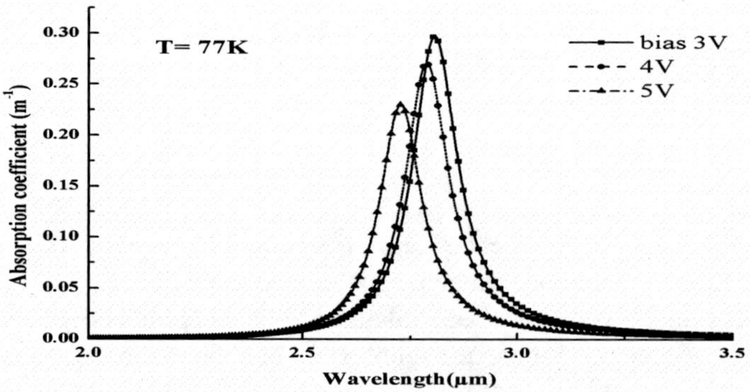

Fig. 2: Calculated Absorption Coefficient as a Function of Optical Wavelength under Different Applied Biases

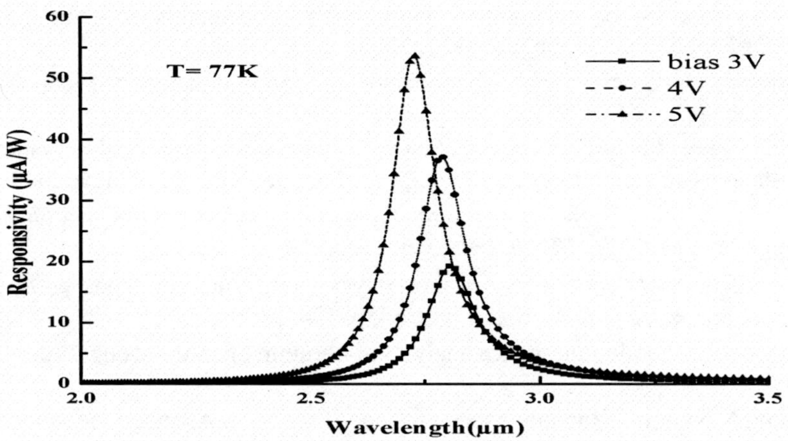

Fig. 3: Calculated Responsivity as a Function of Optical Wavelength under Different Applied Biases

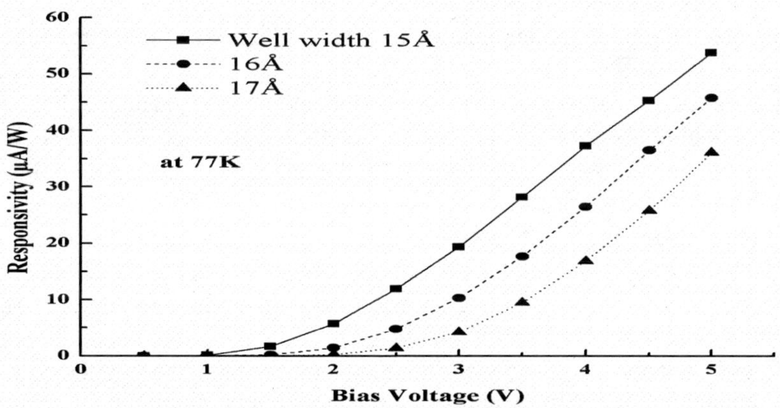

Fig. 4: Responsivity Calculated at 77K as a Function of Bias Voltage under Different Well Width

CONCLUSIONS

We have presented a theoretical model that can be used to analyze Eigen energies, the band profile, and absorption coefficient for GaSb/GaAs type-II QW structures under different applied bias.

REFERENCES

[1] Levine, B.F., Quantum-well infrared phtodetector, *J. Appl Phys.,* Vol. 74, pp. R1-R81, Oct. 1993.

[2] Liu, H.C., in Long Wavelength Infrared Detectors, edited by M. Razeghi ~Gordon and Breach, Amsterdam, pp. 1–59, Chap. 1, 1996.

[3] Liu, H.C. *et al.*, P-type quantum well infrared photodetectors covering wide spectrum, *Electronics Letters*, vol. 38, pp. 909–911, 2002.

[4] Rogalski, A., Quantum well photoconductors in infrared detector Technology, *J. Appl. Phys*., vol. 93, pp. 4355–4391, 2003.

[5] Hsieh, H.W. *et al.* Analysis of GaAs/GaSb/GaAs structures under optical excitation considering surface states as an electron reservoir, *J. Appl. Phys*., vol. 105, p. 103515, 2009.

[6] Lin, W.H., *et al.*, High-Temperature Operation GaSb/GaAs Quantum-Dot Infrared Photodetectors, *IEEE Phot. Tech. Lett*., vol. 23, pp.106–108, 2011.

[7] Chidley, E.T.R. *et al.*, GaAs/GaSb strained-layer heterostructures deposited by metalorganic vapor phase epitaxy, *Appl. Phys. Lett.* vol. 54, p. 1241, 1989.

[8] Yano, M. *et al.*, Photoluminescence analysis of ultrathin quantum wells, *Proc. SPIE* 1283, p. 221, 1990.

[9] Ledentsov, N.N. *et al.*, Radiative states in type-II GaSb/GaAs quantum wells, *Phys. Rev. B* vol. 52, p. 14058, 1995.

[10] Lo, M.C. *et al.*, Discrete monolayer light emission from GaSb wetting layer in GaAs, *Appl. Phys. Lett.* vol. 90, p. 243102, 2007.

[11] Hatami, F. *et al.*, Radiative recombination in type-II GaSb/GaAs quantum dots, *Appl. Phys. Lett.* vol. 67, p. 656,1995.

[12] Matsuda, K. *et al.*, Two-exciton state in GaSb/GaAs type II quantum dots studied using near-field photoluminescence spectroscopy, *Appl. Phys. Lett.* vol. 90, p. 013101, 2007.

[13] Qteish, A. and Needs, R.J., Calculations of the electronic structure of highly strained GaAs/GaSb/GaAs heterostructures, *Phys. Rev. B*, 42, p. 3044, 1990.

[14] Nakai, T. and Yamaguchi, K., Analysis of Sb–As Surface Exchange Reaction in Molecular Beam Epitaxy of GaSb/GaAs Quantum Wells, Jpn. *J. Appl. Phys.*, Part 1vol. 44, p. 3803, 2005.

[15] Liu, G. *et al.*, Optical gain of strained GaAsSb/GaAs quantum-well lasers: A self-consistent approach, *J. Appl. Phys.* vol. 88, p. 5554, 2000.

[16] Vatania, S. and Gildenblat, G., Airy's functions implementation of the transfer-matrix method for resonant tunneling in variably spaced finite superlattices, *J. Quantum Electronics,* Vol. 32, pp. 1093–1105, 1996.

EFFECT OF CARRIER TRAPPING ON THE EFFICIENCY OF Si/SiGe HETEROJUNCTION SOLAR CELL

SANTOSH K. CHOUDHARY AND MUKUL KUMAR DAS

Electronics Engineering, Indian School of Mines, Dhanbad
E-mail: ersantosh05@yahoo.co.in

ABSTRACT

In this paper, we present a numerical device model that describes current voltage characteristics of a SiGe/Si heterojunction solar cell. Model includes effect of heterointerface carrier confinement and doping concentration dependent carrier mobility. Effect of Ge-content in SiGe alloy on the performance of solar cell has been observed and a nonlinear variation of overall efficiency with Ge-content is obtained. Nearly 18.40% efficiency under AM1.5 is obtained for 30% Ge content in SiGe alloy.

INTRODUCTION

Performance of high efficiency III-V multijunction solar cells are becoming a viable technology in the past several years, satisfying the ever-increasing power requirements in daily life application. Efficiency of solar cell is mainly limited by the capability of absorbing whole solar spectrum. Multijunction solar cell consists of individual cells with materials of different energy band-gaps, each absorbing a different band of the solar spectrum, usually connected together in series. However, design complexity of multijunction solar cell is one of its drawbacks. State of the art is to use heterojunction solar cell alternatively for better utilization of solar spectrum. Recently, heterojunction solar cells have attracted a lot interests among researchers around the world due its several advantages over conventional p-n homojunction solar cells. Firstly, short wavelength spectral response can be enhanced by choosing material of large bandgap, secondly, series resistance can be lowered by controlling the doping concentration of

V. Rajendran, R. Yuvakkumar, K. Thyagarajah and K.E. Geckeler (eds.)
Applications of Nano Materials: Electronics, Energy and Environment, pp. 149–154 (2012).

large band gap material without affecting its light transmission characteristics and thirdly, high radiation tolerance can be achieved by increasing the thickness of large band gap material.

III-V based materials become common choice for both multijunction and heterojunction solar cells because of their superior high performance. In the last decade, a great activity has been carried out worldwide to achieve a commercial Photovoltaic (PV) product based on III-V solar cell modules for terrestrial and daily life applications. However, concerning cost of the device, silicon is an obvious choice of materials for its low cost, high reliability and CMOS compatibility. But Si is not a suitable material for absorbing longer wavelengths of solar radiation. New technologies like SiGe alloy answer quite well to solve this trade-off.[1–3] It has been seen that in SiGe heterojunction solar cell, significant absorption can be achieved at longer wavelengths by controlling Ge-composition in SiGe alloy. However, accurate modelling of heterojunction solar cell needs consideration of effects of lattice mismatch and distinctive carrier dynamics e.g., carrier confinement at heterointerfaces. In this work we have developed a device model for $Si_{1-x}Ge_x/Si$ heterojunction solar cell considering the effect of heterointerface carrier confinement and mobility dependent carrier mobility.

MODEL DESCRIPTION

Si/SiGe heterojunction solar cell is considered in our model. The Si/SiGe heterojunction consists of 50 nm n-type SiGe-epilayer which is sandwitch between bottom layer of p-type Si (100 nm) and top capping layer of n-type Si (5 nm). The whole structure is grown on the p-type Si substrate. n-Si capping layer is grown on SiGe surface to increase the luminescence efficiency due to the localization of excitons and increase of the carrier concentration in the Si/SiGe heterojunction. An ITO layer is formed on top of the Si-capping layer which can act as antireflection coating and light is allowed to incident through it. On incident of light on the device [shown in Figure 1(a)], carriers are generated within the device and these photogenerated carriers are separated out either by drift or diffusion process. However, due to the presence of heterointefaces, carrier dynamics is unnatural e.g. holes are trapped in the valence band offset at some stage in its propagation which is shown in Figure 1(b) where energy band diagram of the structure in thermal equilibrium is exposed. $E_{g,SiGe}$ and $E_{g,Si}$ ($E_{g,SiGe} < E_{g,Si}$) are the energy band gaps of SiGe and Si respectively and ΔE_C, ΔE_V are conduction band discontinuity and valence band discontinuity respectively. Light with energy greater than $E_{g,SiGe}$ but less than $E_{g,Si}$ will pass through the first semiconductor, which acts as a window, and be absorbed by the second semiconductor. Carriers created in the depletion region (width w) will be collected by drift process due to the presence of electric field in the region and carriers generated within the diffusion length both sides of the junction (L_p and L_n) will be collected by diffusion process. In our calculation, thickness of capping layer is neglected.

One dimensional steady state continuity equation for holes in n side under low injection condition can be written as,

$$G_n - \frac{n_p - n_{po}}{\tau_n} - \frac{1}{q}\frac{dJ_n}{dx} = 0 \qquad \text{... (1)}$$

where, G is generation rate of carriers, n_p is density of electrons, n_{po} is equilibrium electron density, τ_n is minority electron recombination life time and J_n is current density due to electrons. The current density equation for electrons under diffusion can be written as

$$J_n = -qD_n \frac{dn_p}{dx} \quad \text{... (2)}$$

These two equations can be solved under proper boundary condition to get expression for electron density and hence diffusion current density (J_{diff}) for electrons. Similarly diffusion current density for holes can be obtained by solving self consistent continuity equation and current density equation for holes. For drift current component i.e. current due to carriers generated in the depletion region, we have considered electric field is very high in this region and all the generated carriers are collected without any recombination and, hence, expression for drift current density (J_{dr}) can be written as

$$J_{dr} = qF(\lambda)(1-R)\Big[e^{-\alpha_{SiGe}(\lambda)x_j}\left(1-e^{-\alpha_{SiGe}(\lambda)w_{SiGe}}\right) + e^{-\alpha_{SiGe}(\lambda)(w_{SiGe}+x_j)}\left(1-e^{-\alpha_{Si}(\lambda)w_{Si}}\right)\Big] \quad \text{... (3)}$$

Where F is the no. of incident photons/cm^2/s per unit bandwidth, R is the fraction of these photons reflected from the surface, $\alpha_{SiGe}(\lambda)$ is the absorption coefficient of SiGe, $\alpha_{Si}(\lambda)$ is wavelength dependent absorption coefficient of n Si region, w_{SiGe} is the depletion width of n^+ SiGe region, w_{Si} is the depletion width of p Si region.

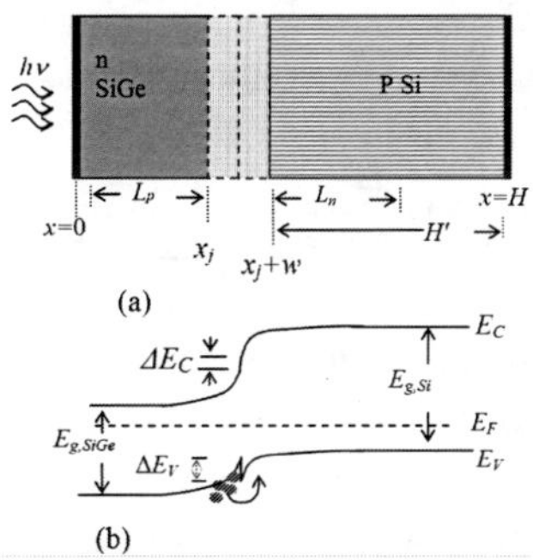

Fig. 1: (a) Schematic Structure of SiGe/Si Solar Cell, (b) Energy Band Diagram

However, due to the presence of band offsets in the device structure, electrons will be trapped at the conduction band discontinuity and holes will be trapped at valence band discontinuity during their propagation which is shown in Figure 1(b). In our model, we have considered that the trapped carriers will be coming out from trap by thermionic emission process which is governed by,

$$e_h = Be^{\Delta E_V/kT} \quad \text{... (4)}$$

where, e_h is the emission rate of holes, k is Boltzman constant and T is room temperature in°K and B is constant. temperature in °K and B is constant. Similar expression for emission rate of electrons can be obtained as

$$e_e = Be^{\Delta E_C/kT} \quad \text{... (5)}$$

During trap/confinement, carriers spent longer time in the offsets and hence some of them may be recombined and due to this recombination, total photocurrent will be decreased by a factor. The factor for current component due to holes (K_h) and due to electrons (K_e) can be expressed respectively as,

$$K_h = \frac{e_h}{e_h + e_{rp}} \quad \text{... (6)}$$

and

$$K_e = \frac{e_e}{e_e + e_{rn}} \quad \text{... (7)}$$

where, e_{rp} and e_{rn} are band to band recombination rates of holes and electrons respectively.

Hence, the total photocurrent density $J(\lambda)$ can be expressed as,

$$J(\lambda) = J_{ni}(\lambda) + K_h J_{pi}(\lambda) + J_{dr}(\lambda) \quad \text{... (8)}$$

Where,

$$J_{nl}(\lambda) = -\frac{q \times F(\lambda) \times e^{-\alpha_{SiGe}(\lambda)d_1 - \alpha_{Si}(\lambda)x_3}}{1 - \alpha_{Si}^{-2}(\lambda)L_n^{-2}} \left\{1 - \frac{1}{\alpha_{Si}(\lambda)\, L_n}\left[\frac{(S_n L_n/D_n)\left(\cosh(H'/L_n) - e^{-\alpha_{Si}(\lambda)x_4}\right) + \sinh(H'/L_n) + \alpha_{Si}(\lambda) L_n e^{-\alpha_{Si}(\lambda)x_4}}{(S_n L_n/D_n)\sinh(H'/L_n) + \cosh(H'/L_n)}\right]\right\}$$

$$J_{pl}(\lambda) = -\left(\frac{q \times F(\lambda)}{1 - \alpha_{SiGe}^{-2}(\lambda)L_p^{-2}}\right) \times \frac{\left\{\left(\frac{S_p}{\alpha_{SiGe}(\lambda)D_p}\right) + 1 - \left[\left(\frac{S_p}{\alpha_{SiGe}(\lambda)D_p}\right)\cosh\left(\frac{x_{jSiGe}}{L_p}\right) + \left(\frac{1}{\alpha_{SiGe}(\lambda)L_p}\right)\sinh\left(\frac{x_{jSiGe}}{L_p}\right)\right] \times e^{-\alpha_{SiGe}(\lambda)x_{jSiGe}}\right\}}{(S_p L_p/D_p)\sinh(x_{jSiGe}/L_p) + \cosh(x_{jSiGe}/L_p)}$$

where, S_p (Sn) is surface recombination velocity, D_p (Dn) is diffusion coefficient and L_p (Ln) diffusion length of holes (electrons). Here, we have neglected effect of electron trapping in the diffusion component of photocurrent because the conduction band offset is negligible in comparison with valence band offset. Also, the effect of carrier trapping for both electron and hole are neglected in J_{dr} because contribution of drift current is very small in the total photocurrent.

Photocurrent at particular wavelength is calculated using eqn.8 and over all photocurrent (J_{ph}) is calculated by integrating it over full wavelength range of solar spectrum. Now, diode current and hence V-I curve under illumination is obtained by calculating reverse saturation current of the diode and using the formula

$J = J_s\left(e^{\frac{qV}{kT}} - 1\right) - J_{ph}$ where, J_s is reverse saturation current density of diode and V is applied bias.

SIMULATED RESULT

We have solved continuity and current density equations for the structure considering hetero-junction effect, Shockely-Read-Hall and Auger recombination mechanisms. Position dependent carrier generation rate in each layer is considered in solving above equations. Optical constants e.g. refractive index and absorption coefficients of the materials for individual layers are taken from literature. Diode V-I characteristics under AM1.5 solar spectrum is plotted and maximum of V.I product is extracted. Hence, efficiency of the cell is calculated and plotted in Figure 2 for both the cases considering trapping phenomenon and without considering it. Thicknesses of n-SiGe and p-Si layers are fixed at 50 nm and 100 nm respectively in this plot. It is clear from figure that efficiency increases with increase in Ge-mole fraction (x) following a nonlinear variation in case of without trapping. But in case of result considering carrier trapping, efficiency initially increases nonlinearly with x and after reaching a peak value it decreases with further increase in x. This is due to the fact that with increase in x, more carriers (hole here) are trapped in the valence band offset and they spent some time there. During this stay they recombine with the electrons and as a result of that overall photocurrent due to holes decreases. Peak in efficiency reaches nearly at $x = 0.2$ which confirms that the result is in good agreement with the experimental data.[1]

Another device parameter, thickness of the absorbing layers have great role on the efficiency. However, thickness of the SiGe layer is limited by its critical thickness and hence it is kept constant within the limit. Efficiency of the cell with the thickness of p-Si layer is observed and is plotted in Figure 3. Plot is taken for three different values of Ge-content and considering fixed value of SiGe layer at 50nm. It is clearly seen that efficiency increases nonlinearly with thickness for a particular x. This is due to the enhanced absorption of solar light. The thickness of Si-layer is varied keeping lesser than the diffusion length of electrons in Si and hence efficiency increases with increase in thickness due to enhanced absorption of light. For very low value of thickness, efficiency increases very sharply because all the photogenerated carriers are separated by diffusion without almost any recombination loss.

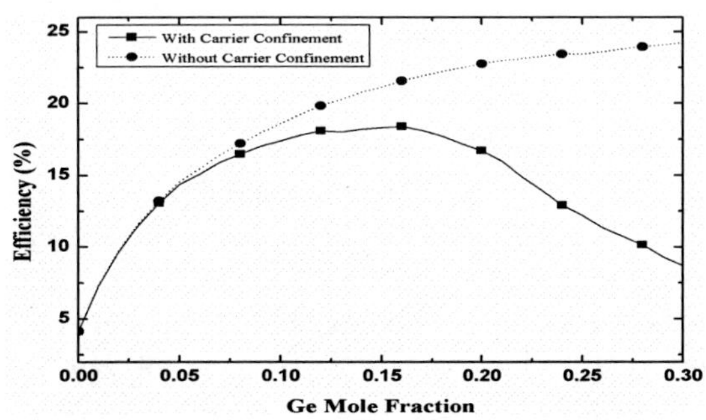

Fig. 2: Variation of Overall Efficiency of Si/SiGe Heterojunction Solar Cell with Ge-Content for Model Considering Carrier Confinement and without Considering Carrier Confinement

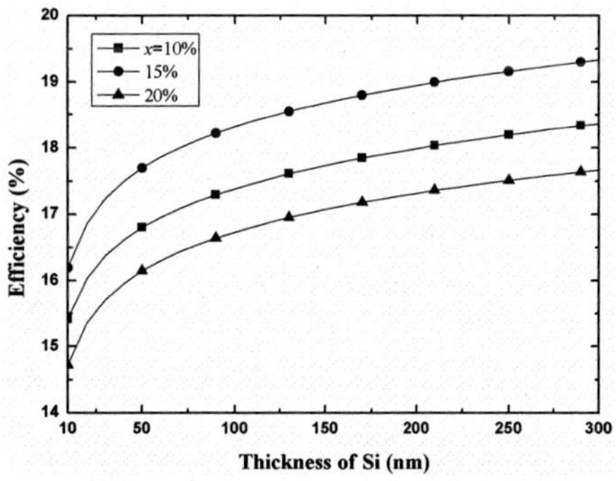

Fig. 3: Plot of Efficiency as a Function of Thickness of Si-layer for Different x

CONCLUSIONS

Efficiency of $Si_{1-x}Ge_x/Si$ heterojunction solar cell considering the effect of carrier trapping at heterointerfaces is investigated. Some optimum values of Ge-content have to be chosen for enhanced efficiency. Variation of efficiency for different Si-layer thickness has also been studied.

REFERENCES

[1] Chang, S.T., *et al.*, "Si/SiGe hetero-junction solar cell with optimization design and theoretical analysis" *Thin Solid Films*, Vol. 519, pp. 5022–5025, 2011.

[2] Liao, M.H. and chen, C.H., "The Investigation of Optimal Si-SiGe Hetro-structure Thin-Film Solar Cell With Theoretical Calculation and Quantitative Analysis" *IEEE Transation on Nanotech.* Vol. 10, no–4, 2011.

[3] Liao, M.H., Cheng, T.-H., "2.0 μm electroluminescence from $Si/Si_{0.2}Ge_{0.8}$ type II heterojunctions", *JAP,* Vol. 103, p. 013105, 2005.

[4] Das, Mukul K. and Das, N.R., "Calculating the Responsivity of a Resonant-Cavity-Enhanced $Si_{1-x}Ge_x/Si$ Multiple Quantum Well Photodetector" *J. App. Physics, AIP*, Vol.105, pp. 093118–1 - 093118–8, 2009.

[5] Liao, Milton M.H., "An investigation on the ight-emmision mechanism of metal-insulator-semoconductor light-emiting-diodes with different SiGe quantum well structures", *Appl. Phys. Lett.,* Vol. 94, p. 241908, 2009.

[6] People, R., "Physics and Application of Ge_xSi_{1-x}/Si Strained- Layer Heterostructures" *IEEE J. Quant. Elec.* Vol. QE-22, no-9, 1986.

[7] Das, Mukul K. and Das, N.R., "On Optimum Designs of a RCE Si/SiGe/Si MQW Photodetector for Long Wavelength Applications" *Optical and Quantum Electronics, Springer*, Vol. 41, pp. 539–549, 2009.

[8] Griggs, Melissa J., *et al.*, "p-n Junction Heterojunction Device Physics Model of a Four Junction Solar cell" *Proc. of SPIF,* Vol. 6339, pp. 63390D–1 -63390D–7, 2006.

[9] Koster, L.J.A., *et al.*, "Device model for the operation of polymer/fullerene bulk heterojunction solar cells" *Physical Review*, Vol. B 72, p. 085205, 2005.

[10] Sze, S.M., "*Physics of Semiconductor Devices*" 2nd Edition, 1981.

[11] Pallab Bhattacharya, "*Semiconductor Optoelectronic Devices*" 2nd Edition, 1999.

[12] King, R.R., *et al.*, "High-Efficiency Space and Terrestrial Multijunction Solar Cells Through Bandgap Control In Cell Structures" *IEEE Conf. Photovoltaic Specialists Conference*, pp. 776–781, 2002.

[13] Aiken, D.J. *et al.*, "Consideration of High Bandgap Subcells for Advanced Multijunction Solar Cells", *4th IEEE, World Conference Photovoltaic Energy Conversion*, Vol.1, pp. 838–841, 2006.

METAL-SEMICONDUCTOR-METAL PHOTODETECTOR WITH ENHANCED TE-POLARIZATION TRANSMISSION

AYMAN KARAR[1*], CHEE LEONG TAN[2], KAMAL ALAMEH[1,4*], YONG TAK LEE[2,3,4] AND AMPALAVANAPILLAI NIRMALATHAS[5]

[1]Electron Science Research Institute, Edith Cowan University, AU
[2]School of Photonics Science, Gwangju Institute of Science and Technology (GIST), Gwangju, Republic of Korea
[3]Department of Information and Communications, GIST, Republic of Korea
[4]Department of Nanobio Materials and Electronics, World Class University (WCU), GIST, Republic of Korea
[5]Department of Electrical and Electronic Engineering, The University of Melbourne, AU
E-mail: akarar@ecu.edu.au; k.alameh@ecu.edu.au

ABSTRACT

In this paper, we use the finite difference time-domain (FDTD) method to optimize the TE-polarized light transmission of a metal-semiconductor-metal photodetector (MSM-PD) employing a dielectric waveguide on top of metal nano-gratings. Simulation results demonstrate that the funneling transmission of the TE-polarized light through the nanoslit of the structure is highly dependent on the structure geometries such as the waveguide and nano-grating heights. We also demonstrate that adding a dielectric waveguide layer on top of the nano-metal grating supports both the TM- and TE polarizations, and enhances the light transmission for TE-polarization approximately 3-times.

INTRODUCTION

Metal-semiconductor-metal photodetectors (MSM-PDs) have several attractive features, such as high speed, ease of fabrication and monolithic integration with VLSI circuitry, which make

V. Rajendran, R. Yuvakkumar, K. Thyagarajah and K.E. Geckeler (eds.)
Applications of Nano Materials: Electronics, Energy and Environment, pp. 155–160 (2012).

them an excellent candidate for application in high-speed optical interconnects, high-speed sampling, and ultra-high speed optical fibre communications.[1, 2] However, the surface reflectivity and the shadowing effect of the metal fingers prevent conventional MSM-PDs from achieving external quantum efficiency greater than 50% for equal electrode width and spacing. Recently several MSM-PD structures, driven by TM-polarized light, based on nano-patterning the metal fingers have been reported demonstrating substantial transmission enhancement through the excitation and guidance of surface plasmon polaritons (SPPs) into the photodetector slits.[3, 4] Since there is no cut-off wavelength for the fundamental TM-polarized slit mode, it is possible to achieve extraordinary transmission with almost any subwavelength slit width. However, since SPPs are TM-polarized mode, only the TM-polarization component of the incident light can be resonantly enhanced the transmission through the subwavelength slits. In this case, subwavelength slits act as polarization selector, which means that the penetration of the TE-polarized mode is suppressed. Furthermore, the TE-polarized mode intrinsically has a cut-off wavelength, making such MSM-PD structures polarization sensitive, and hence less attractive for applications requiring polarization insensitive operation.

In this paper, we use the FDTD analysis to optimize a novel MSM-PD structure employing a dielectric thin layer waveguide deposited onto nano-patterned metal fingers. Our simulation results demonstrate 3-times enhancement in light transmission TE-polarized light in comparison with conventional MSM-PDs.

DESIGN OF MSM-PD WITH ENHANCED TE-POARISATION TRANSMISSION

Recently, an MSM-PD structure based on the deposition of a dielectric layer on top of the metal fingers has been proposed by Nikitin *et al.* using the coupled mode method,[5] and subsequently developed by Guillaumée *et al.*, who experimentally demonstrated transmission enhancement for TE-polarized light through a subwavelength slit.[6] However, such a structure is sensitive to two parameters, namely, the height of the dielectric layer as well as the periodicity of the nano-patterned metal grating, which were not fully optimised. In this paper, we investigate and optimize the key parameters of the dielectric-based MSM-PD structure, shown in Figure 1(a), to maximise the transmission of both the TE- and TM- polarized modes. The dielectric waveguide based MSM-PD structure consists of a subwavelength aperture of width X_w sandwiched between linear metal nano-gratings of heights h_g, and a period (Λ). The entire structure is grown on top of a semiconductor substrate. The metal contact is covered with a thin dielectric film with height h_{WG}. The presence of a thin dielectric layer on top of patterned metal gratings allows a TE-polarized incident light to couple to the dielectric waveguide modes and be guided towards the slit. At the same time, the extraordinary transmission of the TM-polarised light is maintained.

A 2D Finite Difference Time Domain (FDTD) software developed by Optiwave Inc was used to simulate the structure shown in Figure 1(a). A mesh step size of 10 nm was used in the simulation, with a time step satisfying the condition $\delta t < 0.1\delta_x/c$, where δ_x is the mesh size and c is the speed of light. This high-resolution sampling yielded solutions that converged at reasonable computation times. The excitation field was modeled as a Gaussian-modulated continuous plane wave in the x-direction. The anisotropic perfectly matched layer (APML) boundary conditions were applied in both the x- and z-directions to accurately simulate the light reflected off both sides, as well as the light reflected off the top and bottom surface of the MSM-PD structure. In all the simulations, the light wave was normally incident on the top surface of the metal nano-gratings site. The gold (Au) dielectric permittivity was defined by the Lorentz-Drude model [3] and the refractive index for the dielectric layer was chosen to be 2.3, mainly to demonstrate the concept of TE-polarized light transmission enhancement.

The TE and TM transmission spectra for two MSM-PD structures (without a dielectric waveguide), one with and the other without metal nano-gratings, are shown in Figure 1(b). An essential difference between the spectra for the TM- and TE-polarizations is already noticeable for a single slit without metal nano-gratings. While for the TM-polarized mode the spectra show interlaced maxima associated with the Fabry-Perot slit waveguide resonance, for the TE-polarized mode resonance maxima are displayed, however, rapid fall-off, due to the slit mode cut-off, is seen at long wavelengths. Note that, no significant difference is seen in the TE-polarized mode transmission spectra for the structure with and without metal nano gratings. However, the TM mode is resonantly enhanced by using the nano-gratings. Figure1(c) shows the measured I-V characteristics for a nano-grating-patterned GaAs MSM-PD illuminated with 6.42 mW of laser power, for two input polarization states that correspond to the maximum, i.e. TM mode (dashed-dot), and minimum, i.e. TE mode (dashed), possible measured photocurrents, respectively. These results indirectly reveal the polarization dependent loss of the plasmonics-based MSM-PD device.

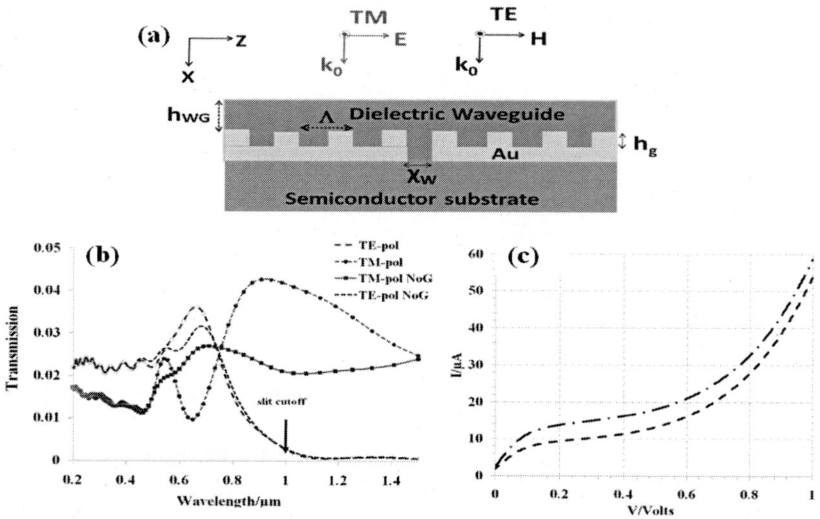

Fig. 1: a) 2-Dimensional Schematic of the Dielectric-Based MSM-PD Structure, **b)** Transmission Spectra of Two Structures without Dielectric Waveguides, for Fingers with and without Metal Nano-gratings, **c)** I-V Characteristics of the MSM-PD with Nano-gratings for TE- and TM-Polarized Input Laser Beams of Power 6.42 mW

RESULTS AND DISCUSSION

Several parameters of the dielectric-waveguide-based MSM-PD structure are optimized. These parameters are the nano-grating height, h_g, nano-grating period, Λ, waveguide height, h_{WG}, duty cycle of the nano-gratings. Each parameter was varied over a certain range of values, while all other parameters were kept constant. Initially, we used the structure shown in Figure 1(a) with metal nano-gratings to optimize the waveguide height h_{WG}. The dielectric waveguide height was varied, while keeping the subwavelength aperture width X_w and h_g constant at 430 nm and 200 nm, respectively. Moreover, the duty cycle, grating period and pitch number were kept at 0.5, 830nm and 7, respectively. Figure 2(a) illustrates the TE-polarized transmission spectra with no dielectric waveguide, 150 nm, 200 nm and 250 nm waveguide height. It is apparent from Figure. 2 (a) that the dielectric waveguide not only affects the amount of the light flux transmitted through the slit, but also the peak resonance wavelength. This indicates that the dielectric layer allows the incident light to resonantly couple to the metal nano-gratings. As seen from Figure 2(a), the resonance peak is highly dependent on the waveguide height and is red-shifted with increasing the waveguide height. Keeping the dielectric waveguide height at 200 nm and the other parameters constant, the simulated TE-polarized transmission spectra is shown in Figure 2(b) for metal nano-grating height from 50 nm to 200nm. As seen in Figure 2(b) while the metal nano-grating height affects the TE-polarized light transmission flux, it has no impact on the resonance wavelength. It is also noticed that the transmission peak increases with decreasing the metal nano-grating height.

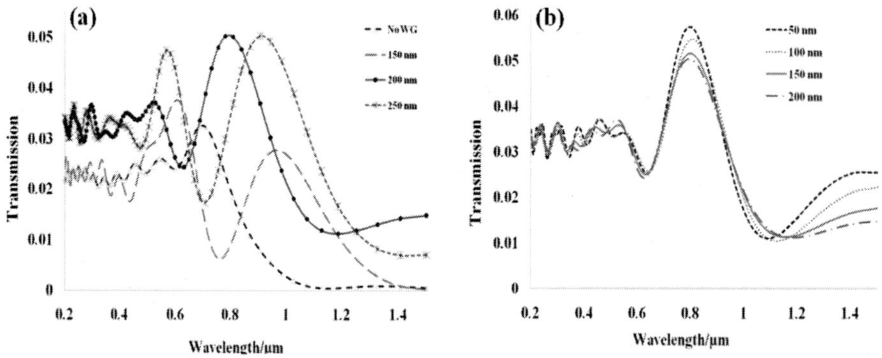

Fig. 2: Simulated TE-polarisation Transmission Spectra for Different **a)** Dielectric Waveguide Height and, **b)** Nano-Grating Height with 430 nm Slit Width

Figures 3(a and b) show the simulated TE-polarisation transmission spectra for different metal nano-grating periods, and duty cycles, for a slit width of 430 nm, a nano-grating height of 50 nm and a dielectric waveguide height of 200 nm. The duty cycle is defined as the ratio of the nano-grating line width to the period, Λ. It is seen from Figure 3(a) that the resonance peak of the TE-polarised light transmission is red-shifted when the periodicity increases. The maximum attainable transmission peak occurs at 730 nm for a nano-grating period of 600 nm. It is noticed from Figure 3(b) that the nano-grating duty cycle has a significant impact on both the transmission peak and the resonance wavelength, and that the resonance wavelength is red-shifted when the duty cycle increases, while the maximum transmission occurs when the duty cycle is 50%.

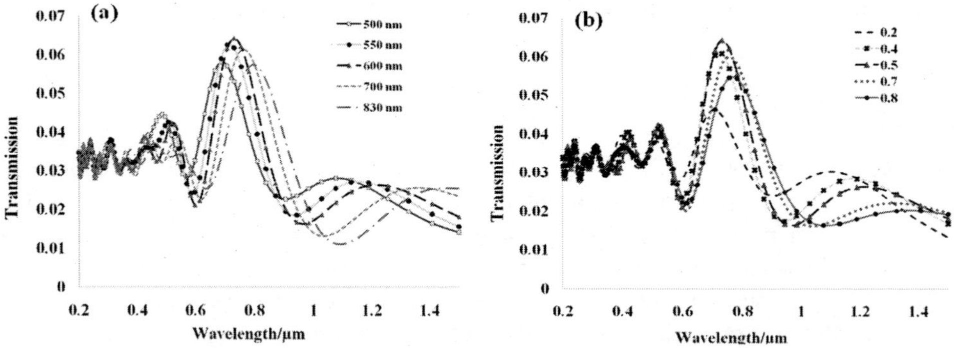

Fig. 3: Simulated TE-Polarisation Transmission Spectra for Different **a)** Periodicity and, **b)** Duty Cycle with 430 nm Slit width, 50 nm Nano-Grating Height and 200 nm Dielectric Waveguide Height

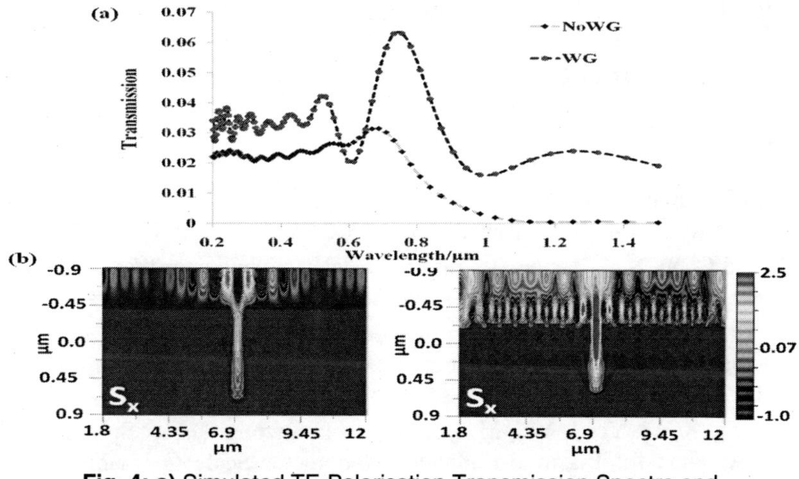

Fig. 4: a) Simulated TE-Polarisation Transmission Spectra and,
b) Simulated Power Distribution without (left) and with (right) Dielectric Waveguide

The TE-polarized light transmission spectra are shown in Figure 4(a), for two optimized MSM-PDs with and without a dielectric waveguide (WG). It is obvious that the dielectric waveguide on top of the metal nano-gratings significantly enhances the TE-polarised light transmission, compared with the conventional MSM-PD device without a dielectric waveguide (NoWG). The resonance wavelength (corresponding to the highest transmission) is 755 nm for Λ = 600 nm and the transmission enhancement is almost 3-times that of a conventional MSM-PD device without a dielectric waveguide. Moreover, as seen from Figure 4(a) the cutoff wavelength is increased when the slit is filled with the dielectric. The simulated S_x poynting vectors, i.e. energy flowing along the x direction, are shown in Figures 4(b), for the MSM-PD without (left) and with (right) a dielectric waveguide. It is worthwhile noting that for an MSM-PD without a dielectric waveguide, the power transmitted into the active area of the semiconductor is insignificant, compared with the power transmitted for an MSM-PD with a dielectric waveguide.

CONCLUSIONS

Finite difference time-domain (FDTD) method has been used to optimize the TE-polarized light transmission of a metal-semiconductor-metal photodetector (MSM-PD) employing a dielectric waveguide and metal nano-gratings. Simulation results have confirmed the dependence of the TE-polarized light through the MSM-PD nanoslit of the structure on the metal nano-grating height and dielectric waveguide height. TE-polarised light transmission enhancement of 3-times has been demonstrated through MDM-PD parameter optimisation.

ACKNOWLEDGEMENTS

We acknowledge the support of Edith Cowan University and the World-Class University Program funded by the Ministry of Education, Science, and Technology through the National Research Foundation of Korea (R31–10026).

REFRENCES

[1] Soole.J. and Schumacher. H., "InGaAs Metal-Semiconductor-Metal Photodetectors for Long Wavelength Optical Communications," *IEEE J. Q. Elec.*, 27, 3,737–752, 1991.

[2] Liu. M.Y. and Chou. S.Y., "Internal emission metal-semiconductor-metal photodetectors on Si and GaAs for 1.3 μm detection," *Appl. Phys. Lett.*, 66, 2673 –2675, 1995.

[3] Karar. A., Das. N., Tan. C.L., Alameh. K., Lee. Y.T., and Karouta. F., "High-responsivity plasmonics-based GaAs metal-semiconductor-metal photodetectors," *Appl. Phys. Lett.*, 99, 33112–33112, 2011.

[4] Ren. F.F., Ang. K.W., Ye. J, Yu., M, Lo. G.Q., and Kwong. D.L., "Split Bull's Eye Shaped Aluminum Antenna for Plasmon-Enhanced Nanometer Scale Germanium Photodetector," *Nano Lett.,* 11, 1289–93, 2011.

[5] Nikitin. A.Yu., Garcia-Vidal. F.J., and Martin-Moreno. L., "Enhanced optical transmission, beaming and focusing through a subwavelength slit under excitation of dielectric waveguide modes," *J. Opt. A: Pure Appl. Opt.,*11, 2009.

[6] Guillaumée. M., Nikitin. A. Yu., Klein. M.J.K., Dunbar. L.A., Spassov. V., Eckert. R., Martín-Moreno. L., García-Vidal. F.J. and Stanley. R.P., " Observation of enhanced transmission for s-polarized light through a subwavelength slit," *Opt. Express*, 18, 9, 9722–9727, 2010.

EFFECT OF ANCHORING ATOMS ON TRANSPORT PROPERTIES OF A CARBON-DIMER BASED MOLECULAR JUNCTIONS: A FIRST PRINCIPLES STUDY

R.M. HARIHARAN AND D. JOHN THIRUVADIGAL

Centre for Materials Sciences and Nanodevices, Department of Physics and Nanotechnology, SRM University, Kattankulathur
E-mail: john.d@ktr.srmuniv.ac.in

ABSTRACT

The conductance of a molecular device is sensitive to the contact geometry between the molecules and the probing electrodes. Combining the density functional theory calculations for molecular electronic structure with a non-equilibrium Green's function method for electron transport, we calculate the molecular conductance of carbon dimer connected between Au leads through two different anchoring atoms Se and Te. The current-voltage characteristics and transmission spectra of two systems are studied. The results exhibit that, depending on the anchoring groups and the subsequent different metal-molecule chemical bonds, the current varies over more than four orders of magnitude under the same bias. Furthermore, the system exhibits negative differential (NDR) effect, when anchored with Te atom. This emphasizes the great importance of the anchoring groups in molecular devices.

INTRODUCTION

Over the past decade, molecular devices have received significant attention owing to their great potential practical applications in atomic-scale circuits.[1] Many interesting physical properties of molecular devices have been predicted theoretically and verified experimentally, such as single-electron characteristics,[2] negative differential resistance (NDR),[3,4] molecular rectification [5,] and field-effect characteristics,[6] etc.

V. Rajendran, R. Yuvakkumar, K. Thyagarajah and K.E. Geckeler (eds.)
Applications of Nano Materials: Electronics, Energy and Environment, pp. 161–166 (2012).

A basic goal in molecular electronics is to measure and control electron transport through a molecule attached to two electrodes. To reach this goal, one must not only synthesize molecules with desired functions, but also design proper molecule-electrode contacts. This is because the contacts play a critical role in the electron-transport properties of molecules.[7,10] One way to optimize the molecule-electrode contacts is to select proper anchoring groups at the two ends of a molecule so that they can bind to the electrodes and in the meantime, provide efficient electronic coupling between the molecule and the electrodes. So the key issue in such molecular electronic device is to find suitable anchoring groups and to construct contact structures that provide both stability and high contact transparency.[9] Therefore, the extensive and profound understanding of these effects in molecular devices is quite necessary. Using DFT+NEGF method, Zeng *et al.*[8] performed transport calculations on carbon dimer sandwiched between Au electrodes, with anchoring groups, H, Cu and S.

In this present work, we focus our concentration on the anchoring group effect in molecular devices. Specifically, we study the transport properties of a carbon-dimer C_2 cluster sandwiched between two Au (111) electrodes. The anchoring groups connecting the cluster and the electrodes are chosen as the atoms Se and Te respectively. Thus, equivalently, we study the transport properties of two different molecules C_2Se_2 and C_2Te_2, since both contain a C_2 unit and they are all of a linear chain structure. Reason for choosing Selenium and Tellurium as anchoring atoms because they serves as efficient alligator clip connection to Au other than sulfur.[9]

Also, we show that asymmetric molecular junctions, such as carbon dimer with appropriately chosen anchoring group, gives rise to I-V curve showing NDR effects.

COMPUTATION METHOD

The simulation model adopted in this work is presented: a C_2 molecule is sandwiched between two Au (111) bulk electrodes via different anchoring atoms. And in the present work, the central sandwiched molecules are C_2Se_2 and C_2Te_2, which are all linear and symmetrical molecules. The C–C and C–Se bond lengths in C_2Se_2 molecule and C–C and C–Te bond lengths in C_2Te_2 molecule are obtained by fully optimizing the respective molecules. For the left and right electrodes, we select a 3 × 3 unit cell in the x and y directions to avoid the interaction between the molecules and their mirror images. The terminating atoms of the molecules are located symmetrically at the hollow sites above the Au (111) surface. By full optimization, we obtain the equilibrium distances between the central molecules and the electrodes.

Our calculations are carried out using the Atomistic Toolkit (ATK 11.2.3 version) [11] program which combines density functional theory (DFT) and Non-equilibrium Green's functions (NEGF) method to simulate the transport properties of two-probe ems. The exchange-correlation functional is described by generalized gradient approximation (GGA) in the form of the Revised Perdew-Burke-Ernzerh (revPBE) in the present calculations.[2] The valence wave functions are expanded by systems. The exchange-correlation functional is described by generalized gradient approximation (GGA) in the form of the Revised Perdew-Burke-Ernzerh (revPBE) in the present calculations.[2] The valence wave functions are expanded by localized numerical atomic orbital (LNAO), with the single zeta plus polarization basis set (SZP) for Au element and the double zeta plus polarization basis set (DZP) for C, H, Se and Te atoms. The Brillouin zone of the leads is sampled by 1 × 1 × 80 k points in the directions of x, y and z (z is the electron transport direction), which is enough to produce the results.

RESULTS AND DISCUSSION

The I-V characteristics of the two probe system can be obtained by using the Landauer–Buttier formula, the current is the integration of the transmission function over the bias window around the Fermi level:

$$I = \frac{2e}{h}\int_{\mu_1}^{\mu_2} T(E,V)[f(E-\mu_1)-f(E-\mu_2)]dE$$

where T (E, V) is the transmission function through the device at energy E and funcion *f*(E − μL/R) are the Fermi distribution functions of electrons in the left/right electrodes. μL/R = μ ± eV/2 are the chemical potentials of the left/right electrodes, with μ the Fermi level of the electrodes in equilibrium and V the bias applied to the device.

In Figure 1, the I–V curves of two systems are presented. As can be seen from this figure, the current for C_2Se_2 linearly increases with the voltage increasing from 0.0 V to 2.0 V, while the current of C_2Te_2 increases in the opposite direction, reaches maximum in 0.58 V and drops to zero at 1.0 V. Further the current for C_2Te_2 rapidly increases from 1.0 V to 2.0 V. This indicates evident NDR behavior, which can be clearly observed in the I–V curve for C_2Te_2 system at zero bias range as shown Figure 1.

On the whole, the amplitude of the I–V curves of two different systems varies over more than four orders of magnitude. For example, when the bias is 1.5 V, the currents are 30,638 nA for C_2Se_2, and 1,36,254 nA for C_2Te_2 respectively. Obviously, NDR behavior and the change in magnitude of current are originated from the contact atoms Se and Te linked to the C_2 cluster to the electrodes. Thus, it shows the importance of different linking atoms which result in such a big difference of the currents through the same molecule with different linking atoms.

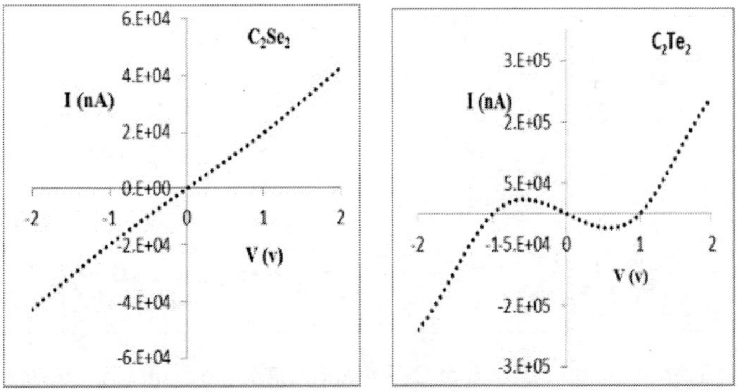

Fig. 1: The I-V Characteristics of C_2Se_2 and C_2Te_2 Molecules Respectively

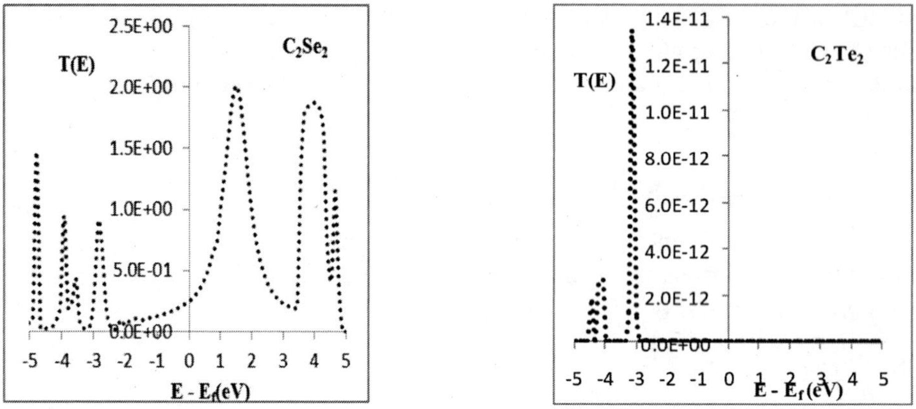

Fig. 2: The Transmission Spectra of C_2Se_2 and C_2Te_2 Molecules at 0V Respectively

In order to understand the difference in I-V characteristics of C_2Se_2 and C_2Te_2 systems, transmission spectrum of those systems at zero bias respectively are shown in Figure 2. As can be seen from Figure 2, the transmission function of C_2Te_2, has very large gap near the fermi level, the transmission spectrum in the zero bias window becomes smaller and smaller, which results in NDR behavior. This NDR behavior may arise due to the the coupling between the Te molecule and electrodes. The transmission coefficient can be related to the molecular orbitals which have been modified by the electrodes when connected to different anchoring groups.[11]

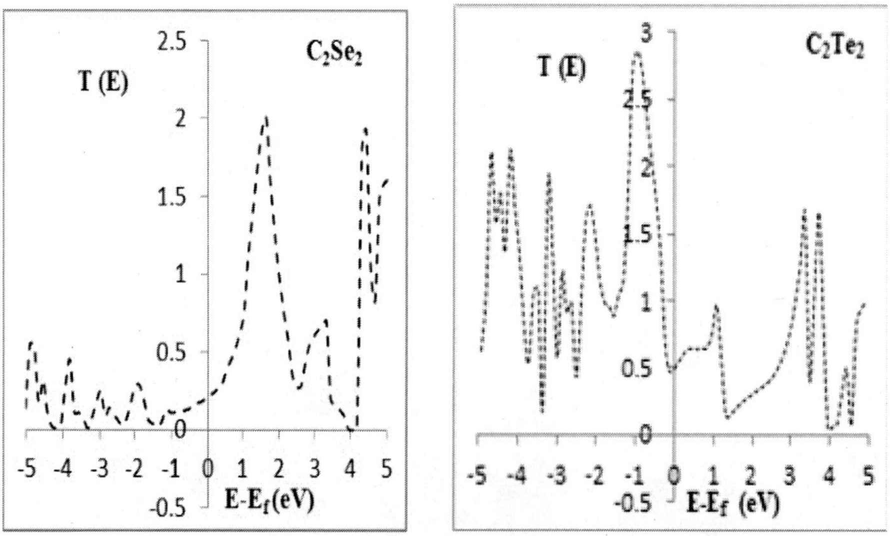

Fig. 3: The Transmission Spectra of C_2Se_2 and C_2Te_2 Molecules at 1V Respectively

Thus when a Te molecule is connected to gold electrodes, the energy levels and spatial distribution of molecular orbitals are modified due to the molecule–electrode interaction and leads to NDR behavior. NDR behavior is not reported in the earlier work on carbon dimer by Zeng *et al.* with three different anchoring groups H, Cu and S.[8] Whereas the transmission function of C_2Se_2, has sharp peaks near fermi level which results in linear conductance of the molecule. The results show that the anchoring group plays a crucial role in NDR effect and overall conductivity of the molecular junctions.

However as can be seen from figure 3, with an increase in the bias, much transmission energy enters into bias windows which results in the increasing of current for C_2Te_2. For the C_2Te_2, there is a large platform around the Fermi level, so the current is very big when the bias voltage is applied to the device. However, we hardly find any new transmission entering into bias window when the bias is applied for C_2Se_2. So a current progresses in the same way as that zero bias for C_2Se_2.

CONCLUSIONS

Using the density functional theory combined with a non-equilibrium Green's function approach, the electron transport properties of molecular junctions containing carbon dimer molecule with two different anchoring groups Se and Te were calculated. The current in these two different systems varies by more than four orders of magnitude under the same bias range. Furthermore, the system exhibits negative differential resistance (NDR) effect, when anchored with Te atom at zero bias range. The results show that the anchoring group and the subsequent different metal-molecule chemical bonds play a significant role in determining the overall conductivity of the molecular junctions.

ACKNOWLEDGEMENTS

This work was supported by the DST- FIST programme (Govt. of India) under Grant No. SR/FST/PSI-010/2010. The work was carried out in the Centre for Materials Sciences and Nanodevices, Department of Physics and Nanotechnology, SRM University.

REFERENCES

[1] Joachim, C., Gimzewski, J.K. and A. Aviram, *Nature,* 2000, 408, 541.

[2] Postma, H.W.C., Teepen, T., Yao, Z., Grifoni and M., Dekker, C., *Science*, 2001, 293, 76.

[3] Chen, J., Wang, W., Reed, M.A., Rawlett, A.M., Price, D.W. and Tour, J.M., *Applied Physics Letters,* 2000, 77, 1224.

[4] Long, M.Q., Chen, K.Q., Wang, L., Zou, B.S. and Shuai, Z., *Applied Physics Letters*, 2007, 91, 233512.

[5] Aviram, A. and Ratner, M.A., *Chemical Physics Letters*, 1974, 29, 277.

[6] Lu, J., Yin, S., Peng, L.M., Sun, Z.Z. and Wang, X.R., *Applied Physics Letters,* 2007, 90, 052109.

[7] Fang Chen, Xiulan Li, Joshua Hihath, Zhifeng Huang, and Nongjian Tao *J. Am. Chem. Soc.,* 2006, 128, 15874–15881.

[8] Wang, R.N., Zheng, X.H., Dai, Z.X., Haoa, H., Song , L.L. and Zeng , Z., *Physics Letters A*, 2011, 375, 657.

[9] Jorge M. Seminario, Angelica G. Zacarias, and James M. Tour *J. Am. Chem. Soc.,* 1999, 121, 411.

[10] Ming-Jun Li, Hui Xua, Ke-Qiu Chen and Meng-Qiu Long, *Physics Letters A,* 2012, 376, 1692.

[11] ATK-VNL Tool Kit 11.2.3 Version, *www.quantumwise.com.*

FABRICATION OF PbSe–ZnO BASED THIN FILMS FOR SOLAR CELL APPLICATIONS

D. Selvakumar and R. Jayavel

Centre for Nanoscience and Technology, Anna University, Chennai
E-mail: rjvel@annauniv.edu

ABSTRACT

A new type of excitonic solar cell based on heterojunctions between PbSe semiconducting thin film and thin ZnO film has been fabricated. These solar cells can generate large photocurrents and higher photovoltages compared to Schottky cells assembled with similar nanocrystal films. The n-type and p-type nanomaterials such as ZnO and PbSe were synthesized by sol-gel method. The synthesized nanoparticles were coated over the ITO substrate by thermal evaporation method. The heterojunction thin films were annealed at 400°C to form p-n junction. Device performance was analyzed using Semiconductor Parameter Analyzer (SPA) under the bright and dark field illumination. The conductivity was estimated from the sheet resistance value by using Four Probe Method. The material coated over the substrate and its crystallinity was identified by using thin film XRD and the nature of material coated was analyzed by Hall Effect measurement system. Thickness of the film coated over the substrate was determined by cross section of FESEM analysis. The solar cell performance was carried out by differentiating the I-V characteristics under bright and dark field light illumination.

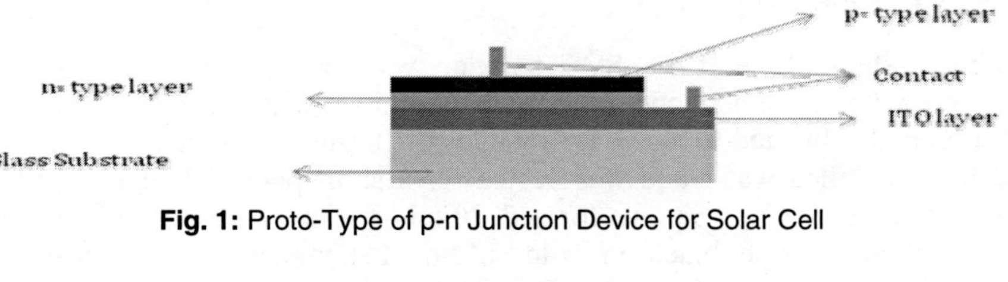

Fig. 1: Proto-Type of p-n Junction Device for Solar Cell

V. Rajendran, R. Yuvakkumar, K. Thyagarajah and K.E. Geckeler (eds.)
Applications of Nano Materials: Electronics, Energy and Environment, pp. 167–172 (2012).

INTRODUCTION

The existing types of organic-based solar cells, including dye-sensitized solar cells (DSSCs), can be categorized by their photoconversion mechanism as excitonic solar cells, XSCs. Their distinguishing characteristic is that the charge generation and separation are simultaneous and this occurs via exciton dissociation at the heterointerface. There is a well-known difference between the photoconversion mechanism in inorganic photovoltaic (IPV) cells and in organic (excitonic) solar cells, XSCs: light absorption in XSCs leads to the production of excitons (mobile excited states), while in IPV cells it leads directly to the creation of free electron-hole pairs. This difference has fundamental, and under-appreciated, consequences for the theoretical description of the photoconversion process, and also for efforts to optimize the performance of XSCs. The conventional cell is much less efficient than the XSC. There are at least two reasons for this: (1) recombination throughout the bulk is more efficient than recombination only at the interface (excitonic layer), given the same rate constants for recombination. (2) The XSC has the photogenerated force, operating in concert with separating the two carriers and driving them toward their respective electrodes; while in the conventional cell, opposes.

MATERIALS AND METHOD

Zinc acetate di-hydrate [$C_4H_6O_4Zn{\cdot}2(H_2O)$], Sodium Hydroxide [NaOH], Selenium powder [Se], Sodium sulfite (sodium sulphite) [Na_2SO_3], Lead acetate tri-hydrate [Pb $(CH_3COO)_2{\cdot}3H_2O$] and Indium Tin Oxide (ITO) substrate all with a high purity from 99.9 to 99.99%, purchased from Merck, were used as starting materials.

Synthesis of ZnO Nanoparticles

Zinc oxide (ZnO) nanoparticles were prepared by sol-gel process; 0.5M Zinc acetate and 0.5M NaOH were dissolved in 50 ml of distilled water separately. NaOH solution was slowly added to zinc acetate solution while stirring vigorously and a milky white solution was obtained. The precipitate of Zn $(OH)^{2-}_4$ was deposited and dried at 100°C to get the fine particles of ZnO. The ZnO target was prepared using ZnO nanoparticles for thermal evaporation.

Synthesis of PbSe Nanoparticles

0.1 M sodium selenosulfate [Na_2SeSO_3] solution was prepared from 0.005 M elemental selenium mixed with 0.2 M aqueous solution of sodium sulfite. The mixture was refluxed for 4 hrs at 120°C and subjected to aging for two days. 1.8 gm of lead acetate was dissolved in 250 ml of triple distilled water and was heated for a short period. 15 ml of 15% ammonia solution was added, a white precipitate formed. 30 ml of selenosulphate solution was added to the above solution become black in color. After 1 hour of heat treatment, 20 ml of selenosulphate solution was added and a dark black precipitate was obtained. The precipitate was then thoroughly washed with hot sodium sulphite solution to remove the selenium ions. The precipitate was dried in a vacuum oven. The PbSe target was obtained for thermal evaporation using PbSe nanoparticles.

Thin Film Diode Fabrication

The vacuum thermal evaporation deposition technique consists of resistance heating (Joule's effect) of the target material to be deposited. The material vapor finally condenses in the form of thin film on the cold substrate surface and pressure is about 10^{-6} or 10^{-5} Torr. Initially ITO substrate was washed with acetone and dried for a few minutes. ZnO film was deposited on ITO substrates at a substrate temperature of 400°C with 2.2 x10^{-5} mbar vacuum and 150 mA current. Prior to deposition, the ITO layer was patterned to produce bottom electrode. The ZnO film was annealed at 450°C under oxygen atmosphere using tubular furnace. PbSe film was deposited over ZnO film from PbSe target under the same environment. The resultant films were subjected to annealing at 400°C as to form p-n junction.

RESULTS AND DISCUSSION

X Ray Diffraction (XRD) pattern for thin films was obtained from Rigaku Ultima III XRD as shown in Figure 2, 3. For the XRD measurement, individual film of 1 cm^2 was used as the sample. Figure 2 shows the XRD pattern of ZnO thin film and the planes present are (100), (002) and (101) and matches with ICDD file number 082029.

From the XRD data shown in Figure 2, it was clearly absorbed that the crystalline nature of the ZnO thin film is optimized at an annealing temperature of 450°C. XRD pattern of PbSe thin film is shown in Figure 3.

The planes corresponding to PbSe film are (200), (220), (311), (400) and (420), which agree with the ICDD file number 652941 and confirms the crystallinity at an optimized sintering temperature of 400°C. XRD data shows that crystallinity and sintering temperature are directly related and responsible for junction formation in device fabrication. The grain size of the films was calculated by Gaussian Fit analysis and the average grain size of ZnO film was 50nm and PbSe film was 30 nm. The grain size of the film increases with increasing in sintering temperature.

The conductivity of the individual thin film of ZnO and Pbse films were measured by Four Probe method. The voltage was applied through the two probes and the current was measured through other two probes based on the sheet resistance present in the film. The sheet resistance was calculated from,

$$R_s = \frac{\rho}{t} = 4.53\frac{V}{I}$$

The conductivity of the film was measured to be 0.7355E-9 Ω/m for ZnO thin film and 1.6025E–6 Ω/m for PbSe film.

The semi-conducting nature of the films was analyzed by Hall effect measurement system (HMS – 3000 VER 3.51.3). The film was inserted in the system and the electric and magnetic fields were applied perpendicular to each other and it experiences a voltage, perpendicular to both the fields. The voltage experienced by the ZnO film has negative direction of z-axis, which intern clearly identified through the negative value of Hall co-efficient due to the presence of electrons as the majority carriers and confirms the n-type nature. Similarly for PbSe film, the voltage experienced is in the positive z-axis indicated by the positive Hall co-efficient and the holes are the majority carriers responsible for p-type nature.

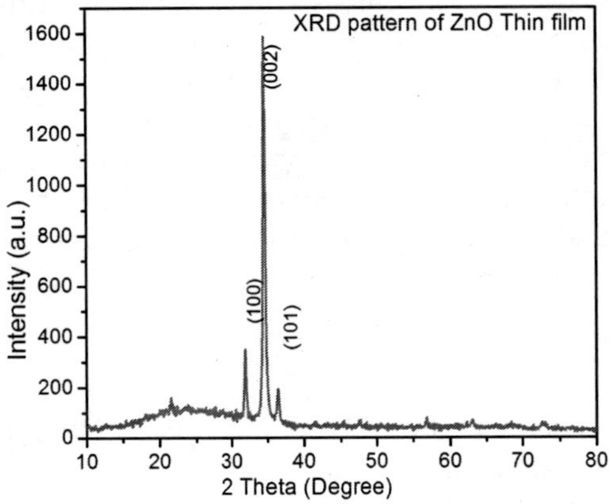

Fig. 2: X Ray Diffraction Pattern of ZnO Thin Film

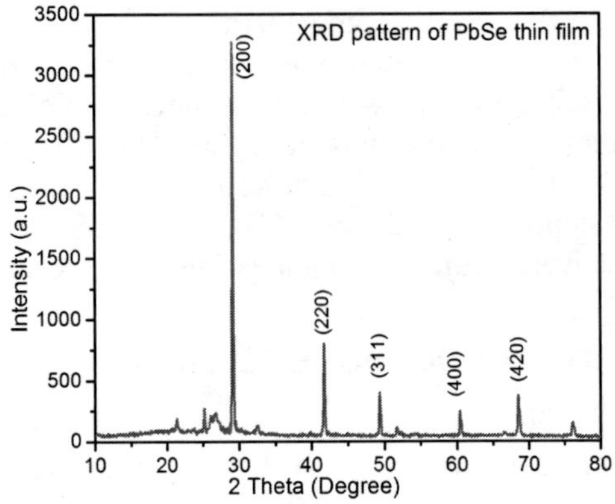

Fig. 3: X Ray Diffraction Pattern of PbSe Thin Film

Field Emission Scanning Electron Microscopy (FESEM) was used for imaging the samples.

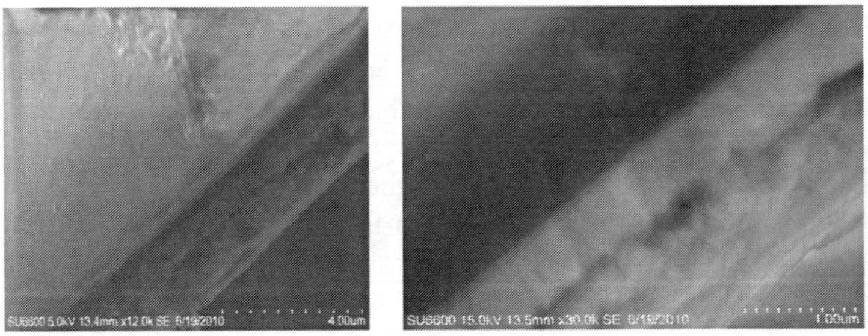

Fig. 4: FESEM Image of p-n Junction Diode from PbSe and ZnO Thin Film

The images (Figure 4) show that PbSe and ZnO layer coated over the ITO glass substrate, which also reveals the growth uniformity. The thickness of the films was measured to be 306 nm for ZnO and 393 nm for PbSe. The sintering effect influences the formation of pn junction between the layers of PbSe and ZnO, whose thickness is 104 nm.

The device performance was measured by semiconductor parameter analyzer (SPA) under bright and dark field light illumination as shown in Figure 5. The diode characteristics were identified in the dark field illumination. In bright filed, the leakage current plays a major role and responsible for photo generation carriers. The PbSe is a low band gap material and absorbs almost all electrons from the ZnO material. There is a further optimization required to improve the performance of the device. The bright filed illumination takes place under different illuminations like 40 w, 100w and 300w. There was no carrier generation at lower watt illumination condition and 300w was the optimized illumination. From the bright field illumination the efficiency was calculated and the device acts as a solar cell.

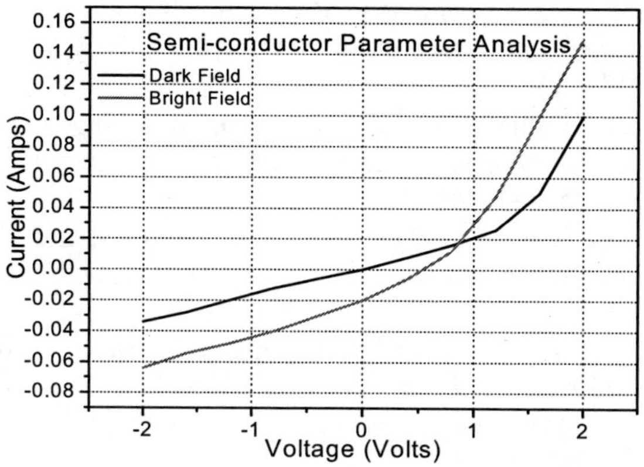

Fig. 5: Semi-Conductor Prameter Analysis of PbSe and ZnO Thin Film Device

CONCLUSION

The solar cell structure was fabricated using ZnO and PbSe thin films as active layers coated over ITO substrate. The low band gap p-type material (PbSe) was used to absorb most of the electrons emitted from the ZnO film, yet further optimization is required to improve the device performance. The crystallinity and the film formation were confirmed by XRD and FESEM analysis. The semi-conducting nature of the material was measured from Hall Effect and four probe method. The device performance was analyzed by using semi-conducting parameter analyzer (SPA), which clearly indicates the solar cell performance under bright and dark field light illumination.

ACKNOWLEDGEMENT

The authors would like to thank Prof. Nandita Dasgupta, IIT Madras and Prof. T. Balasubramanian, NIT Tirchy for extending the characterization facilities.

REFERENCES

[1] N Bouhssira, S Abed, E Tomasella, J Cellier, A Mosbah, M S Aida and M Jacquet, 'Influence of annealing temperature on the properties of ZnO thin films deposited by thermal evaporation', Appl. Surf. Sci., Vol. 252, (2006) pp. 5594–5597.

[2] Brian L Wehrenberg, Dong Yu, Jiasen Ma and Philippe Guyot-Sionnest, 'Conduction in charged PbSe nanocrystal films', J. Phys. Chem. B, Vol. 109, (2005) pp. 20192–20199.

[3] Fei Li, Zhen Li and Fujiang Jin, 'Fabrication and characterization of ZnO micro and nanostructures prepared by thermal evaporation' Physica B, Vol. 403, (2008) pp. 664–669.

[4] Feng Chen, Kevin L Stokes, Weilie Zhou, Jiye Fang and Christopher B Murray, 'Synthesis and properties of Lead Selenide nanocrystal solids', Mat. Res. Soc. Symp. Proc., Vol. 691, (2002) pp. 359–364.

[5] O A Fouad, A A Ismail, Z I Zaki and R M Mohamed, 'Zinc oxide thin films prepared by thermal evaporation deposition and its photocatalytic activity', Appl. Catal. B: Environ., Vol. 62, (2006) pp. 144–149.

[6] Hidero Unuma, Naoki Shigetsuka, Kanji masui, Jiro Nagao and Minoru Takahashi, 'Sintering of chemically synthesized Lead(II) selenide fine powders', J. Am. Ceram. Soc., Vol. 81, No. 2, (1998) pp. 444–446.

[7] Jin-Hong Lee, Kyung-Hee Ko and Byung-Ok Park, 'Electrical and optical properties of ZnO transparent conducting films by sol-gel method', J. Cryst. Growth, Vol. 247, (2003) pp. 119–125.

[8] Joshua J Choi, Yee-Fun Lim, Mitk'EI B Santiago-Berrios, Matthew Oh, Byung-Ryool Hyun, Liangfeng Sun, Adam C Bartnik, Augusta Goedhart, George G Malliaras, Hector D Abruna, Frank W Wise and Tobias Hanrath, 'PbSe nanocrystal excitonic solar cells', Nano Lett., Vol. 9, No. 11, (2009) pp. 3749–3755.

[9] KunWei Li, XiaoTian Meng, Xue Liang, Hao Wang and Hui Yan, 'Electrodeposition and characterization of PbSe films on indium tin oxide glass substrates', J. Solid State Electrochem., Vol. 10, (2006) pp. 48–53.

[10] Kurtis S Leschkies, Timothy J Beatty, Moon Sung kang, David J Norris and Eray S Aydil, 'Solar Cells based on Junctions between colloidal PbSe nanocrystals and thin ZnO films', acsnano, Vol. 3, No. 11, (2009) pp. 3638–3648.

[11] Liping Liu, Qing Peng and Yadong Li, 'Preparation of monodisperse Se colloid spheres and Se nanowires using Na_2SeSO_3 as precursor' Nano Res. Vol. 1, (2008) pp. 403–411.

[12] A Lipovskii, E Kolobkova, V Petrikov, I Kang, A Olkhovets, T Krauss, M Thomas, J Silcox, F Wise, Q Shen and S Kycia, 'Synthesis and characterization of PbSe quantum dots in phosphate glass', Appl. Phys. Lett., Vol. 71, No. 23, (1997) pp. 3406–3408.

[13] A Sashchiuk, L Amirav, M Bashouti, M Krueger, U Sivan and E Lifshitz, 'PbSe nanocrystal assemblies: Synthesis and structural, optical and Electrical characterization', Nano Lett., Vol. 4, No. 1, (2004) pp. 159–165.

[14] Shiyong Gao, Donghua Liu, Dan Xu, Dongmei Li, Youliang Hong, Haiyong Chen, Quanqin Dai, Shihai Kan, Hongdong Li and Guangtian Zou, 'Synthesis of Se nanospheres and PbSe nanoshells in solution under refluxing and stirring', Smart Mater. Struct., Vol. 16, (2007) pp. 2350–2353.

[15] Sumetha Suwanboon, 'Structural and optical properties of nanocrystalline ZnO powder from sol-gel method', ScienceAcia, Vol. 34, (2008) pp. 031–034.

[16] Xinjun Wang, Ke Li, Yutao Dong and Kai Jiang, 'Preparation and Characterization of monodispersed PbSe nanocubes', Cryst. Res. Technol., Vol. 45, No. 1, (2010) pp. 94–98.

[17] Xuan Wang, Qingwen Li, Zhibo Liu, Jin Zhang, Zhongfan Liu and Rongming Wang, 'Low-temperature growth and properties of ZnO nanowires', Appl. Phys. Lett., Vol. 84, No. 24, (2004) pp. 4941–4943.

[18] B.D. Yao, Y F Chan and N Wang, 'Formation of ZnO nanostructures by a simple way of thermal evaporation', Appl. Phys. Lett., Vol. 81, No. 4, (2002) pp.757–759.

COMPOSITES OF SINGLE WALLED CARBON NANOTUBES AND LIQUID CRYSTALS AS SWITCHABLE CONDUCTORS

M. VIJAY KUMAR AND S. KRISHNA PRASAD

Centre for Soft Matter Research, Jalahalli, Bangalore
E-mail: vijayphy@csmr.res.in

ABSTRACT

Realizing large anisotropic electrical conductivity in systems wherein the magnitude of the anisotropy can be switched and thus controlled by employing an external field such as magnetic, electrical or optical fields is of significant interest. An attempt in this direction has been to combine the large electrical conductivities possible with carbon nanotubes (CNT) and the field-switchable orientation of liquid crystals (LC), although stabilization of such composites is a difficult proposition. Recently we have performed measurements wherein at least over reasonable lengths of time the LC-CNT composites are stable without any serious segregation. Here we describe results of electrical conductivity, which establish that the LC-CNT composites have a bright future in attaining switchable electrical insulator-conductor devices, such as for example Mott insulator-conductor systems. The added advantage due to the presence of LC is the phenomenal ease with which large scale devices can be fabricated. The presentation also discusses possible means of increasing the loading factor, which currently is quite low.

INTRODUCTION

The research and development of nano materials in science and technology have received significant interest in the past decade.[1–4] From different dimensions carbon nanotubes (CNT) and liquid crystals have established their own niche world owing to many interesting properties, and a few similarities. Composites of the two materials are expected to yield

V. Rajendran, R. Yuvakkumar, K. Thyagarajah and K.E. Geckeler (eds.)
Applications of Nano Materials: Electronics, Energy and Environment, pp. 173–178 (2012).

extraordinary properties especially with the combination of switchability arising due to liquid crystallanity, and large mechanical and electrical anisotropy of CNTs. The anisotropy of the electric conductivity, being large along the tube axis and small perpendicular to it, has led to proposals of nano-tubes being used in the fabrication of electronic devices.[5–8] Carbon nanotubes occur in two general morphologies, single-wall (SWNT) and multi-wall nanotubes (MWNT), and are normally produced by the techniques of arc-discharge, chemical vapor deposition, or laser ablation.[9–11] For many applications a uniform alignment of CNTs is essential, and in general the ability of providing nanotubes a predetermined direction is of great importance. Additionally, it is desirable to be able to manipulate this direction, for example, by the application of external fields, such as electric, magnetic, or mechanical. Liquid crystals (LC) are anisotropic fluids, located between the isotropic liquid and three-dimensional positionally ordered solid states.[12] Due to molecular self-assembly LCs exhibit orientational order of e.g., rod-shaped molecules, resulting in anisotropic physical properties while maintaining flow properties in their least ordered structure—the nematic phase. The average direction of the long molecular axes called the director n, can easily be controlled by suitable surface treatment or external fields. The phenomenon of reorientation when achieved with an electric field, known as the Freedericksz transition,[13] is exploited in all common LCD applications from cheap wrist watches to the highly sophisticated laptop displays. Attempts have been made [14–20] to employ liquid crystals for imposing alignment on dispersed carbon nanotubes, and to use the Freedericksz transition to manipulate the alignment direction through elastic interactions with the liquid crystal director field. This approach opens the possibility to dynamically and collectively change the orientation direction of ensembles of nanotubes and thus their direction of mechanical and especially electrical anisotropy.

EXPERIMENTAL

For the experiments two liquid crystal compounds were chosen one of which (labelled SPLC) is strongly polar having a nematic (N) to isotropic (I) transition temperature (T_{NI}) at 35°C. The other liquid crystals (WPLC) is a weakly polar one and has a much higher T_{NI} at 123.8°C. Both compounds have a positive dielectric anisotropy $\Delta\varepsilon$ ($= \varepsilon_{\parallel} - \varepsilon_{\perp}$, where $\varepsilon_{\parallel}$ and $\varepsilon_{\perp}$ indicate the dielectric permittivity parallel and perpendicular to the nematic director). The used CNT having an average length of 500 nm and diameter 2 nm was procured from Heji. Several LC-CNT composites comprising low (<0.1wt%) CNT concentrations have been prepared. However, we present here the results on a representative mixture with 0.05wt% concentration. A novel method of preparing the mixture on the lines of the procedure described in an earlier paper [19] is employed. The procedure used results in the prepared composites appearing slightly dark in colour compared to the pure LCs materials with no sedimentation of CNT, a problem seen generally with LC-CNT mixtures. The samples were filled into prefabricated Indium Tin Oxide coated glass cells with the electrode dimension of 0.5cm ×1cm and a cell gap of 25μm through capillary action in the nematic phase. The cell was then observed under a polarizing optical microscope in the crossed polarizer configuration to ascertain the uniform dispersion of CNT. For dielectric and conductivity measurements, the cell was kept inside a homemade heater and placed between the pole pieces of an electromagnet with a maximum field of 1.6 T. The orientation of the cell normal being parallel or perpendicular to the maganetic field direction yields ($\sigma_{\parallel}$,$\varepsilon_{\parallel}$) or ($\sigma_{\perp}$,$\varepsilon_{\perp}$).The measurements were carried out using an impedance analyzer(HP 4294A) at five different frequencies spanning 100Hz –1MHz.

RESULTS AND DISCUSSION

Figures 1(a) and (b) depict the temperature dependence of conductivity in || and ⊥ directions for SPLC as well as WPLC along with their composites with CNT obtained at a frequency of 1 kHz. The composites with both the LC materials exhibit a significant enhancement of the conductivity upon addition of CNT. The magnitude of the enhancement is about 2 orders of magnitude for SPLC whereas it is a further one order higher with WPLC compound. The anisotropy in conductivity is still retained in the composites. Further for the SPLC composite there is a factor of 2.5 increase in the ratio $\sigma_a = \sigma_{\parallel}/\sigma_{\perp}$ with respect to that for the pure material. The WPLC case has the opposite behavior with the ratio σ_a decreasing by a factor of 2 for the composite [For purposes of better comparison the data in the isotropic phase obtained with the two orientations have been matched]. In fact, σ_a becoming smaller than 1 is a feature known to be associated with the appearance of smectic ordering or 1-dimensional periodicity parallel to the director. Since it is clear from microscopy observation that the system remains in the nematic phase, this feature suggests that the additional ordering is only short range. Detailed measurements for mixtures with different concentrations as well studies such as XRD are being planned to understand the difference in the behavior between SPLC and WPLC composites.

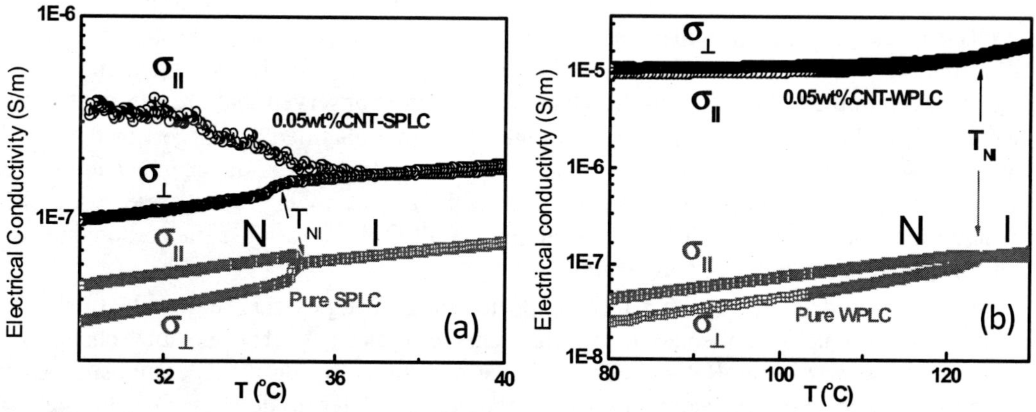

Fig. 1: Electrical Conductivity at 1 KHz as a Function of Temperature ∥ and ⊥ to the Nematic Director for Pure and Composites of (a) SPLC and (b) WPLC Materials

Figures 2(a) and (b) present the thermal behavior of $\varepsilon_{\parallel}$ and $\varepsilon_{\perp}$ for the two pure compounds and their composites. In both the cases the addition of CNT results in a significant increase in both $\varepsilon_{\parallel}$ and $\varepsilon_{\perp}$ as well in the permittivity of the I phase. However, the magnitudes of increase are definitely dependent on the polar character of the host material: for SPLC the increase in, for example the isotropic permittivity if ~ 25%, but for WPLC case it is a factor of 3. A possible reason for the amplified change in WPLC situation could be connected with the low frequency dielectric relaxation that the pure WPLC exhibits. Further work focusing on the dynamics of the director relaxation is expected to shed more light on this feature.

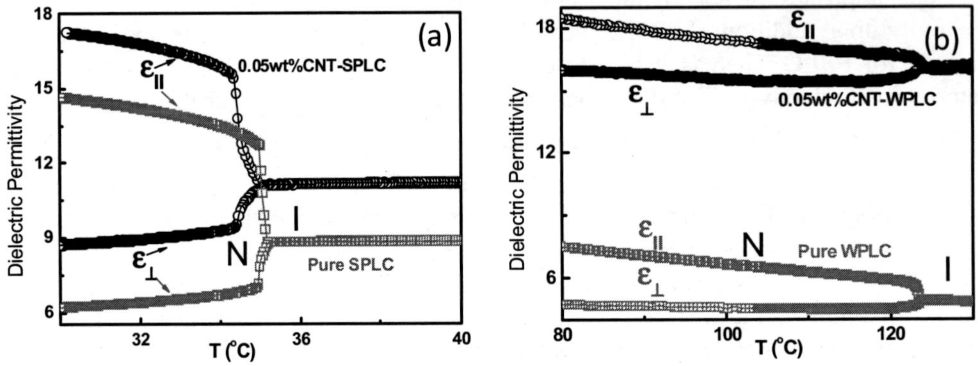

Fig. 2: Thermal Variation of the Dielectric Permittivity at 1 KHz along (‖) and Perpendicular (⊥) to the Director for Pure and Composites of (a) SPLC and (b) WPLC

Another interesting aspect observed is the temperature dependent permittivity behavior in the I phase of the strongly polar material. The pure material exhibits a concave shaped anomaly (CSA) that has been argued to be associated with the antiparallel arrangement of the molecules even in the I phase, but in the vicinity of T_{NI}. It is observed that the temperatures at which the anomaly has a maximum and the highest ε value obtained are larger for the SPLC composite than for the pure SPLC compound. We propose that this is direct indication of the enhancement of the nematic-like short range order in the I phase owing to the presence of CNT particles. Since CSA is not seen for even the pure WPLC, it is not possible to compare the effects between the two types of composites.

Preliminary studies indicate that the composites of both types can be switched using an electric field causing a reorientation of the nematic molecules. The notable changes in conductivity between the parallel and perpendicular directions for the SPLC composite makes it attraction for possible electrically controlled conductivity altering devices. Development of proper systems in which the change in conductivity is substantially higher than seen here would make the LC-CNT composites quite attractive.

CONCLUSIONS

We have reported anisotropic electrical conductivity and dielectric permittivity studies on two liquid crystals and their low concentration composites with SWCNT. The two liquid crystals are chosen such that one of them is strongly polar having its dipole moment parallel to the long axis of the molecule whereas the other one is weakly polar with its dipole in the transverse direction. Both conductivity and permittivity retain their anisotropic nature even in the composites. More importantly the conductivity exhibits large increase (as much as 3 orders of magnitude—for the composite with the weakly polar material. The qualitative behavior of the permittivity is also quite different upon inclusion of CNT. Along with the mentioned electrical switchable character which imparts a field-switchable conductivity value, these materials are quite promising for devices.

ACKNOWLEDGMENTS

The financial support by SERB, DST under the project No. SR/S2/CMP-0029/2011 is gratefully acknowledged.

REFERENCES

[1] Harris, P.J.F., Carbon Nanotubes and Related Structures, Cambridge. University Press, Cambridge, 1999.

[2] Saito, R., Dresselhaus, G. and Dresselhaus, M.S., Physical Properties of Carbon Nanotubes, Imperial College Press, London, 2001.

[3] Rao, C.N.R., Tatishkumar, B.C., Govindaraj, A. and M. Nath, *Chem Phys Chem* 2, 78 2001.

[4] Nalwa, H.S., Ed., Encyclopedia of Nanoscience and Nanotechnology, 15, Volume set (11–25 vols) (American Scientific Publishers, 2011).

[5] Baughman, R.H., Zakhidov, A.A. and W.A. de Heer, Science 297, 787, 2002.

[6] Tans, S.J., Verschueren, A.R.M. and Dekker, C., Nature Londond 393, 49 1998.

[7] Avouris, P., Hertel, T., Martel, R., Schmidt, T., Shea, H.R. and Walkup, R.E., Appl. Surf. Sci. 141, 201 1999.

[8] Thelander, C. and Samuelson, L., *Nanotechnology* 13, 108 2002.

[9] Iijima, S., Nature Londond, 354, 56 1991.

[10] Endo, M., Takeuchi, K., Igarashi, S., Kobori, K., Shiraishi, M. and H.W. Kroto, J. Phys. Chem. Solids, 54, 1841 1993.

[11] Thess, A., Lee, R., Nikdaev, P., Dai, H., Petit, P., Robert, J., Xu, C., Lee, Y.H., Kim, S.G., Rinzler, A.G., Colbert, D.T., Scuseria, G.E., Tomanek, D., J.E. Fischer and R.E. Smalley, Science 273, 483 1996.

[12] See for e.g., S. Chandrasekhar, Liquid Crystals, 2nd Edition, Cambridge University Press, 1992.

[13] Freedericksz, V. and Tsvetkov, V., Phys. Z. 6, 490 1934.

[14] Dierking, I., G. Scalia and P. Moralas, and D. LeClere, Adv. Mater., 16, 865, 2004.

[15] Lagerwall, J.P.F., G. Scalia, M. Haluska, U. Dettlaf, S. Roth, and F. Giesselmann, Adv. Mater., 19, 359, 2007.

[16] Basu, R. and G. Iannacchione, Appl. Phys. Lett., 93, 183105, 2008.
[17] Lee, W., C.Y. Wang, and Y. C. Shin, Appl. Phys. Lett., 85, 513, 2004.
[18] Shah, H.J., H. Pan, and C. Hsieh, Jpn. J. Appl. Phys., Part 1, 45, 6392, 2006.
[19] Jayalakshmi, V. and S. Krishna Prasad, Appl. Phys. Lett. 94, 202106, 2009.
[20] Muklesur Rahman and Wei Lee, J. Phys. D: Appl. Phys. 42 063001, 2009.

LOW TEMPERATURE SYNTHESIS AND THERMOLUMINESCENCE STUDIES OF UNDOPED AND Ce^{3+} DOPED $YAlO_3$ NANOCRYSTALLINE PHOSPHOR

H.B. PREMKUMAR[1,2], D.V. SUNITHA[1], H. NAGABHUSHANA[1*], S.C. SHARMA[1], S.C. PRASHANTHA[2], B.M. NAGABHUSHANA[3] AND R.P.S. CHAKRADHAR[4]

[1]*Centre for Nano Research (CNR), Tumkur University, Tumkur*
[2]*East West Institute of Technology, Bangalore*
[3]*Department of Chemistry, M.S. Ramaiah Institute of Technology, Bangalore*
[4]*National Aerospace Laboratories (CSIR), Bangalore*

ABSTRACT

$YAlO_3$ has a distorted perovskite structure with high effective atomic number (Z_{eff} = 31.4), and can be used in diverse applications such as dosimeter, optical recording media, scintillators and solid state lasers.[1,2] Undoped and Ce-doped (3 mol %) nanocrystalline $YAlO_3$ phosphor, prepared via low temperature combustion method, is described. An efficient phosphor can be prepared by this method in a muffle furnace maintained at 400°C in a very short time (few minutes). The phosphor is characterized by powder X-ray diffraction (PXRD), scanning electron microscopy (SEM), transmission electron microscopy (TEM), UV-Visible measurements etc. Thermoluminescence (TL) measurements have been carried out in undoped and Ce^{3+}doped phosphor after γ-irradiation in the dose range 0.5–6 kGy. The undoped sample has a well resolved glow peak at ~180°C along with shouldered peaks at 126, 240, 288 and 350°C respectively. However, in Ce $^{3+}$doped samples two peaks at 209°C and 308°C was recorded. It is observed that TL intensity increases with increase of γ-dose in both undoped and Ce^{3+}doped samples. The kinetic parameters at different γ-doses namely activation energy (E), order of kinetics (b), frequency factor (s) of Ce^{3+}doped samples was determined using Chens glow peak shape and Luschiks methods.

V. Rajendran, R. Yuvakkumar, K. Thyagarajah and K.E. Geckeler (eds.)
Applications of Nano Materials: Electronics, Energy and Environment, pp. 179–184 (2012).

INTRODUCTION

Yttrium aluminium perovskite YAP ($YAlO_3$) doped with rare earths or transition metals demonstrate very attractive properties being a promising material as a host matrix for laser, phosphors, scintillator, thermoluminescent detector, holographic recording medium and pigments.[1,2] YAP is one of four crystalline phases occurring in the system Al_2O_3–Y_2O_3 [3] presenting an orthorhombic perovskite structure with Y/Al ratio 1:1. Several studies [4] have shown that Ce-doped $YAlO_3$(Ce:YAP) is a good candidate as a g-ray detector. Ce: YAP is an interesting scintillator which exhibits fast scintillation, high light yield and very good mechanical and chemical properties.Recently, TL properties of pure and doped $YAlO_3$ have been studied in single crystals.[5] However, sufficient work on the TL studies of $YAlO_3$ in nano form up to now has not been done. In the present paper, we report $YAlO_3:Ce^{3+}$ nanophosphor prepared by the low temperature solution combustion method for the first time and the phosphor was well characterized by powder X-ray diffraction (PXRD), Fourier transform infrared (FTIR), Scanning Electron microscopy (SEM), Transmission electron microscopy (TEM), UV-Visible spectroscopy (UV-Vis) and Thermoluminescence (TL) studies. From TL, the kinetic parameters for different γ- doses namely activation energy (E), order of kinetics (b),_and frequency factor (s) were determined using Chens glow peak shape method and the results are discussed in detail.

EXPERIMENTAL

$YAlO_3$: Ce^{3+} (3 mol %) nanophosphor was prepared using aluminum and yttrium nitrates in stoichiometric proportions with oxalyl dihydrazide (ODH) fuel . ODH to Nitrate ratio was calculated as described elsewhere.[6] The synthesis was carried out as follows: first, the nitrates and ODH were mixed in a Pyrex dish using double distilled water and mixed uniformly using magnetic stirrer for about 5 min. Thereafter, the Pyrex dish was placed in a furnace preheated at ~ 400 ± 10°C. The mixture underwent dehydration at lower temperatures and decomposition resulting in deflation with the simultaneous evolution of large amounts of gases. After this step, the foam was automatically ignited, giving a voluminous and fluffy product. The whole process lasted within 5 min. The dish was then taken out of the furnace, and the foamy product was crushed into a fine powder and this powder was calcined at 1000°C for 3 hrs.

The phase purity and the crystallinity of the nanophosphors were examined by powder X-ray diffractometer (PANalytical X'Pert Pro) using Cu K_α(1.541 Å) radiation with a nickel filter. The surface morphology of the product was examined by Scanning Electron Microscopy (SEM) (JEOL JSM 840A). Transmission Electron Microscopy (TEM) analysis was performed on a Hitachi H-8100 (accelerating voltage up to 200 kV, LaB_6 filament) equipped with EDS (Kevex Sigma TM Quasar, USA).The FT-IR studies were performed on a Perkin Elmer Spectrometer (Spectrum 1000) with KBr pellets. The UV-VIS absorption of the samples was recorded on SL 159 ELICO UV-VIS Spectrophotometer. TL measurements were carried out at room temperature using Nucleonix TL reader, and using gamma-irradiation as excitation in the dose range 0.5–6 kGy.

RESULTS AND DISCUSSION

The powder X-ray diffraction patterns of the un-doped and 3 mol% Ce^{3+}doped $YAlO_3$ nanopowders are shown in Figures 1 and 2 respectively. The PXRD patterns exhibit dominant diffraction peaks due to the orthorhombic phase of $YAlO_3$ (JCPDS File No. 70–1677) along with $Y_3Al_5O_{12}$ phase.[7] The crystallite size (d) of un-doped and Ce^{3+} doped $YAlO_3$ samples were estimated using Debye-Scherer's formula $D = 0.9\lambda/\beta \text{Cos}\theta$, where D is the average grain size of the crystallites, λ is the incident wavelength, θ is the Bragg angle and β is the diffracted full width at half maximum (in radian) caused by the crystallites.[8] The average size of these nanophosphors is estimated to be approximately 45–80 nm.

Scanning Electron Microscopy (SEM) analysis was carriedout to study the morphology the nanophosphors and is shown in Figure 3. SEM micrograph of $YAlO_3$:Ce^{3+} showed that the particles are, agglomeration, crispy-foam with almost circular in shape. The TEM image further confirms that the $YAlO_3$: Ce^{3+} particles are fused agglomerates with irregular shaped particles, the average size of the particles varies in the range 45–90 nm. This is in good agreement with the value obtained from Scherrer's formula.

The UV-Vis spectra were obtained in the wavelength region of 200–750 nm for Ce^{3+} doped and un-doped $YAlO_3$ sample, the absorption peaks were observed at ~257–304 nm, along with broad absorption at ~459 nm. However, in un-doped sample, the absorption peaks were observed at ~250–287 nm. It is observed that, the absorption peaks are red shifted in Ce^{3+}doped sample. The maximum absorption (257 nm), can arise due to transition between valence band to conduction band. The weak absorption in the UV-visible region is expected to arise from transitions involving extrinsic states such as surface traps or defect states or impurities.[9]

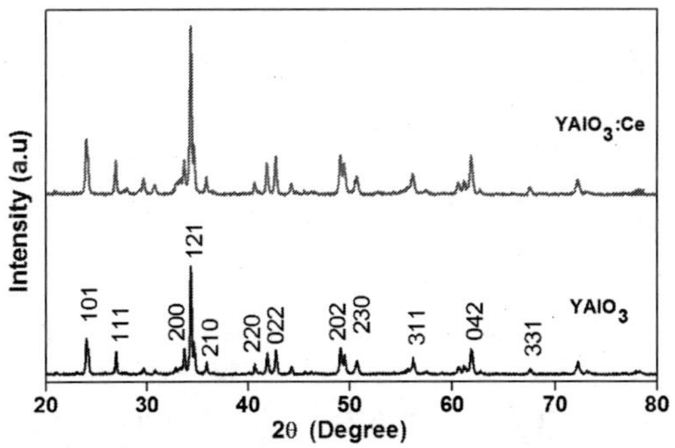

Fig. 1

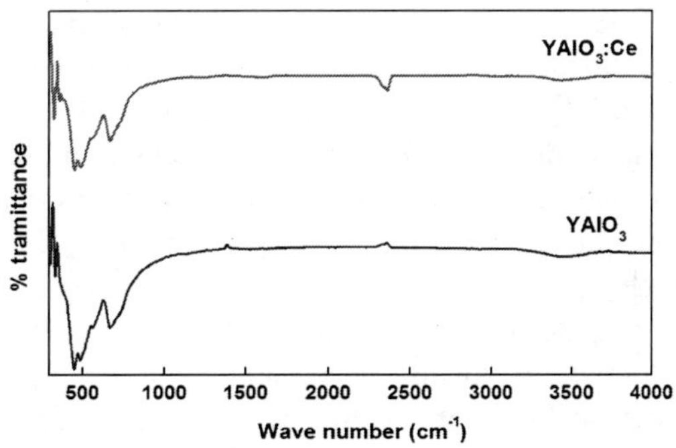

Fig. 2

The band gap in un-doped and Ce^{3+} doped $YAlO_3$ was estimated using Tauc relation [10] and the details are given elsewhere.[11] Band gap energy of 5.55 eV and 6.03 eV were observed for Ce^{3+} doped and un-doped $YAlO_3$ respectively (Figure 6). It has been reported that reduction in E_g for Ce^{3+} doped should arise neither from a quantum size effect nor from crystal structure variation but rather from surface traps states or point defects.[12] The large amount of surface defects can exist on the as-prepared $YAlO_3$ nanostructures due to their high surface area.

A slight increase in E_g can be attributed to the defect states.

Fig. 3

FT-IR spectra of un-doped and Ce^{3+} doped $YAlO_3$ samples are shown in Figure 2. The strong bands between 400–800 cm^{-1} are typical metal–oxygen (Y–O and Al–O) vibrations for the perovskite structure compounds.[13] The broad band around 3450 cm^{-1} and a narrow band at 1635 cm^{-1} were assigned to stretching and bending modes of adsorbed water, respectively.

TL glow curves of 0.5–6 kGy γ-irradiated $YAlO_3$: Ce^{3+}nanophosphor, warmed at a heating rate of 3°C s^{-1} is shown in Figure 4. TL glow curve consists of two TL glow peaks with the main peak at ~209°C and a shouldered peak at 308°C. The variation of TL glow peak intensity with accumulated dose was studied and the TL intensity increases linearly for 209°C and 308°C peaks up to 1 kGy and thereafter TL intensity increases sub-linearly up to 6 kGy. This linear/sub-linearity/saturation may be explained on the basis of track interaction model.[14] It is reported that in this model, the number of generated traps due to irradiation depends upon both the length of tracks in matrix and area of cross-section of tracks. In case of nanomaterials there still exist some particles that would have missed-due to very tiny size-while being targeted by high energy radiation. Thus on increasing the dose these nanoparticles which had earlier been left out from the radiation now generate trapping and luminescent centers. So we do not get saturation in nanomaterials even at higher doses. This same phenomenon may also be explained on the basis of large surface to volume ratio which results in a higher surface barrier energy for nanoparticles. Thus with the increase in doses, the energy density crosses the barrier and a large number of defects produced in the nanomaterials ultimately keep on increasing with the dose till saturation is achieved.

For TL dosimetric applications, a clear knowledge of kinetic parameters such as trap depth (*E*) and frequency factor (*s*) and order of kinetics are essential. Therefore, the obtained curves are deconvoluted using origin 8.1 software. Then the order of kinetics (*b*) the form factor μ_g ($\mu_g = (T_2 - T_m)/(T_2 - T_1)$) which involves T_1 and T_2 and activation energy was calculated using Chen's glow peak shape method.[15]

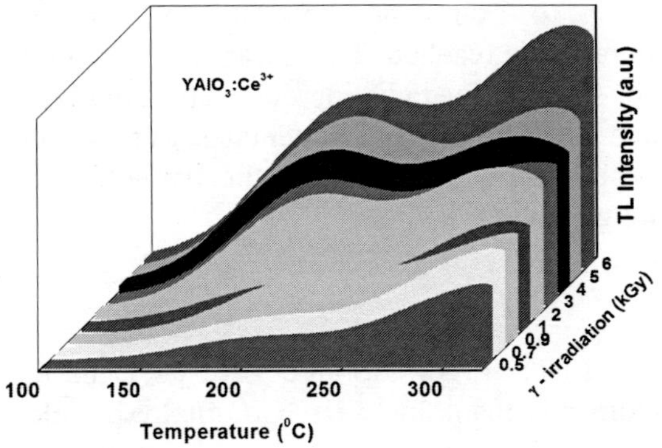

Fig. 4

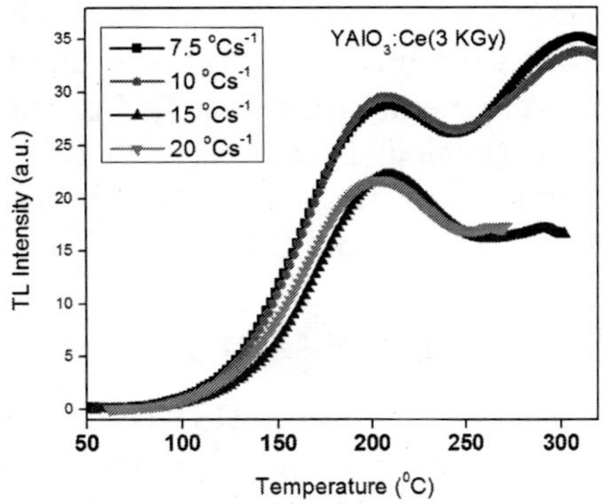

Fig. 5

In order to compare the kinetic parameters Luschiks [16] method has been used to estimate the activation energy (E) and frequency factor (s) using the second order energy equations. The estimated parameters for both the methods are given in Table.

Table

γ-dose (kGy)	*Peak*	T_m*(°C)*	*b* (μ_g)	*Chen's method*	*Luschiks method*
				E_{ave} *(eV)*	E_{ave} *(eV)*
0.5	1	147	2(0.49)	0.986	0.980
	2	226	2(0.49)	1.282	1.240
1.0	1	168	2(0.51)	0.618	0.640
	2	222	2(0.50)	1.145	1.260
3.0	1	180	2(0.50)	0.756	0.720
	2	223	2(0.50)	1.317	1.3 60
	3	284	2(0.49)	1.811	1.990
6.0	1	169	2(0.49)	0.988	0.920
	2	220	2(0.50)	1.8 14	1.8 40
	3	287	2(0.51)	0.887	0.820

CONCLUSIONS

Ce^{3+}-doped $YAlO_3$ perovskite was prepared by a simple and inexpensive combustion method in a few minutes at low temperature (400±10°C). The average crystallite size was estimated to be in the range of 45–90 nm from the broadenings of the XRD peaks and which is in good agreement with TEM results. The crystallites are almost spherical in shape with agglomeration. TL glow curves of 0.5–6 kGy γ-irradiated$YAlO_3$: Ce^{3+}nanophosphor consist of two TL glow peaks with the main peak at ~209°C and a shouldered peak at 308°C. The TL intensity increases linearly for 209°C and 308°C peaks up to 1kGy and thereafter TL intensity increases sub-linearly up to 6 kGy and finally saturation is reached. The linear response up to 0.5–1 kGy might make $YAlO_3$: Ce^{3+}nanophosphor suitable for radiation dosimetry.Finally, it can be concluded that the method reported here is a simple experimental set-up, which consumes surprisingly short time between the preparation of reactants and the availability of the final product, and at low cost coupled with energy saving.

ACKNOWLEDGEMENTS

The author (H.N.) thanks **DST Nano Mission** (Project No. SR/NM/NS-48/2010) New Delhi for sanction of the Project. The author H. B. Premkumar is thankful to **Dr. Channakeshavulu,** Principal, EWIT, for his constant encouragement.

REFERENCES

[1] Loutts, G.B., M. Warren, L. Taylor, R.R. Rakhimov, H.R. Ries, G. Miller, M.A. Noginov, M. Curley, N. Noginova, N. Kukhtarev, H.J. Caulfield, P.Venkateswarlu, Phys. Rev. B 57 (1998) 3706 –3709.
[2] Noginov, M.A., N. Noginova, M. Curley, N. Kukhtarev, H.J. Caulfield, P. Venkateswarlu, G.B. Loutts, J. Opt. Soc. Am. B 15 (1998) 1463 –1468.
[3] Medraj, M., Hammond, R., Parvez, M. A., Drew, R. A. L. and Thompson, W. T., High temperature neutron diffraction study of the Al_2O_3–Y_2O_3 system. *Journal of the European Ceramic Society*, 2006, 26, 3515–3524.
[4] Mares, J.A., M. Nikl, K. Blazek, Phys. Status Solidi A 127 (1991) K605.
[5] Ya. Zhydachevskii, A. Durygin, A. Suchocki, A. Matkovskii, D. Sugak, P. Bilski, S. Warchol, Nucl. Instr. Meth. B. 227 (2005) 545–550.
[6] NeetuAhlawat, SujataSanghi, AshishAgarwal and RajniBala, J. Molec. Struct. 963 (2010) 82–86.
[7] Zhensen Liu, Junfeng Ma, Yong Sun, Zuwei Song, Jingrui Fang, Ye Liu, Chang Gao, Jingang Zhao, Ceram. Int. 36 (2010) 2003–2006.
[8] Klug, P., L.E. Alexander, X-Ray Diffraction Procedure, Wiley, New York, 1954.
[9] Cao, H.Q., X.Q. Qiu, B. Luo, Y. Liang, Y.H. Zhang, R.Q. Tan, M.J. Zhao, Q.M. Zhu, Adv. Funct. Mater. 14 (2004) 243–246.
[10] Tauc, J., Optical Properties of Solids, in: F. Abeles (Ed.), North-Holland, Amsterdam, 1970.
[11] Manjunatha, C., D.V. Sunitha, H. Nagabhushana, B.M. Nagabhushana, S.C. Sharma, R.P.S. Chakradhar, Spectrochim. Acta A (2012) Inpress.
[12] Dhananjaya, N., H. Nagabhushana, B.M. Nagabhushana, B. Rudraswamy, C. Shivakumara, K. Narahari, R.P.S. Chakradhar, Spectrochim. Acta A 86 (2012) 8–14.
[13] Adak, A.K., P. Pramanik, Mater. Lett. 30 (1997) 269–273.
[14] Sahare, P.D., RanjuRanjan, Numan Salah and S.P. Lochab, J. Phys. D: Appl. Phys. 40 (2007) 759 –764.
[15] Chen, R., Y. Krish, Analysis of thermally stimulated processes, New York: Pergamon, 1981.
[16] Lushchik, C.B., Sov. Phys. JETF, **3**(1956) 390–399.

MICROBIAL FUEL CELL APPLICATION USING SULFONATED POLY ETHER ETHER KETONE/ Fe_3O_4 NANOCOMPOSITE ELECTROLYTE

PRABHU NARAYANASWAMY VENKATESAN AND SANGEETHA DHARMALINGAM

Department of Chemistry, Chennai
E-mail: Sangeetha@annauniv.edu

ABSTRACT

Polymer Nano composites are promising alternative polymer electrolyte membranes for fuel cells because of their higher proton transfer, cost reduction and improved thermal, mechanical stabilities and gas barrier properties. E. coli, electrogenic bacteria was used as a biocatalyst in anodic chamber with air cathode using Pt/C catalyst. Four different compositions of SPEEK/Fe_3O_4 membranes (5%, 10%, 15%, 20%) were fabricated by solvent evaporation method. Water absorption, ion exchange capacity of the fabricated polymer nano-composite membranes were studied and also characterized by FT-IR, XRD, Profilometer and SEM. Water absorption of the composites was reduced with increase in the Fe_3O_4 content. The characteristic Fe_3O_4 vibrational stretch at 517 cm^{-1} was confirmed from the AT–IR spectrum. XRD spectra of the fabricated polymer composites showed six characteristic peaks of Fe_3O_4 (2θ = 30.1, 35.5, 43.1, 53.4, 57.0, and 62.6°). The maximum power obtained in a 25 cm^2 Single Chamber Air Cathode Microbial Fuel Cell (SCMFC) was compared with that of commercial available Cation Exchange Membrane (Nafion®).

INTRODUCTION

Exhausted use of fossil fuel in recent years made us to think of alternate energy. Even though other renewal energy sources such as solar, wind, hydrothermal sources are possible it requires more space and are seasonal. Researches in the field of Power generation from renewable sources are extensively going all over the world to overcome energy crisis in near future.[1]

V. Rajendran, R. Yuvakkumar, K. Thyagarajah and K.E. Geckeler (eds.)
Applications of Nano Materials: Electronics, Energy and Environment, pp. 185–190 (2012).

Microbial fuel cells (MFC) are the one among the renewable alternatives where research are going on not only to overcome electricity crisis but also for various factory and domestic waste water treatments[2] MFC is a reliable system that may produce limited power or bio hydrogen from different kinds of electron donors. Simultaneous bioelectricity generation and waste water treatment is considered as one of the most important applications of MFC.

Traditional Microbial fuel cell consists of anodic and cathodic chamber. Substrate oxidation into proton and electron by microorganism (Bacteria) takes places in the anodic chamber. Produced electrons reached the anode surfaces and then reach the cathode through the external circuit. The protons produced in the anodic chamber are moved towards the cathode through the proton exchange membrane (separator) and at the cathode chamber water produced due to electron proton reaction. Oxidizing agents such as Potassium permanganate, potassium dichromate in cathodic increases the performance of Microbial fuel cell. Separators play a major role in increasing the efficiency of microbial fuel cell.[3] In addition to the increase in the efficiency of microbial fuel cell membranes get deformed while MFC was run for longer time (ref.). Magnetite the addition of composites decrease the oxygen cross over and substrate loss during the microbial fuel cell operation. having sulfonated polyether ether ketone having less oxygen mass transfer coefficient (K_O) compared to that of Nafion 117®.[4]

EXPERIMENTAL

The materials that were used in the study were procured commercially from different sources. PEEK (Mol. Wt. 1,00,000) from Victrex while the the sulfonating agent H_2SO_4, D-Glusose, Nutrient broth and N-Methyl Pyrollidone was purchased from Merck. Fe_3O_4 nanoparticles was procured from Sigma Aldrich. The Chemicals were used for the study without any further purification. SPEEK membrane with 5%, 10%, 15%, 20% of Fe_3O_4 were fabricated and characterized with FTIR (Alpha T Bruker Optics) for the presence of functional groups, XRD (Xpert Pro, Pan Analytical) to study the change in crystallinity, SEM (HITACHI S-3400N), were used to study the morphology of the membranes.

SULFONATION OF PEEK

PEEK polymer was sulfonated using sulfuric acid as sulfonating agent. The weighed amount of PEEK was dissolved in concentrated sulfuric acid and magnetically stirred for five hours. After 5 hours, the reaction mixture was poured into cold water and the sulfonated PEEK (SPEEK) was obtained in the form of white precipitate. The SPEEK was washed with deionized water for several times until the pH become neutral. The SPEEK obtained from the above process was dried in a vacuum oven at 80°C overnight. It was then dissolved in a suitable quantity of NMP and cast onto a clean, dry petri dish.

The membrane was obtained by evaporating the solvent in vacuum oven at 80°C for 24 hours. The obtained membranes were pale brown in colour and were peeled off from the dish and stored for further analysis.

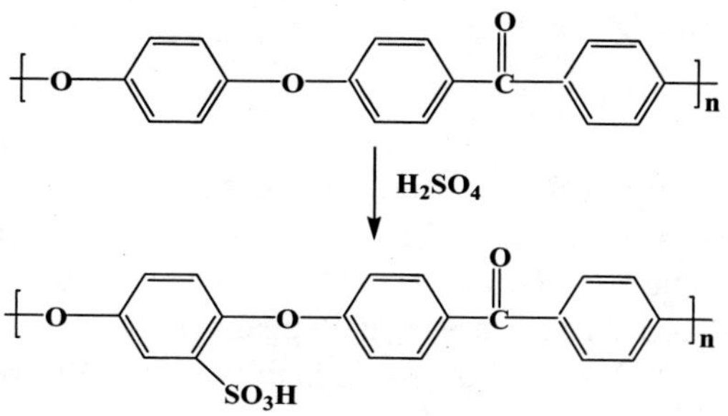

Preparation of Composites

The SPEEK was dissolved in the solvent NMP and different proportions of magnetite was added to the solution and mechanically stirred for 24 hours and then ultrasonicated for about 15 minutes, followed by casting into a clean petri dish to obtain membrane of required thickness. The cast membrane was dried at 80°C for 24 hours.

MFC Construction and Operation

A batch model bottle MFC was fabricated using media bottle (250 ml capacity, Borosil) with 2.5 cm membrane diameter (electrode spacing of 6 cm) and E.coli as bacterium in the anodic chamber.

Membrane electrode assembly (MEA) was prepared by the following method. The carbon cloth was 30% wet proofed and then carbonized using Vulcan X C–72 (Arora Mathey, Kolkata) (3mg/cm2) and heat treated for three hours at 250o C for 2 h. The catalyst coated carbon cloth was then hot pressed on the membrane at 80o C and 1.5 ton pressure for 3minutes to obtain MEA for cathode. The anode chamber was continuously flushed with N2/CO2 (80:20) to maintain anaerobic conditions as well as the pH of the growth medium at 7.

E. coli strains used were from our pre adapted laboratory culture collection. Under strict anaerobic conditions the cells were cultured using bicarbonate buffered defined medium under N2/CO2 (80:20) at 25°C.

Fuel Cell Analysis

Cell voltage and current were observed and recorded using a precision digital multimeter (Model 702, Metravi, India). The circuit was completed with a resistor of 1000 Ω and used to determine the power generation as a function of load. Current was calculated as

$$I = V/R$$

using Ohms's law where, I (mA) is the current, V (mV) is the voltage and R is the external resistance (Ω). Power (P) was calculated as,

$$P = IV$$

The power density curves were obtained after stable voltage was obtained.

RESULTS AND DISCUSSION

FTIR Analysis

The FTIR spectra of the composite membrane were shown in figure 1. The FTIR spectra show the characteristic peaks of Sulfonated PEEK polymer and the presence of composite in the polymeric membrane. The stretch at 1591, 1485, 1382, 1178 and 760 cm^{-1} corresponds to the characteristic peak of Fe_3O_4. The characteristic peaks of SPEEK at 1750 cm^{-1} corresponds to C = O stretch and peak at 1125 cm^{-1} corresponds to O = S = O were also observed.

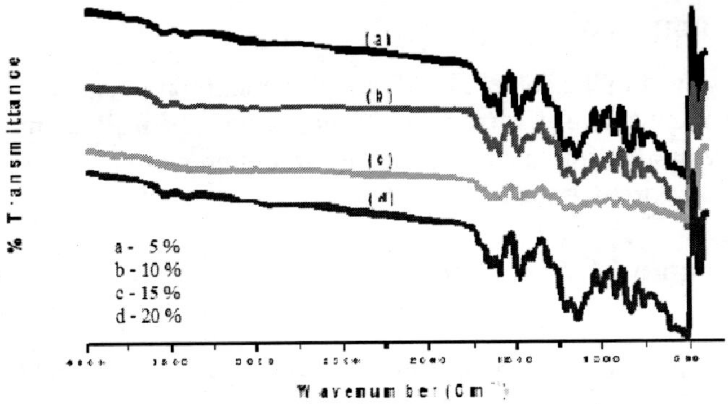

Fig. 1: FTIR Spectra of SPEEK/Fe_3O_4 Composite Membranes

XRD

XRD spectra of the composite membrane were shown in figure.The XRD patterns of composite membranes show the characteristics patterns of Fe_3O_4. The spectra show the characteristic XRD patterns of iron oxide at 2 theta values of 30, 35, 57, 65 and the amorphous nature of the polymer get reduced while increasing the concentration of iron oxide.

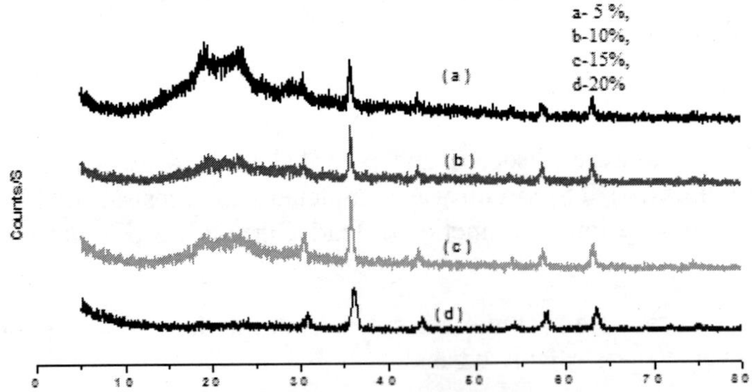

Fig. 2: XRD Patterns of SPEEK/Fe_3O_4 Composite Membranes

SEM

SEM images shows the presence of composites on the surface of SPEEK membrane with uniform dispersion and some aggloremation was observed on higher weight percentage of composite.

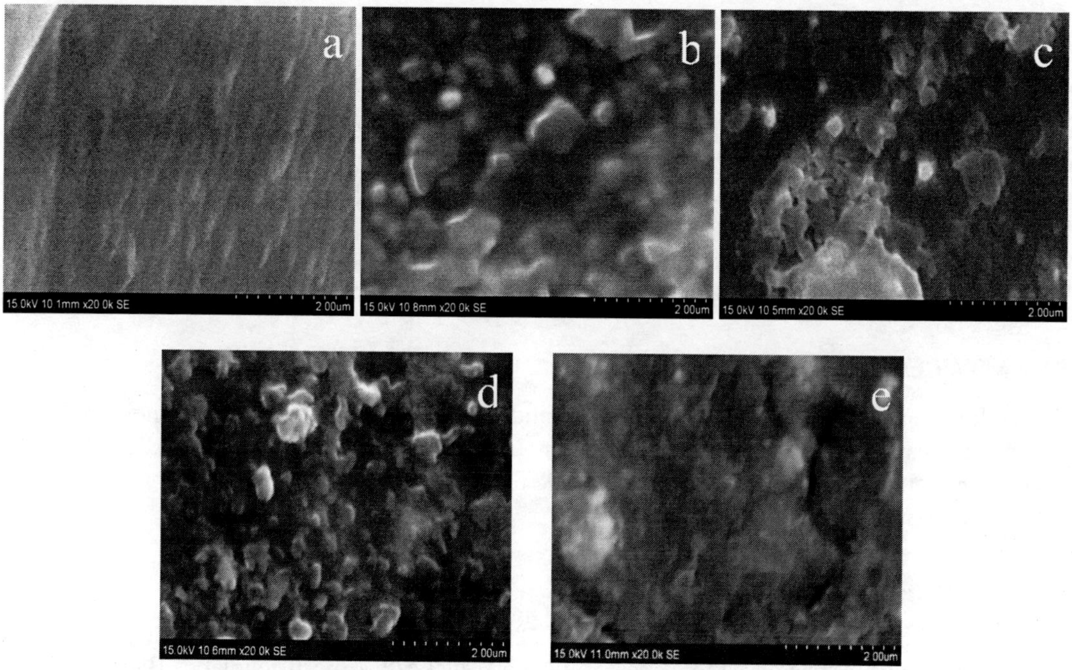

Fig. 3: SEM Images of SPEEK/Fe_3O_4 Composite Membrane

OCP and MFC Performance

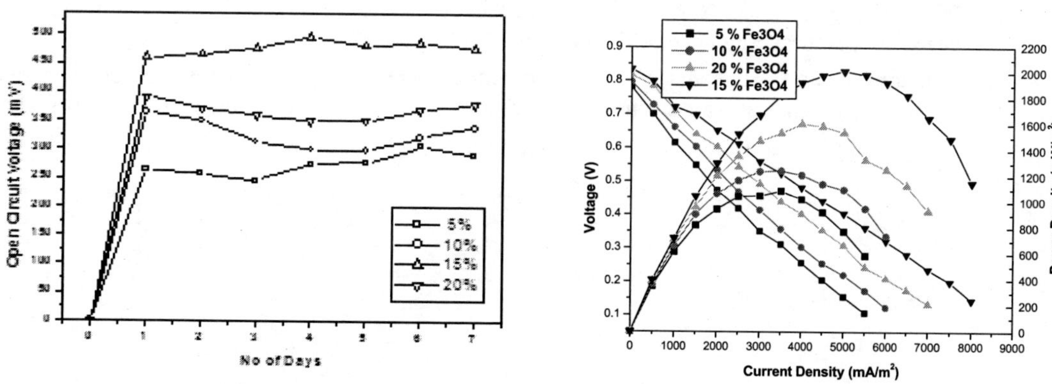

Fig. 4: OCP and Power Density of SPEEK/Fe_3O_4 Composite Membranes

The Open Circuit Potential (OCP)of the fabricated composite membranes was seen over a period of seven days and fuel cell performance was seen after the OCP was get stabilised. The composite membrane with 15% of Fe_3O_4 showed higher OCP (500 mV) and power density (2000 mW/m^2) compared to the other fabricated composite membranes.

CONCLUSION

The fabricated composite membranes are characterized using various techniues such as FTIR, XRD and SEM to analysis the modification in the chemical structure and morphology of the composite membranes. The fabricated membrane with 15% of Fe_3O_4 have higher OCP and power density compared to that of other fabricated membranes with different weight percentages of Fe_3O_4. This was attributed the optimum level of Ion exchange capacity with the composite and proton conductivity of the membrane. The results of this study showed, the low cost, non-fluorinated composite membrane is a promising candidate for PEM MFC.

ACKNOWLEDGEMENT

The authors thank the Department of Science and Technology (DST) India, for their financial support to carry out this work vide letter No. DST/TSG/AF/2010/09, dt.01-10–2010. Instrumentation facility provided under FIST-DST and DRS-UGC to Department of Chemistry, Anna University is sincerely acknowledged.

REFERENCES

[1] Mokhtarian, N., Rahimnejad, M., Najafpour, G.D., Wan Daud, W.R. and Ghoreyshi A.A., Effect of different substrate on performance of microbial fuel cell, African Journal of Biotechnology, 11, 3363–3369, 2012.

[2] Aelterman, P., Rabaey,K., Clauwaert, P. and Verstraete, W., Microbial fuel cells for wastewater treatment, Water Science and Technology, 54, 9–15, 2006.

[3] Li, W.W., Sheng G.P., Liu X.W. and Yu, H., Recent advances in the separators for microbial fuel cells, Bioresource Science and Technology, 102, 244–52, 2011.

[4] Sivasankaran, A. and Sangeetha, D., Development of MFC using sulphonated polyether ether ketone (SPEEK) membrane for electricity generation from waste water, Bioresource Technology 102, 11167–11171, 2011.

EFFECTS OF OXIDE NANOPARTICLES ON DIESEL-ETHANOL BLEND IN CI ENGINE

S. VENKADESWARAN, S.C.G. KIRUBA DANIEL[1], KASI NEHRU[2] AND MUTHUSAMY SIVAKUMAR[1]

Department of Automobile Engineering, Anna University of Technology, Tiruchirappalli
[1]Department of Nanoscience and Technology, Anna University of Technology Tiruchirappalli
[2]Department of Chemistry, Anna University of Technology, Tiruchirappalli
E-mail: muthusiva@gmail.com

ABSTRACT

Effects of SiO_2 and TiO_2 nanoparticles on diesel-ethanol blend are studied based on emission and efficiency characteristics of a Compression Ignition (CI) engine. SiO_2 and TiO_2 nanoparticles were synthesized by sol-gel synthesis and characterized using UV-Vis spectroscopy and Atomic Force Microscopy. The size of the synthesized SiO_2 and TiO_2 nanoparticles were 15–25 nm and 20–30 nm respectively. In these experiments SiO_2 or TiO_2 nanoparticles added to diesel ethanol blends were run in a four stroke diesel CI engine operating at 1500 rpm. The Specific Fuel Consumption (SFC) decreases relatively for upto 1.7 % for diesel-ethanol blend with SiO_2 nanoparticles compared to neat diesel and Brake Thermal Efficiency (BTE) has increased by 5% compared to neat diesel. Furthermore, from the analysis of exhaust gas the smoke absorption coefficient decreases upto 65% for diesel-ethanol blend with SiO_2 nanoparticles. The oxide nanoparticles are found to enhance complete combustion and reduce the exhaust emissions significantly.

INTRODUCTION

In our experiments we have chosen diesel ethanol blends with oxide nanoparticles. Oxide nanoparticles can be easily dispersed in ethanol and also with little amount of ethanol can significantly improve the combustion characteristics of the blend.[4] Diesel ethanol blends do

V. Rajendran, R. Yuvakkumar, K. Thyagarajah and K.E. Geckeler (eds.)
Applications of Nano Materials: Electronics, Energy and Environment, pp. 191–196 (2012).

not require any modification of engine design and also it is easy to produce complete ethanol blends with diesel. Diesel fuel additives are used to lower HC, CO, NO_x emissions and to enhance oxidation rates. Oxide nanoparticles shows promise in improving emission reductions due to enhanced oxidation rates.[5]

Generally diesel engines are more reliable and economical compared to petrol engines. Since the fuel prices are going up, there is an urgent need for a better efficient fuel and also there is a need to reduce the emissions of noxious gases to protect the environment from pollution. The market share of diesel vehicles has increased in many countries. However the problem with diesel engine is its lower thermal efficiency and higher emission.[1–2] In this paper investigation is being carried out to overcome the above said problems by using oxide nanoparticles. Because the nanoparticles are small having very large surface area and their properties are significantly different from microsized particles, they can be utilized as efficient catalyst. In earlier studies, various metals have been introduced in solid propellant formulations, gas propellants and solid fuels.[3] Mainly oxide nanoparticle plays a vital role during combustion due to shorter ignition delay and shorter burn times.

EXPERIMENTAL

All the chemicals used were obtained from Merck (India) and used as such. To synthesize TiO_2 nanoparticles, 50 ml of 9 M NaOH aqueous solution was placed in a two neck flask equipped with a condenser and heated in an oil bath upto 110°C. 10 g bulk TiO_2 powder was added to the above solution and stirred magnetically. After 20 hrs, the reaction mass was cooled down to room temperature, a precipitate was obtained which was separated and washed with miliQ water repeatedly until the conductivity of the supernatant reached 0.8 us/cm. Then the precipitate was soaked in 0.1 M HCl solution for 10 min, after which the TiO_2 nanoparticles was dried by hot air oven at 60°C for 2 hrs and was stored. For the synthesis of SiO_2 nanoparticles, 15 ml of ethanol was added to 50 ml water and sonicated for 10 min. Then 2 ml of Tetraethyl ortho silicate (TEOS) was added to the above solution and continued sonication for 20 min. 1ml of ammonia was added and again sonicated for 60 min. Resultant SiO_2 nanoparticles was obtained by drying the ethanol-water mixture. Both the oxide nanoparticles were characterized by UV-Vis spectroscopy and AFM.

We have used diesel Kirloskar engine for our fuel enhancement studies. The engine specifications are listed in Table 1. An eddy current dynamometer was coupled with the engine. The engine exhausts were measured by smoke meter. In order to study the effects of adding oxide nanoparticles to diesel-ethanol blends, tests were conducted at 1500 rpm. Initially all tests were carried out with diesel fuel. The nanoparticles are added to diesel-ethanol blends with the help of stirring. Here we have chosen three types of combinations are as follows:

1. Diesel + Ethanol
2. Diesel + Ethanol + TiO_2 nanoparticles
3. Diesel + Ethanol + SiO_2 nanoparticles

To minimize the effects of different fuels on each other, the engine was allowed to operate for 15 min to clean the fuel system before testing each fuel. In each test mode the engine was allowed to operate for a few minute to reach stable conditions. After stable conditions the time taken for a run is noted. From this, we can calculate the Specific Fuel Consumption (SFC), Brake Thermal Efficiency (BTE) and by using smoke meter, smoke absorption coefficient was calculated.

Table 1: Specification of the Diesel Engine

Brake Power	3.7 kw
Speed	1500 rpm
Bore	80 mm
Stroke	110 mm
Ignition	Compression ignition
Cooling	Water cooling
Loading System	Electrical dynamometer
No. of cylinder	Single

RESULTS AND DISCUSSION

The synthesized TiO_2 and SiO_2 nanoparticles exhibited significant peaks at 436 and 213 nm (Figure 1) respectively in UV-Vis spectroscopy. Similarly Atomic Force Microscopy images confirmed the presence of nanoparticles in the size range of 20–50 nm (Figure 2). The synthesized nanoparticles were added to ethanol for uniform dispersion in diesel.

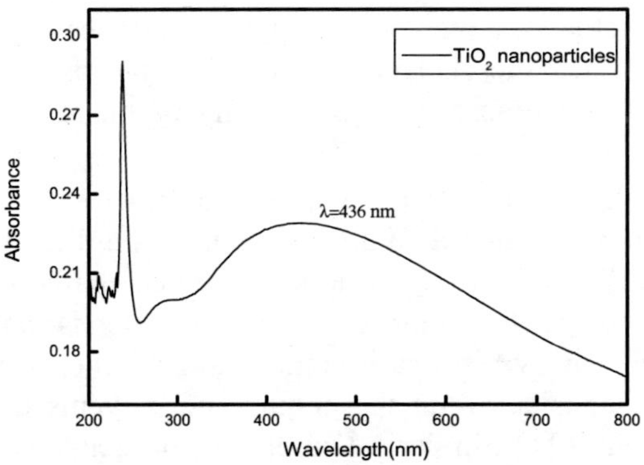

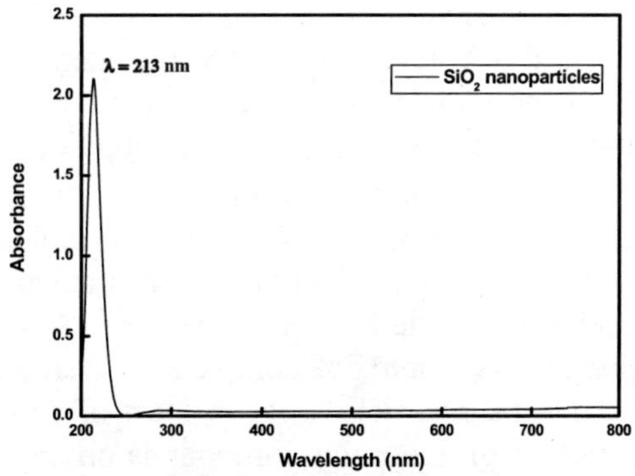

Fig. 1: UV- Vis Absorption Spectrum of Synthesized Titanium Dioxide Nanoparticles Exhibiting Peak at 436 nm and Silicon Dioxide Nanoparticles Exhibiting Peak at 213 nm

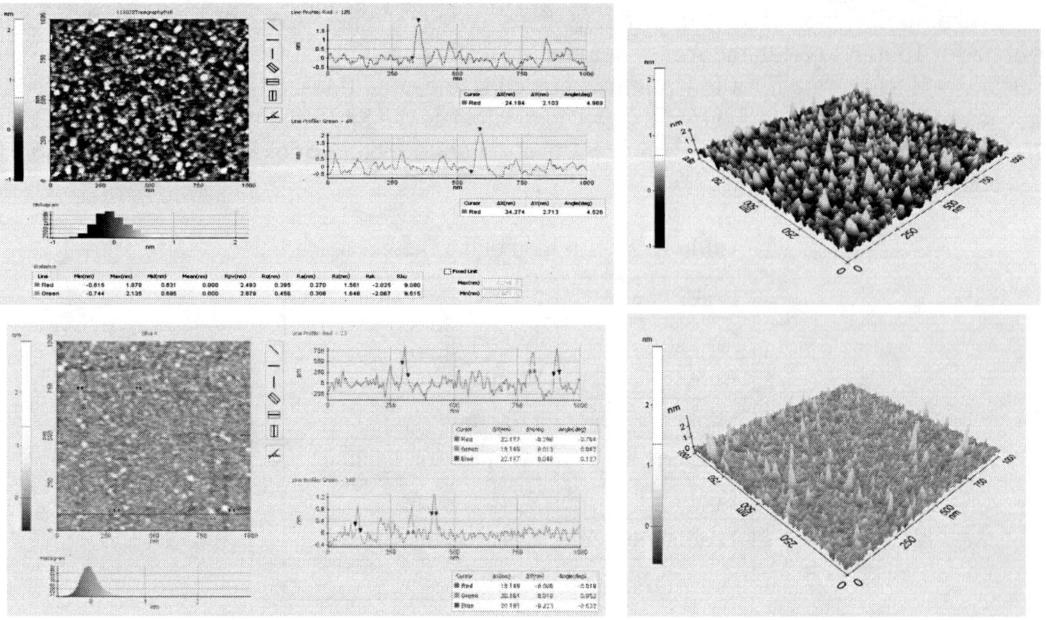

Fig. 2: Atomic Force Microscopy Image (2D and 3D) of TiO_2 and SiO_2 Nanoparticles having Size Distribution of 15–40 nm

SFC and BTE can be calculated from engine speed and fuel flow speed from burette to engine. Figure 3A shows the variation of SFC with Brake Power for different loads. Diesel-ethanol blends has decreased SFC level upto 0.72% compared to neat diesel an diesel-ethanol blends with TiO_2 nanoparticles has decreased SFC level upto 0.68% compared to neat diesel lower compared to diesel-ethanol blends. Diesel-ethanol blends with SiO_2 nanoparticles have decreased SFC level upto 1.7% when compared to neat diesel. Thus diesel-ethanol blends with SiO_2 nanoparticles gives lesser fuel consumption compared to diesel-ethanol blend and diesel-ethanol blends with TiO_2 nanoparticles.

Figure 3B shows variation between BP to BTE for different loads. The diesel-ethanol blends have enhanced BTE of 1.47% compared to neat diesel. Whereas the titanium dioxide nanoparticles with diesel-ethanol has enhanced BTE of 1.17% but lesser when compared to diesel-ethanol blends. However the SiO_2 nanoparticles with diesel-ethanol blends has enhanced BTE upto 5% compared to neat diesel. This was primarily due to activation energy of oxide nanoparticles acts to burn of carbon deposits within the engine and prevents the deposition of non polar compounds on engine walls.[1] Similarly Diesel-ethanol blends has decreased Smoke Absorption Coefficient upto 30% and also TiO_2 nanoparticles with diesel-ethanol blends decreased upto 46% compared to neat diesel as seen in Figure 3C. But SiO_2 nanoparticles with diesel-ethanol blends decreased 65% compared to neat diesel since oxide nanoparticles acts as an oxygen donating catalyst and provides oxygen for the oxidation of CO and resulting in reduction in emissions.

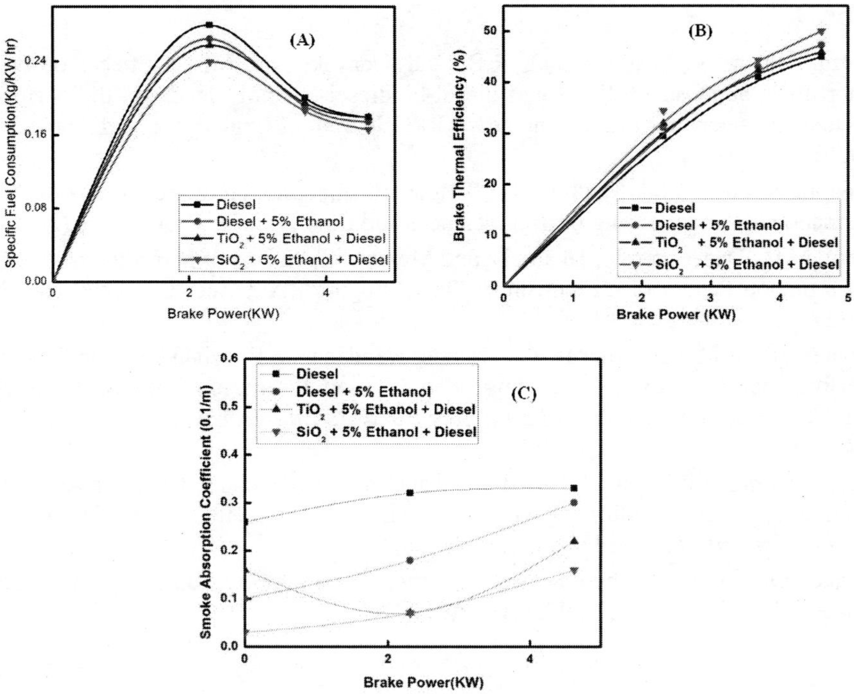

Fig. 3: Graphs showing Specific Fuel Consumption (A), Brake Thernal Efficiency (B) and Smoke Absorption Coefficient (C)

CONCLUSIONS

The performance and emission characteristics of neat diesel and diesel-ethanol blends with the addition of SiO_2 nanoparticles and TiO_2 nanoparticles are investigated on four stroke diesel engine. The engine tests were done by diesel-ethanol blend, TiO_2 nanoparticles with diesel-ethanol blends and SiO_2 nanoparticles with diesel-ethanol blends. Smoke Absorption Coefficient investigated with the help of smoke meter. The above all variations compared and results are as follows:

- SFC of engine was decreased by 1.7 % in SiO_2 nanoparticles with diesel-ethanol blends compared to neat diesel, TiO_2 nanoparticles with diesel-ethanol blends and diesel-ethanol were decreased slightly.
- BTE of engine increased by 5% with the help of SiO_2 nanoparticles with diesel-ethanol blends compared to neat diesel. The TiO_2 diesel-ethanol blends increased by 1.17% and diesel-ethanol blends increased by 1.47% compared to neat diesel.
- Smoke Absorption Coefficient was decreased by SiO_2 nanoparticles with diesel-ethanol upto 65% compared to neat diesel. The SiO_2 nanoparticles with diesel-ethanol blends decreased by 46% and diesel-ethanol blends gives 30% decrement of smoke absorption coefficient.

REFERENCES

[1] Mozhi Selvan, V. Arul, Anand, R.B. and Udayakumar, M., " Effects of cerium oxide nanoparticle addition in diesel and diesel-biodiesel-ethanol blends on the performance and enission characteristics of a CI engine", ARPN Journal of Engineering and Applied Sciences, 4 (2009).

[2] Nagaprasad, K.S. and Madhu, D., "Effect of Injecting Hydrogen Peroxide into diesel", International Journal of Engineering Sciences and Emerging Technologies, 2 (2012) 24–28.

[3] Shafii, M.B., Daneshvar, F., Jahani, N. and Mobini, K., "Effect of Ferrofluid on the Performance and Emission Pattern of a Four-Stroke Diesel Engine", Advances in Mechanical Engineering, (2011).

[4] Bawankure, D. Mayur, Potekar, A. Prashant, Nandane, A. Bhushan and Gandhewar, R. Vivek, "Performance Evaluation of Compression Ignition Engine by Diesel-Ethanol-Methylester blend", *International Journal of Emerging trends in Engineering and Development*, 2 (2012) 147–154.

[5] Jung, Heejung, Kittleson, B. David and Zachariah, R. Michael, "The influence of a Cerium additive on ultrafine diesel particle emissions and kinetics of oxidation", *Journal of Combustion and Flame,* 142 (2005), 276–288.

[6] Nalcaci, Orkun ovez, "The removal of Carbon Monoxide by iron oxide nanoparticles in car exhaust", Thesis, Dokuz Eylül University (2007).

EFFECT OF TRANSITION METAL DOPING ON STRUCTURAL, OPTICAL AND ELECTRICAL STUDIES OF SnO_2 NANOPARTICLES

P. Baraneedharan, C. Siva[2], K. Nehru[3] and M. Sivakumar[4]

Department of Nanoscience and Technology, Anna University of Technology, Tiruchirappalli
[1]Department of Chemistry, Anna University of Technology, Tiruchirappalli
E-mail: muthusiva@gmail.com

ABSTRACT

The intentional incorporation of atomic impurities into semiconductor materials is a common routine for the purpose of tailoring properties such as band gap and electric conductivity to realize specific applications. A simplistic chemical route was adapted to synthesis transition metal doped SnO_2 nanoparticle synthesis. The doped nanomaterials are characterized using electron microscopy and atomic force microscopy for their surface morphology. The structural analysis was carried out using x-ray diffraction pattern. The electrical and optical properties were investigated for different concentration of transition metal doping.

INTRODUCTION

A variety of pure and doped semiconductor metal oxides have been investigated for their morphological, electrical and optical properties.[1] Among them SnO_2, a wide band gap n-type semiconducting material attracted keen interest because of its electrical and optical properties like transparent for visible light and reflective for IR .In addition the lower cost and environmentally friendly in nature gives them more consideration than other metal oxides.[2]

It is obvious that the routine way to modify or tailor the characteristics of a material is that by intentional introduction of dopant to the crystal structure. Doping of metals like Aluminium, Copper, Ferrite, Cobalt in to the lattices of metal oxide can guide to increase in

V. Rajendran, R. Yuvakkumar, K. Thyagarajah and K.E. Geckeler (eds.)
Applications of Nano Materials: Electronics, Energy and Environment, pp. 197–202 (2012).

surface area of SnO_2, stabilize the surface and promote reduce in grain size.[3] The doping of a variety of cations of transition metals into different sub lattices of metal oxides render diverse kinds of optical, electrical and structural properties.[4] Several methods have been developed by wide group of researchers for metal doping into SnO_2 nanostructures includes sol-gel, co-precipitation, citrate gel route, microwave synthesis route and spray pyrolysis which yield particles with high level of homogeneity.[20] Chemical co precipitation method offer many advantages over other methods such as freedom from contamination, simplicity, lower cost, higher homogeneity and good control over particle size.[5] Numerous cation metal doping like Al, Er3+[7], La[8], Eu3 + [9], Ag[10] have been made in SnO_2 crystal structure and its electrical, optical, morphological studies were carried out by various group. But only a few reports presented on binary metal doping on SnO_2 nanostructures and their property studies.

In this work, we present a simple co-precipitation route to prepare Cu^{2+}, Co^{2+} co doped SnO_2 nanostructure. Besides, the optical, electrical, morphological properties are also studied in detail.

EXPERIMENTAL

Pure and Co^{2+}, Cu^{2+} co doped SnO_2 powders were prepared by co-precipitation chemical route. All chemicals used were analytical grade and used without further purification. $SnCl_2.2H_2O$, Co $(NO_3)_2.6H_2O$, Cu $(NO_3)_2.3H_2O$ have been used as initial precursor material for synthesis of SnO_2 and Co^{2+}, Cu^{2+} co doped SnO_2 powders.

In typical synthesis procedure citric acid was added into 50 ml of double distilled water with magnetic stirring until pH turn into 2. Now $SnCl_2.2H_2O$ and appropriated amounts of Co $(NO_3)_2.6H_2O$ and Cu $(NO_3)_2.3H_2O$ were added. After 30min of stirring appropriate amount of ammonia solution was added drop wise to the above solution under magnetic stirring. The resulting solution was stirred for 2hrs to form a white precipitate, which was then washed several times with water and dried in the oven. The final product further calcinated at 250°C for 2hrs resulting in Co^{2+}, Cu^{2+} co doped SnO_2 powder.

Crystalline and structural analysis of SnO_2 and Co^{2+}, Cu^{2+} co doped SnO_2 powder were determined by X-ray diffraction with Cu- Kα radiation λ = 1.54060 Å ranging from 20°C to 80°C.Morphological analysis was carried using Scanning electron microscope. Optical absorbance spectra of samples were recorded using UV-Vis spectrophotometer (Jasco V-680) in the wavelength range of 200–800 nm

RESULTS AND DISCUSSION

Structural Analysis

The XRD patterns of pure SnO_2 and Co^{2+} Cu^{2+} co doped SnO_2 nanoparticles shown in Figure 1.

The diffraction pattern can be ascribed to tetragonal structure of SnO_2 structure for all the samples. No other diffraction peaks obtained from copper oxides and cobalt oxides or any impurities for the structure indicates the incorporation of both Cu^{2+} and Co^{2+} ions to the SnO_2

lattices. The peak intensity decreases and FWHM increases with the addition of Cu^{2+} and Co^{2+} ions indicating that the sample size decreased.[11] The incorporation of ions lead to the degradation in the crystal lattice showing that though dopant incorporated to Sn^{4+} sites it produces crystal defects around the dopant and charge imbalance created .There is a slight change in lattice parameter value of Cu^{2+} and Co^{2+} doped SnO_2 when compared with pure SnO_2 stating that they could occupy regular lattices of SnO_2 sites due to the closer ionic radii of Cu^{2+} and Co^{2+}. The average grain size calculated using Scherrer formula,

$$d = K\lambda / \beta\cos\theta$$

Where, K is constant taken usually as 0.9, λ be the wavelength of x-ray radiation, β be the full width half maximum height at the diffraction angle. The calculated average crystalline size is 30 nm.

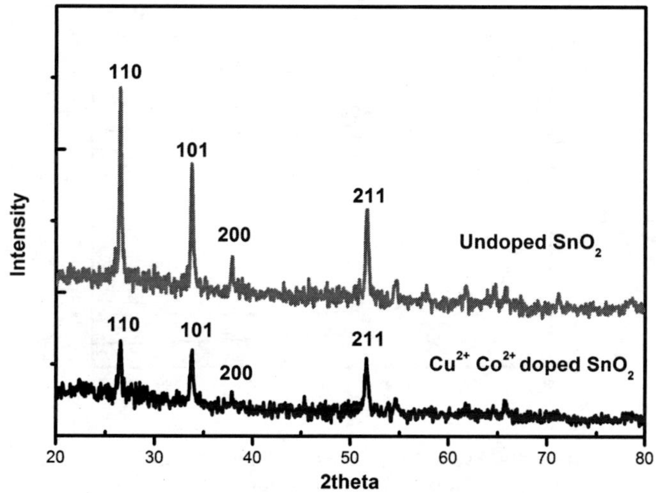

Fig. 1: XRD Pattern of SnO_2 and Co2+ Cu2+ co Doped SnO_2 Nanoparticles Corresponding to Tetragonal Structure

Optical Studies

Generally the size and band gap structure are directly correlated. When there is an increase in size the bang gap decreases and vice versa. The optical absorption spectra of SnO_2 and Cu^{2+} and Co^{2+} doped SnO_2 with absorption at 300 nm and 350 nm shown in Figure 2.

With the blue shift in absorption when compared with bulk SnO2, the absorption is assigned to the direct transition of electrons in the SnO_2 structure.[25] The obtained band gap value is larger than the value of bulk SnO_2 (3.6eV). The measured transmittance value is used to calculate the absorption coefficient using the formula,

$$\alpha = 2.303 \log (1/T)/t$$

Where, t is the thickness of the sample.

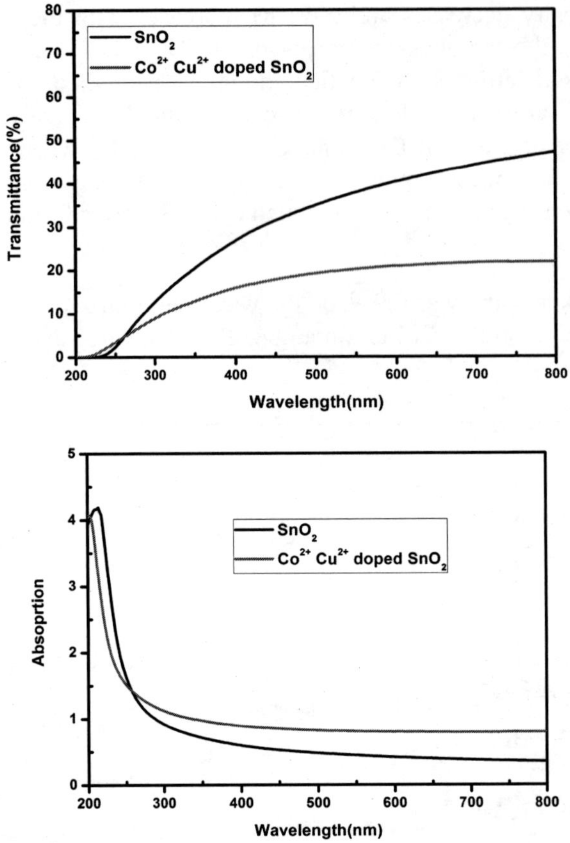

Fig. 2: Transmittance and UV-Vis Absorption Spectra of SnO_2 and Co2+ Cu 2+ Co Doped SnO_2 Nanoparticles

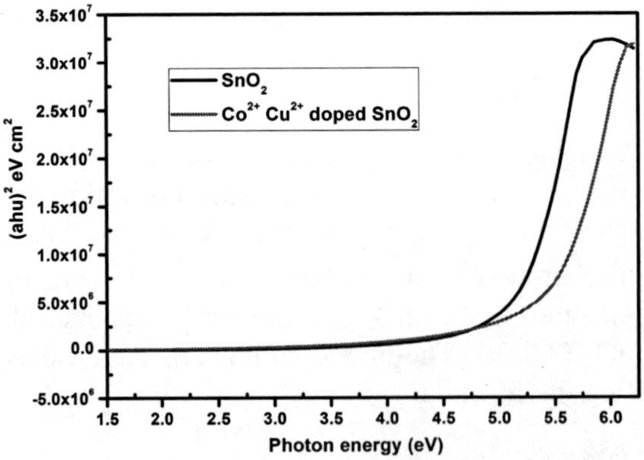

Fig. 3: $(\alpha h\upsilon)^{1/2}$ vs. Photon Energy Plot of SnO_2 and Co2+ Cu2+ Co Doped SnO_2 Nanoparticles

The optical band gap Eg was evaluated from the transmission spectra and the optical absorption coefficient near the absorption edge is given by,

$$h\upsilon\alpha = A\,(h\upsilon - Eg)^{1/2}$$

Where A is constant, Eg the optical band gap, h is the Planks constant and υ the frequency of incident d. The band gap was estimated by plotting $(\alpha h\upsilon)^{1/2}$ vs $h\upsilon$ as shown in Figure 3 and extrapolating the linear portion near the onset of absorption edge to the energy axis. The energy gap value calculated from the graph is 4.1 eV and 3.6 eV.

It can be seen that the incorporation of Cu and Co lead to an increase in band gap value. This increase can be attributed to alloying effect of parent compound with impurity doped. In addition few suggested merging of sub bands with the conduction and valance bands forming continuous bands.

CONCLUSION

A simple chemical route has been used to synthesize SnO_2 and Cu^{2+} and Co^{2+} doped SnO_2 nanoparticles. The XRD pattern reveals that the samples are in tetragonal structure with no impurity peaks and with ~ 30 nm in size. The intensity of the diffraction peak gets reduced on doping with Cu^{2+} and Co^{2+} on SnO_2 crystal structure. The optical studies carried out using absorption and transmittance spectra. The Tauc's plot obtained from transmittance spectrum discloses there is a decrease in band gap due to doping of Cu^{2+} and Co^{2+} which may be due to alloying effect. Hence the doping Cu^{2+} Co^{2+} can be used as a method to tailor the band gap of a material to our desired range.

REFERENCES

[1] Ahmed, A.S., Muhamed, S.M., Singla, M.L. and Tabassum, S., *J. Lumin* 131 (2011) 1.

[2] Matin, B.M., Mortazavi, Y., Khodadadi, A.A., Sens. Actuators B: Chem 151 (2010) 140.

[3] Karthikeyan, K., Amaresh, S., Kalpana, D., Kalaiselvan, R., Lee, Y.S., *J. Phys. Chem. Solids* 73 (2012) 363.

[4] Ashiq, M.N., Saleem, S., Malana, M.A., Rehman, A.U., *J.Alloy.Compd.* 486 (2009), 640.

[5] Gul, I.H., Ahmed, W. and Maqsood, A., *J. Magn. Mater* 320 (2008) 270.

[6] Mohagheghi, M.M.B. and Saremi, M.S., *J. Appl. Phys,* 37 (2004), 1248.

[7] Sambasivam, S., Kim, S.B., Jeong, J.H., Choi, B.C., Lim, K.W., Kim, S.S. and Song, T.K., *Curr. Appl. Phys,* 10 (2010), 1383.

[8] Fu, C., Wang, J., Yang, M., Su, X., Xu, J. and Jiang, B., *J. Non-Cryst. Solids*, 357 (2011) 1172.

[9] Fu, Z., Yang, H.K., Moon, B.K., Choi, B.C. and Jeong, J.H., *Curr. Appl. Phys,* 9 (2009) 1360.

[10] Liu, C.H., Zhang, L. and He, Y.J., *Thin Solid Films,* 304 (1997) 13.

[11] Lang, J., Li, X., Yang, J., Yang, L., Zhang, Y., Yan, Y., Han, Q., Wei, M., Liu, X., Wang, R., *Appl. Surf. Sci.* 257 (2011), 9574.

[12] Gajendran, J. and Rajendran, V., *J. Optoelectron. Adv.* M 5 (2011) 44.

HIERARCHICAL TiO_2 NANOWIRE OVER FABRIC PLATFORM: POTENTIAL CANDIDATE FOR WEARABLE FIELD EMITTER AND PHOTOCATALYST

SOUMEN MAITI[1], UDAY N. MAITI[1,2] AND KALYAN K. CHATTOPADHYAY[1]

[1]*Thin Film and Nanoscience Lab., Jadavpur University, Kolkata, West Bengal, India*
[2]*KAIST, South Korea*
E-mail: sou.maiti@gmail.com

ABSTRACT

Hierarchical titanium oxide nanostructure over conducting carbon fabric is realized via hydrothermal route which shows great potential as wearable field emitter and photocatalyst. In response to external applied electric field, the branched hierarchy exhibits excellent electron emission ability with turn-on field as low as 2.9V/μm. Additionally it also displays excellent environmental remediation ability through catalytic degradation of hazardous dye (orange II) under UV illumination.

INTRODUCTION

Questionable durability of carbon nanotube in harsh condition [1] enforced the researcher in quest of other cold cathode material and the pursuit of alternative was achieved by the realization of oxide based semiconductor nanostructures [2,3] which overcome most of the bottlenecks of CNT as a field emitter. Furthermore, semiconductors nanostructures over foldable substrate ignite the visualization of electronic devices with distinctive portability and flexible features.

TiO_2, a kind of functional material has already registered their widespread applicability in several fields notably in catalytic and solar cell application. But the field emission performance of TiO_2 is less explored till date. Herein wearable field emitter and photocatalyst with rutile hierarchical TiO_2 nanowires deposited via environment friendly hydrothermal route over conducting, foldable carbon fiber was realized.

V. Rajendran, R. Yuvakkumar, K. Thyagarajah and K.E. Geckeler (eds.)
Applications of Nano Materials: Electronics, Energy and Environment, pp. 203–208 (2012).

EXPERIMENTAL

Ultrasonically cleaned and dried carbon cloth substrates were dip coated with TiO_2 sol and heated in air at 400°C for 1 hour. The seeded substrates were placed within a Teflon autoclave containing a solution of water, hydrochloric acid and titanium but-oxide and kept at 180°C for 4 hours. Thereafter the autoclave was cool down to room temperature and product was collected via washing the cloth with DI water. For branched growth of nanowires, the same process commence a new over nanowire coated fabrics instead of bare one with deposition time temperature and time kept at 110°C and 6 hours respectively. Finally the hierarchical TiO_2 nanostructure was obtained by washing and drying cloth.

RESULTS AND DISCUSSION

Figure 1(a-d) displays the FESEM image of the synthesized nanostructure grown on the carbon cloth substrate. Large scale uniformity of the synthesized product is clearly evident from the low magnification as shown in Figure 1a. Further close inspection shows that the nanowires grow over the entire cylindrical surface of each individual fiber of the cloth with their tip radiating outwards from the axis of the fiber. High magnification image in Figure 1b illustrates the actual dimension of ~200 nm in diameter and that top of each nanowire is

Fig. 1: FESEM Images of the Nanowire Array and Hierarchical Nanowire (a, c) Low Magnification and (b, d) Higher Magnification Respectively

highly corroded to form sharp spikes. Further modification by hydrothermal treatment crates a drastic change in morphology. Multiple branches were appeared over the nanowire surface (Figure 1c) to form forest like complex nano-hierarchical morphology. The dimension of these secondarily grown rods is ~150 nm in length and ~10 nm in diameter.

To gain clear information on the morphology and the crystallinity, HRTEM investigation of synthesized product was carried out. Figure 2a shows the TEM images of an individual nanowire. Careful observation of the top edge of the nanowire confirms that the corroded top of the nanowire consists of individual ultrathin spikes (~15 nm in diameter). Parallel lattice fringes with lattice spacing of 0.32 nm corresponds to the inter-planar distance of rutile TiO_2 (110) crystal face indicating [001] as the growth direction, parallel to C axis. (Figure 2b) Further the secondarily grown multiple nano-branches along the backbone of the TiO_2 nanowire are evident from Figure 2c whereas the HRTEM images in the Figure 2d mimic the similar crystallinity with same lattice spacing as of the host nanowire.

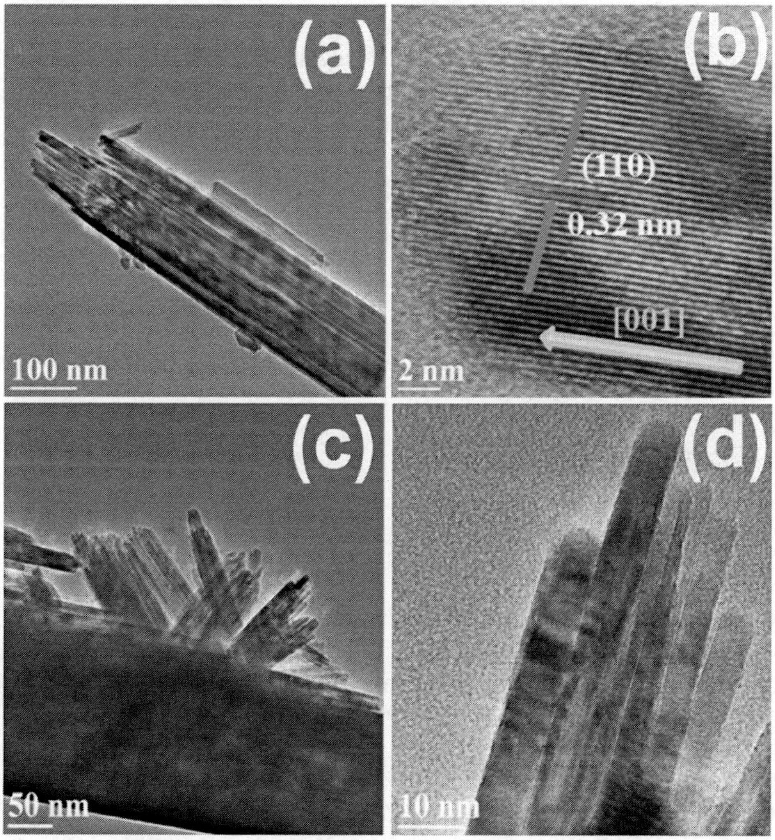

Fig. 2: TEM Images (a, c) and HRTEM Images (b, d) of the Nanowire and Hierarchical Nanowire Respectively

Figure 3a shows the plot of the electron emission current density (J) as a function of applied electric field strength (E). The plot reflects the noticeable difference in the FE characteristics for the two distinct morphologies. The turn-on field and the threshold field (corresponding to current density 10 μA cm^{-2}, 1 mA cm^{-2} respectively), were found to be 2.9 V/μm and 6.98 V/μm only for nano-hierarchy which is significantly low compare to the same for the nanowires (4.2 V/ μ m, 8.89 V/μm) respectively. According to the Fowler-Nordheim (FN) formalism, the dependence of emission current density (J) on the applied electric field (E) follows the relations $J = (A\beta^2 E^2 / \phi)\exp(-B\phi^{3/2} / \beta E)$ where A and B are constants, φ is the work function of the emitter (4.5 eV for titanium oxide). "β" the geometrical field enhancement factor of the emitter strongly depends on several parameters such as nanostructure geometry, emitter tip separation, substrate conductivity etc.[3] The calculated β value from the FN plot (Figure 3b) comes out to be 4365 and 2548 for the nano hierarchy and the nanowire array respectively. In response to external applied electric field, the recorded much better electron emission ability of the branch hierarchy as compared to nanowire is attributable to sharp tip geometry of the branches over nanowire backbone. But the nanostructure geometry cannot be solely accredited for the registered high β value for both as the underlying conducting carbon cloth plays a vital role also as unanimously confirmed by Maiti *et al.*[4]

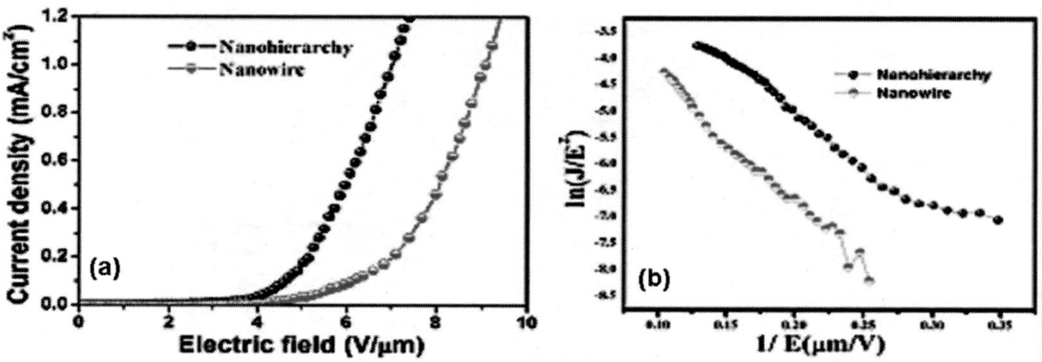

Fig. 3: (a) J–E Plots for Samples under Investigation and (b) F–N Plots Corresponding to J–E Curves

Time dependent evolution of the UV absorption peak corresponds to the hazardous orange II dye in presence of UV illumination as a consequence of the photocatalytic degradation of the hierarchical rutile TiO_2 sample was displayed in Figure 4a. The intensity of the characteristics absorption peak gradually decreases and finally diminishes after 180 min for the hierarchical sample whereas the same occurs after 230 min for the nanowire array (not shown in figure). From the figure it also clear that there is no appreciable degradation of the dye over TiO_2 either in the absence of the UV irradiation or in the absence of the catalyst. The relative decrement of the dye concentration(C/Co) as a function of time was plotted in Figure 4b where Co and C are initial (t = 0) and final (t) concentration respectively.

Under UV light valance band electron promoted to the conduction band leaving behind a hole in the valance band. This photogenerated hole along with the electron via some intriguing chemical reaction with the adhering water generates highly reactive hydroxyl radical which finally oxidizes the dye. Catalyst with high surface area should have the extra possibility of encountering electron/hole with water/oxygen thereby producing active hydroxyl radical finally results in complete degradation of dye. Thus the hierarchical nanostructure on behalf of their high surface area presented excellent environmental remediation ability as compared to the nanowire.[5]

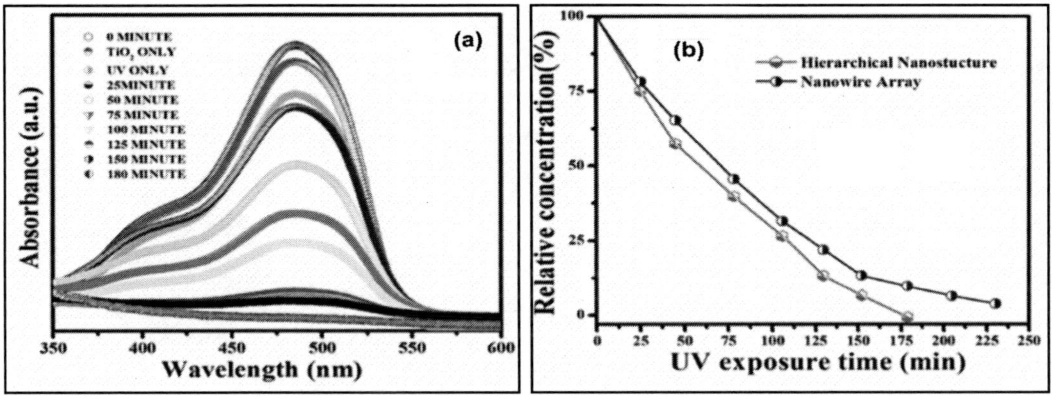

Fig. 4: (a) Temporal Evolution of UV-VIS Absorption Spectra Corresponding to Orange II for Hierarchical Nanowire (b) Relative Concentration Versus UV Exposure Time for Both

CONCLUSION

In summary, we have successfully synthesized forest like TiO_2 hierarchical nanostructure on carbon cloth in hydrothermal route which not only serve the purpose of flexible field emitter but also it registered its usability as wearable photocatalyst also.

REFERENCES

[1] Dean, K.A., Burgin, T.P. and Chamala, B.R., Evaporation of carbon nanotubes during electron field emission, Appl. Phys. Lett., 2001, 79, 1873.

[2] Alivov, Y., Klopfer, M. and Molloi, S., Enhanced field emission from clustered TiO_2 nanotube arrays, Appl. Phys. Lett., 2011, 99, 063104.

[3] Maiti, U.N., Maiti, S., Thapa, R and Chattopadhyay, K.K., Flexible cold cathode with ultralow threshold field designed through wet chemical route, Nanotechnology, 2010, 21, 505701.

[4] Maiti, U.N., Maiti, S., Das, N.S. and Chattopadhyay, K.K., Hierarchical graphene nanocones over 3D platform of carbon fabrics: A route towards fully foldable graphene based electron source, Nanoscale, 2011, 3, 4135.

[5] Maiti, U.N., Maiti, S. and Chattopadhyay, K.K., An ambient condition, one pot route for large scale production of ultrafine (<15 nm) ZnO nanowires from commercial zinc exhibiting excellent recyclable catalytic performance: Approach extendable to CuO nanostructures K.K., CrystEngComm, 2012, 14, 640.

EFFECT OF Li^{+} CO-DOPING ON THE LUMINESCENCE PROPERTIES OF ZnO: Tb^{3+} NANOPHOSPHORS

PARTHA P. PAL AND J. MANAM

Deptartment of Applied Physics, Indian School of Mines, Dhanbad
E-mail: phys.ppal@gmail.com

ABSTRACT

Li^{+} co-doped ZnO:Tb^{3+} nanocrystals were synthesized via room temperature chemical co-precipitation method and the effect of Li^{+} co-doping was studied. The samples are characterized by means of X-ray diffraction (XRD), Fourier transform infrared spectroscopy (FTIR), Photoluminescence (PL) spectroscopy and Thermoluminescence studies. The as synthesized samples were found to have hexagonal wurtzite structure. The peaks are well matched with the JCPDS data card No. 79–0206. The average grain size of ZnO:Tb^{3+} calculated by Debye-Scherrer formula is found to be in the range 10–25 nm. Though XRD study shows no change in the XRD peak pattern of the samples, probably a slight shift of the peaks towards lower angle can be seen for the Li^{+} co-doped sample. It is found that Photoluminescence intensity is enhanced due to little amount of Li^{+} co-doping on the sample. The X-induced thermoluminescence studies gives a glow peak at 320ºC for the Li^{+} co-doped sample which is a little lower angle shifted and twice intense than the TL glow peak of Tb^{3+}-doped sample at 355ºC.

INTRODUCTION

Study of rare earth doped ZnO has been a subject of interest in the recent past. Recently nanocrystalline ZnO is being synthesized in its various shapes like nanorods, nanospheres, nanoflowers etc. and it is being doped with rare-earth (RE) ions to get enhanced luminescence properties which is being used in various optoelectronic devices like vacuum fluorescent display

---

V. Rajendran, R. Yuvakkumar, K. Thyagarajah and K.E. Geckeler (eds.)
Applications of Nano Materials: Electronics, Energy and Environment, pp. 209–214 (2012).

(VFD), field emission display (FED) and electroluminescent display (ELD) etc.[1–5] The intra-shell transition between the 4f shells of the RE ions make the luminescence properties very attractive. But it is very difficult to dope RE ions in ZnO matrix because, firstly, the ionic Radii of RE ions (e.g. Tb^{3+}~0.92Å) is much larger than that of Zn^{2+} ions (~0.74 Å) and secondly the insufficiency of charge compensating material while RE^{3+} ions substitutes the Zn^{2+} ions in the ZnO host matrix. The charge may be compensated from the local defect sites but it will be much convenient to provide a charge compensating material for the formation of stable compound. The alkali metals like Na, K, Li etc. are the most suitable candidates in this regard.

In this work, undoped ZnO, Tb^{3+} doped ZnO and Tb^{3+} doped ZnO with Li^+ codopant, were synthesized successfully by co-precipitation method. The luminescence properties of these samples were systematically studied by thermoluminescence (TL) and photoluminescence (PL) study separately while the structural properties were studied by X-ray diffraction (XRD). In case of PL study also the intensity of the peak for Li^+ codoped sample is found higher than that of ZnO:Tb^{3+} and the undoped sample. Moreover, the Tb^{3+} related peak at 544 nm is found to be more prominent in the Li^+ co-doped sample.

EXPERIMENTAL

Undoped, Tb^{3+} doped and Li^+ co-doped ZnO were prepared by co-precipitation method. Undoped ZnO sample has been prepared by mixing equal volume of 0.05 M ethanolic $ZnCl_2$ solution with 0.20 M ethanolic NaOH at room temperature. Firstly, suitable amount of $ZnCl_2$ and NaOH was separately dissolved in ethanol to prepare 0.05 M $ZnCl_2$ solution and 0.20 M NaOH solution at room temperature. Then the two solutions were mixed and stirred by a magnetic stirrer for 6 hours continuously at room temperature. During this process NaOH reacts with $ZnCl_2$ to form ZnO. The mixed sample was then given for filter with a 125 mm Whiteman 42 filter paper. After complete filtration the residue on the paper is collected and dried. It was then again washed in ethanol and dried to collect and kept for characterization. For co-doping Li^+ and Tb^{3+} ions, requisite amount of LiCl and $TbCl_3$ [Sigma-Aldrich, 99.9%] salt was added before the mixing of 0.05 M $ZnCl_2$ solution and 0.20 M NaOH solution. The samples are annealed at 200°C for two hours.

The powder sample was characterized by Bruker D8 Focus XRD measuring instrument in a wide range of Bragg angle 2θ (10° ≤ 2θ ≤ 80°) with a Bruker D8 Focus XRD measuring instrument with Cu target (K_α = 1.5406 Å) at a scanning rate of 2°/min at the Dept. of Applied Physics, ISM. The FTIR spectrum (KBr pellet) was taken in the region 400 to 4000 cm^{-1} on 'FTIR Spectrum RX I', (Perkin Elmer, Switzerland) spectrometer in the dept. of Applied Chemistry, ISM. The PL emission and excitation studies were carried out by commercially available personal computer based Hitachi FL 2500 Fluorescence spectrophotometer with 150W xenon lamp in the wavelength range 350 nm to 700 nm. The thermoluminescence studies were carried out by using personal computer based Thermoluminescence Analyzer System (TL 1007) supplied by Nucleonix Systems private ltd. India.

RESULTS AND DISCUSSION

XRD Studies

The room temperature XRD patterns of the undoped, ZnO:Tb^{3+} and Li co-doped ZnO:Tb^{3+} powder samples has been shown in the Figure 1.

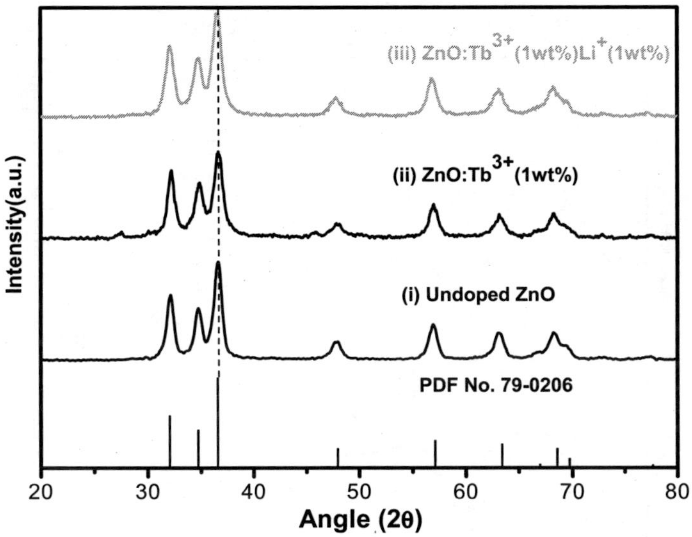

Fig. 1: Room Temperature XRD Pattern (i) Undoped ZnO, (ii) ZnO:Tb^{3+}(1wt%), (iii) Li^+ (1wt%) Co-doped ZnO:Tb^{3+}(1.0 wt%) along with the Standard XRD Peaks for ZnO (JCPDS No. 79–0206)

The XRD study was done in a wide range of Bragg angle (20°–80°). The XRD pattern of all the samples shows ZnO hexagonal wurtzite structure. The peaks are well matched with the JCPDS data card No. 79–0206. The average grain size of ZnO:Tb^{3+} calculated by Debye-Scherrer formula is found to be in the range 10–25 nm. The XRD study shows no change in the XRD peak pattern which shows that Li^+ ions are well incorporated in the crystal lattice; probably a slight shift of the peaks towards lower angle can be seen for the Li^+ co-doped sample indicating an increase in lattice parameters due to Li^+ doping.

FTIR Studies

The 'FTIR' spectra of the undoped ZnO, Tb^{3+} doped ZnO and Li^+ co-doped ZnO:Tb^{3+} was also studied systematically and the results are shown in Figure 3. The FTIR spectra of the undoped ZnO sample consist of four large absorption bands. The large band located at 475 cm^{-1} attributes to the Zn-O stretching in ZnO lattice. Two others bands locating at 1545 and 1431 cm^{-1} are attributed to asymmetric and symmetric C=O stretching vibration modes. The presence of C=O is may be due to the atmospheric CO_2. The wide absorption band at 3438 cm^{-1}

corresponds to the stretching mode of OH group.[6] These bands are present in the Tb^{3+}-doped sample and Li^{+} co-doped $ZnO:Tb^{3+}$ sample also. It can be observed that there is no change in the Zn-O and -OH peak position due to the Li^{+} but the peak for Zn-O stretching is found to be more prominent for the Li^{+} co-doped sample and there is a shift in peak for C=O. The increase in peak sharpness is also noticed indicating a better crystal quality for the Li^{+} doping. But, no peak due to the incorporation of Tb^{3+} ion only in the Tb^{3+}-doped sample is observed, which may be due to the low concentration of the doping ions.

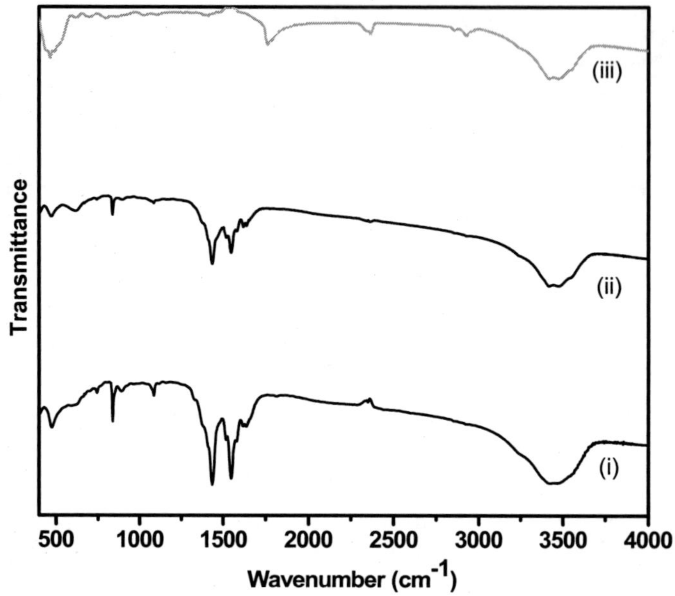

Fig. 2: FTIR Spectra of the (i) as Synthesized ZnO, (ii) Annealed ZnO and (iii) Annealed $ZnO:Tb^{3+}$ (1wt%) Nanorods

Photoluminescence Studies

The room temperature photoluminescence spectra of the Li^{+} co-doped $ZnO:Tb^{3+}$ and $ZnO:Tb^{3+}$ sample are shown in the figure. Figure 2(b) shows the PL emission spectra of the Li^{+} co-doped $ZnO:Tb^{3+}$, $ZnO:Tb^{3+}$ and the undoped ZnO sample. The photoluminescence intensity is found to be increased due to inclusion of Li^{+}. While all the sample gives a broad blue peak near 435 nm due to the host lattice, the Tb^{3+}-doped sample shows a blue-green emission peak near 534 nm which is due to the transition $^5D_4 \rightarrow {}^7F_5$ of Tb^{3+}ions.[7] This peak is found to be more prominent and more intense in Li^{+} co-doped sample at the position of 554 nm. The corresponding excitation spectra have been shown on Figure 2(b) for the confirmation of the study above. The charge compensation by the Li^{+} ions providing a chemical stability and the improved crystal quality has a great influence on the enhancement of the PL peak intensities in the Li^{+} co-doped sample.

Thermoluminescence Studies

The thermoluminescence study of the co-doped, RE doped and undoped samples were carried out in the temperature range 27–400°C. Figure 3 shows the TL glow curves of the undoped, Tb^{3+} doped and Li^+ co-doped ZnO:Tb^{3+} samples. The Li^+ co-doped sample shows an intense peak at 320°C, while the Tb^{3+}-doped sample gives much more shifted peak at 355°C. The undoped sample shows peak at 185°C. The TL intensity of the Li^+ co-doped sample is found to be almost twice than that of ZnO:Tb^{3+} sample. That means the inclusion of Li^+ ion in the lattice increase the intensity about two orders of magnitude. The shifting of the TL peak towards the lower angle due to the inclusion of Li^+ may be due to the influence on inducing the enhancement of lower temperature peak.[8]

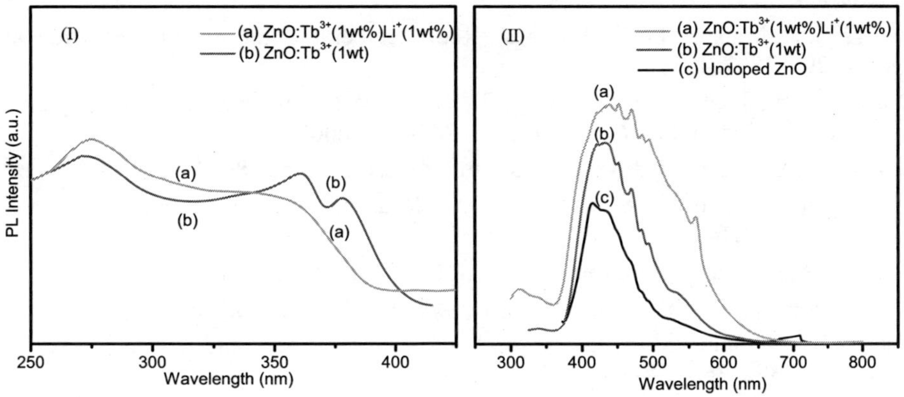

Fig. 3: Room Temperature (I) Excitation Spectra of the (a) Li^+ Codoped ZnO:Tb^{3+} and (b) Tb^{3+}doped ZnO at E_x = 540 nm and (II) Emission Spectra of the (a) Li^+ Co-doped ZnO:Tb^{3+} (b) Tb^{3+}doped ZnO and (c) Undoped ZnO Sample at E_m = 275 nm

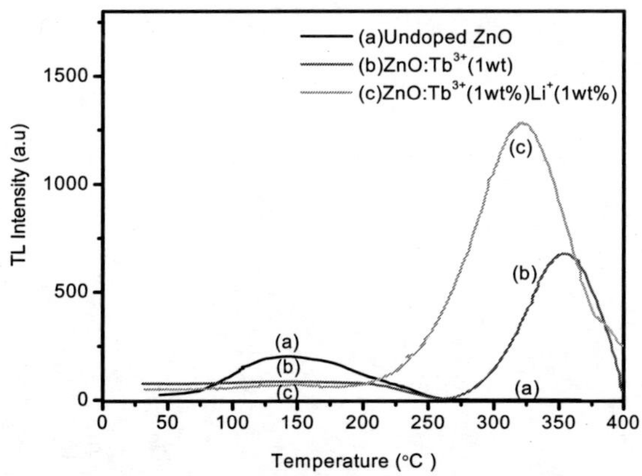

Fig. 4: X-irradiated TSL Glow Curve for Li^+ Co-doped ZnO, ZnO:Tb^{3+} and Undoped ZnO Irradiated for 5 mins

CONCLUSIONS

Li^+ co-doped $ZnO:Tb^{3+}$ nanoparticles were synthesized by precipitation method using ethanol as solvent and its effect on Luminescence properties are discussed. It is found that Li^+ ion can influence lattice parameters. Both the PL and TL peak intensities are improved due to charge compensation by the Li^+ ions. The TL glow peak is found to be low angle shifted and the sample can be well used in display devices and Thermoluminesence dosimetry.

REFERENCES

[1] Nakanishi, Y., Miyake, A., Kominami, H., Aoki, T., Hatanaka, Y., Shimaok and G., Preparation of ZnO thin films for high-resolution field emission display by electron beam evaporation, *Appl.Surf. Sci.* 142, 233–236, 1999.

[2] Huang, M.H., Mao, S., Feick, H., *et al.*, Room temperature ultraviolet nanowire nanolasers, *Science,* 292, 1897–1899, 2001.

[3] Zhang, J., Feng, H., Hao, W. and Wang, T., Luminescence of nanosized ZnO/polyaniline films prepared by self-assembly, *Ceram. Int.* 33, 785–788, 2007.

[4] Hellemans, A., Laser Light From a Handful of Dust, Science 284, 24–25, 1999.

[5] Yi, L., Hou, Y., Zhao, H., *et al.*, "Photo- and electro-luminescence properties of ZnO: Zn thin film, Displays 21,147–149, 2000.

[6] Maensiri, S., Laokul, P. and Promarak, V., *J. Cryst. Growth,* 289, 102, 2006.

[7] Zhang, J., Feng, H., Hao, W. and Wang, T., *Mat. Sci. Eng. A,* 425, 346–348, 2006.

[8] Rojas, S.S., Yukimitu, K., de Camargo, A.S.S., Nunes, L.A.O. and Hernandes, A.C., *J. Non-Cryst. Solids* 352, 3608, 2006.

EXTRACTION OF Fe_2O_3 FROM THE INDUSTRIAL WASTE FOR PHOTONIC APPLICATIONS

C. BABEELA AND T.C. SABARI GIRISUN

Department of Physics, Bishop Heber College (Autonomous), Tiruchirappalli
E-mail: sabari.girisun@gmail.com

ABSTRACT

The industrial waste (EOFD) was obtained from nearby industries, was subjected to microwave irradiation (900 W, 473 K, 20 minutes) to recover the valuable metals in the form of metal oxides by thermal method. From Powder X-ray diffraction analysis, Fe was identified as a major phase in raw sample and Fe_2O_3 was identified as a major phase in irradiated sample. SEM analysis shows that the particle size of the irradiated powder was reduced five times when compared to the raw powder. The recorded EDAX pattern shows clearly that increase in the oxygen content after irradiation, confirming the conversion of metal into metal oxides. The irradiated powder was dispersed in ethanol and the optical characteristic of the sample was measured by UV-Vis spectrophotometer. The electrical study was done by two probe method. Thus, the metal oxides derived could be helpful to reduce the burden on the environment, increase the availability of raw materials that are useful for the photonic applications.

INTRODUCTION

In the past decade, steel making process is the important sector in furnace industry. Steel is an alloy that consists mostly of iron and has carbon content between 0.2% and 2.1% by weight, depending on the grade. Iron has been found in the earth's crust only in the form of an ore i.e. combined with other elements such as oxygen and sulphur. Electric Oil Furnace Dust (EOFD) is a solid waste generated during steel making process in electric oil furnace. Fe, Si, Ni, Mo, Cu, V, Cr, S and C are the heavy metals present in the EOFD waste. Due to the increase demand of steel throughout the world (1,057,000,000 tons/ annum), the dust produced per

V. Rajendran, R. Yuvakkumar, K. Thyagarajah and K.E. Geckeler (eds.)
Applications of Nano Materials: Electronics, Energy and Environment, pp. 215–220 (2012).

year is 21,140,000 tons. This EOFD was placed in a hazardous waste landfill. Due to the large number of the heavy metals in EOFD, the environmental suffer deep impacts like soil, ground water and ecology pollution. The industrial waste severs as the most destructive threat to the health and environment. At the same time, by thermal treatment it is possible to convert all the metals into its metal oxide form and thereby the environmental problem can be avoided. Out of this metal oxides, Fe_2O_3 is of great technological interest due to their wide applications in environmental degradation of metals and alloys, microwave devices as well as high density recording media, magnetic and nonlinear optics, optoelectronic devices, photo electrochemical solar cells, gas sensors and photo catalytic hydrogen fuel production. The present communication deals with the structural, electrical and optical properties of the industrial waste (EOFD) after microwave treatment.

EXPERIMENTAL

To recover the valuable metals in the form of metal oxides, the scales were initially grinded using mortar for five hours to powder the sample. The grinded powders were initially washed with acetone and ethanol to remove any possible organic impurities present in it. The purified and grinded raw materials were subjected to microwave irradiation using LG MC7148 microwave oven. The powders were irradiated to radiation at maximum intensity of 900W and maintained at temperature of 473 K for 20 minutes. The structural and phase of the powdered material was analyzed using Powder X-ray Diffraction (XRD, Rigaku DMAX-111A) using CuKα as a source having wavelength 1.5418 Å and with a step size 0.02 s^{-1} in the range of 10 to 75°. The microstructural and elemental analysis of the materials was carried out using SEM-EDAX. The optical characterization of the powder was done by using UV-Visible Spectrophotometer (Varian, Cary-50). And the electrical characterization of the powder was carried out by two probe technique.

RESULTS AND DISCUSSION

X-ray Diffraction

The powder XRD analysis was carried out for raw and irradiated samples to make a basic confirmation about the various metal and metal oxides present in the system. From the recorded XRD pattern, Fe was identified as a major phase in the raw and irradiated sample. The major peak at 42° which corresponds to ferrous metal in the raw material turns into ferrous oxide (36°). The other minor peaks corresponds to V_2O_5, Cr_2O_3, Mn_2O_3, MoO_3 and SiO_2. Table 1, clearly shows that the raw material that are rich in heavy metals like Fe turns into oxide form like Fe_2O_3 and Fe_3O_4. The lattice parameter of Fe_2O_3 was found to be a = 5.023, b = c = 13.708 Å [JCPDS No: 89–8104] with rhombohedral phase.

Table 1: XRD Data for Raw and Irradiated Material

d-values		*Assignment*
Experimental	*Literature*	
Raw material		
2.1504	2.1404	V 1 1 0
2.1397	2.1347	Fe 1 0 0
2.1243	2.1275	Cr_2O_3 1 1 1
1.6829	1.6803	Mn 3 2 1
1.2826	1.2822	SiO_2 3 1 3
Irradiated		
2.6915	2.6919	Fe_2O_3 1 0 4
2.6728	2.6784	SiO_2 111 0
2.5157	2.5115	Fe_2O_3 1 1 0
2.4584	2.4511	Fe_3O_4 2 2 2
1.7354	1.7332	Fe_3O_4 4 2 2

Scanning Electron Microscopy

The SEM images provide the direct information about the size and typical morphologies of the powder. The surface morphology of raw and irradiated samples is as shown in Figure 2. SEM analysis clearly shows that the particle size of the irradiated powder was reduced five times when compared to the size of raw powder. However there is no major change in the morphology of the samples.

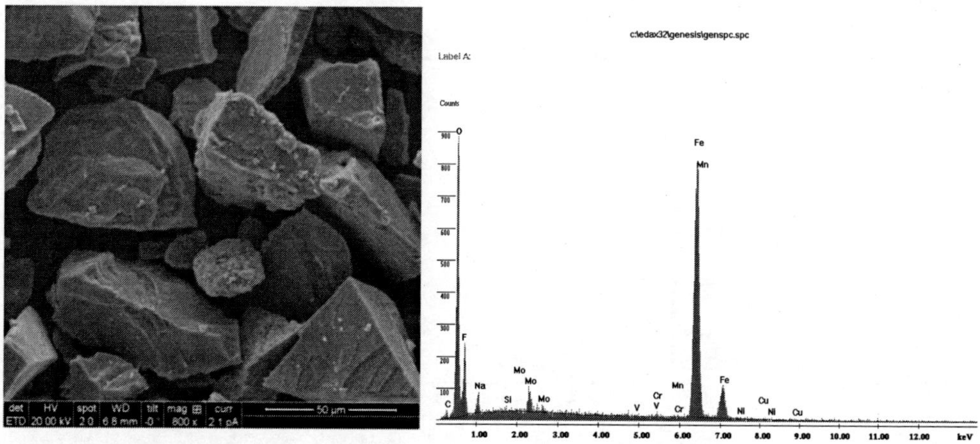

Fig. 2: SEM Image and EDAX Pattern of the Irradiated Powder

The recorded EDAX pattern also shows that ferrous as a major element composition with 64 wt% and 59 wt% for the raw and irradiated powder respectively. Most interestingly, the peak at 0.5 keV which corresponds to the oxygen element shows a sound variation in wt% due to microwave irradiation. The increase in the oxygen content after irradiation confirms the conversion of metal into metal oxides i.e. Ferrous turns ferrous oxides. The other minor peaks corresponds to the other heavy metals like Cu, Si, Mo, V, Cr, Ni, Mn etc.

Optical Properties

As the samples of interest contain heavy metals, the irradiated powder is almost insoluble in all solvents. Hence the irradiated powder was dispersed in (0.01 mg in 10 ml) ethanol and the absorbance pattern was recorded as a function of wavelength in the range of 190 to 1100 nm in steps of 1 nm (Figure 3).

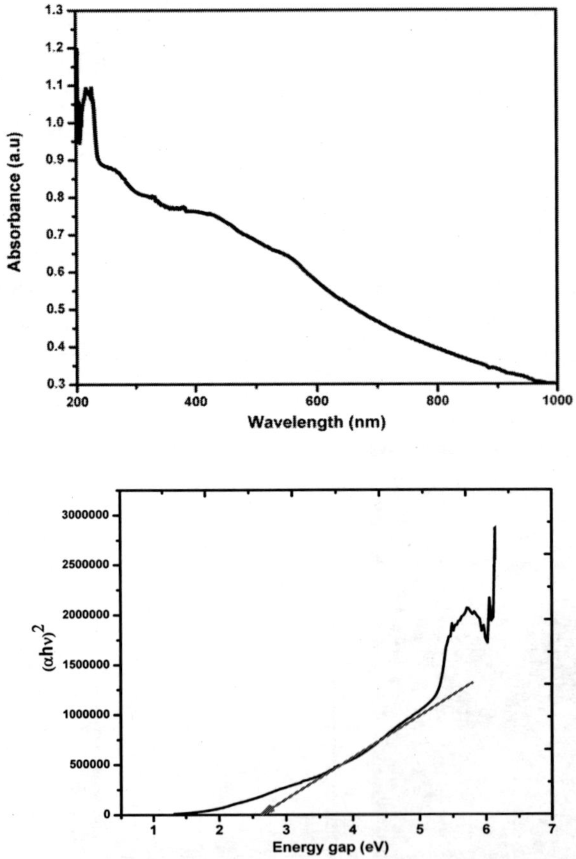

Fig. 3: Absorbance and Tauc's Plot of the Irradiated Sample

The absorbance maximum of the sample lie at 226 nm and beyond that absorbance decreases throughout the visible region. By Tacu's plot, the optical band gap of the sample was found to be 2.6 eV and the literature comparison shows that it may corresponds to the band gap of ferrous oxide material. Thus it can be inferred that the material with good absorbance and moderate bandgap can be used as a potential candidate for photonic applications.

Electrical Properties

The electrical study was done by two probe method. The irradiated sample was initially made into a pellet of thickness 4 mm and was subjected to electrical studies. The change in resistivity of the powder as a function of temperature is as shown in Figure 4. Resistivity decreases as temperature increases, which shows the semiconducting nature of the sample. By Arrhenius plot, the electrical band gap was estimated to be 3 eV.

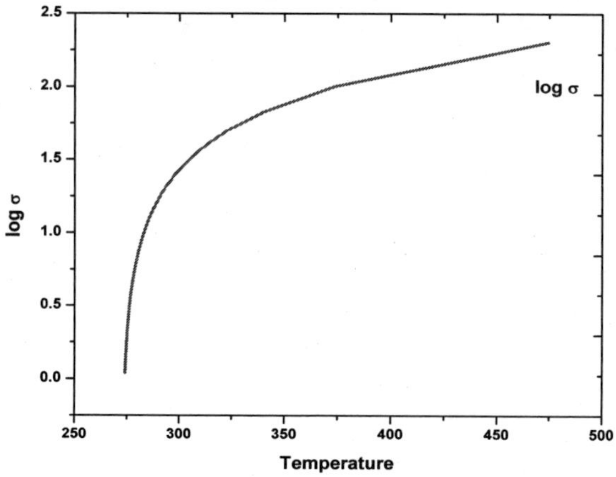

Fig. 4: Arrhenius Plot of the Irradiated Sample

CONCLUSION

The industrial waste (EOFD) collected from the nearby industries of Tiruchirappalli, was subjected to microwave irradiation (900 W, 473K, 20 minutes) to recover the valuable metals in the form of metal oxides. Basic confirmation of the element present in the dust was carried out for raw and irradiated samples, and XRD analysis confirms the formation of metal oxides. In particular Fe and Fe_2O_3 were identified as major peak in the raw and irradiated dust respectively. SEM reveals that the size of the raw powder was reduced after microwave irradiation. EDAX pattern clearly shows that the increase in the oxygen content after irradiation confirming the conversion of metal into metal oxides. The irradiated industrial dust, with good absorbance and moderate bandgap is a potential candidate for photonic applications. Thus it can be concluded that, the industrial dust which serves as threat to environment, can be reused as useful raw materials for various photonic applications like solar cells, gas sensors, optical limiters.

REFERENCES

[1] Shinde, S.S., Moholkar, A.V., Kim, J.H., , R.K., Rajpure, Surface and Coatings Technology, 205, 3567, 2011.
[2] Pickles, C.A., Journal of Hazardous Materials, 179, 309, 2010.

HIGHLY DENSE AND ALIGNED GROWTH OF MULTI WALL CARBON NANOTUBES USING AQUEOUS $FeCl_3$ AS CATALYST

MOHAN LAL, BHARAT BHUSHAN SHARMA[1], POORNENDU CHATURVEDI, JASWANT SINGH RAWAT AND PARTAP KUMAR CHAUDHURY

Solid State Physics Laboratory, DRDO, New Delhi
[1]D.A.V. Institute of Engineering and Technology, Jalandhar, Punjab
E-mail : mohan.lal@hotmail.com

ABSTRACT

Growth of Carbon Nanotube (CNT) was achieved using a very efficient and economical process of catalyst deposition. Aqueous $FeCl_3$ salt was used as the catalyst for CNT growth. The iron nanoparticles were formed over the substrate by pre-treatment of the samples. Removal of Chlorine was ensured during the confirmed by the EDAX studies. The growth was carried out using C_2H_2 as the source gas by the CVD technique. The grown CNTs were highly dense and aligned. The alignment of CNTs were achieved in both Horizontal and Vertical direction on the same substrate.

INTRODUCTION

Carbon Nanotube were discovered by S. Iizima in 1991 [1] and because of their remarkable properties CNTs have found applications in number of fields like nanocomposites,[2] field emission display,[3,4] gas storage [5] and gas sensor.[6,7] The high quality multi wall nanotubes (MWNT) can be synthesized by different method like arc-discharge method,[8,9] laser ablation method,[10,11] chemical vapor deposition (CVD) method [12, 13] but CVD technique is preferred for the selective growth of CNTs. In CVD technique transition metals such as iron (Fe), Cobalt (Co) and Nickel (Ni) are used as catalyst for the growth of CNTs.[14] These metal films over the substrate crack into small metal particles which catalyze the growth of CNTs. So control on the areal density, size and regularity of the catalyst particles is a key element to achieve uniform growth of CNTs.[15]

V. Rajendran, R. Yuvakkumar, K. Thyagarajah and K.E. Geckeler (eds.)
Applications of Nano Materials: Electronics, Energy and Environment, pp. 221–226 (2012).

In this paper we are demonstrating aligned growth of Carbon Nanotube (CNT) using a very easy and efficient process of catalyst deposition. We have used aqueous $FeCl_3$ salt as catalyst for the growth of carbon nanotubes (CNTs). This technique does not require any sophisticated metal deposition system for catalyst deposition. Till now the reported results using salts as the catalyst over Si wafer have mostly shown randomly oriented or low density growth of CNTs.[16,17] We were able to obtain aligned and dense growth of CNTs using the same method.

EXPERIMENTAL

Preparation of Catalytic Film on Silicon Surface: The $Fecl_3$ salt was used as precursor for the formation of catalytic Iron nanoparticles. The 4 mg composition of $Fecl_3$ salt was dissolved in 27 ml of DI water. The Si wafer was dipped in aqueous $FeCl_3$ solution and heated at 80–90°C temperature in the oven. After drying a film of $FeCl_3$ was formed over the substrate as shown in the Figure 1.

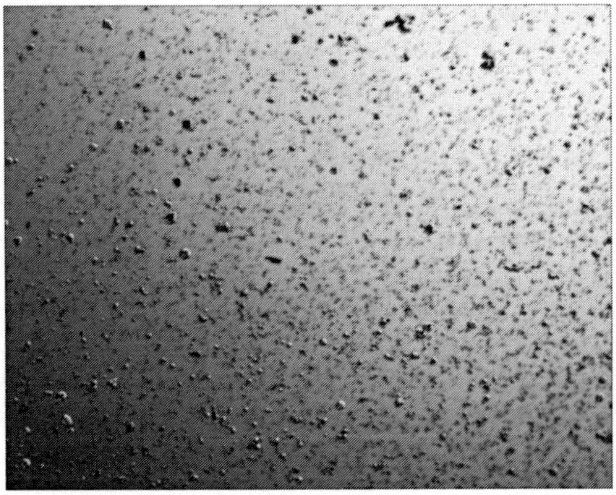

Fig. 1: Optical Image of $FeCl_3$ Deposited Si

Pretreatment and Nanoparticles Formation

The as prepared samples were treated with the H_2 and NH_3 at 900°C for 15 minutes to reduce the $FeCl_3$ films over the wafer. This reduction step removes the chlorine present in the film and aids in the formation of nanoparticles.

Growth of VA-CNTs

The samples with iron Nanoparticles were put to growth in CVD reactor. Growth was done using C_2H_2 as the carbon source gas at a temperature of 900°C and for duration of 10 minutes. H_2 and NH_3 were also flowing simultaneously into the reactor to reduce the iron particles and avoid any oxide formation.

RESULT AND DISCUSSION

The Figure 1 shows the optical image of the film of $FeCl_3$ formed over the substrate. The Figure 2 shows the FESEM micrograph of the sample after pretreatment of the sample. In the image Fe-nanoparticles are visible which have been formed after the reduction of catalyst film in the presence of NH_3 and H_2. A very dense and homogenous iron nanoparticles distribution was achieved and this is required in order to obtain the very dense growth of CNTs. The size distribution of Fe nanoparticles from the FESEM image has been shown in the form of histogram in Figure 3 and we can see that the maximum concentration of particle size is around 20 nm. Figure 4(a) and 4(b) shows the Energy dispersive using x-ray (EDAX) of as deposited $Fecl_3$ salt on silicon substrate and after the pretreatment respectively. From the spectrum it can easily be concluded that there is no chlorine present over the sample after the pretreatment of the samples. The removal of chlorine is important because otherwise it will not allow the formation of the iron nanoparticles. We can also observe the oxygen peaks in the spectrum. These peaks are here because of the oxidation of the film in the environment.

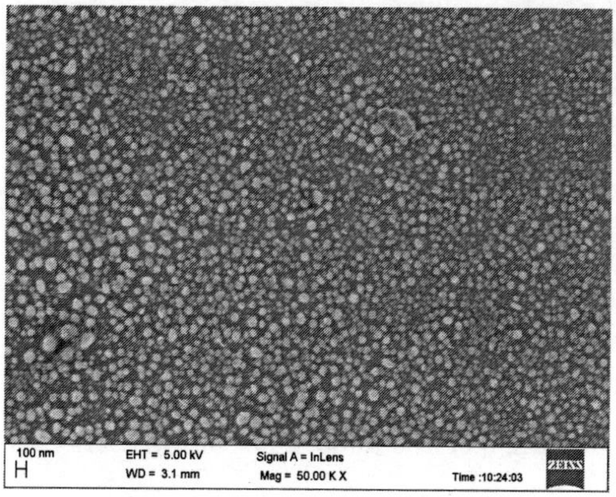

Fig. 2: FESEM image Iron Nanoparticles

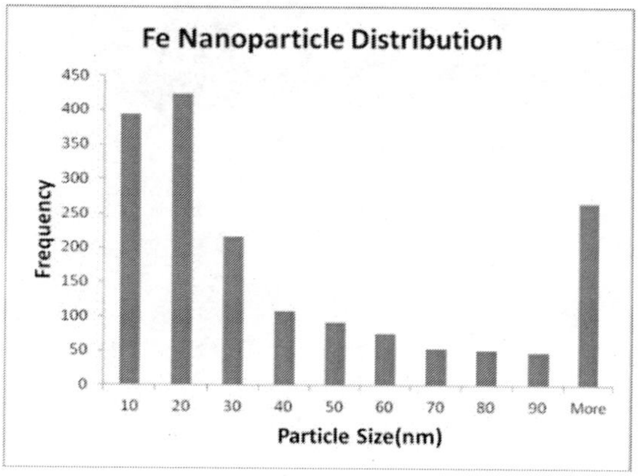

Fig. 3: Distribution of Fe-Nanoparticles

The Figure 5 shows the SEM micrograph of the sample after the growth. We can observe that the CNT grown are vertically oriented in direction and having a length of approx 15 micron. Figure 6 shows a scratch on the substrate from which we can observe how dense the grown CNTs are. In the inset we can see the CNTs have formed a mat all over the substrate. Figure 7 shows the SEM image of one of the edge of the sample we can see that CNTs were also grown on the edges of the sample and were having the horizontal direction of the growth. Again the length in both the orientation is in the same range which is very much expected. So the growth rate of our CNTs was 1.5 micron/minute. From Figure 8 it can be observed that the diameter of the grown CNTs in the range of 60 nm. This is expected from the iron particle distribution of our pretreated samples.

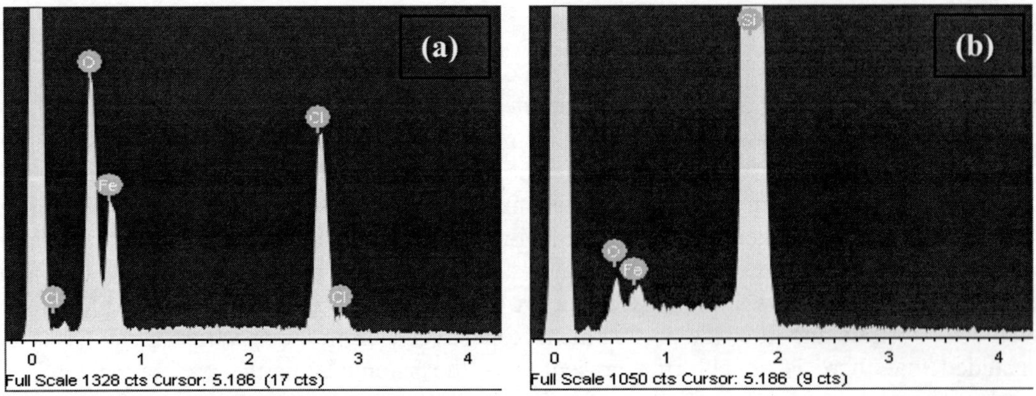

Fig. 4: EDAX Spectrum of Pre and Post Annealed Samples

Fig. 5: SEM Image of Grown CNTs

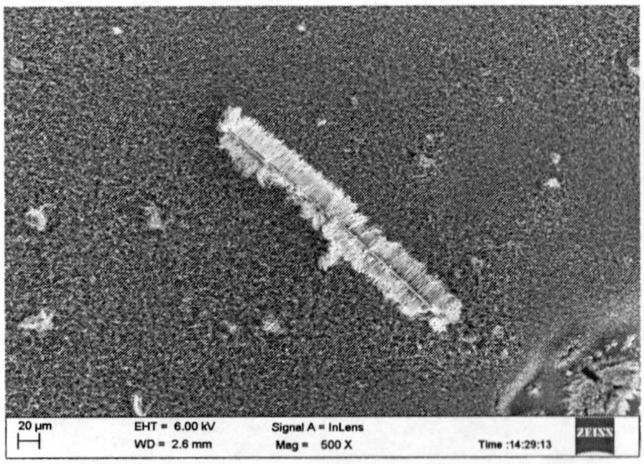

Fig. 6: Mat Like Structure of Grown CNTs

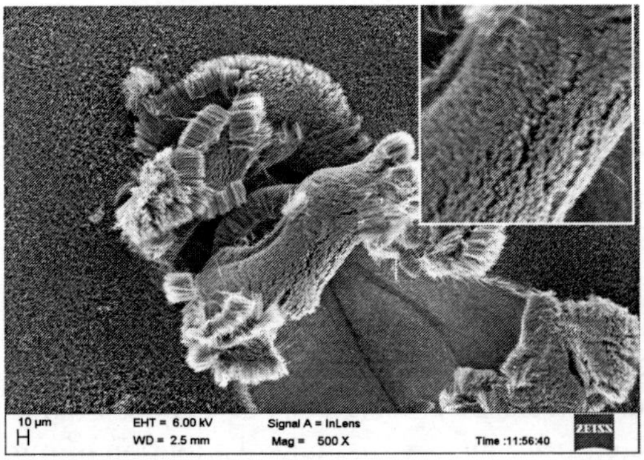

Fig. 7: Horizontally Aligned CNTs from Edge of Sample

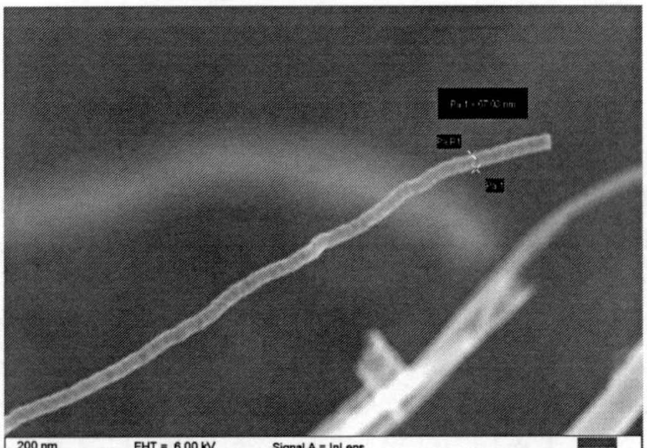

Fig. 7: Horizontally Aligned CNTs from Edge of Sample

CONCLUSION

Highly dense and aligned CNTs were grown using aqueous $FeCl_3$ as the catalyst and C_2H_2 as the source gas. The CNTs were grown in both orientations i.e. Horizontal and Vertical over a single substrate. The technique does not require any sophisticated metal deposition system or lithography technique for catalyst deposition and orientation of the CNTs.

ACKNOWLEDGEMENT

We are thankful to the Dr. Ashok Kapoor and Mr. Anand in Characterization Group at SSPL for FESEM and EDAX characterization of the samples.

REFERENCES

[1] Iijima, S., Helical microtubules of graphitic carbon. *Nature*, 1991, 354 (6348), pp. 56–58.

[2] Dalton, A.B., *et al.*, Super-tough carbon-nanotube fibres. *Nature*, 2003, 423 (6941), pp. 703–703.

[3] Bonard, J.-M., *et al.*, Field emission from single-wall carbon nanotube films. *Applied Physics Letters*, 1998, 73(7), pp. 918–920.

[4] Saito, Y., *et al.*, Field emission of carbon nanotubes and its application as electron sources of ultra-high luminance light-source devices. *Physica B: Condensed Matter,* 2002, 323(1–4), pp. 30–37.

[5] Dillon, A.C., *et al.*, Storage of hydrogen in single-walled carbon nanotubes, *Nature,* 1997, 386 (6623), pp. 377–379.

[6] Ting, Z., *et al.*, Recent progress in carbon nanotube-based gas sensors. Nanotechnology, 2008, 19(33), p. 332001.

[7] Baughman, R.H., Zakhidov, A.A. and de Heer, W.A., Carbon nanotubes—the route toward applications. *Science*, 2002, 297 (5582), pp. 787–92.

[8] Gamaly, E.G. and Ebbesen, T.W., Mechanism of carbon nanotube formation in the arc discharge. *Physical Review B*., 1995, 52(3), pp. 2083–2089.

[9] Keidar, M. and Waas, A.M., On the conditions of carbon nanotube growth in the arc discharge. *Nanotechnology*, 2004, 15(11), p. 1571.

[10] Scott, C.D., *et al.*, Growth mechanisms for single-wall carbon nanotubes in a laser-ablation process. *Applied Physics A: Materials Science & Processing*, 2001, 72(5), pp. 573–580.

[11] Puretzky, A.A., *et al.*, Investigations of single-wall carbon nanotube growth by time-restricted laser vaporization. *Physical Review B*., 2002, 65(24), p. 245425.

[12] Kukovitsky, E.F., *et al.*, CVD growth of carbon nanotube films on nickel substrates. *Applied Surface Science*, 2003, 215(1–4), pp. 201–208.

[13] Hart, A.J., *et al.*, Uniform and selective CVD growth of carbon nanotubes and nanofibres on arbitrarily microstructured silicon surfaces. Nanotechnology, 2006, 17(5), p. 1397.

[14] Esconjauregui, S., Whelan, C.M. and Maex, K., The reasons why metals catalyze the nucleation and growth of carbon nanotubes and other carbon nanomorphologies. *Carbon*, 2009, 47(3), pp. 659–669.

[15] Mukul, K., Carbon Nanotube Synthesis and Growth Mechanism, 2011.

[16] Cui, R., *et al.*, Comparison between Copper and Iron as Catalyst for Chemical Vapor Deposition of Horizontally Aligned Ultralong Single-Walled Carbon Nanotubes on Silicon Substrates. *The Journal of Physical Chemistry C*, 2010, 114 (37), pp. 15547–15552.

[17] Fu, Q., Huang, S. and Liu, J., Chemical Vapor Depositions of Single-Walled Carbon Nanotubes Catalyzed by Uniform Fe_2O_3 Nanoclusters Synthesized Using Diblock Copolymer Micelles. *The Journal of Physical Chemistry B*, 2004, 108(20), pp. 6124–6129.

EFFECT OF HYDROXYL PASSIVATION ON BAND GAP OF SILICON NANOWIRES

SANA KAUSAR, SHIRISH JOSHI AND SYED MUJAHID HUSAIN

Department of Physics, Govt. M.V.M. College Bhopal, Department of Electronics JNCTS, Bhopal

ABSTRACT

The electronic properties of very thin Silicon nanowires passivated with hydrogen and hydroxyl group of hexagonal shape was studied by ab-initio method using the generalized gradient approximation. We replaced all the hydrogen atom with hydroxyl and then optimzed the structures of hydroxyl-terminated [111] Si nanowire .We calculated the electronic band structure and compared them with the hydrogen-terminated Si nanowires. Remarkable change in the electronic band structure occured between hydrogen terminated and hydroxyl terminated Si nanowire, that is, the band gap of [111] Si nanowire reduces on passivation with hydroxyl atoms.

INTRODUCTION

Recently nanowires become the most studied nanomaterials due to continuous discoveries and their superior physical properties in comparison to larger bulk materials. These physical properties arise in a variety of scientific and engineering feilds like mechanics, magnetism,, thermal sciences, optics and others.[1,2] Due to these superior physical properties nanowires are viewed as the potential building blocks of future nanoelectronics such as field effect transistors (FET), single electron transistors structured on the scale of 10 nm, tunable optoelectronics and quantum information technology, etc. By reducing device size, the effects of defect states can be readily occurred: such as space charge limited current and hopping transport. Also, the interface property is a significant factor in nanoelectronic device design. In order to solve these problems, a comprehensive investigation of electrical and optical

V. Rajendran, R. Yuvakkumar, K. Thyagarajah and K.E. Geckeler (eds.)
Applications of Nano Materials: Electronics, Energy and Environment, pp. 227–232 (2012).

properties becomes extremely important.The mechanical properties of nanowires have extensively studied to gain fundamental information about the strength of nanomaterials, and how free surfaces, which are prevalent in nanomaterials, contribute to the unique properties of nanowires.[3,4] Semiconductor nanowires have shown promising results for next generation integrated electronics .The current state of understanding on the mechanical behaviour and properties of nanowires has come from both experiment [5] and atomistic calculations. This collection of information has established that, in general, nanowires have significantly higher yield stresses and strains than bulk materials, that single defects can cause catastrophic losses of strength in single-crystal nanomaterials, and that the mechanical properties are strongly dependent on the axial orientation of the nanowires.

BAND GAP OF HYDORGEN PASSIVATED SINWS ALONG [1 1 1] DIRECTION

We import the crystal structure of bulk Si and Create a [111] surface then Specify the surface [111] and the number of layers the surface is converted to crystal structure by builting a vacuum layer of 4.0Å thickness we Find out the periodic lattice length in the Z direction, which is 9.406 angstrom here. All the the redundant atoms in the Z direction are deleted the structure according to the periodic lattice length is Rebuild then a supercell in the X-Y plane is build the redundant atoms in the X-Y plane is deleted. After Rebuilting the cell shape the nearest atom length between neighbor cells is checked. If the length is less than 10 angstrom, that time the cell size is changed. Now the final structure of Si nanowire along [111] direction is shown in Figure 1.In this study, we build silicon nanowire with nearly hexagonal cross section, which has 54 atoms in its primitive unit cells and are grown along [111] direction. The top (upper) View of silicon nanowire along [111] direction is shown in Figure 1.Yellow spheres stand for silicon atoms and white spheres for hydrogen. Hydrogen passivation is used to eliminate the defect states within the band gap so that the band structure of the chain should briefly feature the characteristics of silicon. for eliminating the inter-layer interaction a vacuum layer the thickness of which is 4 Å is built between two neighboring chains, there is the lattice constant C_o and all the atomic positions are fully optimized. Upon relaxation, the structure of the ideal bare nanowire get reconstructed. Furthermore, to allow possible reconstructions involving two unit cells, structure optimization of an ideal nanowire is performed. In the later optimization, the energy per Si atom and the atomic structure did not change from the single cell optimization. Table 1 lists the studied cases with the atomic numbers within them, the area of the cross section and band gap. The lattice constants along the chain axes were optimized, involving variations within 3.5%. We have performed first-principles plane wave calculations[6,7] within DFT [9] using ultrasoft pseudopotentials.[7,8] The exchange correlation potential has been approximated by generalized gradient approximation GGA using Perdew Burke Ernzerhof exchange correlation functional[11] both for spin-polarized and spin unpolarized cases. For partial occupancies, we use the Methfessel-Paxton smearing method[12] The adopted smearing width is 0.1 eV for the atomic relaxation and 0.05 for the accurate band structure analysis and density of states calculations here total energy/atom convergence tolerance is 0.1000E-05 eV. All structures have been treated within a super cell geometry using the periodic boundary conditions. The lattice constants of the tetragonal super cell in the *x-y* plane are taken as *asc* = 23.10918, *bsc* = 22.661879 Å and *csc* = 9.616978 Å along the *z* axis. For the double unit cell calculations, the lattice constant is taken as *csc* = 2*co* to prevent the interactions between the nearest neighbor impurity atoms located in adjacent cells and the number of atom in a super cell is given in Table 1. In the selfconsistent potential and total energy calculations, the Brillouin zone of SiNW is sampled in the **k** space within the Monkhorst-Pack scheme[10,13] by 1x1x40 mesh points as determined by the convergence tests. A plane wave basis set with kinetic energy in the range from 300 to 400 eV has been used depending on the structure of SiNWS. All atomic positions and lattice parameters are optimized by using the BFGS where total energy and atomic forces are minimized. The convergence or energy is chosen as 10^{-4} eV between two ionic steps, and the maximum force allowed on each atom is 0.3000E-01eV/A. The band structure around the Fermi of this hydorgen passivated silicon nanowire is displayed in Figure 1. Upon passivation with H and subsequent optimization, most of the peak positions coincide with those of the ideal SiNW.

Table 1

H	*si*	*O*	*Lattice Parameters (Å)*	*Cell Angles*	*Area of Cross Section (Å)²*	*Band Gap (eV)*
26	28		a = 21.874769 b = 22.477034 c = 9.518816	alpha = 90.000003 beta = 92.272954 gamma = 90.000005	41.470582	2.596
26	28	26	a = 27.276257 b = 26.051662 c = 9.701105	alpha = 87.147725 beta = 90.207746 gamma = 88.013101	41.470582	1.823

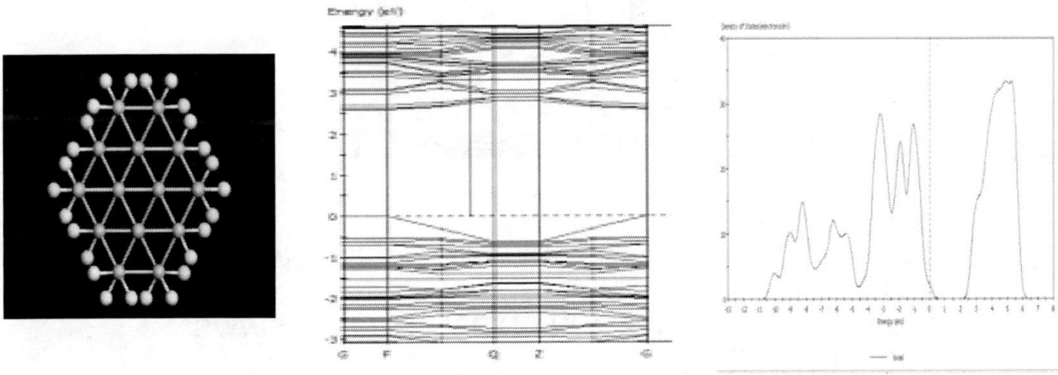

Fig. 1

Yellow sphere represents the silicon atom and white spheres represent hydrogen atoms. And in the middle energy band structure of SiNWS are presented. On the left, Density of states of SiNWs are presented.

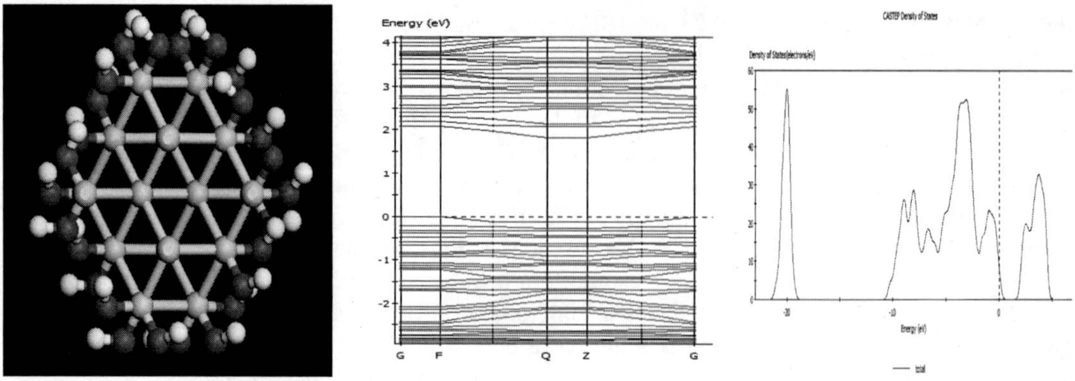

Fig. 2

Yellow sphere represents the silicon atom and white spheres represent hydrogen atoms and red sphere represent oxygen atoms. And in the middle energy band structure of SiNWS are presented. On the left, Density of states of SiNWs are presented.

Fig. 3

Optimesed Silicon nanowire (little distortion of chemical bonds close to the surfaces).

EFFECT OF HYDROXYL PASSIVATION ON BAND GAP OF SINWS ALONG [1 1 1] DIRECTION

Until now we have discussed Si nanowire with hydrogen full-termination.But, in the experimental usage dopant may be adsorbed on the Si nanowire surface. The adsorbents could change the electronic structures and other related properties of Si nanowire. Here, we examine the influence of the adsorbents on the band structure of Si nanowires along [111] direction. Starting with the hydrogen terminated Si nanowires. We replaced all the hydrogen atom with hydroxyl and the structures of hydroxyl-terminated [111] Si nanowire is shown in Figure 2. In Figure 2 yellow bolls represent silicon atoms, white ones for hydrogen and red bolls represent the oxygen atoms. During relaxation the bond lengths are modified, and the bond angles change a little. Most of the silicon wire, however, is still preserved. That is, the hexagonal cross section and the hydrogen saturation on the surface still keep the characteristics of those in the hydrogen-terminated Si nanowires. The optimization did not change the configurations. The crystalline core was almost the same as the hydrogen terminated Si nanowires, with just a little distortion of chemical bonds close to the surfaces as shown in Figure 3. Employing DFT calculation with Perdew Burke Ernzerhof functional, we studied the adsorbed [111] Si nanowires systematically. We found that the total energies of the systems vary from –4903.690 eV to –4903.674 eV, showing a very small discrepancy between each other. It indicated that the stability of hydorogen passivated and hydroxyl adsorbed systems is almost the same.Consequently, the electronic band structures of these nanowires are shown in Figure 1 and Figure 2 we found noticeable difference between them from the comparison. We calculated their electronic band structures and compared them with the former hydrogen-terminated Si nanowires. Remarkable change in the electronic band structure occured between hydrogen terminated and hydroxyl terminated Si nanowire, that is, the band gap of [111] Si nanowire reduces on passivation with hydroxyl atoms.The band gap varied from 2.596 eV to 1.283 eV.

RESULTS

Our results show that SiNWs passivated with hydroxyl has less band gap in comparison to the Si nanowire passivated with hydrogen. The electronic band structures of these nanowires are shown in Figures 1 and 2. The electronic band structure shows that the hydroxyl passivation effect the physical properties of the whole wire, such as it modified the electronic band structure. We calculated the electronic band structure and compared them with the hydrogen-terminated Si nanowires. Remarkable change in the electronic band structure occured between hydrogen terminated and hydroxyl terminated Si nanowire, that is, the band gap of [111] Si nanowire reduces on passivation with hydroxyl atoms.The band gap varied from 2.596 eV to 1.283 eV.

REFERENCES

[1] Lieber, C.M. (2003). Nanoscale science and technology: building a big future from small things *MRS Bull.,* 28, 486–91, 15, 353–89.

[2] Lieber, C.M. and Wang, Z.L. (2007). Functional nanowires *MRS Bull.,* 32, 99–104.

[3] Diao, J., Gall, K. and Dunn, M.L. (2003). Surface-stress-induced phase transformation in metal nanowires *Nat. Mater.,* 2, 656–60.

[4] Cammarata, R.C. (1994). Surface and interface stress effects in thin films, *Prog. Surf. Sci.,* 46, 1–38.

[5] Wu, B., Heidelberg, A. and Boland, J.J. (2005). Mechanical properties of ultrahigh-strength gold nanowires, *Nat. Mater.,* 4, 525–9.

[6] Payne, M.C., Teter, M.P., Allen, D.C., Arias, T.A. and Joannopoulos, J.D., *Rev. Mod. Phys.*, 64, 1045 [1992].
[7] Numerical computations have been carried out using VASP software. G. Kresse and J.
[8] Vanderbilt, D., *Phys. Rev. B* 41, R7892 [1990].
[9] Holmes, J.D., Johnston, K.P., Doty, R.C. and Korgel, B.A., *Science* 287, 1471 [2000].
[10] Monkhorst, H.J. and Pack, J.D., *Phys. Rev* B 13, 5188[1976].
[11] Perdew, J.P., Chevary, J.A., Vosko, S.H., Jackson, K.A., Pederson, M.R., Singh, D.J. and Fiolhais, C., *Phys. Rev. B*, 46, 6671 [1992].
[12] Methfessel, M. and Paxton, A.T., *Phys. Rev. B* 40, 3616 [1989].
[13] Monkhorst, H.J. and Pack, J.D., *Phys. Rev. B,* 13, 5188 [1976].

AN INSIGHT INTO BHARADWAJA'S VYMANIKA ŚASTRA IN THE PREPARATION OF VATADHARANA LOHA

PADMASHREE ANAND, R.S. KULKARNI, E.S. DWARAKADASA[1] AND S. VYNATEYA[2]

CMRTU, RVCE Campus, Mysore Road, Bangalore
[1]Metallurgy (Retd.), IISc., Bangalore
[2]MTD, CPRI, Bangalore
E-mail: padmashreeanand@gmail.com

ABSTRACT

Throughout history, there are records scattered in Sanskrit literature about Indian scientists to be the best materials scientists since they knew how to use the available materials without affecting the balance of materials in nature or the environment. In several instances, proper links regarding sources and references follow up work, practice and procedures are missing despite the uniqueness and possible utility being entirely new to modern science. Maharshi Bharadwaja's ancient work on 'Vymanika Śastra' is one of the most significant in this regard. A few attempts have been made to decipher and follow details given in this treatise to repeat the experiments and recreate some of the property—specific materials.

The purpose of this research is to re-enact the making of one such material described in the text, reproduce those conditions and check whether there is validity of the claimed properties. An attempt was made to reproduce 'Vatadharana Loha',—a light and strong material as mentioned in the scripture. The methodology adopted, analysis of the microstructures of the samples prepared and certain properties of Vatadharana Loha obtained, are described in this paper.

INTRODUCTION

Vymanika Śastra, a part of the unknown work, 'Yantra Sarvasva'(encyclopedia of machines) authored by Maharshi Bharadwaja consists of nearly 6000 lines in lucid Sanskrit dealing with

V. Rajendran, R. Yuvakkumar, K. Thyagarajah and K.E. Geckeler (eds.)
Applications of Nano Materials: Electronics, Energy and Environment, pp. 233–238 (2012).

the construction of an aircraft in about 600 aphorisms embedded in 8 chapters. Most of the scientific information in ancient India has been recorded in Sanskrit language but in the encrypted form. The unique feature in addressing knowledge of ancient Indian origin is in the methodology adopted, distinctly different from the approach of modern scientific methods. The best approach to check the veracity of the claims made in the text is to reproduce at least some of the materials mentioned therein.

EXPERIMENTAL WORK

The property—specific material selected for the study was Vatadharana Loha—a light, soft and strong material. The deciphering of terminologies for the Vatadharana loha was done and an attempt was made to reproduce it whose composition is given in Table 1.

Table 1: Composition of Vatadharana Loha

Sanskrit Name	Modern Chemical Equivalent	Proportion
Simhaasyakam	Iron rust	Equal parts
Shaarana	Juice of jute??	
Surya	Copper	
Varchula	Salt Petre (KNO_3)	
Temperature : 102 kakshyas*		

*Kakshya—appears to be a non-linear scale of temperature.

The coal fired blower type furnace was allowed to attain a temperature of 900°C and A1 (~1 kg) SiC crucible was kept inside the furnace. For each trial, when the crucible reached the temperature of the furnace, the charge [Table 2] was placed in the crucible with occasional stirring to obtain homogeneity and the crucible was allowed to attain a temperature of ~1350°C so as to melt the charge. Shaarana probably means juice of jute, and if it is so, the purpose of addition of it probably is to facilitate the reduction of iron oxide to iron. So mother in law's tongue (a highly fibrous plant with the biological name Sansevieria trifasciata) was used in place of juice of jute in Trial # 4 and 5. The molten metal was poured into a metal mold (Trials 1, 2 and 4) and into the sand mold (Trial 3 and 5).

Table 2

Vatadharana Loha	*Trial # 1*	*Trial # 2*	*Trial # 3*	*Trial # 4 & 5*
Raw materials	Weight in grams	Weight in grams	Weight in grams	Weight in grams
Iron rust	25	55	75	55
Juice of jute??	Not available	Not available	Not available	110g of Mother in law's tongue
Copper	200	110	150	110
Salt Petre (KNO_3)	200	110	150	110

RESULTS

Chemistry, conductivity and density were checked (Table 3).

Table 3

Trial #	*Chemical Composition*	*Electrical Conductivity Sm^{-1}*	*Electrical Conductivity (IACS)*	*Thermal Conductivity W/mK*	*Density g/cm^3*
1	Cu 97.39%; Fe 2.61%	1.867×10^7	32.2%	137.22	7.95
2	Cu 97.09%; Fe 2.91%	1.229×10^7	21.2%	90.33	7.81
3	Cu 96.79%; Fe 3.21%	1.036×10^7	17.9%	75.64	7.68
4	Cu 97.15%; Fe 2.85%	1.387×10^7	23.91%	101.26	7.75
5	Cu 97.29%; Fe 2.71%	1.519×10^7	26.2%	110.9	7.87

The samples were prepared for metallographic examination in the standard way and the microstructure was examined (Figures 1–5).

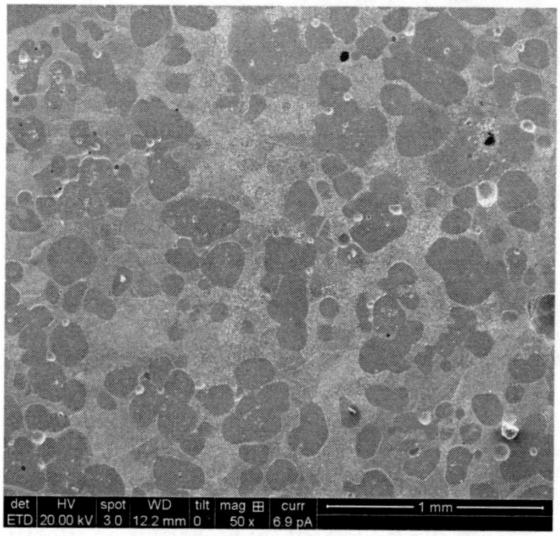

Fig. 1: Trial # 1

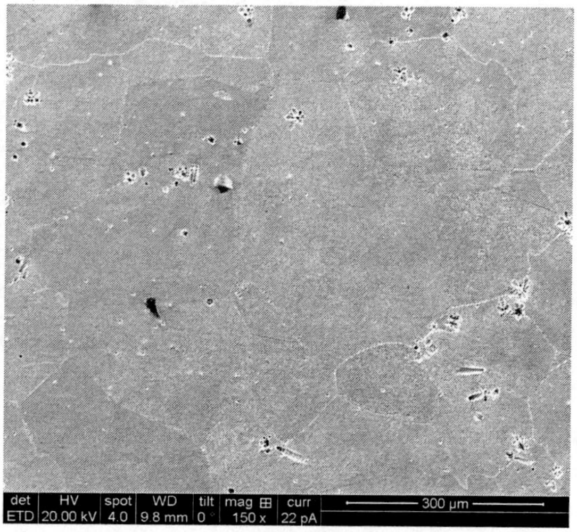

Fig. 2: Trial # 2

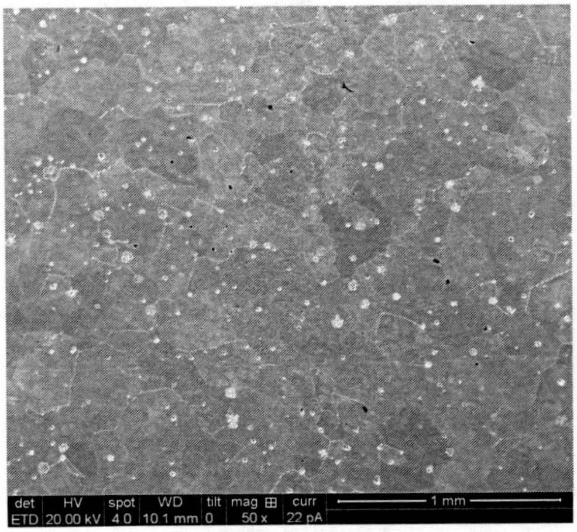

Fig. 3: Trial # 3

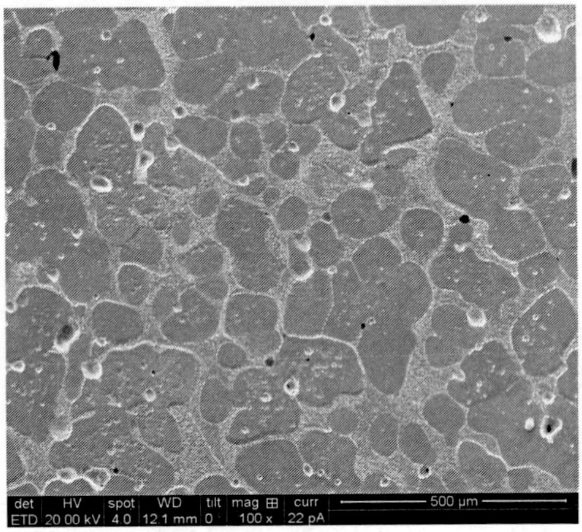

Fig. 4: Trial # 4

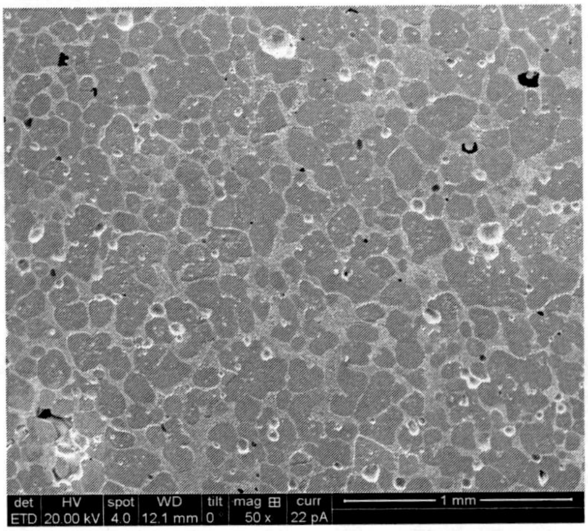

Fig. 5: Trial # 5

DISCUSSION

Figure 1 indicated iron powder particles were dispersed in the matrix of copper. Fine grains of copper were observed throughout the microstructure. Pores were observed distributed uniformly across the microstructure which may have resulted due to the gas generated by addition of KNO_3. The single phase structure in Figure 2 shows that iron has dissolved in copper and grains of alpha phase of copper. Many pores have formed flower like group structures. In Figure 3, the single phase structure indicates that iron has dissolved in copper and there are grains of alpha phase of copper. Pores are more when compared to Trial 1 and 2. It is evident from the microstructure (Figure 4), iron has dissolved in copper and grains of alpha phase of copper are present. Also the pores are evidently more when compared to previous trials. In the microstructure of Trial 5 (Figure 5), iron powder particles were dispersed in the matrix of copper. The uniform distribution of pores in the structure is much evident and increase in the number of pores is probably due to the addition of Mother in law's tongue in both Trials 4 and 5. The results (Table 3) indicate that with increase in the percentage of iron in the product, electrical conductivity, thermal conductivity and density of the product decreases. There was also a limitation on the temperatures that could be reached as also the rates of heating. With the availability of quickly heating modern furnaces, it is possible that something is amiss.

CONCLUSION

This research is a step towards integration of ancient and modern science and technology. It is expected that the present experiments will lead to a greater clarification and lend credibility about the work reported in ancient literature. This research sheds light on the fact that KNO_3 can be used to produce porosity in copper alloys and thus further research in this regard may culminate in the preparation of copper foam which will be a light material and thus can find lots of applications in aerospace industry.

BIBLIOGRAPHY

[1] Shastra, Vymanika and Bharadwaja, Maharshi, Translated into English and edited, G.R. Josyer, Coronation Press, Mysore (1973).

[2] Sastra, Brihad Vimana (Ed) Swami Brahmamuni Parivrajaka Gurukul Kandgi, Dayanand Bhavan, New Delhi (1959).

[3] Vimana in Ancient India, D.K. Kanjilal, Sanskrit Pustak Bhandar, Calcutta (1985).

[4] Sheth, Naren, "Was Maharshi Bharadwaja a Solar Scientist?", in 'Science and the Vedas', (Ed.) Ishwarbhai Patel, Maharshi Academy of Vedic Sciences (1984) 36–39.

[5] VIMANA Aircraft of Ancient India and Atlantis, David Hatcher Childress, Adventure Unlimited Press (1994), 103–200.

[6] High Technology in Ancient Sanskrit Literature—A Preliminary Report, CSR Prabhu, Samskritavignanavaibhavam, June (1999).

[7] Briefly ……., Hinduism Today, April,1997.

[8] Dongre, N.G., Malaviya, S.K. and Rao, P. Ramachandra, The study of various materials described in Amshubodhini of Maharshi Bharadwaja, Final Report, NML, Jamshedpur, (2000)

[9] Amshubodhini, Maharshi Bharadwaja, pp..69-77

[10] Kawadkar, M.K., Bharatiya Bouddhika Sampada; May, Aug, Nov (2000); Feb, May, Aug (2001); Feb, Aug (2002)

[11] Damodar Joshi, Rasaratnasamuccaya by Vagbhata, edited with English translation, (1991–1992)

[12] Dongre, N.G., Malaviya, S.K. and Rao, P. Ramachandra, Prakasha Stambhanabhida Loha of Maharshi Bharadwaja, *Journal of History of Sciences,* 33 (4) (1998), pp. 273–280.

[13] Didolkar, V.K. and Kawadkar, M.K., A novel infrared transparent glass material as per 'Amshubodhini' (2002).

[14] Dina Nath Mishra, Wonder that was India, The Pioneer, Feb 06 (2005).

[15] M.P. Rao *et al.*, Vymanika Shaastra Rediscovered – Project Study Report, AR & DB, New Delhi (2002).

[16] Vimana Ancient flying Machine, http://www.crystalinks.com, Ellie Crystal

[17] Ancient and Lost Civilizations, http://www.crystalinks.com, Ellie Crystal

[18] Proof Advanced Ancient Indian Civilization Existed, http://www.abovetopsecret.com

[19] Airplanes in Ancient India, http://www.SciForums.com

TRANSITION METAL (Co, Cu AND Ni) OXIDES-DOPED ZnO-PPy NANOCOMPOSITE ON PT MODIFIED ELECTRODES FOR NO_x DETECTION

SUBASH PRAKASH, SEENIVASAN RAJESH[1], SARKARAI RAJASINGH, CHANDRAN KARUNAKARAN[1] AND VEERAPANDY VASU[2]

Department of Physics, VHNSN College, Virudhunagar, Tamil Nadu
[1]Biomedical Research Laboratory, Department of Chemistry, VHNSN College, Virudhunagar, Tamil Nadu
[2]School of Physics, Madurai Kamaraj University, Madurai, Tamil Nadu
E-mail: mms_prakash@yahoo.co.in

ABSTRACT

Semiconductor oxides are important materials in gas sensing systems because of their sensitivity and selectivity to specific gases. This research investigates the effect of transition metal viz., Co-, Cu- and Ni- oxides (TMOs) doped ZnO on the NO_x gas-sensing ability at room temperature. The electrochemical NO_x sensors were designed and developed by forming the nanocomposite of TMOs-doped ZnO-polypyrrole (PPy) on Pt electrode. For the morphological characterization, scanning electron microscopy was used to determine the effect of various dopants on the grain structure. Energy dispersive X-ray analysis was used to verify the composition of the TMOs-doped ZnO. Further, the electrochemical behavior of the TMOs-doped ZnO-PPy-Pt electrodes as sensor investigated by cyclic voltammetry exhibited the well defined characteristic redox peaks for the dopants. The TMOs-doped ZnO-PPy-Pt electrodes showed an enhanced electrocatalytic activity towards NO_x compared to the undoped ZnO-PPy-Pt electrode. A linear response over the concentrations ranging from 0.9 to 2000 μM, 1 to 2000 μM and 0.8 to 2500 μM, with the detection limit of 0.9 μM, 1 μM and 0.8 μM for NO_x was observed at the Co/ZnO-PPy-Pt, Cu/ZnO-PPy-Pt and Ni/ZnO-PPy-Pt electrodes respectively. Here, the TMOs-doped ZnO-PPy nanocomposite facilitated the electron transfer provided an electrochemical approach for the determination of NO_x at room temperature with good sensitivity, good reproducibility and retained stability over a long period.

V. Rajendran, R. Yuvakkumar, K. Thyagarajah and K.E. Geckeler (eds.)
Applications of Nano Materials: Electronics, Energy and Environment, pp. 239–246 (2012).

INTRODUCTION

Nitric oxide (NO) plays a vital role in the normal vascular biology and pathophysiology.[1] NO, a neuronal signaling molecule in the central and peripheral nervous system can be synthesized by mammalian cells. Further it acts as a physiological messenger and cytotoxic agent.[2] It is very unstable with a half-life of 2–30 seconds and rapidly reacts with molecular oxygen to form NO_x (NO and NO_2).[3] Therefore, sensitive and accurate methods for the determination of NO_x are required.

Many methods for the determination of NO_x have been proposed,[4] such as electron paramagnetic resonance,[5] spectrophotometry [6] and chemiluminesence.[7] Among them, electrochemical techniques are proved to be powerful tools due to their advantages of simplicity, portability, inexpensive and fast analysis.[8] Recently, the ZnO-In_2O_3 films have been widely used for detecting NO_x gas.[9] However, the sensor lacking high sensitivity and operating at high temperature (200–500°C) is a persistent issue.[10] So, it is necessary to develop a highly sensitive electrochemical sensor by modification of ZnO performing at room temperature. In order to improve the stability, sensitivity and selectivity towards specific gases at room temperature, small amounts of additives/dopants are added into the pure MOS gas sensors.

Here in this present work, the development of NO_x sensors is based on the use of ZnO thin films doped with different TM *viz.,* Co, Cu and Ni oxides as additives/catalyst. According to our knowledge, this is the first report describing the TMOs-doped ZnO-PPy-Pt electrodes for NO_x sensing applications. The conducting polymer of polypyrrole (PPy) matrix increases the rates of reaction occurring at the electrode surface by increasing the surface area and also mediating the direct electron transport.[11]

EXPERIMENTAL

Reagents

Pyrrole, sodium nitrite ($NaNO_2$), and diethylenetriaminepentaacetic acid (DTPA) were purchased from Sigma-Aldrich Company, USA. $Co(NO_3)_2$ and $Ni(NO_3)_2$ were purchased from Merck limited, USA. $Cu(NO_3)_2$ was purchased from Loba chemie. Pyrrole was used without purification.

Instrumentations

All electrochemical experiments were performed on CHI 2000A electrochemical workstation (CHI, USA) with a conventional three electrode system which consisted of an Ag/AgCl wire as reference electrode, a Pt wire as auxiliary or counter electrode, and a TMO-doped ZnO-PPy-Pt as the working electrode. The size of Pt wire working electrode was 0.5 mm in diameter. The morphological scanning electron microscopic (SEM) images and energy dispersive X-ray Analysis (EDX) spectrum were obtained using a FEI Quanta FEG 200-High Resolution Scanning Electron Microscope (FEI Co. Netherlands).

Fabrication of TMOs-Doped ZnO-PPy-Pt Electrodes

Prior to the experiment, the Pt electrode of 0.5 mm diameter was polished with alumina powder (size 0.05 and 1.0 micron) and washed twice with double distilled water. The cleaned Pt electrode was pretreated with 1 M H_2SO_4 by scanning the potential from –0.2 to +1.45 V for 10 cycles and then washed with double distilled water. This pretreatment was employed before each electropolymerization of pyrrole for the effect of removing adsorbed species and altering the micro structure and roughness of the electrode surface. ZnO entrapped PPy matrix on Pt electrode was prepared by a two-step procedure. First, PPy films were polymerized on Pt electrode according to the published report.[12] Briefly, the irreversible oxidation of 0.4 M pyrrole in 0.1 M KCl occurred in the potential range from 0 to +0.9 V vs. Ag/AgCl electrode for 10 complete cycles under argon atmosphere. The optimum of 10 cycles for pyrrole electropolymerization was used because of its increased conductivity and surface area as reported earlier.[13] Then, pure ZnO was electrochemically deposited on PPy-Pt electrodes from the solution containing 0.1 mM $Zn(NO_3)_2$ and 0.1 mM KCl as supporting electrolyte by applying the potential from 0 to –0.9 V vs Ag/AgCl.[31] $Co(NO_3)_2.6H_2O/Cu(NO_3)_2.6H_2O/$ $Ni(NO_3)_2.6H_2O$ was added to a solution of $Zn(NO_3)_2$ (0.1 mM) in 20 ml of methanol-ethanol (3:1 by volume) mixture respectively. The resulting solution was allowed to evaporate at room temperature. TMOs-doped ZnO was electrochemically deposited on PPy-Pt electrodes from the above resulting substance and 0.1 mM KCl as supporting electrolyte by applying the potential from 0 to –0.9 V vs Ag/AgCl.[14]

Sample Preparation

NO_x gas was prepared according to the following equation. NO_x gas was generated based on the stoichiometric reaction in the presence of air.[15]

$$NaNO_2 + H_2SO_4\ NO_x\uparrow + H_2O + Na_2SO_4 \quad \text{... (1)}$$

Since the conversion of NO_2^- to NO_x is stoichiometric, the final concentration of NO_x generated is equal to the concentration of $NaNO_2$ in the solution.[16]

RESULTS AND DISCUSSION

Characterization of TMO-Doped ZnO-PPy-Pt Electrodes

The surface morphological images of Co/ZnO-PPy-Pt, Cu/ZnO-PPy-Pt and Ni/ZnO-PPy-Pt electrodes are shown in Figure 1. The SEM morphological images indicated the co-existence of doped TMOs on ZnO-PPy nanocomposite. The low density ZnO material is favorable for the catalytic application. ZnO strongly adhere onto the network structure provided by PPy. They combine each other very closer, forming the layered structured of staggered, which is important to tightly attach the doped ZnO-PPy nanocomposite film on the surface of the Pt electrode and increase the electron transfer rate.

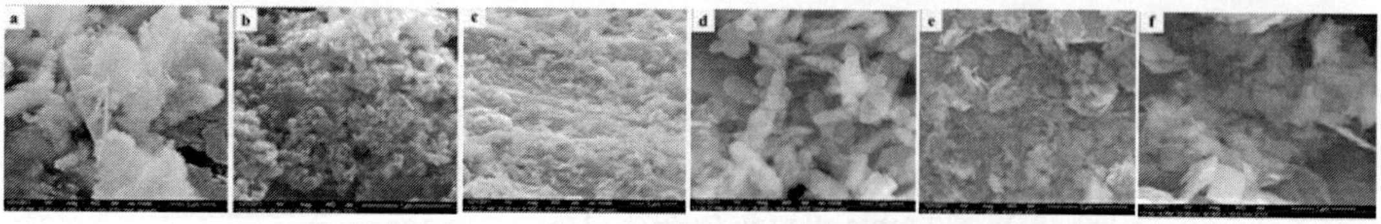

Fig. 1: SEM images of (a) Co/ZnO-PPy-Pt (5 μm), (b) Co/ZnO-PPy-Pt (1 μm), (c) Cu/ZnO-PPy-Pt (5 μm), (d) Cu/ZnO-PPy-Pt (1 μm), (e) Ni/ZnO-PPy-Pt (5 μm) and (f) Ni/ZnO-PPy-Pt (1 μm) electrodes (different magnification)

A representative EDX spectrum was performed for the Co/ZnO-PPy-Pt, Cu/ZnO-PPy-Pt and Ni/ZnO-PPy-Pt electrodes respectively and shown in Figure 2 (a, b and c). The peaks associated with Zn, C, O, Co, Cu and Ni were seen in the EDX spectrum, which confirm that the presence of TMOs-doped ZnO-PPy nanocomposite.

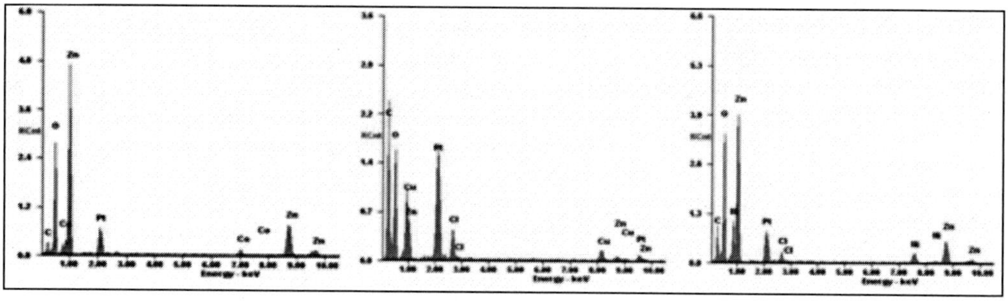

Fig. 2: EDX Spectrum of (a) Co/ZnO-PPy-Pt Electrode, (b) Cu/ZnO-PPy-Pt Electrode, (c) Ni/ZnO-PPy-Pt Electrode

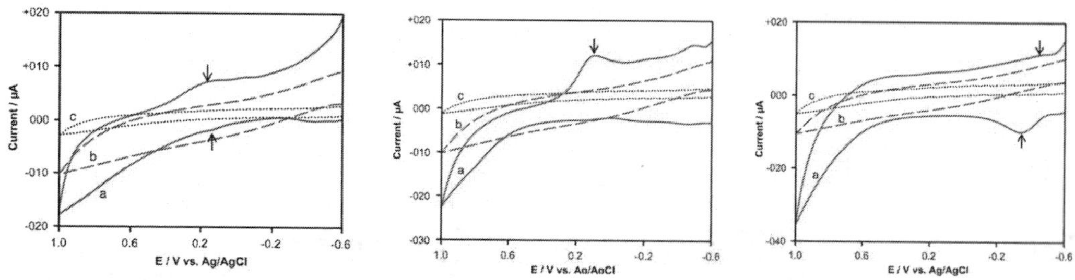

Fig. 3: (i) Typical CV response of (a) Co/ZnO-PPy-Pt (solid), (b) ZnO-PPy-Pt (dashed) and (c) PPy-Pt (dotted) electrodes in 0.1 M PBS (pH 7.0); **(ii)** Typical CV response of (a) Cu/ZnO-PPy-Pt (solid), (b) ZnO-PPy-Pt (dashed) and (c) PPy-Pt (dotted) electrodes in 0.1 M PBS (pH 7.0); **(iii)** Typical CV response of (a) Ni/ZnO-PPy-Pt (solid), (b) ZnO-PPy-Pt (dashed) and (c) PPy-Pt (dotted) electrodes in 0.1 M PBS (pH 7.0); scan rate: 50 mVs^{-1} vs. Ag/AgCl.

The cyclic voltammetric studies were performed for the Co/ZnO-PPy-Pt, Cu/ZnO-PPy-Pt and Ni/ZnO-PPy-Pt electrodes in 0.1 M PBS (pH 7.0) at a scan rate of 50 mVs^{-1} and these voltammograms are shown in Figure 3(i), 3(ii) and 3(iii). The Co/ZnO-PPy-Pt electrode

(Figure 3(i)) exhibited a pair of redox peaks at 0.12 V (anodic peak) and 0.15 V (cathodic peak) indicating the presence of Co. In Figure 3(ii), the quasi reversible peak obtained at 0.10 V confirms the doping of Cu element in the Cu/ZnO-PPy-Pt electrode. A pair of redox peaks were obtained at –0.34 V (anodic peak) and –0.47 V (cathodic peak) indicating the presence of doped Ni in Ni/ZnO-PPy-Pt electrode (Figure 3(iii)). But, there are imperceptible redox peaks exhibited by ZnO-PPy-Pt and PPy-Pt electrodes clearly confirming the presence of dopants.

Electrochemical Response of TMOs-Doped ZnO-PPy-Pt Electrode for NO_x Sensing

The typical CV responses of the TMOs-doped ZnO-PPy-Pt, ZnO-PPy-Pt and PPy-Pt electrodes in 0.1 M PBS in the applied potential range from –1.0 V to +1.0 V at a scan rate of 50 mVs^{-1} in the presence of NO_x obtained from 500 μM NO_2^- reaction with H_2SO_4. A new irreversible anodic oxidation peak appeared at about +0.89 V (Figure 4(i) a) for the Co/ZnO-PPy-Pt electrode. For the Cu/ZnO-PPy-Pt electrode anodic oxidation peak appeared at about +0.90 V (Figure 4(ii) a). At about +0.88 V (Figure 4(iii) a) a new irreversible anodic oxidation peak appeared for the Ni/ZnO-PPy-Pt electrode.

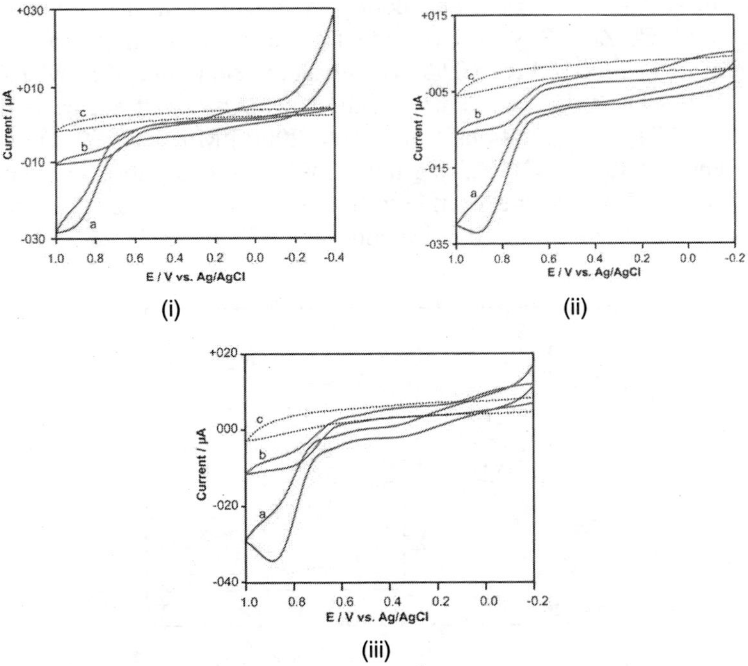

Fig. 4: (i) Typical CV response of (a) Co/ZnO-PPy-Pt (b) ZnO-PPy-Pt and (c) PPy-Pt electrodes in the presence of 500 μM NO_x in 0.1 M PBS (pH 7.0); **(ii)** Typical CV response of (a) Cu/ZnO-PPy-Pt (b) ZnO-PPy-Pt and (c) PPy-Pt electrodes in the presence of 500 μM NO_x in 0.1 M PBS (pH 7.0); **(iii)** Typical CV response of (a) Ni/ZnO-PPy-Pt (b) ZnO-PPy-Pt and (c) PPy-Pt electrodes in the presence of 500 μM NO_x in 0.1 M PBS (pH 7.0); scan rate: 50 mVs^{-1} vs. Ag/AgCl.

But for the ZnO-PPy-Pt electrode, there was only broad anodic oxidation peak response observed (Figure 4(i) b, 4(ii) b, 4(iii) b). And for the PPy-Pt electrode, there was no anodic oxidation peak observed (Figure 4(i) c, 4(ii) c, 4(iii) c). The above results clearly revealed that the enhanced electrocatalytic properties only for the Co/Cu/Ni doped ZnO-PPy nanocomposite. Thus, the TMOs-doped ZnO-PPy nanocomposite combined the advantages of polymer as p-type semiconductors and high electron mobility of inorganic n-type doped semiconductors. In addition, in the absence of H_2SO_4 under the same condition an anodic oxidation peak of NO_2^- observed at about +0.68 V for the TMOs doped ZnO-PPy-Pt electrode. This result clearly shows that the current response observed in TMOs-doped on ZnO-PPy-Pt electrode is only for NO_x.

Effect of NO_x Concentrations

Figures 5–7 exhibit the typical CVs obtained for different concentrations of NO_x using the Co/ZnO-PPy-Pt, Cu/ZnO-PPy-Pt and Ni/ZnO-PPy-Pt electrodes in 0.1 M PBS at 50 mVs^{-1}. A progressive increase in the current response was observed around at +0.88 V with the increase in the concentration of NO_x from 0 to 2000 μM.

The observed anodic peak currents vs. NO_x concentrations for the Co/ZnO-PPy-Pt, Cu/ZnO-PPy-Pt and Ni/ZnO-PPy-Pt electrodes are plotted as shown in Inset Figures 5, 6 and 7. The calibration curve of Co/ZnO-PPy-Pt electrode thus exhibits a linear range for the NO_x concentrations from 1 to 2000 μM ($r^2 = 0.9904$) with a detection limit of 1 μM and sensitivity of 0.022 μA μM^{-1} cm^{-2}. The calibration curve of the Cu/ZnO-PPy-Pt electrode thus exhibits a linear range for the NO_x concentrations from 0.9 to 2000 μM ($r^2 = 0.9931$) with a detection limit of 0.9 μM and sensitivity of 0.023 μA μM^{-1} cm^{-2}. The calibration curve of the Ni/ZnO-PPy-Pt electrode thus exhibits a linear range for the NO_x concentrations from 0.8 to 2500 μM ($r^2 = 0.9967$) with a detection limit of 0.8 μM and sensitivity of 0.023 μA μM^{-1} cm^{-2}.

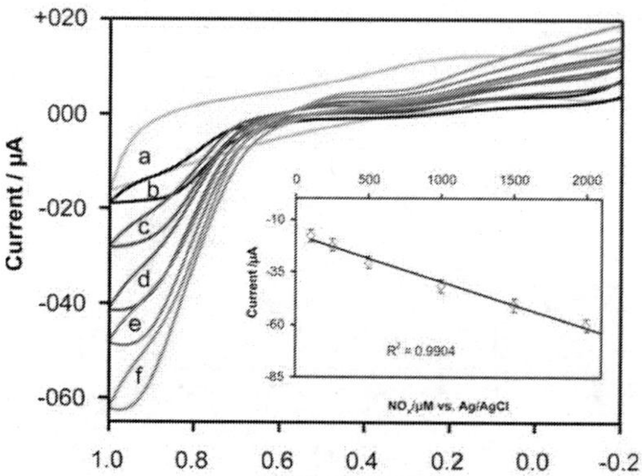

Fig. 5: Typical CV response of Co/ZnO-PPy-Pt electrode in 0.1 M PBS containing (a) control, (b) 250 μM, (c) 500 μM, (d) 1000 μM (e) 1500 μM and (f) 2000 μM NO_x at scan rate of 50 mVs^{-1} vs. Ag/AgCl. Inset. A linear calibration plot of anodic peak currents against NO_x concentrations (each point represents the three repetitive measurements).

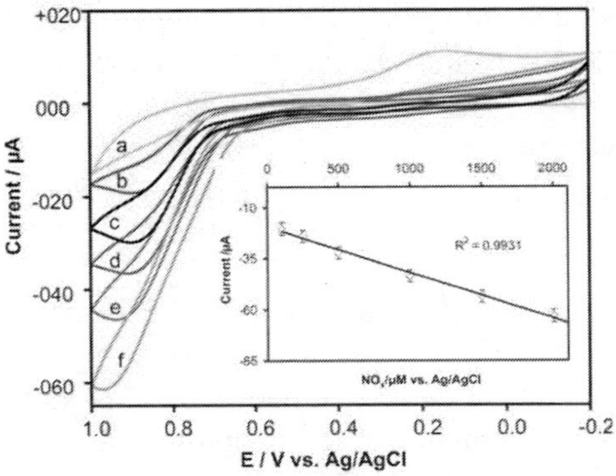

Fig. 6: Typical CV response of Cu/ZnO-PPy-Pt electrode in 0.1 M PBS containing (a) control, (b) 250 µM, (c) 500 µM, (d) 1000 µM (e) 1500 µM and (f) 2000 µM NO_x at scan rate of 50 mVs^{-1} vs. Ag/AgCl. Inset. A linear calibration plot of anodic peak currents against NO_x concentrations (each point represents the three repetitive measurements).

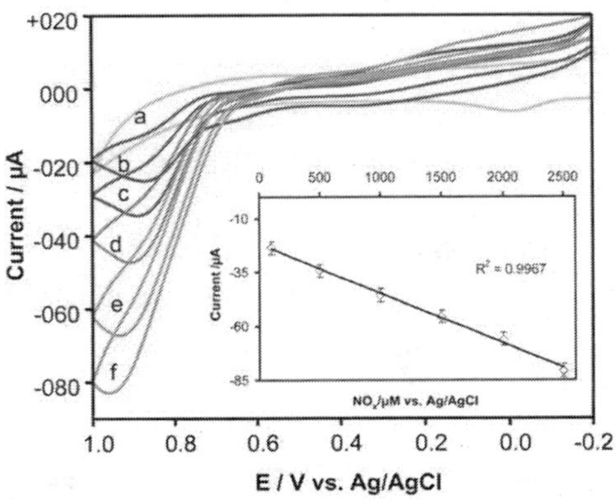

Fig. 7: Typical CV response of Ni/ZnO-PPy-Pt electrode in 0.1 M PBS containing (a) control, (b) 250 µM, (c) 500 µM, (d) 1000 µM, (e) 2000 µM and (f) 2500 µM NO_x at scan rate of 50 mVs^{-1} vs. Ag/AgCl. Inset. A linear calibration plot of anodic peak currents against NO_x concentrations (each point represents the three repetitive measurements).

Selectivity, Stability and Reproducibility of TMOs Doped ZnO-PPy-Pt Electrodes

The reproducibility of the present method for the determinations of NO_x was investigated in 0.1 M PBS containing 500 µM NO_x by repetitive determination in triplicate. There were no changes appeared on the CVs, the results showed good reproducibility of the modified electrodes. Further studying the stability of the sensors, the electrocatalytic responses of NO_x were recorded three times daily and the current responses were found to be constant for above 30 days.

The study of interferences on the electrode response is very important so as to ensure appropriate preparation by without their effects. In this sense, various possible interfering substrates were used to evaluate the selectivity of the sensor for NO_x. The time vs. current response of the sensor for the addition of 500 μM each of ascorbic acid and uric acid into the 0.1 M PBS containing 500 μM NO_x was studied chronoamperometrically, there were no significant changes in the current.

CONCLUSION

In this work, we have prepared the TMOs doped ZnO-PPy modified Pt electrodes for the determination of NO_x. The modified electrodes gave a significant activity towards the electrochemical oxidation of NO_x at room temperature. This study demonstrates that TMO doped ZnO-PPy nanocomposites enhanced and promoted the direct redox electron transfer process compared with the pure ZnO-PPy nanocomposite electrode. The TMO doped ZnO-PPy nanocomposite combined advantages of PPy polymer as p-type semiconductors and high electron mobility of inorganic n-type doped semiconductors, perhaps result in good sensitivity, reproducibility, good stability and selectivity towards NO_x.

REFERENCES

[1] Moncada, S., Palmer, R.M. and Higos, E.A., *Pharmacol. Rev.* 43, 109–142, 1991.

[2] Koshland, D.E., Science 258, 1861–1864, 1992.

[3] Ignarro, L.J., Fukuto, J.M., Griscavage, J.M., Rogers, N.E. and Byrns, R.E., *Proc. Natl. Acad. Sci.* USA, 90, 8103–8107, 1993.

[4] Moorcroft, M.J., Davis, J. and Compton, R.G., *Talanta* 54, 785–803, 2001.

[5] Arroyo, C.M. and Kohno, M., *Free Radical Res. Commun.* 14, 145–155, 1991.

[6] Kelm, M., Feelisch, M., Spahr, R., Piper, H.M., Noack E. and Schrader, J., *Biochem. Biophys. Res. Commun.*, 154, 236–244, 1988.

[7] Aoki, T., *Biomed. Chromatogr.* 4, 128–130, 1990.

[8] Fan, C., Li, G., Zhu. and Zhu, J, D., *Anal. Chim. Acta.*, 423, 95–100, 2000.

[9] Lin, C.-Y., Fang, Y.-Y., Linb, C.-W., Tunney, J.J. and Hoa, K.-C., *Sens. Actuators B* 146 28–34, 2010.

[10] Santos, J., Serrini, P., O'Beirn, B. and Manes, L., *Sens. Actuators B* 43, 154–160, 1997.

[11] Albert, K., Lewis, N., Schauer, C., Sotzing, G., Stitzel, S., Vaid, T. and Walt, D., *Chem. Rev.* 100, 2595–2626, 2000.

[12] Cioffi, N., Ditaranto, N., Torsi, L., Picca, R. A., De Giglio, E., Sabbatini, L., Novello, L., Tantillo, G., Bleve Zacheo, T. and Zambonin, P.G., *Anal. Bioanal. Chem.*, 382, 1912–1918, 2005.

[13] Nagami, N., Umakoshi, H., Shimanouchi, T. and Kuboi, R., *Biochem. Eng. J.,* 21, 221–227, 2004.

[14] Lee, D.U., Pradhan, D., Mouawia, R., Hwan Oh, D., Heinig, N.F., Tong Leung, K. and Prouzet, E., *Chem. Mater.,* 22, 218–225, 2010.

[15] Zhou, H., Xu, Y., Chen, T., Suzuki, I. and Li, G., *Anal. Sci.*, 22, 337–340, 2006.

[16] Zhang, X., *Frontiers in Bioscience* 9, 3434–3446, 2004.

PARTICIPATION OF NANOCRYSTALLINE TiO_2 SURFACE IN THE ELECTRON TRANSFER BETWEEN SEMICONDUCTOR SOLID AND ADSORBED COBALT(III)-*R*py COMPLEX

A.S. GANESHRAJA AND K. ANBALAGAN

Department of Chemistry, Pondicherry University, Kalapet, Pondicherry
E-mail: kanuniv@gmail.com; asgchem84@gmail.com

ABSTRACT

Cis-[Co^{III}(tn)$_2$(Rpy)Br]Br_2, (R = 4-CN, H, 4-Bz, 4-Me, 4-Et, and 4-MeNH), in aqueous 2-propanol exhibit varying adsorption characteristics and led to surface compound formation. UV (λ = 254) excitation of the nano-TiO_2//cobalt(III)-(Rpy) surface compound resulted in interfacial electron transfer reaction (IFET). The IFET has been found to dependent upon the coordination environment of the complex, more precisely due to Rpy ligand. The latter being electron donor/acceptor, alters the photoefficiency of formation of the Co^{II}_{aq} species. In addition, the proposed mechanism of interfacial electron transfer reaction includes the formation of Co^{II} ion implanted nanocrystalline TiO_2. This Photoreduction was found to be solvent controlled. Photoefficiency of the formation of Co^{II}_{aq} was spectrally analyzed simultaneously Co^{II}:TiO_2 was isolated from the photolyte solution. The isolated solid was subjected to FTIR, DRS, PXRD, and SEM-EDX instrumental analysis. It is concluded that the removal of metal ion in the form of complex is coordination structure dependent, hence, seems more specific in removal efficiency and in doping the anatase lattice. Semiconductor nanocrystalline solid such as TiO_2 has focus as effective photocatalysts in removal of metal ion and in doping of semiconductor matrix.

V. Rajendran, R. Yuvakkumar, K. Thyagarajah and K.E. Geckeler (eds.)
Applications of Nano Materials: Electronics, Energy and Environment, pp. 247–254 (2012).

INTRODUCTION

Understanding the photophysics of transition metal complexes attached to semiconductor surfaces is essential for the design of artificial systems for solar energy conversion. In particular, Cobalt(III)-pyridyl complexes have attracted a great deal of attention as a promising class of compounds with long-lived charge-separated states and rich photochemical properties. However, their electronic excitations and photoconversion mechanisms remain only partially understood. This paper builds upon our recent work[1] and addresses the study of TiO_2 nanoparticles sensitized with *cis*-$[Co^{III}(tn)_2(R\text{py})Br]Br_2$ (where *R* = *4*-CN, *H*, *4*-Bz, *4*-Me, *4*-Et, and *4*-MeNH) adsorbates. Our study includes the characterization of the electronic excitations, electron injection time scales, and interfacial electron transfer mechanisms,[2] photoefficiency of interfacial electron transfer (IET) and photo degradation of 2-propanol. Here, we focus on TiO_2 surfaces modified by Co(III)-pyridyl complexes attached by nitrogen linkers. Emphasis is given to the characterization of the electronic excitations and injection time scales as determined by the nature of the molecular adsorbates and the attachment modes.

EXPERIMENTAL

Materials and Methods

$CoCl_2.6H_2O$ (99%), *R*-pyridine (*R*py), ferric chloride, potassium oxalate, sodium acetate, ammonium thiocyanate, nanocrystalline titanium dioxide (surface area = 200–220 m^2/g and particle size = 25 nm) and DMSO-d_6 (NMR solvent) were purchased from Sigma Aldrich. 1, 3-diamino propane (LR), pyridine, 1,10-phenanthroline and all other chemicals were purchased from Himedia and SD. Fine Chemicals (India). All the solvents and 1, 3-diamino propane were purified by distillation and water was triply distilled over alkaline $KMnO_4$ in an all glass apparatus. Analytically pure crystals of *cis*-$[Co^{III}(tn)_2(R\text{py})Br]Br_2$ adsorbates (where *R* = *4*-CN, *H*, *4*-Bz, *4*-Me, *4*-Et, and *4*-MeNH) were synthesized by a modified procedure[3] and recrystallized.

Photocatalytic Activity Measurement

The photocatalytic activity of nano-TiO_2 was evaluated by measuring the efficiency of reduction of $Co^{III}(tn)_2(R\text{py})Br^{2+}$ complex in neat water/aqueous 2-propanol solutions under 254 nm light irradiation. The photoefficiency was calculated by estimating photogenerated Co^{II} by Kitson's method[4]. Photoreduction was carried out in a reactor vessel using 254 nm low pressure mercury vapor lamp, which were housed in a fume hood (Lab Guard), covered with black polythene sheet to prevent extraneous light. The photoreactor is double walled quartz vessel, in which, photolyte mixture (100 mg of nano-TiO_2 crystals in 80 mL of *cis*- $Co^{III}(tn)_2(R\text{py})Br^{2+}$ (1.68×10^{-3} M) solution and 1M $NaNO_3$) was taken in the inner jacket and cool water in the outer jacket. Prior to irradiation, the suspension of the catalyst was achieved by ultrasonic treatment and continued to stir magnetically in dark to attain cobalt(III) complex ion adsorption/desorption equilibrium on the catalyst. The photolyte suspension was dosed with 254 nm light for definite irradiation periods (0, 2–16 min). About 4 mL aliquots were sampled and centrifuged to remove the semiconductor particles and then spectrally analyzed. In order to diminish the experimental error, the experiments were repeated at least two-three times for the same sample and calculated the mean value. Photoefficiency of Co(II) formation in terms of percentage, PE (%) was computed using the formula: PE (%) = $[(A_t - A_i)/A_t] \times 100$ where A_i and A_t are the absorbances of the photolysed solutions initially and at definite time interval 't' respectively.

RESULTS AND DISCUSSION

Photocatalytic Reduction of $Co^{III}(tn)_2(Rpy)Br^{2+}$

Cobalt(III)-*R*py complex is a good UV light absorber, however, decomposes on exposure to light over a long period of time due to ligand to metal charge transfer (LMCT) bands (λ = ~356 nm). Surprisingly, addition of nano-TiO_2 (anatase) provoked the complex to degrade more efficiently and a higher amount of cobalt(II) was generated. Figure 1 is a repetitive scan spectra of $Co^{III}(tn)_2(4\text{-}Etpy)Br^{2+}$ complex observed at definite irradiation time intervals; there is blue shift in the absorption maxima at $\lambda = 357 \rightarrow 351$ nm and red shift at $\lambda = 512 \rightarrow 518$ nm. Absence of isosbestic point in the repetitive scan spectra indicates the decomposition of the complex during reduction process suggesting perturbation of the Co^{III} centre due to interfacial electron transfer.

Table 1: Photoefficiency P.E (%) of formation of Co(II) upon λ = 254 nm irradiation of *cis* $Co^{III}(tn)_2(Rpy)Br^{2+}$ in aqueous 2-propanol at 298 K. ionic strength 1M $NaNO_3$, complex concentration = 1.68×10^{-3} M

Rpy		*PE (%) Water/Pr^{i}OH (w/w %)*							*pKa (Rpy)*
		100/0	*95/5*	*90/10*	*85/15*	*80/20*	*75/25*	*70/30*	
4-CN	–	14±0.5	18±1.0	19±1.0	21±0.7	26±0.8	28±0.9	32±0.6	1.90
	nano-TiO_2	17±0.7	20±0.9	24±0.5	29±0.8	31±0.6	37±1.0	41±1.4	
H	–	15±0.8	21±0.7	29 ±1.0	34±0.5	38±1.4	40 ±0.9	42±1.2	5.25
	nano-TiO_2	18 ±1.0	25±0.7	34±0.6	39 ±1.0	45±0.5	51 ±0.9	55 ±0.5	
4-Bz	–	12±0.8	19 ±1.0	24 ±0.6	28±0.7	32 ±0.8	38±0.5	41±1.2	5.59
	nano-TiO_2	18 ±0.6	21±0.8	29±0.7	31±0.5	36 ±1.0	44 ±0.9	57±1.4	
4-Me	–	12 ±1.0	19±1.0	26±0.5	29 ±0.8	34 ±0.9	38±0.8	49±1.4	6.02
	nano-TiO_2	19±0.7	26 ±0.5	33 ±1.0	38 ±0.6	46 ±0.8	54±1.4	57 ±0.9	
4-Et	–	18 ±1.2	24±0.6	28±1.0	31±0.5	36±1.4	41±0.8	49±0.5	6.02
	nano-TiO_2	21±0.6	28±0.5	35 ±1.0	39±0.9	49 ±0.8	58±0.5	64±0.7	
4-MeNH	–	19±0.8	24 ±0.5	29±0.5	32±1.0	46 ±1.0	52±0.8	53±0.9	9.70
	nano-TiO_2	26±0.8	31±0.6	37±0.7	44±0.5	54±1.0	63±0.9	69±0.8	

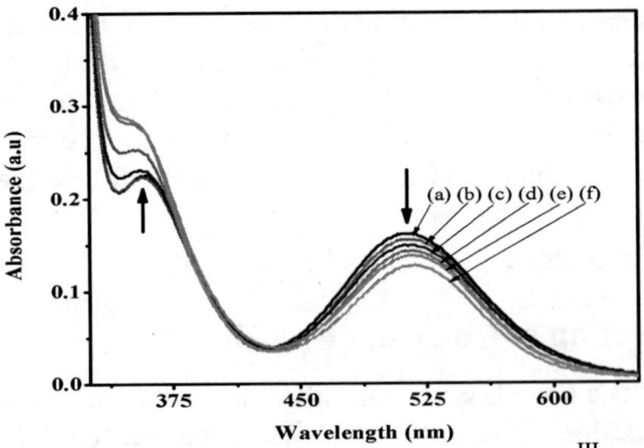

Fig. 1: Repetitive scan spectra recorded for the photocatalysed $Co^{III}(tn)_2(4\text{-Etpy})Br^{2+}$ complex ion in nano-TiO_2 suspension in neat water at various time intervals: (a) 0, (b) 2, (c) 4, (d) 18, (e) 12 and (f) 16 min respectively. Complex concentration = 1.68×10^{-3} M, ionic strength = 1 M $NaNO_3$, pH ~ 7 and at 298 K.

Photo-Oxidation of 2-Propanol to Acetone by ^1H NMR

The photoreduction was systematically followed by NMR measurements which indicate the growth of acetone peak progressively appearing in signal intensity as a function of dosage of light. Table 2 exhibits ^1H NMR signals due to 2-propanol appearing at δ = 1.0 to 1.1 ppm (-CH_3) and δ = 3.8 to 3.93 ppm (-CH) before the initiation of photoreduction and after definite time of irradiation periods. There is a steady growth of signal due to photo released acetone and the signal intensity increases with respect to dosage of light. That is, new ^1H NMR signal at δ = 2.04 ppm indicates the formation of acetone and the signal strength (Table 2) has increased considerably with increase in time of irradiation. In fact, the integrated intensity of the acetone signals gradually increase, whereas the -CH and -CH_3 (2-propanol) signals gradually decrease. The origination of acetone is from the oxidation of 2-propanol resulting due to scavenging of the valence band holes by alcohol: nano-TiO_2 (h^+) + $(CH_3)_2CHOH \rightarrow CH_3COCH_3$.

Table 2: ^1H NMR data for the generator of acetone upon λ = 254 nm irradiation of $Co^{III}(tn)_2(4\text{-Etpy})Br^{2+}$ at 298 K in DMSO-d_6. Photocatalyst nano-TiO_2, ionic strength 1N $NaNO_3$, complex 1.86×10^{-3} M. 'A' and 'P', respectively, denote acetone and 2-propanol

Irradiating Time (min)	*2x -CH₃(d) P δ ppm*		*-CH(m) P δ ppm*							*2x -CH₃(s) A δ ppm*
0	0.98	0.99	3.72	3.73	3.75	3.61	3.78	3.79	3.81	–
8	0.98	0.99	3.72	3.73	3.75	3.76	3.78	3.79	3.81	2.02
16	0.98	0.99	3.72	3.73	3.75	3.76	3.78	3.79	3.81	2.02
45	0.98	0.10	3.71	3.73	3.75	3.76	3.77	3.79	3.80	2.02

Mechanism of Photoinduced Electron Transfer Reaction

Surface compound formation:	nano-TiO_2 + $Co^{III}(tn)_2(Rpy)Br^{2+}$ $\rightarrow$ nano-TiO_2// $Co^{III}(tn)_2(Rpy)Br^{2+}$	(1)
Photoexcitation:	nano-TiO_2 + hν $\rightarrow$ nano-TiO_2 (e^-, CB) + nano-TiO_2 (h^+, VB)	(2)
Charge trapping:	(e^-, CB) $\rightarrow$ (e^-, tr)	(3)
	(h^+, VB) $\rightarrow$ (h^+, tr)	(4)
Electron-hole recombination:	(e^-, tr)/(e^-, CB) + (h^+, tr)/(h^+, VB) $\rightarrow$ (e^-, tr) + heat	(5)
Electron transfer:	nano-TiO_2 (e^-, CB)/(e^-, tr) + ($Co^{III}{}_{ad}$) $\rightarrow$ $Co^{II}{}_{surf}$ + $Co^{II}{}_{aq}$	(6)
Hole transfer:	nano-TiO_2 (h^+, VB)/nano-TiO_2 (h^+, tr) + $(CH_3)_2CHOH$ $\rightarrow$ CH_3COCH_3	(7)

Where (e^-, tr) and (h^+, tr) represent charge trapped surface states while $Co^{III}{}_{ad}$, $Co^{II}{}_{surf}$, and $Co^{II}{}_{aq}$ represent surface compound, surface implanted species and aqueous species respectively. It is unlikely that the photoreduction of Co(III) proceeds through multielectron process, but it should occur in one electron step. The predominant reduction path of the metal centre is as given in eqs. 1–7 with a limited contribution from charge transfer population states like: σ(N) → dσ*(Co) and σ(Cl) → Co.

*p*Ka Dependent Photoreduction

In this investigation, a linear dependence of photoefficiency(%) of $Co^{III}(tn)_2(Rpy)Br^{2+}$/$TiO_2$ suspension with respect to *R*py ligand in terms of acidity constant (*p*Ka) as shown in Figure 2

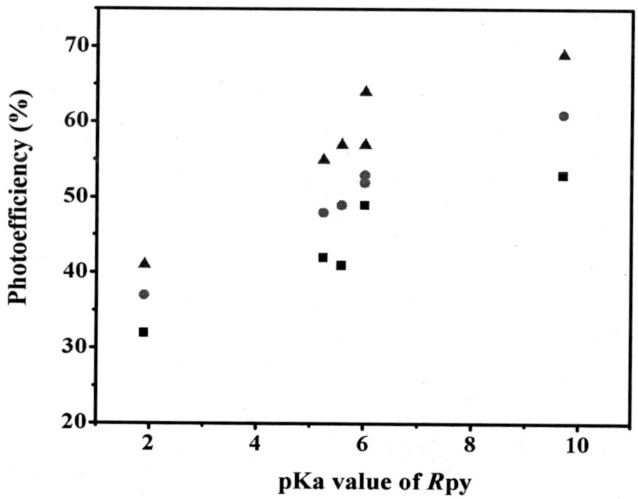

Fig. 2: Linear Plot of Photoefficiency *vs.* *p*Ka Value of Aryl Amine (*R*) in $Co^{III}(tn)_2(Rpy)Br^{2+}$ with the Presence of ■ Absence of Nano-TiO_2, ● Presence Poly-TiO_2, ▲ Presence Nano-TiO_2 in 70/30 (w/w)% water/Pr^iOH

is observed. Linear regression analysis of PE% versus *p*Ka yields slope indicating that the electron donating group in *R*py (*p*Ka > 5.25) enhances photo efficiency (%) and that electron withdrawing group in *R*py (*p*Ka = 1.90) reduces the photo catalytic behavior. Therefore, it can be concluded that coordination environment of the metal centre of a transition metal ion can greatly influence the photochemical character of the complex.

Characterizations of Cobalt-Doped Nano-TiO_2

The isolated solid were subjected to spectral, macrostructural, microstructural, and surface morphological analyses using instrumental methods as FTIR, DRS, PXRD, and SEM-EDX instrumental analysis. Transmittance peaks observed between 450–600 cm^{-1}, which can be assigned to the Ti-O-Ti bond as shown in Figure 3(A). Both the pure and Co-doped nano-TiO_2 showed a characteristic band at 511 cm^{-1} corresponding to a Ti-O bond of the anatase phase.Ti-O vibration tends to shift to lower energy region as Co content in nano-TiO_2. The DRS of undoped anatase showed 100% reflectance, and doped samples absorb more effectively in the range of 480–600 nm. Nano-TiO_2 with low cobalt density absorbs at ~563 nm, and the λ_{max} shifts towards ~574 nm (Figure 3(B)). The red shift attributed to the 3d Co^{2+} electrons into the conduction band of nano-TiO_2.

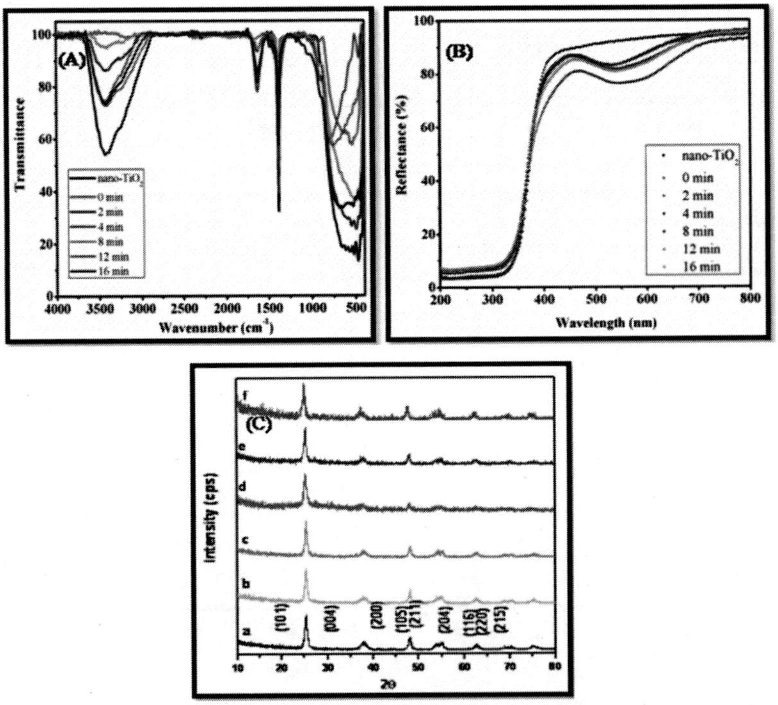

Fig. 3: (A) Diffuse Reflectance Spectra, (B) FT-IR Spectra and (C) PXRD Pattern of the Pure Nano TiO_2 and x%Co: Nano-TiO_2

The XRD patterns indicate phase pure anatase[6] form and confirm the absence of phase transformation to other related crystalline phases like rutile and brookite. This indicates that only a small amount of the elements are substituted into the Ti^{4+} sites, while interstitially incorporated Co(II) cannot be discounted as shown in Figure 3(C). SEM images, which illustrate that the particles mainly belong to anatase phase, loosely agglomerated, spherical with a few hundred nm in size (Figure 4(A)). Homogeneous and continuous surface structure with well dispersed spheres appears as the dopant density in Co/TiO_2 is increased (Figure 4(B)(C)).

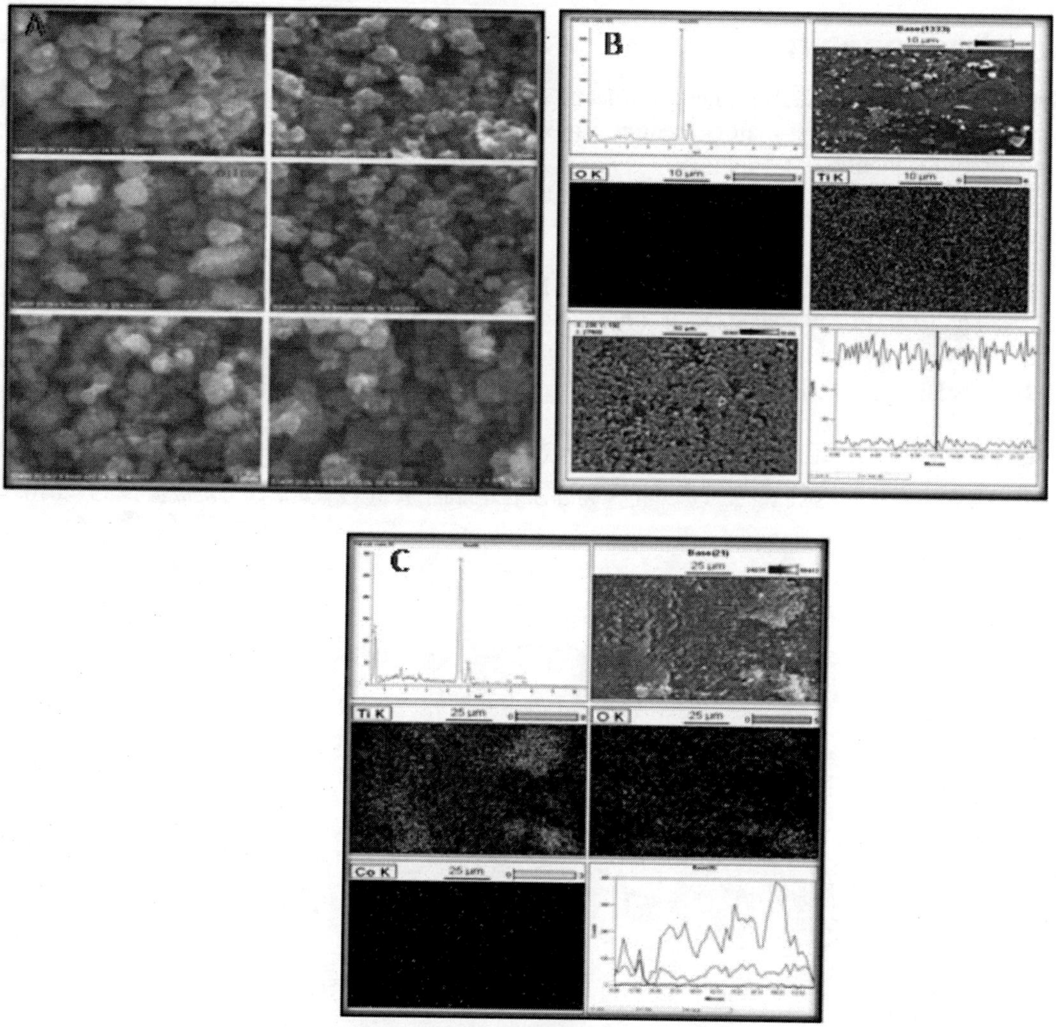

Fig. 4: (A) SEM Images and (B), (C) EDX Spectra, Mapping, Line Spectra of Pure Nano-TiO_2 and x%Co:nano-TiO_2 Particles

CONCLUSION

We have demonstrated that *cis*-$[Co^{III}(tn)_2(R\text{py})Br]Br_2$ complexes are photocatalytically reduced into Co^{II}_{sol}. The photoreduction is found to be solvent dependent. The photoproduced acetone illustrates the scavenging effect of isopropanol. Interfacial electron transfer reaction mechanism is important to understand the functioning and construction nano-TiO_2 photocatalysed photochemical energy conversion systems. It is concluded that the removal of metal ion, hence, seems more specific in removal efficiency and in doping the anatase lattice.

ACKNOWLEDGMENT

KA is thankful to the CSIR (sanction order: No. 01(2570)/12/EMR-II/3.4.2012), New Delhi for financial support through major research project. The authors are thankful to CIF, Pondicherry University for instrumental facility.

REFERENCES

[1] (a) Anbalagan, K., Ganeshraja, A. S., and Maharaja Mahalakshmi, C., *Met. Chem. Phys.* 134, 747, 2012. (b) Anbalagan, K., *J. Phys. Chem.*, C 115, 3821, 2011.

[2] Kelly, C.A. and Meyer, G.J., *Coord. Chem. Rev.*, 211 295, 2001.

[3] Anbalagan, K., Maharaja Mahalakshmi, C. and Ganeshraja, A.S., *J. Mol. Struct.,* 1005 45, 2011.

[4] Kitson, R.E., *Anal. Chem.,* 22, 664, 1950.

[5] Ohko, Y., Hashimoto, K. and Fujishima, A., *J. Phys. Chem.,* A101, 8057, 1997.

[6] Santhosh, S.M., Swetha S., Geetha, S. and Balakrishna, R., *Cent. Eur. J. Chem.*, 8, 453, 2010.

STRUCTURAL AND PHOTOLUMINESCENCE ANALYSIS OF Zn_2SiO_4:Eu^{3+} and Zn_2SiO_4: Eu^{3+}+Li^+ RED LUMINESCENT POWDER PHOSPHORS BY SOL-GEL METHOD

B. CHANDRA BABU AND S. BUDDHUDU

Department of Physics, Sri Venkateswara University, Tirupati, Andhra Pradesh
E-mail: chandrababuphd@gmail.com

ABSTRACT

By employing a sol-gel method, Zn_2SiO_4:Eu^{3+} and Zn_2SiO_4:Eu^{3+}+Li^+ powder phosphors have been synthesized to characterize structural and optical properties from the measurement of their XRD, FT-IR, EDAX and photoluminescence (PL) spectra. Photoluminescence spectral analysis and red emission decay curve of Eu^{3+}:Zn_2SiO_4 powder phosphors with and without the presence of Li^+ ion have been studied. The reason behind the observance of a bright red emission from the Eu^{3+} due to the presence of Li^+ under the UV source (366 nm) has successfully been explained. These phosphors could find potential applications as luminescent screen materials.

INTRODUCTION

In the past few years, particular interest has been focused on inorganic luminescent materials, which find practical applications in different display devices such as flat panel displays, plasma display panel, field emission display (FED), LEDs, sensors and lasers.[1–2] Thus rare earth ions are better luminescent materials because of their possessing sharp and intense emissions due to their 4f intra shell transitions. Among the lanthanides Eu^{3+} has attracted significantly as an luminescent active ion with a red emission $^5D_0 \rightarrow {}^7F_2$ transition at 616 nm, based on host nature in a non-centro symmetric site.[3] In order to enhance the red emission intensity of Eu^{3+}, certain other rare earth or transition metal ions are usually co-doped as sensitizers. In order to improve the brightness of phosphors, different kinds of fluxes are also used

V. Rajendran, R. Yuvakkumar, K. Thyagarajah and K.E. Geckeler (eds.)
Applications of Nano Materials: Electronics, Energy and Environment, pp. 255–262 (2012).

to enhance the luminescence. Specially for Li^+ ion doping into the host lattice of phosphors could enhance the luminescence efficiency, even in a smaller quantity of its presence.[4] Sol-gel method is one of the most important soft-chemistry techniques to prepare phosphors with higher uniformity in particle size distribution, non-aggregation and higher photoluminescence intensity.[5] In recent years, silicate based phosphors have attracted a great deal due to its providing stable crystal structure, good mechanical strength and high thermal stability by the tetrahedral silicate $(SiO_4)^{2-}$ group in the hosts.[6] The reason is that the Li^+ ions would serve as a self-promoter for a better crystallization to reduce the defect around Eu3+ ions in the zinc orthosilicates (Willemite α-Zn_2SiO_4, Rhombohedral crystal structure) host. In the present work, we report the results concerning structural characterization, and PL spectral analysis.

EXPERIMENTAL

Spectral pure chemicals were used in the present work. Zn_2SiO_4: Eu^{3+} and Zn_2SiO_4: Eu^{3+}+Li^+ sol-gel powder phosphors were prepared by a sol-gel method [7] by using TEOS $(SiOC_2H_5)_4$ (99%) as SiO_2 source, zinc nitrate hexahydrate ($Zn(NO_3)_2$ $6H_2O$) as Zn source, europium nitrate ($Eu(NO_3).6H_2O$) as the dopant Eu^{3+} ion and $LiNO_3(Li^{+)}$ as co-dopant ion, high pure ethyl alcohol (EtOH) as the solvent and 0.2 ml of HCl as the acid catalyst. The crystalline structures of the prepared samples were investigated on a 3003TT SEIFERT X-ray diffraction system with a CuKα line (λ = 1.5418Å) of radiation and FTIR spectrum of samples was recorded on a FTIR spectrophotometer by using a KBR pellet. The photoluminescence spectra of Zn_2SiO_4:Eu^{3+} and Zn_2SiO_4:Eu^{3+}+Li^+ samples were recorded on a YVON Fluorolog-3 Fluorimeter with a Xe-arc lamp of power 450W as an excitation source. Lifetimes of red emissions were measured with a phosphorimeter attachment to the main system fitted with a Xe-flashlamp (25W).

RESULTS AND DISCUSSION

The phase purity and the crystal structure of samples were characterized by powder X-ray diffraction. The XRD patterns of the samples Zn_2SiO_4, Zn_2SiO_4:Eu^{3+} and Zn_2SiO_4:Eu^{3+}+Li^+, all the samples were obtained after annealing at 1000°C for 2 h as shown in Figure 1. The diffraction peak positions and sample relative intensities are in agreement with the single phase characteristic peaks of α-Zn_2SiO_4 crystal phase [JCPDS Card No. 79–2005 with space group (148)].

No second phase was detected at the current doping and co-dopant levels indicating that the Eu^{3+} ions and Li^+ ions are obviously homogenously mixed and effectively doped in the host lattice. However, the XRD profile of Zn_2SiO_4: Eu^{3+} co-doped Li^+ sample indicates that, the intensity of (1 0 2) peak has been increased, which indicates an enhancement in crystal quality of the Zn_2SiO_4, within the range of X-ray some values at higher θ values represents splitting L^+ ions dopant can introduce some bounded electron—hole pair acts as and intermediates for the transfer of excitation energy.[8] Figure 2 shows an EDAX elemental analysis of Eu^{3+} doped Zn_2SiO_4 powder phosphors reveals the presence of elements in the sample studied, that was sintered at 1000°C for 2 h.

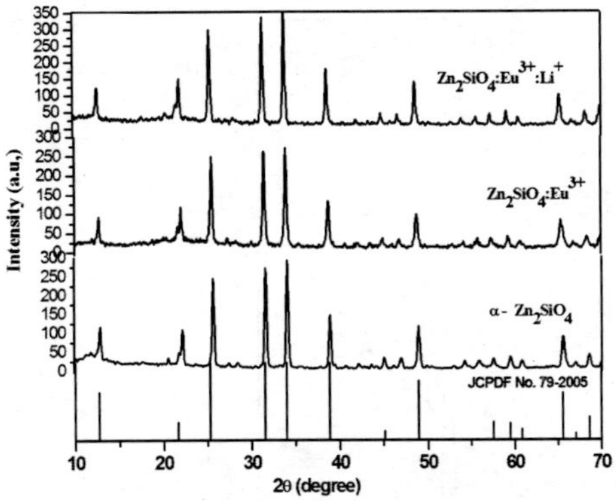

Fig. 1: XRD Diffraction Profiles of (a) JCPDS No:79–2005, (b) Willemite α-Zn_2SiO_4, (c) Zn_2SiO_4: Eu^{3+} and (d) Zn_2SiO_4:Eu^{3+}+ Li^+ Powder Phosphors

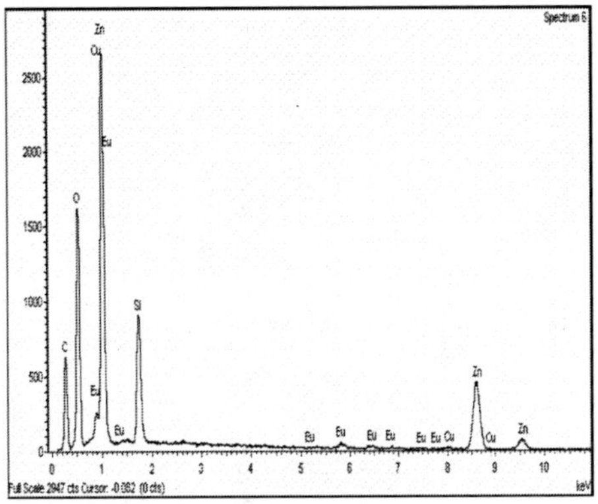

Fig. 2: EDAX Profile of Zn_2SiO_4: Eu^{3+} Powder Phosphor

The host and Eu^{3+}: Zn_2SiO_4: Eu^{3+}+ Li^+: Zn_2SiO_4 powders related FT-IR profiles show common bands as shown in Figure 3. A strong broad band at 3435 cm^{-1} and 1631cm^{-1} could provide an evidence of molecular water in the structure which are assigned to O-H stretching vibrations, (Symmetric and ant symmetric) and H-O-H bending mode.[9] The characteristic strong vibrational modes typical of this materials can be observed: 870 cm^{-1} (υ_1 SiO_4); 911, 935 and 977 cm^{-1} (υ_3 SiO_4), 460 cm^{-1}, (υ_4 SiO_4), 577 cm^{-1} (υ_1 ZnO_4), and 616 cm^{-1} (υ_3 ZnO_4); where υ_1 stands for totally symmetric stretching, υ_3 is asymmetric stretching and υ_4 - asymmetric deformation.[10]

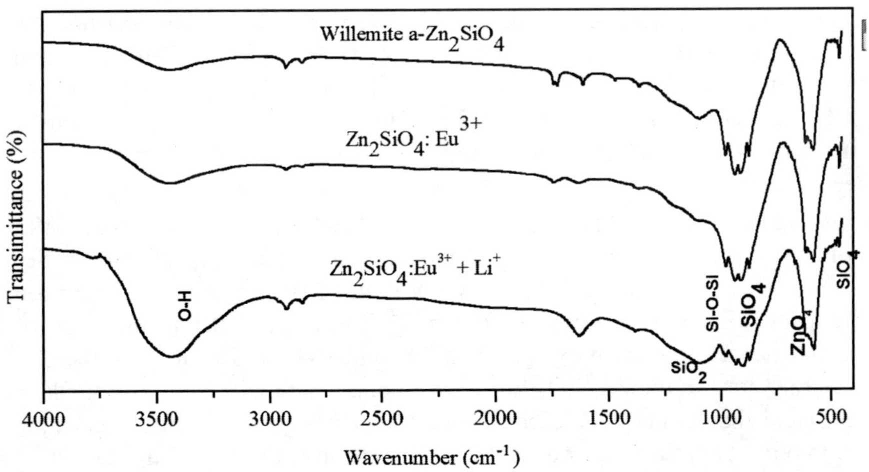

Fig. 3: FT-IR Profiles of (a) Willemite α-Zn_2SiO_4, (b) Zn_2SiO_4: Eu^{3+} (c) Zn_2SiO_4:Eu^{3+}+ Li^+

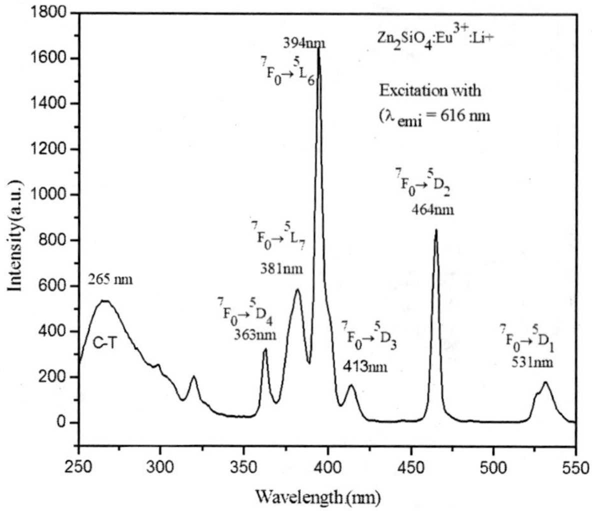

Fig. 4: Excitation Spectrum of Zn_2SiO_4: Eu^{3+}:Li^+ Powder Phosphors

The appearance of the vibrations of SiO_4 and ZnO_4 groups clearly suggests the formation of the Zn_2SiO_4 phase. The IR spectroscopic data analysis confirms that the IR peaks 505, 836 and 911 cm^{-1} are due to Zn_2SiO_4: Eu^{3+}. These results are also in agreement with the XRD analysis, which indicates that a single-phased phosphor has been obtained.

Figure 4 gives the excitation spectrum of $Zn_2SiO_4:Eu^{3+}$+ Li^+ powder phosphors sintered at 1000°C for 2h. The excitation spectrum (monitored at 616 nm) showing a broad band at 265 nm attributed to the Eu^{3+}- O^{2-} charge transfer band (CTB) transition, and the sharp lines in the excitation spectrum in the wavelength region of 350 to 560 nm, at 363 nm, 381 nm, 394 nm, 413 nm, 464 nm, and (527 and 530) nm are all because of f-f transitions of Eu^{3+} ions within its $4f^6$ configuration that forms $^7F_0 \rightarrow {}^5D_4$, $^5G_2$ and $^5L_7$, $^5L_6$ and $^5D_3$, $^5D_2$ and $^7F_1 \rightarrow {}^5D_1$ and also found from the same excitation spectrum of $Zn_2SiO_4:Eu^{3+}$ with a low intensity compared to the $Zn_2SiO_4:Eu^{3+}$: Li^+ powder phosphors. Only the prominent excitation peaks at 394 nm and 464 nm have been selected for the measurement of emission spectra of Eu^{3+} samples.

The emission spectra of $Zn_2SiO_4:Eu^{3+ \text{ and}}$ $Zn_2SiO_4:Eu^{3+}$+Li^+ powder phosphors which are excitated with 394 nm and 464 nm separately which display the characteristic emission transitions $^5D_0 \rightarrow {}^7F_{j\,=\,0,\,1,\,2,\,3\&4}$ of Eu^{3+} as shown in Figures 5(a)&(b) for a comparison purpose. Based on the spectral features, it is evident to identify $Zn_2SiO_4:Eu^{3+}$:Li^+ powder phosphor shows prominent emission richness upon excitation at 394 and 464 nm. Because of, the Zn^{2+} ions are substituted by Eu^{3+} ions in the $Zn_2SiO_4:Eu^{3+}$ phosphors, which will lead to some defects due to different radius and charges.[11] Therefore, the defects may affect the luminescence properties to some extent. In order to reduce the defects, Li^+ ions were added into the $Zn_2SiO_4:Eu^{3+}$ participles as charge compensators. It is well known that the Li^+ ions are prone to be incorporated into the Zn_2SiO_4 lattice because the effective ionic radius of Li^+ ions (59 pm) is smaller than that of Zn^{2+} (74 pm).[12]

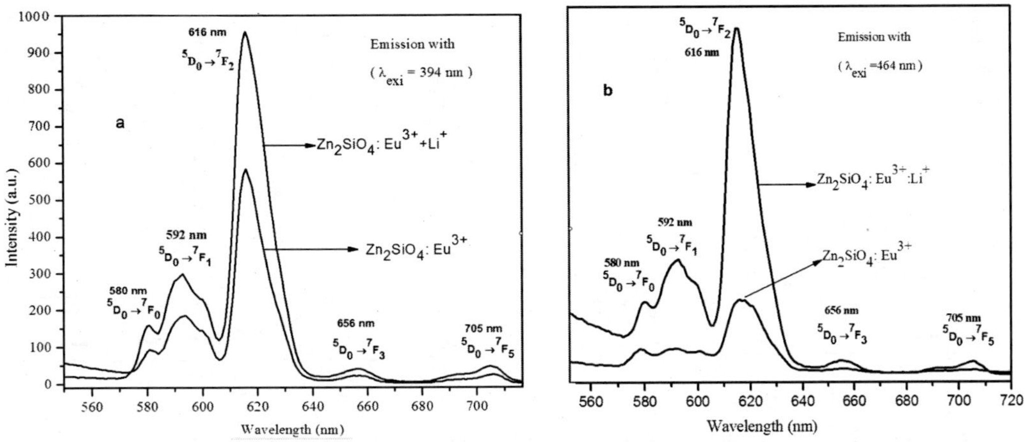

Fig. 5: (a &b) Emission Spectra of Both the Red Luminescent Powder Phosphors

When the Li^+ ions are introduced into the host lattices, defective structure is formed. Therefore, the transition probability of Eu^{3+} ions will be enhanced greatly by the defective structure. However, the PL emission intensity has not been increasing monotonously with an increase in the Li content. It illustrates that only a few Li^+ ions are regarded as charge compensators. When the Li^+ content is higher than Eu^{3+} content, the excess amount of Li^+ ions could occupy the Zn^{2+} sites, resulting in the generation of extra negative charges. In addition, it causes the formation of oxygen vacancies for a charge neutrality in host lattices. But the oxygen vacancies destroy the crystallinity of Zn_2SiO_4 phosphors and leading to the luminescence quenching.[11–13]

The emission spectra of Zn_2SiO_4: Eu^{3+}+ Li^+ consists of lines mainly located in the wavelength range form 550–750 nm with emissions from the excited $^5D_0$ level to the lower lying ^7F_J = 0, 1, 2, 3, 4 levels of the $^4f_6$ configuration of Eu^{3+} as shown in Figure 6(a). The emission intensity is greatly enhanced by Li^+ ions doping, from PL figures, it can be notice that the ultraviolet light (394 nm) excitation provides a strong and bright red emission compared with the visible light (464 nm) excitation. The intensities of different $^5D_0 \rightarrow {}^7F_j$ transition and the splitting of these emissions depend on the local symmetry of the crystal field of Eu^{3+} ion. The emission spectrum mainly exhibits five peaks at 589 nm, 592 nm, 616 nm, 656 nm and 705 nm which correspond to $^5D_0 \rightarrow {}^7F_{0,\ 1,\ 2,\ 3,\ and\ 4}$ transitions from the excited Eu^{3+} to the ground state, respectively. The strong red emission $^5D_0 \rightarrow {}^7F_2$ transition at 616 nm, is attributed due to a forced electric dipole transition ($^5D_0 \rightarrow {}^7F_2$) and it is also a hypertensive transition satisfying the selection rule of ($\Delta J = \pm 2$) and clearly indicates that Eu^{3+} occupies asymmetric site in the lattice.[14] Particularly, the $^5D_0 \rightarrow {}^7F_0$ (579) emission is a strictly forbidden transition (non-degenerated) with a selection rule of ($\Delta J = 0$), if Eu^{3+} ion occupies as inversion symmetry site in the crystal lattice, thus we may suggest the Eu^{3+} ion occupies the site of the type Cs. The $^5D_0 \rightarrow {}^7F_1$ orange emission (592 nm) transition is a parity-allowed magnetic dipole transition ($\Delta J = \pm 1$) and is usually taken as a references transition because the crystal field does not

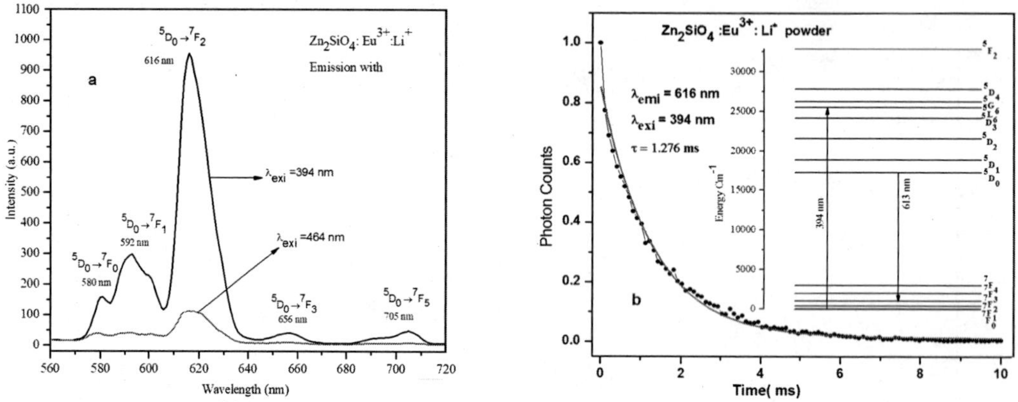

(a) (b)

Fig. 6: (a) Emission Spectra at Excitation 394 nm and 464 nm and (b) Decay Curve of the Red Emission (616 nm) at Excitation 394 nm of $Zn_2SiO_4:Eu^{3+}$: Li^+ Powder Phosphors

considerably alter the intensity of this.[15] The three peaks at 589.2 nm, 592.6 nm and 597.1 nm in accordance with 2J+1 = 3 form $^7F_1$ state suggests that there is only one site occupied by Eu^{3+} ions. If Eu^{3+} is embedded at a site without inversion symmetry, the electric dipole transition of $^5D_0 \rightarrow ^7F_2$ should be dominant, while at a site with inversion symmetry, the magnetic dipole transition of $^5D_0 \rightarrow ^7F_1$ will be preponderant. Therefore, the luminescence intensity ratio of $^5D_0 \rightarrow ^7F_2$ to $^5D_0 \rightarrow ^7F_1$, known as the asymmetry ratio, has been used to measure the degree of distortion from inversion symmetry of the local chemical environment surrounding doping Eu^{3+} in the host.[16]

The fluorescence decay time of $^5D_0 \rightarrow ^7F_2$ emission (616 nm) for $Zn_{2-2x}SiO_4$: xEu^{3+}: xLi^+ doped with Eu^{3+}(x = 0.05 mol %) and co-doped Li^+ (0.05 mol %) ions under excitation at 394 nm has been studied at room temperature as shown in Figure 6(b) and inset fig show corresponding energy level diagram. The measured life time (τ means) for the $^5D_0 \rightarrow ^7F_2$ transition has been determined from the decay curve, which is represented by the following equation. $I = I_0 A e(-t/\tau)$; Where A is the amplitude, I and I_0 are the luminescence intensities at time 0 and t respectively, and τ =1.276 μs as the life time of the red emission of Zn_2SiO_4: Eu^{3+} + Li^+ phosphor.

CONCLUSION

In summary, we have successfully synthesized Zn_2SiO_4: Eu^{3+} and Zn_2SiO_4: Eu^{3+}+Li^+ powder phosphors by using a sol-gel method and are analyzed. Based on the measured excitation spectrum, we have identified two strong excitation bands (394 nm and 464 nm) upon using these two excitation bands, we have measured the emission spectra in ensuring their suitability as more promising red luminescent phosphors. The results clearly indicate that the phosphors containing the Li^+ ion also in $Zn_2SiO_4:Eu^{3+}$ that has revealed a brighter red color emission and therefore it could be suggested that $Zn_2SiO_4:Eu^{3+}:Li^+$ could be found as an ideal coating material with significant importance for its use in luminescent screens of electronic systems. Besides evaluating its luminescence efficiency its structural properties have also been investigated from the measurement of its XRD and FTIR spectra.

ACKNOWLEDGEMENT

One of us (BCB) would like to thank the UGC, New Delhi for the award of a RG National Fellowship to him to carryout the present work.

REFERENCES

[1] Yen, W.M., Shionoya, S. and Yamamoto, H., Practical Applications of Phosphors, Taylor and Francis Group, LLC, CRC Press, 2007.

[2] Zhang, Q.Y., Pita, K. and Kam, C.H., Sol-gel derived zinc silicate phosphors films for full-color display applications, *J. Phys. Chem. Solids,* 64, 33–338, 2003.

[3] Shionoya, S. and Hirano, M., Spin-Wave Sideline in the Luminescence Spectra of Eu^{3+} ion in $KMnF_3$, *Phys. Lett. A*, 26, 533–534, 1968.

[4] Haijie Jin, Hao Wu, and Lianhuan Tian, Improved Luminescence of Y_2MoO_6:Eu^{3+} by doping Li^+ ions for light-emitting diode applications, J. Lumin., 132, 1188–1191, 2012.

[5] Zhang, H.X., Kam, C.H., Zhou, Y., Han, X.Q., Buddhudu, S., *et al.*, Luminescence of Eu^{3+} and Tb^{3+} doped Zn_2SiO_4 nanometer powder phosphors, Thin Solid Films 370, 50–53, 2000.

[6] Naik, Y.P., Mohapatra, M., Dahale, N.D., Seshagiri, T.K, Natarajan, V., Godbole, S.V., Synthesis and luminescence investigation of RE^{3+} (Eu^{3+}, Tb^{3+} and Ce^{3+}) – doped lithium silicate (Li_2SiO_3), *J. Lumin.*, 129, 1225–1229, 2009.

[7] Chandra Babu, B. and Buddhudu, S., Synthesis and characterization of Ni^{2+}:Zn_2SiO_4 nanophosphors by sol-gel method, DAE-SSPS-Dec.2011, *AIP Conf. Proc.*1447,1325, 2012.

[8] Chandra Babu, B., Naresh, V., Jaya Prakash, B. and Buddhudu, S., Structural, Thermal and Dielectric Properties of Lithium Zinc Silicate Ceramic Powders by Sol-Gel Method Ferroelectrics Letters Section, 38:4–6, 114–127, 2011.

[9] Patra, A., Gary A. Baker B. and Sheila N. Baker, Synthesis and Luminescence study of Eu_{3+} in Zn_2SiO_4 nanocrystals, Opt. Mater. 27, 15–20, 2004.

[10] Lin,.C.C. and Shen, P., Sol-gel synthesis of zinc orthosilicate, *J. Non-Cryst. Solids*, 171, 281–289, 1994.

[11] Yang Wu, Yongsheng Wang, Dawei He, Ming Fu, Zhenmin Chen and Yue Li., Spherical Zn_2SiO_4: Eu^{3+} @ SiO_2 phosphor particles in core–shell tructure: synthesis and characterization, *J. Lumin.*, 130, 1768–1773, 2010.

[12] Hongli Liu, Yuying Hao, Hua Wang, Junfu Zhao, Ping Huang and Bingshe Xu, Luminescent properties of R^+ doped Sr_2SiO_4:Eu^{3+} ($R^+ = Li^+$, Na^+ and K^+) red-emitting phosphors for white LEDs, *J. Lumin.* 131, 2422–2426, 2011.

[13] Sabikoglu, I., Ayvacıklı, M., Bergeron, A., Ege, A. and Can, N., Photoluminescence investigations of Li_2SiO_3:Ln (Ln = Er^{3+}, Eu^{3+}, Dy^{3+}, Sm^{3+}) phosphors, *J. Lumin.*,132, 1597–1602, 2012.

[14] Blasse, G., Grabmaier, B.C., Luminescent Materials Springer, Verlag Telos, ISBN0387580190, 1994.

[15] Ana Paula, A., Marques and Marcos Takashi, S., Tanaka and Elson Longo and Edson R. Leite and Ieda Lucia Viana Rosa, J. Fluoresc., 21, 893–899, 2011.

[16] Ying Liu, Zhou-Guang Lu, Ying-Ying Gu and Wei Li, Hydrothermal-assisted ion exchange synthesis and photoluminescence of Li^+ and Eu^{3+} co-doped $NaLa(WO_4)_2$ as near-UV type red phosphors, *J. Lumin.* 132, 1220–1225, 2012.

DESALINATION EFFICIENCY OF THE NOVEL POLYSULFONE BASED BIPOLAR MEMBRANE FOR CHENNAI SEAWATER TREATMENT

KRISHNAVENI VENUGOPAL AND SANGEETHA DHARMALINGAM

Department of Chemistry, Anna University, Sardar Patel Road, Chennai, Tamil Nadu
E-mail: sangeetha@annauniv.edu

ABSTRACT

The most crucial problem that afflicting people around the world is global water scarcity which was mainly due to the rapid growths in population and economy that have resulted in greater demand on the quantity and quality of drinking water, leading to catastrophic water shortage in arid and water-stressed region areas. In the present investigation, we have synthesized cationic and anionic functionalized polysulfone based bipolar membranes with PVA as intermediate layer. The structure and characteristic properties of these membranes were studied by FTIR, ^{1}HNMR, XRD, TGA-DSC, SEM, IEC and water absorption. The efficiency of the membrane performance was evaluated using lab made fabricated BPMED stack. The effect of feed (Chennai Sea water) on the process efficiency such as concentration of acid and alkali production, current efficiency and energy consumption were evaluated. During treatment the transport number of co-ions, conductivity of the solution and change in potential with respect to time at constant current density were also recorded and compared for both the synthesized and commercially procured membranes.

INTRODUCTION

The most crucial problem that afflicting people around the world is global water scarcity which was mainly due to the rapid growths in population and economy that have resulted in greater demand on the quantity and quality of drinking water, leading to catastrophic water shortage in arid and water-stressed region areas. Of the technologies developed, desalting seawater

V. Rajendran, R. Yuvakkumar, K. Thyagarajah and K.E. Geckeler (eds.)
Applications of Nano Materials: Electronics, Energy and Environment, pp. 263–270 (2012).

to produce clean water for drinking, irrigation, industrial and urban development emerged as the most sustainable approach.[1] Membrane technologies are getting more and more attention nowadays due to their reliable contaminant removal without production of any harmful by-products, especially in water and wastewater treatment processes. Electrodialysis (ED) is a viable technology for the desalination of brackish water, seawater and solutions containing organics. ED is a more cost effective technique than reverse osmosis and has the potential to treat the waste stream generated by pressure driven membrane processes that are commonly used in desalination and water reuse applications.[2] Bipolar membrane electrodialysis (BPMED) has greatly extended the potential of electrodialysis in recent years and is effective in resolving secondary salt pollution.

Bipolar membrane (BPM) is a composite ion exchange membrane consisting of a layered ion exchange structure of a cation exchange layer (CEL) joined to an anion exchange layer (AEL) through an intermediate layer (IL). Synthetic bipolar membranes are receiving much attention nowadays because of their potential for application in different industries. In fact, the excellent efficiency of the bipolar membrane technology in converting selective industrial wastes (acids, alkali's or salts) into valuable products that can be recycled and reused has been proved.[3] When applied to treat salt containing wastewater, BPMED can split salt into acid and base and overcome salt pollution and discharged without treatment. The main objective of the present study is to fabricate a BPM from sulfonated and quaternized polysulfones (SPSu and QPSu) with PVA as IL and to characterize it. Polysulfone was chosen for the study since it possesses excellent physical and chemical stabilities apart from the fact that it can be readily functionalized into a charged polymer. In this present study, the desalination of Chennai sea water using PSu based BPM with PVA as IL was evaluated in terms of current efficiency, energy consumption and some important parameters such as membrane stack potential, pH, electric conductivities in concentrated and diluted solutions, transport number of co-ions and acid-base concentrations were examined in both synthesized and commercial PSDVB based membrane systems.

EXPERIMENTAL

Chemical Reagents and Materials

Polysulfone (PSu, Mw – 35,000) was purchased from Aldrich, Chlorosulfonic acid (CSA) from Spectrochem, India, while N, N'-dimethylformamide (DMF) was obtained from Merck. Chloroform, Triethyl amine (TEA), Trimethyl chloro silane ($ClMe_3Si$) and stannous chloride ($SnCl_2$) were purchased from Sisco Research Laboratory Pvt. Ltd. (SRL). Polyvinyl alcohol (PVA, Mw ~ 14,000) and Paraformaldehyde were purchased from Central Drug House (CDH). All the chemicals were of AR grade and used as received. Commercially available Ion Exchange Membranes such as strong acid cation exchange membrane (CMI-7000S), strong base anion exchange membranes (AMI-7001S) were purchased from Membranes International INC, New Jersey, USA and BPM made of PSDVB (Gel Polystyrene cross linked with divinylbenzene) were from Arun Electro chemicals, Chennai.

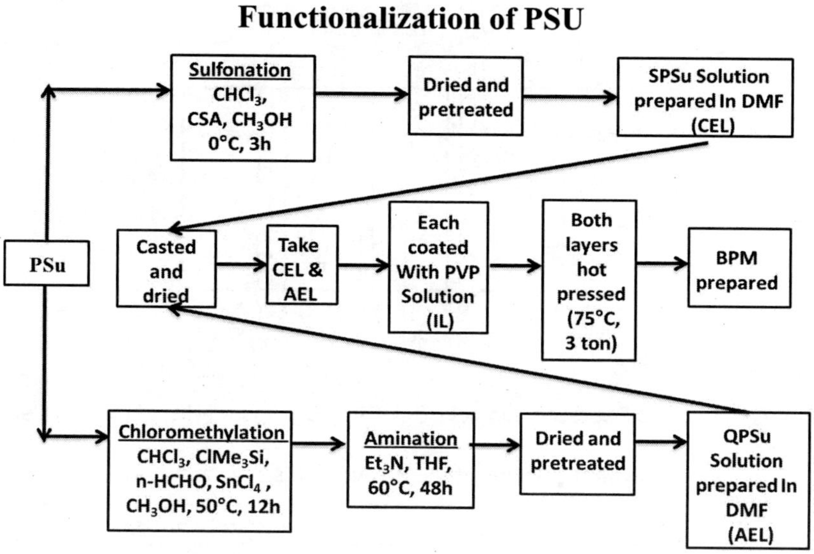

Membrane Characterization

The sulfonation and quaternization of the prepared membranes was confirmed with a FTIR Perkin Elmer spectrometer and 1HNMR Bruker Advance III 500 MHz multinuclear FT-NMR spectrometer. The cross sectional morphologies of the membranes were studied using Hitachi S-3400 N model Scanning Electron Microscope. The ion exchange capacities (IEC) of the CEM and AEM were determined by titration method. The percentage of water absorbed was calculated using the following relation,

$$\text{Water absorption (\%)} \frac{\text{Weight of the wet membrane} - \text{Weight of the dry membrane}}{\text{Weight of the dry membrane}} \times 100$$

Membrane Stack Configurations of ED

BPMED experiments were carried out using lab-scale electrodialysis equipment. The BPMED unit consisted of a DC power supply and an electrodialysis stack. The cell consisted of two pieces of BPM, CEM and AEM with BP-A-C-BP configuration as depicted in the Figure 1 to effect water dissociation and convert the salt into their corresponding acid and base.[4] The desalination experiments were performed in a recirculation batch mode at a constant current density of 5 mA/cm^2 for a run time of 8h at room temperature. The 0.5 M NaCl solution was taken in the electrolyte compartment as electrode rinse solution. The membrane stack configuration for the commercial membranes (CMI, AMI and PSDVB-BPM) was the same as that of synthesized membranes. The details about the BPMED system are furnished in the Table 1.

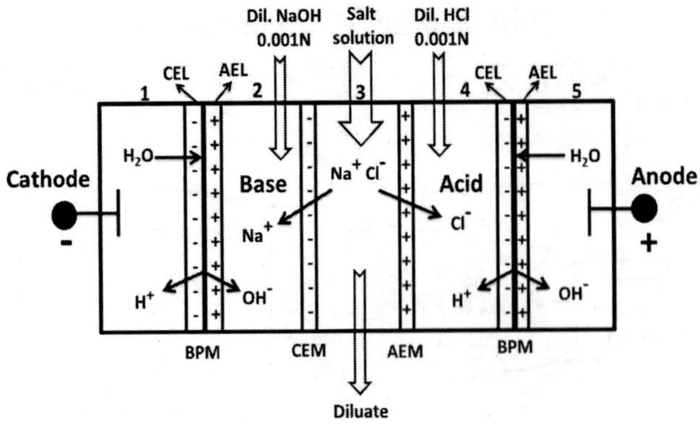

Fig. 1: Schematic Representation of BP-A-C-BP Configured Electrodialysis Cell

Table 1: Details of the Electrodialysis System

Parameter	*Values/Material Used*
Number of Compartments	5
Volume of each chamber	160 mL
Anode	Titanium base metal coated with Ti-Ru-Pd oxides
Cathode	Stainless steel
Current	0–1.5 A
Voltage	0–50 V
Effective membrane area	120 cm^2

The current efficiency,[5] energy consumption [6] and transport number of ions [5] of the process were analyzed using the equations

$$\textit{Current Efficiency } \eta(\%) = \frac{FV\Delta C}{nIt} \times 100$$

$$\textit{Energy Consumption } (kWh) = \int_0^t UIdt$$

$$\textit{Transport Number of ions } t_{\pm}^{m} = \frac{nF\,(C^0V^0 - C^iV^i)}{It}$$

Where F is the Faraday constant (96500 C/mole); V is the total volume (cm^3); ΔC is the change in concentration of acid or base (mol/lit); n is the stoichiometric number (n = 1 in this case); I is the applied current (A); t is the time (sec); U is potential (V); t is the operation time (h); $t_{\pm}^{m}$ is the transport number of an ion; C^0 and C^i are the initial and final ion concentration in the anodic/cathodic compartment while, V^0 and V^i are the initial and final volume of the anodic/ cathodic compartment; t is the time (sec).

RESULTS AND DISCUSSION

FTIR Analysis

The infrared spectrum of PSu, SPSu, CMPSu and QPSu are shown in Figure 2. The various vibration modes of each polymer and their observed frequency ranges are listed in Table 2.

Table 2

Membrane type & mode of vibration	Frequency range
PSu	
(a) Aromatic ν_s(C–H) & δ(C–H)	3026 cm^{-1} and 756 cm^{-1}
(b) Aliphatic ν_{as}(C–H) and ν_s(C–H)	1411 cm^{-1} and 1387 cm^{-1}
(c) Aromatic ν_s(C=C)	Doublet at 1589 cm^{-1}
(d) $\nu_s(SO_2)$ & $\nu_{as}(SO_2)$	1163 cm^{-1} and 1320 cm^{-1}
(e) ν(C–S)	692 cm^{-1}
(f) ν_s(C–O–C) & ν_{as}(C–O–C)	1113 cm^{-1} and 1298 cm^{-}
SPSu	
(a) ν(O–H) – broad intense band	3440 cm^{-1}
(b) $\nu_s(SO_2)$ & $\nu_{as}(SO_2)$	1175 cm^{-1} and 1488 cm^{-1}
(c) ν_s(S–O)	716 cm^{-1}
(d) $\nu_s(SO_3)$ & $\nu_{as}(SO_3)$	1015 cm^{-1} and 1295 cm^{-1}
CMPSu	
(a) ν(C–Cl)	856 cm^{-1}
QPSu	
(a) $\nu_{as}(NR_3^+)$ & $\delta_{as}(NR_3^+)$ (broad, intense)	3373 cm^{-1} and 1582 cm^{-1}
(b) ν(C–N)	1320 cm^{-1}

^{1}H NMR Analysis

The ^{1}H NMR spectra of PSu, SPSu and QPSu are shown in Figure 2. From the NMR study the functionalization of PSu was confirmed.

Morphological Studies

The morphology of various ion exchange membranes were studied by SEM. The cross sectional view of PSu, SPSu, QPSu and BPM in Figure 2 (a,b,c and d) showed that each membrane exhibited a homogeneous and uniform defect-free surface while three distinct regions were evident in Figure 2d. The top layer (seen in Figure 2d) was the sulfonated polymer while the middle one represented the intermediate PVA and the last layer (third layer) represented the quaternized polymer material.

Membrane Water Absorption and Ion-Exchange Capacities

Water uptakes of PSu, SPSu, QPSu and BPM membranes results are shown in Figure 2. The IEC value for the prepared SPSu and QPSu were found to be 1.467 and 1.567 meq/g respectively.

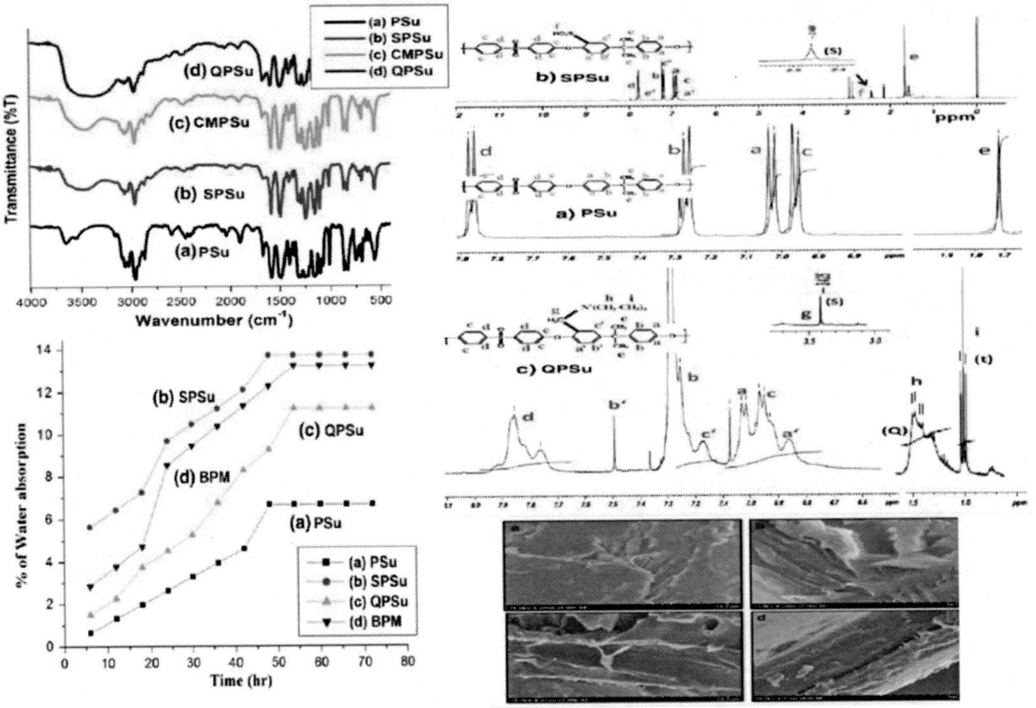

Fig. 2: Water Absorption; FTIR, ^{1}H NMR Spectrum; Cross Sectional SEM Micrographs of (a) PSu, (b) SPSu, (C) QPSu and (d) BPM

Process Parameters of BPMED Systems

The results shows that PSu based prepared membranes showed higher production of both acid and base having concentrations of about 0.6 M and 0.7 M respectively with time, as compared against 0.4 M and 0.5 M of acid and base concentrations, respectively obtained with commercially procured PSDVB based membrane.. The concentration of both acid and base was found to increase linearly with increment of time. The transport numbers of co-ions were calculated for different time intervals. The developed BPM showed a current efficiency of 63% and energy consumption of 0.129 kWh in comparison with 45% and 0.14 kWh in the case of commercial BPM respectively. It revealed that hydrophilic nature of IL enhanced the water dissociation efficiency and reduced energy consumption because both depended on acid/base formation in Anodic/Cathodic Compartment by water dissociation at IL of BPM. From the results, it was obvious that the current efficiency increased gradually with increase in concentration of the concentrate compartment (acid compartment and base compartment) which was observed for various time intervals. At a constant current density of 5 mA/cm^2, energy consumption value was also found to be higher for commercially procured PSDVB based BPM when compared with prepared BPM membrane. Thus, the energy consumption value was lower for the prepared membranes than the commercial membrane. The pH in the acid compartment decreased to nearly acidic pH whereas in the base compartment, it increased to nearly basic pH with increase in time which suggested the high production of acid and base. In the case of feed compartment, pH of the salt solution increased initially (pH ~ 7.6), which reflected the leakage of OH- ions predominating in the initial stage at the feed compartment. The decrease in pH observed during the end of electrodialysis was due to the predominating concentration of H^+ ions transported from the acid compartment during the later stages of the experiment to the feed compartment. Consequently, OH^- ions from base compartment pass to the feed compartment to balance the concentration of H^+ ions in the feed solution. However, at the end of the electrodialysis, the pH of the feed solution was found to be slightly acidic in nature (pH of about 7–6). Whereas the pH in the electrolyte compartment decreased to acidic pH as a result of leakage of H^+ ions from the acid compartment which was evidently much higher when compared to the leakage of OH^- ions from the base compartment. The electrical conductivity of concentrated solutions (acid and base) of the two systems increases with the electrodialysis time which was the indicative of the higher production of the acid and the base in the respective compartments. In contrast, the conductivity of the solution in the dilute compartment decreased with time indicating the effective removal of the salt from the sample solution. Hence, it can be inferred that the effective production of acid and base or in otherwise, the removal of salt from the solution was found to be higher in the case of PSu based prepared membranes than commercially procured membranes.

CONCLUSIONS

The present study showed that the process using electrodialysis with fabricated novel PSu based bipolar membranes which resulted in greater efficiency of removal of salt from sea water as compared with the PSDVB based commercial membranes. This was due to the catalytic effect of PVA on water splitting, hydrophilicity of PVA and hydrogen bonding of PVA. During treatment the conductivity of the solution in various compartments was also measured and confirmed the salt removal along with acid and alkali production. Hence, the results obtained confirmed that the synthesized BPM is a suitable material for desalination of sea water.

ACKNOWLEDGEMENTS

Financial support from the Board of Research in Nuclear Science (BRNS), Mumbai, India (Letter. No. 2010/37C/1/BRNS/826, Dated: 28–06–2010), DST-FIST, DRS-UGC to Department of Chemistry, Anna University were gratefully acknowledged.

REFERENCES

[1] Misdan, N., Lau, W.J. and Ismail, A.F., Seawater Reverse Osmosis (SWRO) desalination by thin-film composite membrane—Current development, challenges and future prospects, Desalination., 287, 228–237, 2012.

[2] Banasiak, L.J., Bruggen, B.V. and Schafer, A.I., Sorption of pesticide endosulfan by electrodialysis membranes, *Chemical engineering journal.*, 166, 233–239, 2011.

[3] Zhang, X., Lu, W., Ren, H. and Cong, W., Sulfuric acid and ammonia generation by bipolar membranes electrodialysis: transport rate model for ion and water through anion exchange membrane, *Chemical and Biochemical Engineering Quarterly*., 22, 1–8, 2008.

[4] Badruzzaman, M., Oppenheimer, J., Adham, S. and Kumar, M., Innovative beneficial reuse of reverse osmosis concentrate using bipolar membrane electrodialysis and electro chlorination processes, *J. Membr Sci.,* 326, 392–399, 2009.

[5] Michael Rajesh, A., Kumar, M. and Shahi, V.K., Functionalized biopolymer based bipolar membrane with poly ethylene glycol interfacial layer for improved water splitting, *Journal of Membrane Science,* 372, 249–257, 2011.

[6] Shi, S., Lee, Y.H., Yun, S.H., Hung, P.X. and Moon, S.H., Comparisons of fish meat extract desalination by eleectrodialysis using different configurations of membrane stack, *Journal of food engineering,* 10, 1417–423, 2010.

TRANSPARENT NANO COMPOSITE THIN FILMS FOR THIN FILM TRANSISTORS AND OPTO-ELECTRONIC DEVICES

S. Sathish and B. Chandar Shekar

Department of Physics, Kongunadu Arts and Science College,
G-N Mills, Coimbatore, Tamil Nadu
E-mail: chandar.bellan@gmail.com; sathishphy@yahoo.co.in

ABSTRACT

Transparent nanocomposite PVA-TiO_2 and PMMA-TiO_2 thin films were prepared by a easy and cost effective dip coating technique. Al/PVA-TiO_2/Al and Al/PMMA-TiO_2/Al sandwich structures were prepared to study the dielectric behavior. The thickness of the films was measured by capacitance method, weighting method and cross checked by electronic thickness measuring instrument (Tesatronic TTD-20). The functional groups, structural, morphology, electrical and optical properties were studied by Fourier Transform Infrared Spectroscopy (FTIR), X-Ray Diffraction (XRD), Scanning Electron Microscope (SEM) and UV-Vis-NIR Spectrophotometer. The FTIR study confirms the formation of metal-oxide (Ti-O) bond in the prepared films. The XRD pattern indicated that the prepared films were amorphous in nature. Very smooth surface without any pits and pin holes were observed from SEM analysis. The dielectric behavior showed that the value of dielectric constant increases with increase of temperature. The transmittance of the films were high above 80% in the visible region and the optical band gap was found to be at around 4 eV by using Tauc's plot. The observed amorphous structure, dielectric behavior and high optical transmittance with wide band gap of the prepared nano composite thin films indicated the feasibility of utilizing these films in thin film transistors and opto-electronic devices.

V. Rajendran, R. Yuvakkumar, K. Thyagarajah and K.E. Geckeler (eds.)
Applications of Nano Materials: Electronics, Energy and Environment, pp. 271–278 (2012).

INTRODUCTION

Nanostructured materials have recently attracted tremendous interest because of their potential uses in the development of nanodevices. Now-a-days, the use of organic-inorganic composites polymeric materials has generated widespread interest in developing new materials. In composites, polyvinyl alcohol (PVA) additive matrix added may have a potential role against thermal and environmental degradation and also is being given a great deal of attention due to its attractive film forming, physical properties, good processability, biocompatibility and good chemical resistance. To further enhance the properties of this polymer, the addition of a inorganic material is advantageous forming a composite, which could be lightweight, flexible and exhibit good moldability.[1] Recently lot of research work on nanocomposites based on polymer/inorganic is going on around the globe for its applications in photonic, Fuel cell, supercapacitors, Photovoltaic, strain sensing, EM detector, thin film transistors, gas sensor, medical, linear optics, storage devices and infrared detector applications.[2–15]

For commercial applications, polyvinyl alcohol is selected because its properties still appear to be unique in terms of the price/performance ratio. PMMA is a high-strength, amorphous polymer possessing good dimensional stability and outdoor wearing properties. Owing to these characteristics, PMMA is among the most heavily studied polymers for nano and micro composite fabrications. New technologies are demanding materials with improved structural, electrical and chemical modifications of traditional PVA and PMMA for the development of nanodevices. In our previous work, we have reported the preparation and characterization of virgin PVA and PMMA thin films prepared by solution growth techniques.[16,17] The present work deals with the effect of addition of TiO_2 on the PVA and PMMA matrix and their structural, morphological, electrical and optical properties of the composite films prepared by dip coating technique.

EXPERIMENTAL

Preparation of Thin Films

Initially, PVA solution was prepared by adding 50 ml of distilled water with 0.5g of PVA powder and this mixture was magnetically stirred at 85°C for 2 hrs. Similarly the PMMA solution was prepared by adding 50 ml of benzene with 0.5g of PMMA powder and this mixture was magnetically stirred for 1 hr. After this, 0.5 g of TiO_2 powder is added to both PVA and PMMA solutions and stirred for 2 hrs followed by repeated ultrasonication until get homogeneous dispersion. A precleaned glass substrate was dipped into it for deposition of composite films by dip coating technique. Al/PVA-TiO_2/Al and Al/PMMA-TiO_2/Al sandwich structure was prepared to study the electrical properties. Both bottom and top electrodes were deposited by vacuum evaporation technique and intermediate thin film was deposited by dip coating technique.

Characterization

The deposited films were characterized by using Fourier Transform Infrared Spectroscopy (FTIR), X-Ray Diffraction (XRD), Scanning Electron Microscopy (SEM), Energy Dispersive X-Ray diffraction (EDAX), UV-Vis-NIR Spectrophotometer and LCR meter.

RESULTS AND DISCUSSION

Fourier Transform Infrared Spectroscopy (FTIR)

The FTIR spectra of composite PVA-TiO_2 and PMMA-TiO_2 thin films recorded in the range of 4000 cm^{-1} – 400 cm^{-1} are shown in Figure (1a,b). A relatively broad and intensive band is observed between 3200cm^{-1} – 3450cm^{-1}, indicating the presence of polymeric association of the free hydroxyl groups and bonded OH stretching vibrations. The wide absorption band centered at 3430cm^{-1} is due to Ti-OH for PVA-TiO_2, where as 3465 cm^{-1} for PMMA-TiO_2 films.[10,18] The bands observed at 2900 cm^{-1}, 1051cm^{-1} and 702 cm^{-1} suggests the presence of C-H stretching, C-O stretching vibrations of inplane and outplane bending vibrations for PVA-TiO_2 composite film. The peaks at 2998 cm^{-1} and 2953cm^{-1} due to the stretching vibration of CH_2 and CH_3. The peak at around 1730 cm^{-1} is due to the stretching vibrations of C = O for PMMA-TiO_2.

In general the IR absorption bands of PVA are quite broad and severally overlapped in the 600–1500 cm^{-1} region due to TiO_2 crystallites embedded in the matrix of polymeric material. The band at 1432 cm^{-1}, 1326 cm^{-1}, 1232 cm^{-1} indicating the symmetric bending mode $v_\delta(CH_2)$, wagging vibration of CH_2 and CH respectively for PVA-TiO_2 composite film.[19]

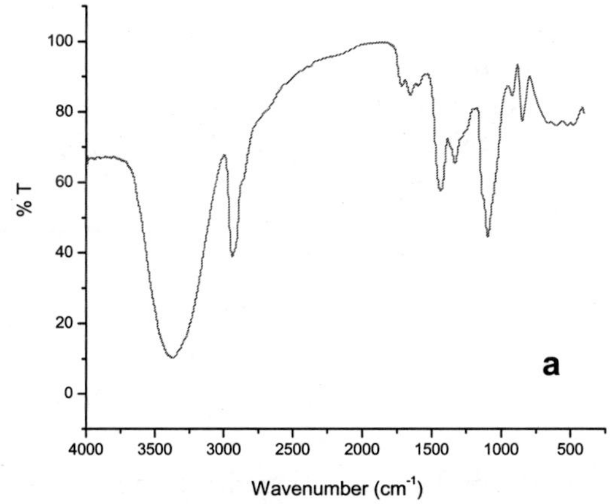

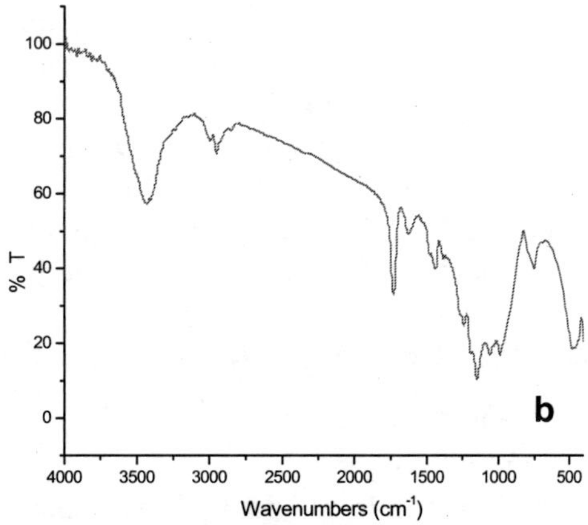

Fig. 1: FTIR Spectra of a) Composite PVA-TiO_2 and b) Composite PMMA-TiO_2 Films

The new bands at 1732 cm^{-1}, 1450 cm^{-1}, 1270 cm^{-1},1245 cm^{-1},1193 cm^{-1}, 1149 cm^{-1} due to stretching vibrations of C = O, C (= O)-O and C-O-C respectively for PMMA-TiO_2 composite film.[20]

The several overlapping bands observed at 800–400 cm^{-1} due to metal oxide (M-O) vibration, indicating the Ti-O-Ti vibrations.[21] The band at around 890 cm^{-1}, 750 cm^{-1} 640 cm^{-1} and 535 cm^{-1} for PVA-TiO_2 and at 800 cm^{-1}, 680 cm^{-1}, 630 cm^{-1}, 560 cm^{-1} and 510 cm^{-1} for PMMA-TiO_2 film is corresponds to Ti-O stretching modes.[22,23]

X-Ray Diffraction (XRD)

XRD pattern of composite PVA-TiO_2 and PMMA-TiO_2 films are shown in Figure (2a,b). It is seen that the XRD pattern of both films are identical and observed several broad humps.

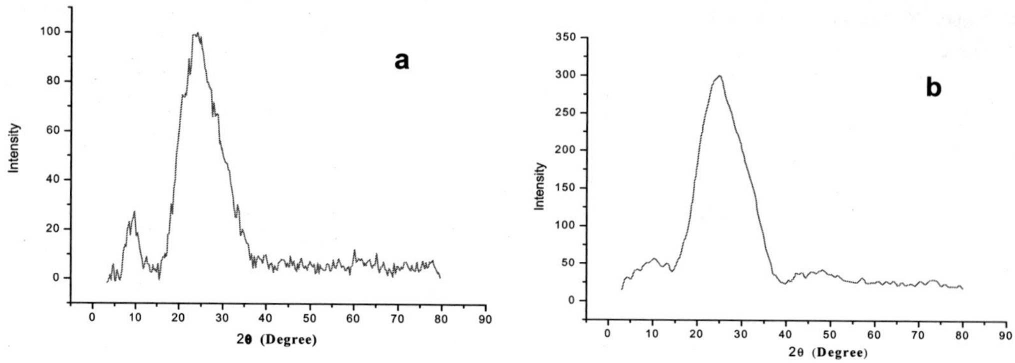

Fig. 2: XRD Pattern of a) Composite PVA-TiO_2 and b) Composite PMMA-TiO_2 Films

The observed broad humps in the XRD spectrum indicated the presence of crystallites of very low dimension.[17] A major broad hump obtained at $2\theta = 20–30°$ is due to polymer matrix such as PVA and PMMA indicated a increase in the amount of amorphous contents.[11,16,24] The XRD pattern of TiO_2 nanoparticles showed three small crystalline characteristic peaks at $2\theta = 25°$, 38° and 47° for PVA-TiO_2 film, where as $2\theta = 26°$,38° and 46° for PMMA-TiO_2 film corresponding to 101, 004 and 200 planes indicating the tetragonal crystal planes of anatase phase.[25,26] The absence of any intense peaks through out the spectrum indicated the predominantly amorphous nature of the film.[27]

Scanning Electron Microscope (SEM)

The scanning Electron Micrograph of composite PVA-TiO_2 and PMMA-TiO_2 films are shown in Figure (3a, b).

The morphology obtained for composite films is similar to the powder filled polymer matrix composite material.[5,15,28] From the SEM images, it can be seen that the various shapes and sizes of the nanoparticles embedded in the film surface. The surface morphology of the composite thin films shows that cluster nanoparticles of TiO_2 are distributed over the film surface and also there are no cracks and pin holes on the surface.

(a)

(b)

Fig. 3: SEM Micrograph of a) Composite PVA-TiO_2 and b) Composite PMMA-TiO_2 Films
The SEM Images Revealed that all Films have Very Different Morphologies of Cluster Like Shape

UV-Vis-NIR Spectrophotometer

The transmittance spectra of composite films were taken in the wavelength range of 300–2500 nm. Figure (4a) shows the transmittance spectra of composite PVA-TiO_2 and PMMA-TiO_2 films. Higher transmittance of about 89 % was obtained for PMMA-TiO_2, where as transmittance of about 83% for PVA-TiO_2 film. The transmittance increases with wavelength for both composite films. The observed increase of transmittance with wavelength is due to the increase in crystalline size associated with higher densification of the films.[29] The observed higher transmittance for composite films are in accordance with the earlier investigation on other composite films.[5,11,20,30,31] The band gap energy has been estimated from $(\alpha h\nu)^2$ versus $(h\nu)$ plot as shown in Figure (4b). The band gap is about 3.77 eV for PVA-TiO_2 composite film, where as 3.78 eV for PMMA-TiO_2 composite film.

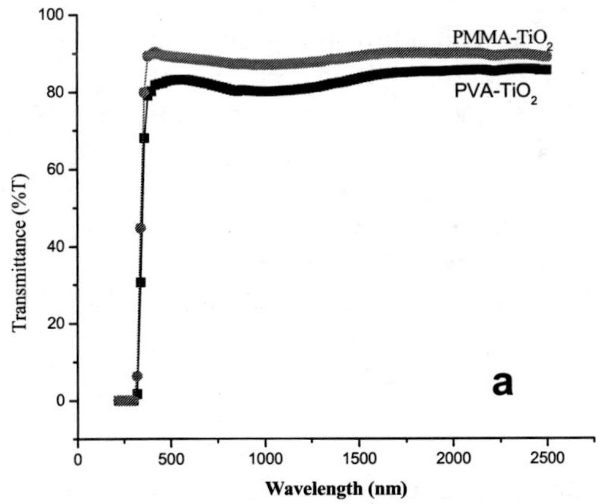

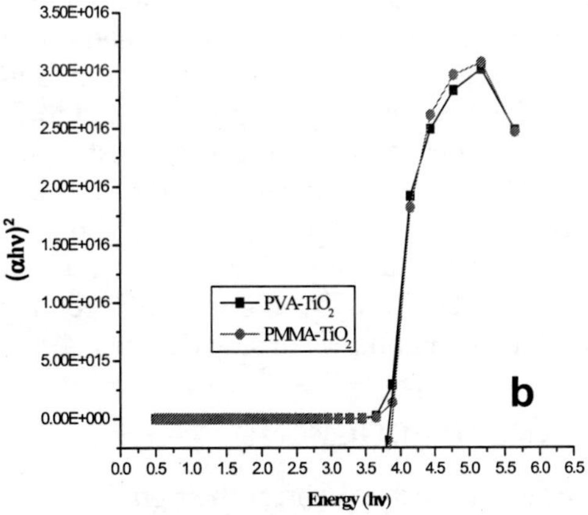

Fig.4: a) UV-Vis-NIR Transmittance Spectra and b) $(\alpha h\nu)^2$ Vs $h\nu$ of Composite PVA-TiO_2 &PMMA-TiO_2 Films

Dielectric Behaviour

The temperature dependence of dielectric constant of composite PVA-TiO_2 and PMMA-TiO_2 films for 1 kHz is as shown in Figure (5). From the figure, it is seen that the dielectric constant of the film increases with temperature [32] and revealed maximum at 200°C is due to the film densification and crystallization of a minor amorphous phase in the as grown film during the heat treatment.[33] This phenomenon may be caused by interfacial orientation and space charge polarization effect and thus increased dielectric constant at high temperature.[34]

Dielectric constant of about 26.8 obtained for PMMA-TiO_2 film where as PVA-TiO_2 film showed a dielectric constant of about 24.6 at 1 kHz. The composite films showed a tremendous increase in the dielectric constant value as compared to their virgin polymer films.

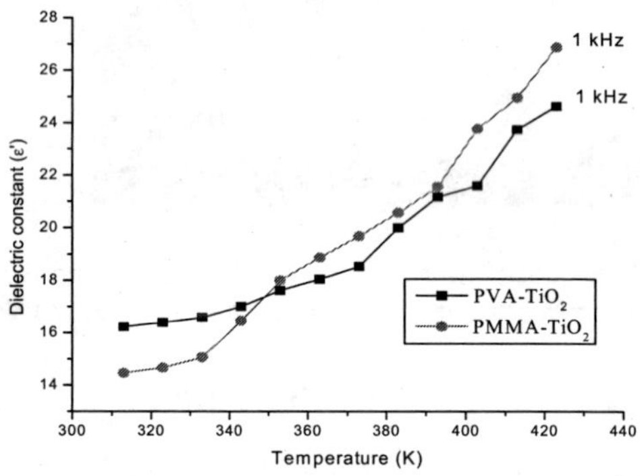

Fig. 5: Dielectric Constant Vs Temperature of Composite PVA-TiO_2 and PMMA-TiO_2 Films at 1 kHz

CONCLUSION

Transparent composite PVA-TiO_2 and PMMA-TiO_2 thin films were prepared by simple and cost effective dip coating technique. The FTIR specta showed several overlapping bands between 800–400 cm^{-1} due to Ti-O-Ti vibration. The existence of little nanocrystalline titanium dioxide in PVA and PMMA matrix was confirmed by XRD, which also indicates the predominantly an amorphous nature of the film. A cluster of TiO_2 nanoparticles were well dispersed in the polymer matrixes and cluster like shape with a uniform grain sizes. A higher optical transmission (>80%) was obtained in the entire UV-Vis-NIR region for both composite films. The optical band gap was found to be higher (377.eV and 3.78 eV) indicating the blue shift. The determined dielectric constant values of 24.6 for PVA-TiO_2 and 26.8 for PMMA-TiO_2 films indicate the enhanced dielectric constant than their respective virgin PVA and PMMA films. Thin films of the composites showed excellent transparency and good dielectric behavior with stability. These outstanding properties of transparent composite PVA-TiO_2 and PMMA-TiO_2 films by dip coating technique make them potential candidates for thin film transistors and optoelectronic devices.

ACKNOWLEDGEMENT

The authors wish to express their gratitude to the Jawaharlal Nehru Memorial Fund (JNMF), New Delhi, India, for financial support to this research work under JNMF Scholarship for doctoral studies.

REFERENCES

[1] Zein, S.H.S., Boccaccini, A.R., and Zainal, W.N.H.W., Effect of Manganese Oxide on the Properties of Polyvinyl Alcohol Multi-Walled Carbon Nanotube Composites as Model Supercapacitors Material, World Appl Sci J., 6(6), 737–747, 2009.

[2] Tintu,R., Saurav,K., Sulakshna,K., Nampoori, V.P., Radhakrishnan,P., and Thomas,S., $Ge_{28}S_{60}Sb_{12}$/PVA Composite Thin Films for Photonic Applications, J non-oxide glasses, 2, 167–174,2010.

[3] El-Tooney, M.M., Synthesis of GMA/PVA Cast Membrane Reinforced with Alumina for Fuel Cell Applications, Nat Sci., 9(12), 160–172, 2011.

[4] Singhal, A., Kaur, M., Dubey, K.A., Bhardwaj, Y.K., Jain, D., Pillai, C.G.S., and Tyagi, A.K., Polyvinyl Alcohol-In_2O_3 Nanocomposite Films: Synthesis, Characterization and Gas Sensing Properties, RSC Adv., 2, 7180–7189, 2012.

[5] Saikia, D., Saikia,P. K., Gogoi, P.K., Saika, P., Synthesis, Characterization and Photovoltaic Application of Silver Doped CdS/PVA Nanocomposite Thin Films, Digest J Nanomater Biostruct., 6, 589–597,2011.

[6] Loh, K.J., and Chang, D., Zinc Oxide Nanoparticles-Polymeric Thin Films for Dynamic Strain Sensing, J. Mater. Sci., 46, 228–237, 2011.

[7] Yahya, N., Akhtar, M.N., Masuri, A.F., and Kashif, M., Synthesis and Characterization of ZnO-CNT Filled PVA Composite as EM Detector, J Appl Sci.,11(8),1303–1308, 2011.

[8] Chen,F.C., and Shien, H.P.D., Low Voltage Organic Thin Film Transistor with Polymeric Nanocomposite Dielectrics, Org Elect., 7, 435–439, 2006.

[9] Karami, H., Investigation of Sol-Gel Synthesized CdO-ZnO Nanocomposite for CO Sensing, Int. J. Electrochem. Sci, 5, 720–730, 2010.

[10] Gu,G., Zhang, Z., and Dang,H., Preparation and Characterization of Hydrophobic Organic-Inorganic Composite Thin Films of PMMA/SiO_2/TiO_2 with Low Friction Coefficient, Appl Surf Sci., 221, 129–135,2004.

[11] Su, W., Wang, S., Wang, X., Fu, X., and Weng, J., Plasma Pre-Teatment and TiO_2 Coating for PMMA for the Improvement of Antibacterial Properties, Surf Coat Technol., 205, 465–469, 2010.

[12] Arlindo, E. P.D.S., Lucindo, J.A., Bastos, C.M.D.O., Emmel, P.D., and Orlandi, M.O., Electrical and Optical Properties of Conductive and Transparent ITO@PMMA Nanocomosites, J. Phys. Chem. C, 116, 12946–12952, 2012.

[13] Stefanescu, E.A., Tan,X., Lin, Z., Bowler, N., and Kessler, M.R., Multifunctional Fiberglass-Reinforced PMMA-$BaTiO_3$ Structural/Dielectric Composites, Polym, 52, 2016–2024, 2011.

[14] Sreeja, R., Jhon, J., Aneesh, P.M., and Jayaraj, M.K., Linear and Nonlinear Optical Properties of Luminescent ZnO Nanopartices Embedded in PMMA Matrix, Opt Commun, 283, 2908–2913, 2010.

[15] Shan, Y., Zhou, Y., Cao,Y., Xu, Q., Ju, H., and Wu, Z., Preparation and Infrared Emissivity Study of Collagen-g-PMMA/In_2O_3 Nanocomposites, Mater Lett.,58, 1655–1660, 2004.

[16] Chandar Shekar B., Sathish, S., and Sathyamoorthy, R., Morphology, Structure and Conduction Properties of PVA Nano Scale Thin Films for Field Effect Organic Thin Film Transistors Applications, Int J Nanotechnol Appl., 5, 297–308, 2011.

[17] Chandar Shekar, B., Na,M.K., Lee, J., and Rhee, S.W., Dielectric, Thermally Stimulated Discharge Current, Pyroelectric and Surface Morphology of PMMA Thin Films Prepared by Isothermal Immersion, Mol. Cryst. Liq. Cryst., 424, 43–53, 2004.

[18] Yasumori, A., Shinoda, H., Kameshima,Y., Hayashi, S., and Okada, K., Photocatalytic and Photoelectrochemical Properties of TiO2-Based Multiple Layer Thin Film Prepared by Sol-Gel and Reactive-Sputtering Methods, J. Mater. Chem., 11, 1253–1257, 2001.

[19] Shehap, A.M., Thermal and Spectroscopic Studies of Polyvinyl Alcohol/Sodium Carboxy Methyl Cellulose Blends, Egypt. J. Solids., 31, 75–91, 2008.

[20] Hu, Y., Gu, G., Zhou, S., and Wu, L., Preparation and Properties of Transparent PMMA/ZrO_2 Nanocomposites using 2-Hydroxy Ethyl Methacrylate as a Coupling Agent, Polym., 52, 122–129, 2011.

[21] Vasquez, J., Lozano, H., Lavanen, V., Cantu, M.L., Romero, P.G., Ana, M.A.S., Benavente, E., and Gonzalez, G., High-Yield Preparation of Titanium Dioxide Nanostructures by Hydrothermal Conditions, J Nanosci Nanotechnol, 9, 1103–1107, 2009.

[22] Choudhury, A., Dielectric and Piezoelectric Properties of Polyetherimide/$BaTiO_3$ Nanocomposites, Mater Chem Phys Lett., 121, 280–285,2010.

[23] Mitsionis, A., Vaimakis, T., Trupalis, C., Todorova, N., Bahnemann, D., Dillert, R., Hydroxyapatite/Titanium Dioxide Nanocomposites for Controlled Photocatalytic NO Oxidation, Appl Catal B: Environ., 106, 398–404,2011.

[24] Singh, P., Kumar, R., Virk, H.S., and Prasad, R., Modification of Optical, Chemical and Structural Response of Polymethyl Methaacrylate Polymer by 70 MeV Carbon Iron Irradiation, Indian J Pure Appl Phys., 48, 321–325, 2010.

[25] Lu, C.H., Wu,W.H., Kale, R.B., Synthesis of Photo Catalytic TiO_2 Thin Films via the High Pressure Crystallization Process of Low Temperature, J Hazard Mater,147, 213–218, 2007.

[26] Pourjafar, S., Rahimpour, A., and Jahanshahi, M., Synthesis and Characterization of PVA/PES Thin Film Composite Nanoparticles for Better Performances and Surface Properties, J Ind Eng Chem., 18, 1398–1405,2012.

[27] Kanarlou, H., and Rafizadeh, S., Investigation of Optical and Structural Properties of Titanium Very Thin Layers Under Heat Process, Australian J Basic Appl Sci., 5, 1182–1187, 2011.

[28] Lagashetty, A., Havanoor, V., Basavaraja, S., and Vengataraman, A., Synthesis of MoO_3 and its Polyvinyl Alcohol Nanostructured Film, Bull. Mater. Sci., 28, 477–481, 2005.

[29] Agbo, P.E., and Nnabuchi, M.N., Core-Shell TiO_2/ZnO Thin Film: Preparation, Characterization and Effect of Temperature on Some Selected Properties, Chal Lett., 8, 273–282,2011.

[30] Augustine, M.S., and Jayalekshmi, S., Solvent Dependent Optical Properties of Spin Coated Transparent Nanocomposite Films of Oleic Acid Capped ZnO Quantum Dots in PVA matrix, International Conference on Advances in Polymer International., India, 148–152, Feb 2010.

[31] Ju, Y., Li, L., Wu, Z., and Jiang, Y., Effect of Oxygen Partial Pressure on the Optical Property of Amorphous Titanium Oxide Thin Films, Energy Procedia,12, 450–455, 2011.

[32] An, M.H., Composition, Structural and Electrical Configurations on DC-Magnetron Sputtered TiO_2 Thin Films, J Korean Phys Soc, 47, 847–851, 2005.

[33] M.J. Alam and D.C. Cameron, Preparation and Characterization of TiO_2 Thin Films by Sol Gel Method, J. Sol. Gel. Sci Technol, 25, 137–145, 2002.

[34] Hsiao, Y.J., and Chang, Y.S., High Dielectric Permittivity of Li and Ta Doped NiO Ceramics, J Phys D: Appl Phys, 40, 863–868, 2007.

CELLULOSIC COMPOSITES FOR EFFLUENT WASTE WATER TREATMENT

NEETA DHAVALE, GAURI SHELAR, RENUKA R GONTE[1] AND K. BALASUBRAMANIAN[2]

Department of Chemistry, Fergusson College, FC Road, Pune
[1]Department of Applied Chemistry, Defence Institute of Advanced Technology, Girinagar, Pune
[2]Department of Materials Engineering, Defence Institute of Advanced Technology, Girinagar, Pune
E-mail: balask@diat.ac.in

ABSTRACT

Novel cellulose composite were synthesized in the form of micro porous beads. Cellulose acetate was used as binder to bind the agro wastes- sawdust and sugarcane molasses. Amount of biomaterial in each type of composite beads was varied. These composite beads were investigated for removal of industrial dyes (direct blue and direct green) from mimicked wastewater systems in batch process. The adsorption behavior of composite beads were tested by performing equilibrium studies. These composite showed effective dye adsorption (~ 85%), which oriented from the high surface area and presence of large number of hydrophilic functional groups. Adsorption capacity was found to be in the order Cellulose -Sugarcane molasses > Cellulose-Sawdust. The beads were found to be highly stable in aqueous medium and preparation technique was simple. These non-cross-linked dye adsorbed beads were further considered as master batches for development of lacquers used for coatings on various substrates. These biocomposite beads can thus find practical application in various industries.

INTRODUCTION

Discharge of industrial effluents directly into the environment has become a worldwide problem. Dyes form one of the prime pollutants released in the natural water bodies, which are difficult to treat. Dyes have synthetic origin and complex molecular structure, which makes them

V. Rajendran, R. Yuvakkumar, K. Thyagarajah and K.E. Geckeler (eds.)
Applications of Nano Materials: Electronics, Energy and Environment, pp. 279–286 (2012).

highly stable in atmosphere and difficult to biodegrade.[1] Colored organic effluents are produced in various industries such as textiles, rubber, paper, plastic, cosmetics, etc.[2] Dyes are ionic, aromatic organic compounds, composed of aryl rings, with delocalized electron systems. The presence of auxochromes results in stronger alteration of the maximum absorption of the compound and provides a strong bonding affinity.[1] Dyes can have acute and/or chronic effects on aquatic organisms, depending on its concentration and exposure time; high visibility of dyes makes their presence as abnormal coloration on surface waters.[2] Discharge of dyes even in a small amount can affect the aquatic life and food web; the complicated molecular structures make it difficult to treat and interfere with municipal waste treatment operations. Dyes from wastewater can undergo chemical and biological changes, consume dissolved oxygen from the stream and destroy aquatic life; can cause allergic dermatitis and skin irritation.[2] Thus, it becomes important to remove these pollutants from the wastewaters before their final disposal.[3]

Various conventional methods of removing dyes include co-agulation, oxidation–ozonation, filtration, reverse osmosis, membrane filtration, biological treatment, electrochemical processes, etc.[4] However, these methods are not widely used due to the high cost and economic disadvantages. In contrast, an adsorption technique is by far the most versatile and widely used method in terms of initial cost, simplicity of design, ease of operation, flexibility and insensitivity to toxic substances. However, selection of solid substrate for adsorption depends on its adsorption capacity and commercial accessibility.[5] Researchers have exploited many low cost, biodegradable and effective adsorbents obtainable from natural resources for the removal of different dyes from aqueous solutions at different operating conditions. Recent research has been oriented towards development of cost effective, easily and abundantly available high adsorption capacity natural biopolymer.[6] These polymers present themselves as attractive candidates due to their physicochemical characteristics, high reactivity, chemical stability and excellent selectivity towards dye molecules, due to the presence of reactive functional groups in polymer chain and high surface area.[7] Cellulose due to its excellent physical and chemical properties in terms of stability, water absorbency and dye removal ability is considered as important adsorbent.[8] Sawdust and sugarcane molasses are naturally occurring biopolymer with large number of hydrophilic groups like -OH, -COOH, etc. along with lignocelluloses, hemicelluloses groups, which enhance their capacity to bind to various substrates and absorb large amount of water.[9]

The present work, aims at development of cellulose composite beads for removal of dyes (Papricol Direct Blue (DB) and Papricol Direct Green (DG)) from mimicked waste water systems. Sawdust and sugarcane molasses were used in composite formulation. Details of DB and DG dyes are summarized in Table 1. Various parameters affecting the adsorption process were studied. It was found fascinating to consider such dye-loaded beads as master batches to impart coloration to the paper and as lacquers for coating various substrates. It was extremely interesting to observe the utilization of largely accessible agricultural waste to separate dyes from industrial wastewater and recycle the same for additional practical applications with very simple and cost efficient method.

EXPERIMENTAL

Cellulose acetate, sodium bicarbonate, acetone, N,N-Dimethylacetamide obtained from Qualigens INDIA and used as received. Sawdust (SD), sugarcane molasses (SM) were purchased from local suppliers, cleaned, vacuum dried and thrashed to fine dust prior to use.

- *Synthesis of Biocomposite Beads:* Cellulose acetate was dissolved 2:1 acetone NN dimethylacetamide to form homogeneous solution. SD, SM and $NaHCO_3$ were added to this solution and the mixture was precipitated in water bath in form of spherical beads. These beads were purified by water washing and dried in oven.
- *Dye Solution:* Direct blue and direct green dyes were used as received. Stock solutions (500 mg/l) were prepared in double distilled water and diluted as required.

Batch Adsorption Experiments

Batch adsorption experiments were carried out using Cellulose acetate (C), Cellulose-acetate-sugarcane molasses (C-SM), Cellulose-acetate-sawdust (C-SD). pH of the solution was maintained at 6.8. Conc. of dye was monitored spectrophotometerically (Systronics 166 spectrophotometer). The effect of adsorbent dose on uptake efficiency and adsorption capacity were tested using 20 ml of 100 mg/l solution. Equilibrium adsorption isotherm studies were conducted using 0.1g adsorbent and 20 ml (10, 50, 100 mg L^{-1}) dye solution. The adsorption efficiency (%Ad) and adsorption capacity (qe) were calculated as

$$Ad\,(\%) = ((C_0 - C_t)/C_0) \times 100 \qquad \ldots (1)$$

$$qe = ((C_0 - C_t)/W) \times V \qquad \ldots (2)$$

where qe is the amount of the dye adsorbed (mg L^{-1}) on the adsorbent at time t, C_0 is the initial concentration of dye, C_t is the concentration of dye at a given time t, V is the volume of the solution used (L), and W is the weight of adsorbent used.

Table 1: Physical Properties of Industrial Dyes Used in Study

Dye	*Structure*		*Mol. Wt.*	*CAS No.*	λ_{max} *(nm)*
DB	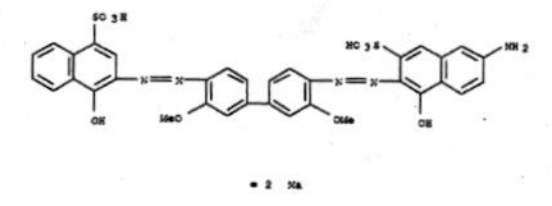	$C_{34}H_{25}N_5Na_2O_{10}S_2$	773.69	6449–35–0	620
DG	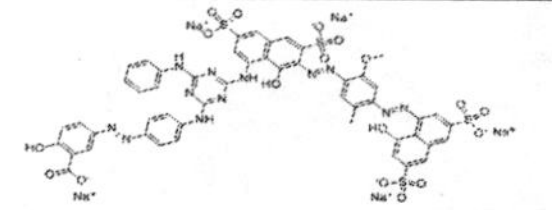	$C_{50}H_{33}N_{12}Na_5O_{18}S_4$	1333.07	6388–26–7	583

RESULTS AND DISCUSSION

The size of composite beads was found to be dependent on diameter of nozzle used for precipitating the solution. Cellulose acetate exhibited excellent binding ability and maximum of 50% SD and 20% SM could be incorporated in bead formulation. These composite beads were found to exhibit good mechanical strength, chemical stability and thermal stability upto 80°C. The advantages render by the non crosslinking was utilization of dye adsorbed beads as master batches for various purpose. The scanning electron microscope image of the C-SM composite shows the presence of 0.30μm to 300μm sized pores on the surface. The low magnification SEM image represents the average bead diameter (3–4 mm) and varying pore diameter (Figure 1a, b).

- *Effect of adsorbent dose:* The effect of composite dose was evaluated using 20ml 100 mg/L dye solution, varying sorbent dose from 0.1g to 0.5g. It was observed that dye adsorption increased drastically with increase in bead dosage due to the increased surface area and adsorption sites (Figure 2a). Further increase in dose showed negligible enhanced adsorption.
- *Effect of Contact Time:* The time dependent behavior of dye adsorption was examined by varying the contact time between adsorbate and adsorbents in the range of 15min to 7 days. Equilibrium adsorption was achieved in 120 min (Figure 2b) after which rate of adsorption became dependent on rate at which the dye molecules were transported from the bulk liquid phase to the actual adsorption sites. The maximum adsorption of DB and DG was observed onto C-SM beads in one week. An initial sharp increase in adsorption gradually slows down due to saturation of adsorption sites. It was further observed that adsorption of DG was comparatively low due to the highly complexed globular structure, which interferes and weakens the interaction between active functional groups and dye.
- *Effect of Initial Concentration of Dye:* The effect of dye concentration was studied varying the concentration from 10ppm to 100mgL^{-1}. Decrease in adsorption with increase in conc. is evident due to the presence of excess amount of DB and DG in the solution (Figure 2c and 2d). Thus at high dye concentration, maximum adsorption occurs at surface which results in decrease in adsorption efficiency.

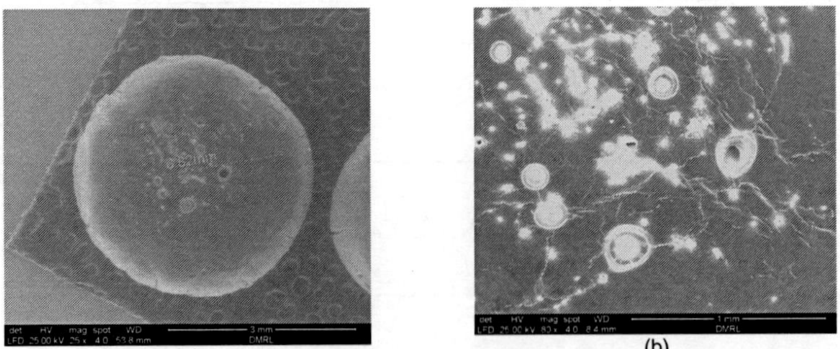

(a) (b)

Fig. 1: SEM (a) Low Magnification (b) High Magnification

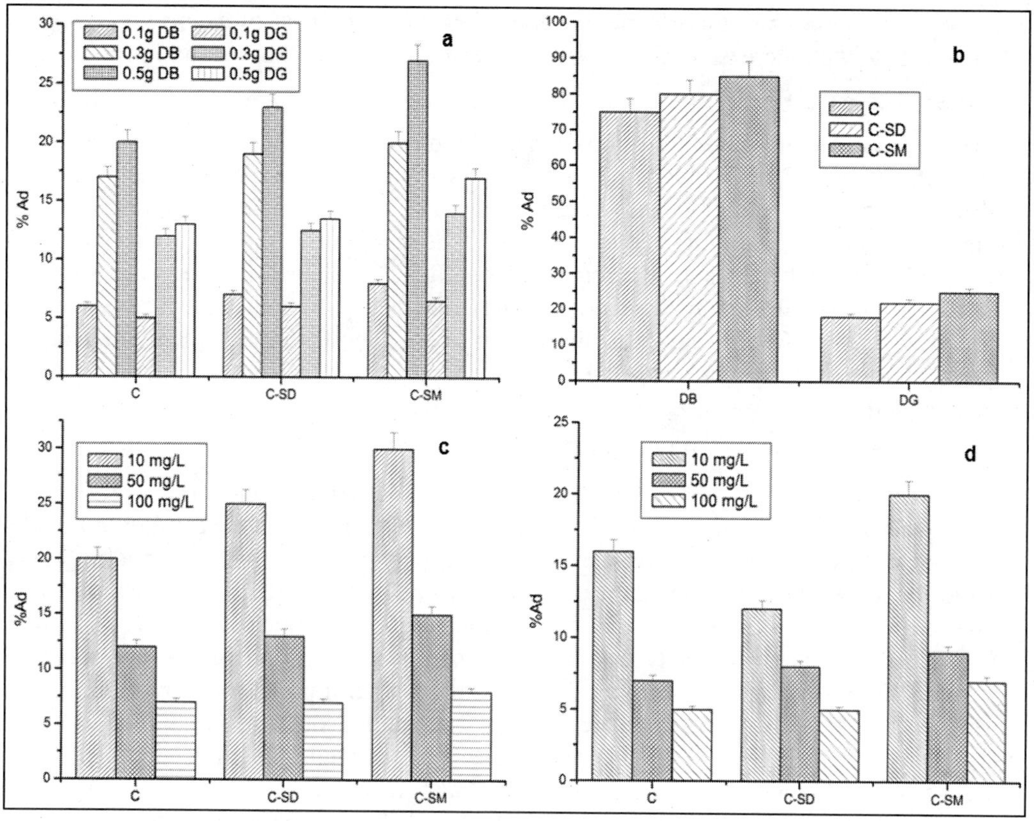

Fig. 2: a) Effect of Dose (b) Effect of Time (c) Effect of DB Conc. (d) Effect of DG conc.

- *Adsorption Isotherm:* Adsorption isotherms are important to describe the adsorption mechanism for the interaction of dye on the adsorbent surface for the design of an adsorption process and to determine the efficiency of adsorption. The isotherm yields certain constants whose values express the surface properties and affinity of the sorbent.
- *Langmuir Isotherm Model:* The basic assumption of the Langmuir theory is that the uptake of dye occurs on a homogeneous surface by monolayer adsorption. The linearized form of Langmuir equation is:

  $$Ce/qe = 1/(qmax\ KL) + Ce/qmax \quad \ldots (3)$$

  where Ce is the equilibrium concentration (mg/L), qe the amount of DB and DG dyes adsorbed (mg/g), qmax is qe for a complete monolayer (mg/g), KL is a constant related to the affinity of the binding sites (L/mg).
- *Freundlich Isotherm Model:* Freundlich isotherm theory states that the ratio of the amount of solute adsorbed onto a given mass of sorbent to the concentration of solute in

the solution is not constant at different concentrations. The heat of adsorption decreases in magnitude with increase in the extent of adsorption. This isotherm equation is as:

$$\ln qe = 1/n \ln Ce + \ln KF \quad \ldots (4)$$

where qe, and Ce are the equilibrium concentrations of dye in the adsorbent (mg/g) and liquid phases (mg/L), respectively, KF and n are the Freundlich constants which are related to adsorption capacity and intensity respectively. A favorable adsorption tends to give Freundlich constant *n* a value between 1 and 10. Strong interaction was observed with C-SM beads. The high values for the regression coefficients obtained for the composite beads clearly indicate that the Freundlich model can describe the equilibrium adsorption isotherm.

- *Temkin Isotherm Model:* Temkin isotherm, considers the effects of the heat of adsorption that decreases linearly with coverage of the adsorbate and adsorbent interactions. The Temkin isotherm has been used in the form as follows:

$$qe = (RT/b_T) \ln (a_T Ce) + (RT/b_T) \ln Ce \quad \ldots (5)$$

where R is gas constant 8.314 $X10^{-3}$ kJ $mol^{-1}K^{-1}$. T is absolute temperature K, b_T is the Temkin constant related to the heat of adsorption ($kJmol^{-1}$) and a_T is the equilibrium binding constant corresponding to the maximum binding energy (L/g). Typical bonding energy range for ion exchange mechanism is reported to be in the range of 8–16$kJmol^{-1}$ while physisorption processes are reported to have adsorption energies less than –40$kJmol^{-1}$. The Temkin isotherm suggests that the chemisorptive adsorption of the dye molecule. Figure 3 shows the various isotherm plots and the isotherm parameters are mentioned in Table 2. The overall results indicate that the Langmuir isotherm, Freundlich isotherm, Temkin isotherm fits well for cellulose composite beads while the adsorption can be best explained by Freundlich model which explains that the process of adsorption is Physicochemical involving the -OH, -COOH groups of the composite beads.

Table 2: Equilibrium Isotherm Parameters for Adsorption of Direct Blue and Direct Green Dye

Isotherm Model		*C* DB	*C* DG	*C-SD* DB	*C-SD* DG	*C-SM* DB	*C-SM* DG
Langmuir	qmax	0.7845	0.8381	0.8222	0.8254	0.9245	0.9353
	KL	0.0254	0.0407	0.0269	0.0662	0.0270	0.0811
	RL	0.7973	0.7106	0.7875	0.6016	0.7870	0.5519
	R^2	0.9763	0.9955	0.9900	0.9674	0.9245	0.9965
Freundlich	n	1.8949	1.9009	1.8638	2.2707	1.9135	2.4381
	KF	0.0511	0.0670	0.0532	0.1065	0.0631	0.1427
	R^2	0.9992	0.9774	0.9998	0.9674	0.9996	0.9646
Tempkin	a_T	0.2869	0.3531	0.2906	0.5392	0.3028	0.6724
	R^2	0.9786	0.9885	0.9901	0.9736	0.9816	0.9694

Reutilization of the Dye Adsorbed Beads

The reutilization of the polymer beads is a key factor in improving process economics. The dye-adsorbed beads were separated from dye solution and dissolved in acetone to obtain coloured cellulosic mixture, which was used for coating on glass surface. Homogenously coloured coating on glass plate was possible as shown in Figure 4. Depending on the quantity of dye being adsorbed onto the beads, the intensity of colour coating can be varied.

CONCLUSIONS

Novel porous biocomposite beads were synthesized using agricultural by-products for removal of industrial dyes from mimicked solutions. The C-SM composite exhibited maximum efficient adsorption (86%) due to presence of hydrophilic –COOH and –OH groups along with lignocellulosic groups. Removal of dye was observed to be a time dependent process, which attained its saturation level in a time span of 7 days. The equilibrium isotherms were well described by Freundlich model which indicates the two-stage adsorption; monolayer adsorption at first stage and later being chemisorption. The poor rate of adsorption in the second stage resulted from slow penetration of dye molecules into the interior of the beads. The dye-adsorbed beads could be utilized as master batches for formation of lacquers for coating purpose and is proposed to be effectively used in preparation of low-grade colour paper. Thus the novel composite beads designed in the present work is distinguished by significant adsorption capacity, reuse ability, stability in aqueous solutions and above all its economic cost and simplicity of preparation. Such beads thus represent as a potential candidate for practical industrial application.

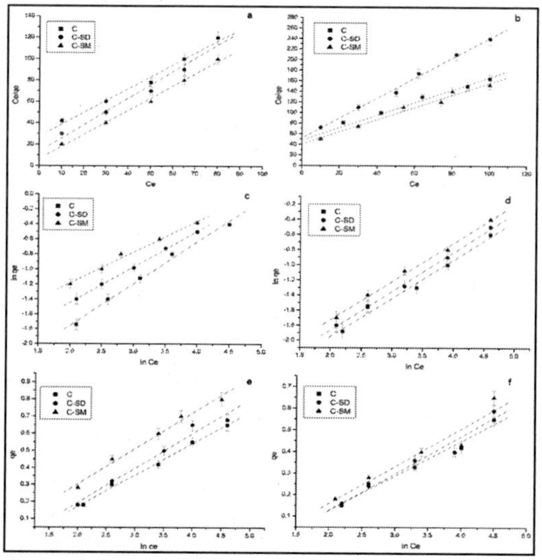

Fig. 3: Isotherm Plot Langmuir Model (a) DB (b) DG; Freundlich Model (c) DB (d) DG; Temkin Model (e) DB (f) DG

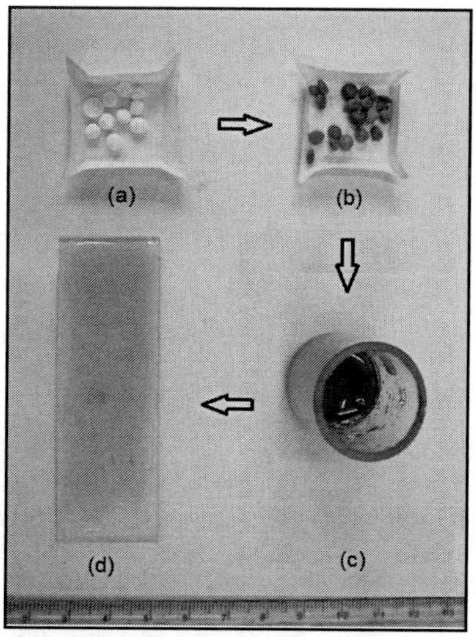

Fig. 4: Reutilization of Dye Adsorbed Beads for Glass Coating

ACKNOWLEDGEMENTS

Authors would like to thank Dr Prahlada, Vice Chancellor, DIAT (DU) for his encouragement in this research activity. The authors would also like to thank Dr. Partha Ghosal, Scientist 'E', Defence Metallurgical Research Laboratory, Kanchanbagh, Hyderabad for the SEM micrographs. The authors would also like to thank the members of characterization center at Fergusson College, Pune.

REFERENCES

[1] Velmurugan P, Rathina kumar V, Dhinakaran G, (2011) International Journal of Environmental Sciences, 1, 7.
[2] Klimiuk E, Gusiatin Z, Kabardo K, (2006) Polish Journal of Environmental Studies, 15–1, 95.
[3] Mohammed H. J., Kadhim B. J., Mohammed Aseel S.H., (2011) Desalination, 8–2, 739.
[4] Abdel-Halim E. S., Al-Deyeb Salem S., (2011) Carbohydrate Polymers, 84, 454.
[5] Siew-Teng Ong, Pei-Sin Keng, (2011) Water, 3, 157.
[6] Gupta V.K., Suhas, (2009) J. Environmental Science, 90, 2313.
[7] Abdel-Halim E. S., Al-Deyeb Salem S., (2011) Carbohydrate Polymers, 84, 454.
[8] Luo X., Zhang L., (2009) Journal of Hazardous Materials, 171, 340.
[9] Bai Y-X., Li Y-F, (2006), Carbohydrate Polymers, 64, 403.

THE INFLUENCE OF Mn SUBSTITUTION ON THE STRUCTURAL, ELECTRICAL AND MAGNETIC PROPERTIES OF $CoFe_2O_4$ FERRITE

S.P. YADAV, S.S. SHINDE, A.A. KADAM AND K.Y. RAJPURE

Electrochemical Materials Laboratory, Department of Physics, Shivaji University, Kolhapur
E-mail: rajpure@yahoo.com

ABSTRACT

The influence of Mn substitution on the structural, electrical and magnetic properties of $CoFe_2O_4$ ferrite is reported. The ferrite system was $CoMn_xFe_2O_4$ ($0 \leq x \leq 0.5$) synthesized by the self propagating auto-combustion route. X-ray diffraction patterns of prepared samples confirm the spinel cubic crystal structure showing (311) characteristic reflection and lattice constant of cobalt ferrite increases with the increase of Mn content. The microstructural analysis was carried out by SEM. Dielectrical properties of Co-ferrite have been studied with frequency and temperature. Room temperature magnetization measurements showed that for the substitution of Co by Mn, there is an initial increase in the saturation magnetization (Ms) and remanence (Mr) for $x = 0.0$ to $x = 0.3$ and decrease for $x = 0.4$ The coercivity of the samples increases with increase of Mn content.

INTRODUCTION

Recent studies on $CoFe_2O_4$ have shown that it can be a suitable material for developing new technologies in the areas of strategic importance. An enhanced magneto-optical effect and a large magnetoelectric effect in cobalt ferrite containing composites are some examples.[1–2] Its intrinsic robustness, high negative magnetostriction and ability to work at higher frequencies by minimizing the eddy current losses makes cobalt ferrite as suitable and economical material for developing oxide based magnetostrictive materials. The magnetoelastic behavior of cobalt ferrite can be utilized in stress sensing applications which in turn can be employed in automobiles to enhance the fuel efficiency by nearly 5%. Efforts are continuously being made

V. Rajendran, R. Yuvakkumar, K. Thyagarajah and K.E. Geckeler (eds.)
Applications of Nano Materials: Electronics, Energy and Environment, pp. 287–294 (2012).

to enhance the mechanical robustness, to reduce the magnetomechanical hysteresis at elevated temperatures and to improve the motor coefficient in order to enhance the stress sensitivity at lower magnetic field strengths by Mn substitution, using metallic binders, etc.[3] It is showed that manganese substituted cobalt ferrites are excellent candidates for stress sensors due to a large magnetomechanical effect and high sensitivity to stress.[4] Nowadays, highly magnetostrictive and resistive ferrite phase becomes an essential part of ME composites so, attempt has been made to prepare the fine particles of $Co_{1-x}Mn_xFe_2O_4$ ($0 \leq x \leq 0.4$) system for the first time by an efficient self propagating auto-combustion route and report their structural and magnetic properties.

EXPERIMENTAL

The $Co_{1-x}MnxFe_2O_4$ ($0 \leq x \leq 0.5$) samples were prepared by the combustion of aqueous solutions of redox mixture containing stoichiometric amounts of corresponding metal nitrates and glycine in a pyrex dish. The equivalence ratio, Φe (O/F) of the redox mixture for the combustion is calculated by using the total oxidizing and reducing valencies of the oxidizer (O) and fuel (F), which serve as numerical coefficients so that Φe (O/F), becomes unity and the heat released is at its maximum.[5] According to the principles used in propellant chemistry, the oxidizing and reducing valencies of various elements are considered as follows: C = 4, H = 1, O = −2, N = 0, M = 2, 3, etc. Thus, typically for ferrites i.e. $M^{2+}Fe^{2+}3O_4$ the oxidizing valency of a divalent metal nitrate, $M(NO_3)_2$ becomes −10; for trivalent $Fe(NO_3)_3$, it is −15, which should be balanced by the total valences in the fuel; glycine (H_2NCH_2COOH), which add up to +9. Hence for the stoichiometric composition of the redox mixture, in order to release the maximum energy for the reaction, require that $-40 + 9n = 0$ or $n = 4.44$ mol of glycine. A mixture of cobalt nitrate, manganese nitrate and ferric nitrate in stoichiometric molar proportion taken in a pyrex dish, was melted by heating at 80°C. Glycine was added to the melt and the slurry was introduced into a furnace preheated to 400°C. After evaporation of the water content, the mixture frothed and ignited to combust with a flame, giving a voluminous and foamy $Co_{1-x}MnxFe_2O_4$. After formation of as synthesized powder, the pellets were prepared by pressing in a die of 1 cm diameter and 0.1mm thickness with the help of a hydraulic press by applying a pressure of 5 ton/in to form the pellets using PVA as a binder. The pellets were final sintered at 1000°C in the presence of air.

RESULTS AND DISCUSSION

X-Ray Diffraction Studies

Powder X-ray diffraction measurements (XRD) were performed with a Siemens D5000 and Cu Kα radiation. Figure 1 shows XRD patterns of Co-ferrite samples deposited at various Mn substations (x = 0.0, 0.1, 0.2, 0.3, 0.4 and 0.5). All the samples are polycrystalline and shows single phase cubic spinel crystal structure. The samples show all the characteristic reflections of ferrite material with most intense (311) reflection which confirms the formation of cubic spinel structure.[6] The XRD patterns were indexed using the JCPDS card Nos. 22–1086 and 74–2403. The lattice parameter was calculated according to the formula $\sin^2\theta/(h^2 + k^2 + l^2) = \lambda_2/4a^2$, where λ is Cu Ka radiation with 1.5405 A°, θ is diffraction angle. The calculation of the lattice parameters was carried out using most intence reflection typical of a spinel structure material with a cubic cell, corresponding to the (311), (220), and (511) planes. The linear increase in lattice constant with Mn content may be attributed to the replacement of smaller Co^{2+} (0.78 Å) by the larger Mn^{2+} (0.83 Å) ions in the $Co_{1.2-x}Mn_xFe_{1.8}O_4$ system obeying Vegards's law [8–9] The lattice constant is found to increase with addition of Mn in $Co_{1.2-x}Mn_xFe_{1.8}O_4$ as shown in table 1. The theoretical density was calculated according to the formula $D_x = 8M/Na^3$, where M is the molecular mass, N is Avogadro's number, and is the lattice parameter which was calculated from the X-ray diffraction pattern. The percentage porosity of sample was calculate using the formula $P(\%) = (d_x- d_a/d_x)\times100$ where d_x is the x-ray density and d_a the actual density The theoretical density and porosity increases with Mn content as shown in Table 1.

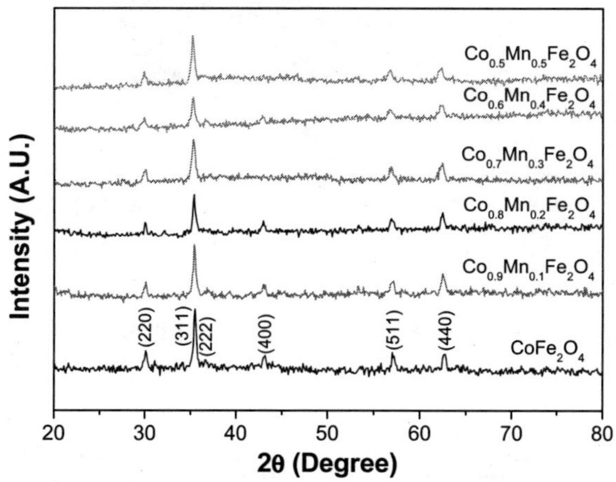

Fig. 1: XRD Patterns of Co-ferrite ($Co_{1-x}MnxFe_2O_4$) Samples Prepared at Various Mn Contents (x = 0.0, 0.1, 0.2, 0.3, 0.4, and 0.5)

Morphological Analysis

The surface morphologies were investigated using scanning electron microscopy. Figure 2 shows scanning electron micrographs of the $Co_{1-x}MnxFe_2O_4$ with x = 0.1, 0.2, and 0.3 respectively. Samples examined by SEM showed that law level of the Mn substitution(x = 0.1, 0.2) gives a uniform microstructure but further increase in Mn substitution intergranular pores increases. The average grain size was calculated by using line intercept method, and its variation with manganese content.[10] It is observed that the average grain size goes on increasing with increase in manganese content.

Fig. 2: SEM Micrographs of Co-ferrite ($Co_{1-x}Mn_xFe_2O_4$) Samples Prepared at Various Mn Contents a) x = 0.1, b) x = 0.2 and c) x = 0.3

Magnetic Properties

The magnetic properties of the material are obtained by B-H hysteresis loop, which is behavior of material in the magnetic field. Figure 3 shows the room temperature B–H hysteresis loop of

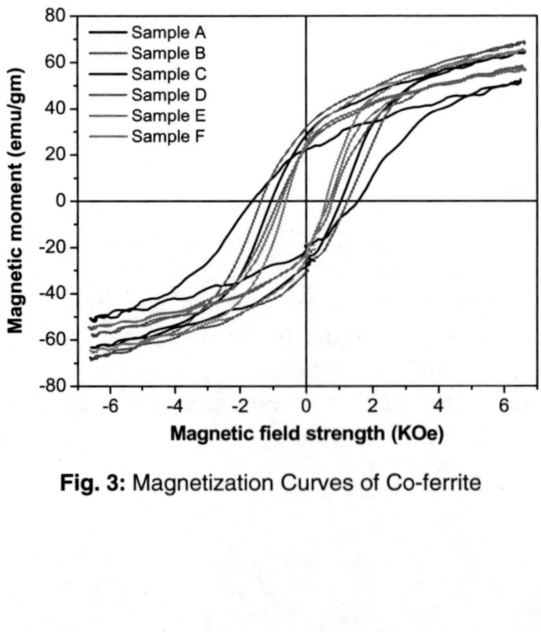

Fig. 3: Magnetization Curves of Co-ferrite

all the compositions of $Co_{1-x}MnxFe_2O_4$ system. All the samples show typical B-H hysteresis loop magnetic behavior, indicating the presence of ordered magnetic of spinel ferrite and the variation of the Ms depend on cation distribution in a spinal lattice.[10] Room temperature magnetization measurements show that for the substitution of Co by Mn, there is an initial increase in the saturation magnetization (Ms) and remanence (Mr) for x = 0.0 to x = 0.3 and decrease for x = 0.4. The coercivity of the samples deceases with increase of Mn content. The various magnetic parameters are as shown in Table 1.

Table 1: Data on Saturation Magnetization, Remanence, Coercivity, Lattice Constant, X-ray Density and Porosity

Ferrites. Composition (x)	*M_r (emu/gm)*	*M_s (emu/gm)*	*Coercivity H_c (Oe)*	*Lattice Constant a (Å)*	*X-ray Density(dx) gm/cm^3*	*Porosity (%)*
x = 0.0	30.68	52.55	1591.81	8.3675	4.5847	8.88
x = 0.1	31.48	58.54	1303.02	8.3722	4.5566	11.11
x = 0.2	35.93	64.77	1050.91	8.3788	4.5334	13.46
x = 0.3	45.88	68.94	820.68	8.3833	4.5134	15.89
x = 0.4	26.79	65.55	734.76	8.3901	4.4841	18.18
x = 0.5	33.30	57.26	603.83	8.3999	4.4524	20.42

Dielectric Properties

The room temperature performance of dielectric properties with frequency has been studied. From analysis of Figure 4, it is clear that higher frequencies showing dispersion of dielectric constant at lower frequencies due to charge transport relaxation time. The large value of dielectric constant is associated with space charge polarization and inhomogeneous dielectric structure viz. impurities, grain structure, defects and pores. The decrease in dielectric constant takes place when jumping frequency of electric charge cannot follow the alternations of applied electric field beyond certain critical frequency. This dielectric dispersion is attributed to Maxwell and Wagner type of interfacial polarization in agreement with Koop's phenomenological theory.[11] Since polarization decreases with increasing frequency and reaches constant values, decrease in dielectric constant with frequency is observed. The dielectric permittivity goes on increasing with the increase of Mn^{2+} concentration in the substituted Co ferrites, showing the usual dielectric dispersion with frequency. Figure 5, shows variation of dielectric loss with frequency for the all samples. For all samples dielectric loss decreases with frequency and almost independent at the high frequency. Figure 6 shows the variation of log (σ_{AC}) with logω^2. The plots are almost linear indicating that the conductivity increases with increase in frequency. It is well known that there are two types of polarons viz. small polarons and large polarons. In the small polaron model the conductivity increases linearly with increase in frequency and in case of large polarons the conductivity decreases with increase in frequency.[12]

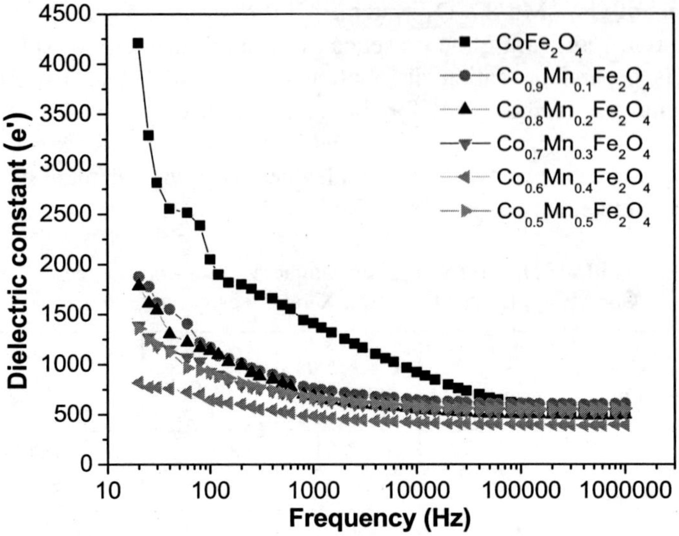

Fig. 4: Variation of Dielectric Constant with Frequency

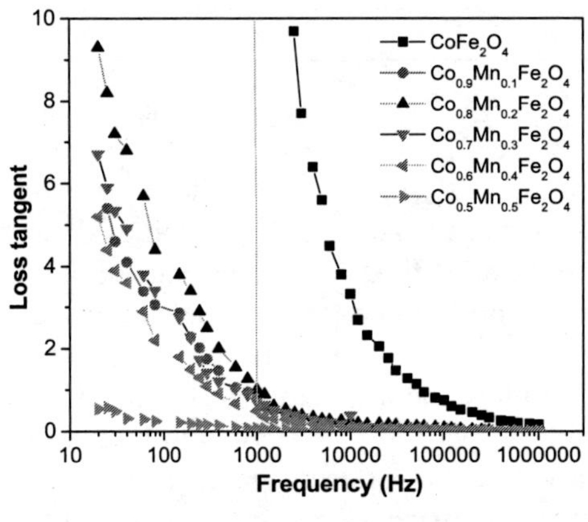

Fig. 5: Variation of Dielectric Loss with Frequency

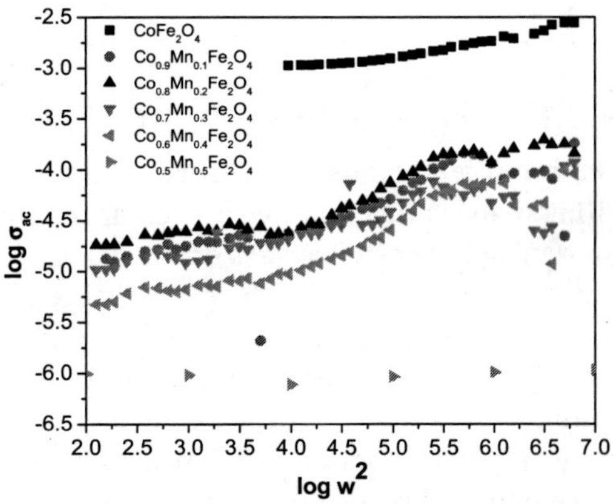

Fig. 6: Variation of log (σ_{AC}) with logω^2

Resistivity Measurement

Figure 7 shows temperature variation of DC resistivity of the samples. The resistivity of the samples decreases with increase in temperature showing the semiconducting nature of the samples. This is due to the increase in thermally activated drift mobility of charge carriers according to hopping conduction mechanism. In such a hopping process the carrier mobility is temperature dependant and is characterized by activation energy. The electrical resistivity is very sensitive to the cation distribution in the ferrites.[13]

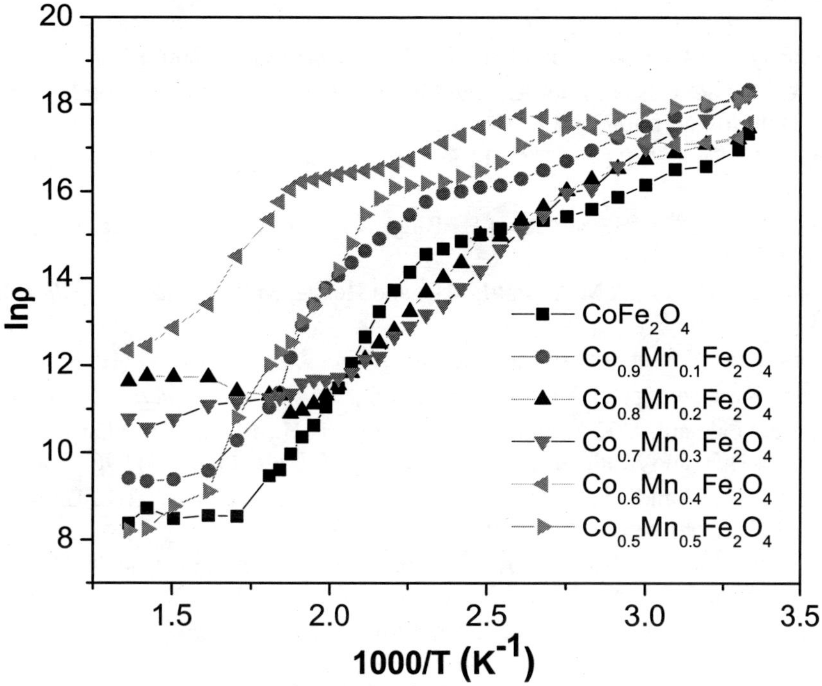

Fig. 7: Variation of DC Resistivity with Temperature for Co-ferrite

CONCLUSIONS

Fine particles of Co-Mn ferrite systems were successfully synthesized by auto-combustion route. The XRD confirms the formation of cubic spinel structure. The lattice parameter increases with increase in Mn content. Also the porosity of the material increases. SEM shows uniform microstructure with average grain size of about 150 nm. The dielectric permittivity goes on increasing with the increase of Mn^{2+} concentration in the substituted Co ferrites, showing the usual dielectric dispersion with frequency. The ac electrical conductivity with frequency shows that the conduction mechanism in the present case is due to small polaron type and it is concentration dependent. DC resistivity measurements lead to conclude that semiconducting behavior for all the samples. From the study it is confirmed that prepared system is suitable for making them good materials for electronic applications with low-cost and high performance.

ACKNOWLEDGEMENTS

Authors are very much thankful to UGC and DST for the financial support through UGC-DSA-I, DST-PURSE and DIST-FST-II programmes.

REFERENCES

[1] Ryu, J., Priya, S., Uchino, K. and Kim, H., *J. Electroceram.* 8 (2002) 107.

[2] Caltun, O.F., Rao, G.S.N., Rao, K.H. and Rao, B. Parvatheeswara, CheolGi Kim, Chong-Oh Kim, I. Dumitru, N. Lupu, and H. Chiriac Sensor Letters Vol.5 (2007)1.

[3] Paulsen, J.A., Ring, A.P., Lo, C.C.H., Snyder, J.E. and Jiles, D.C., *J. Appl. Phys.* 97 (2005) 044502.

[4] Paulsen, J.A., Lo, C.C.H., Snyder, J.E., Ring, A.P., Jones, L.L. and Jile, D.C., *IEEE Trans. Magn.* 39 (2003) 3316.

[5] Prakash, A.S., Khadar, A.M.A., Patil, K.C. and Hegde, M.S., *J. Mater. Synth. Processing* 10 (2002) 135.

[6] Woo Chul Kim, Young Suk Yi and Chul Sung Kim. *J. Appl. Phys* 85 (1999)8.

[7] Chul Sung Kim,Young Suk Yi. Key-Taeck Park,. *Journal of Magnetics.* 5(2000) 111.

[8] Bhatu, S.S., Lakhani, V.K., Tanna, A.R., Vasoya, N.H., Buch, J.U., Sharma, P.U., Tridevi, U.N., Joshi, H.H. and Modi, K.B., *Ind. J. Pure Appl. Phys.* 45 (2007) 596.

[9] Kambale, R.C., Shaikh, P.A., Harale, N.S., Bilur, V.A., Kolekar, Y.D., Bhosale, C.H. and Rajpure, K.Y., *J Alloys and Compounds* 490 (2010) 568.

[10] Caltun, O.F., Rao, G.S.N., Rao, K.H., Rao, B. Parvatheeswara, CheolGi Kim, Chong-Oh Kim, I. Dumitru, Lupu, N. and Chiriac, H., *Sensor Letters*, 5 (2007).

[11] Caltun, O., Chiriac, H., Lupu, N., I. Dumitru, B.P. Rao, *Optoelec, J. Adv. Mat.* 9 (2007) 1158

[12] Koop, J.C.G., *Phys. Rev. B,* 83 (1951) 12.

[13] Kambale, R.C., Shaikh, P.A., Bhosale, C.H., Rajpure, K.Y., Kolekar, Y.D., Smart Mater. Struct. 18 (2009) 115028.

[14] Puri, R.K. and Sayen, V., Proc. ICF-5, India (1989) 245.

SILVER-ACTIVATED CHARCOAL NANOCOMPOSITE FOR MICROBIAL DECONTAMINATION OF WATER

V. SHERLY ARPUTHA KIRUBA, ARUN DAKSHINAMURTHY AND P. MOSAE SELVAKUMAR[1]

Department of Nanotechnology, School of Nanosciences and Technology, Karunya University, Coimbatore, Tamil Nadu
[1]Department of Chemistry, School of Science and Humanities, Karunya University, Coimbatore, Tamil Nadu
E-mail: mosae@karunya.edu; arun@karunya.edu

ABSTRACT

Eco friendly chemical synthesis has various advantages over traditional methods and has become the need of the hour. Here we report a greener method for synthesizing silver nanoparticles from musk melon fruit extract and its characterization using XRD, SEM, particle size analyzer and UV-Visible spectroscopy. Its anti-microbial activity was also studied so as to be used along with activated charcoal as nanocomposite for water purification. The nanoparticles formed were effective against S. aureus and B. subtilis which are indicators of water contamination by humans. The nanocomposite formed was found to be more effective compared to virgin activated charcoal and makes it possible to be used for water filtration against microbial contamination. We conclude that the efficiency of virgin activated charcoal can be improved by compositing it with silver nanoparticles.

INTRODUCTION

Nanoparticles are fundamental building blocks of nanotechnology. The most important and distinct property of nanoparticles is that they exhibit a large surface area to volume ratio, the

V. Rajendran, R. Yuvakkumar, K. Thyagarajah and K.E. Geckeler (eds.)
Applications of Nano Materials: Electronics, Energy and Environment, pp. 295–302 (2012).

reason why they have been effectively used in medicine and bio-medical applications such as drug delivery, water filtration and medical diagnostics. Nanoparticles have also been studied for their anti-microbial properties.[1] Among the various methods for nanoparticle synthesis, 'green synthesis' has proven to be the most eco-friendly method.[2]

Natural disasters like flooding can considerably increase microbial contamination of drinking water[3] and can also result in release of hazardous materials into water resources.[4–6] Treatment of water therefore becomes a priority. Silver nanoparticles play an important role in water treatment due to their anti-microbial activity.

The adsorption technique, based on the transfer of pollutants from the solution to the solid phase, is one of the most efficient water treatment methods.[7] Adsorbents can be used to remove toxic substances such as chlorine-disinfectant residues, organics and particulates.[8–10] Due to its high surface area, porous structure, and special surface reactivity, activated charcoal (AC) has been the most popular and widely used adsorbent all over the world.[11,12] AC is highly inert and thermally stable, and can be used over a broad pH range. It has great capacity of adsorbing various organic compounds and can be easily modified by chemical treatment to increase its adsorption.[13] Although virgin AC can adsorb microbes in water, natural organic matter would rapidly occupy the adsorption sites and promote biofilm growth[14] thereby negatively impacting the quality of treated water.[8,15,16]. Activated charcoal can also remove objectionable taste and odorous compounds such as iodine and chlorine by adsorption and chemical reduction.[17] The major drawback of activated charcoal based adsorption that limits its long-term application is that it requires continuous replacement due to saturation, though suitable for short term acute emergencies.[11] Carbon metabolizing bacteria also pose concerns in treatment of water using AC[18,19] In this work we report the development of a nanocomposite of activated charcoal and 'green' synthesized silver nanoparticles for effective water treatment against microbial contamination.

EXPERIMENTAL

Synthesis and Characterization of Silver Nanoparticles

Silver nanoparticles were prepared using musk melon (*Curcumis melo*) extract, as reducing agent since its rich in L-ascorbate. The fruit pulp was mashed with mortar and pestle and centrifuged at 5000 rpm for 15 mins to pellet the debris. The clear supernatant was used for nanoparticle preparation. 1 mM silver nitrate solution was reduced with 5% musk melon fruit extract. The colloidal solution of silver nanoparticles was characterized with UV-Vis spectroscopy, particle size analyzer, SEM and XRD.

Anti-Microbial Assay

Agar well diffusion method was followed for anti-microbial assays. Two gram negative (*Escherichia coli, Pseudomonas aeruginosa*) and two positive (*Staphylococcus aureus, Bacillus subtilis*) microbes were tested. Ampicillin and tetracycline were used as control antibiotics depending on the strain used. The zone of inhibition was measured and recorded.

Preparation of Activated Charcoal (AC) and its Characterization

Washed and dried coconut shells were treated with sulfuric acid for 2 days and then washed over 15 times with distilled water (centrifuged after overnight sedimentation) till the pH reached 7. It was then dehydrated at 110°C overnight and ground to obtain activated charcoal powder, and was then characterized using SEM and XRD before being used along with the silver nanoparticles as a nanocomposite to test its antimicrobial activity during water filtration.

Preparation of Silver-Activated Charcoal (Ag-AC) Nanocomposite and its Characterization

100 mg of the prepared activated charcoal was added to 100 ml of the silver nanoparticles solution. After mixing briefly, the solution was allowed to settle overnight, filtered the next day and air dried to obtain the nanocomposite. The nanocomposite was characterized using SEM and XRD.

Cartridge Fabrication Using Silver-Activated Charcoal (Ag-AC) Nanocomposite

50 mg of the obtained air dried nanocomposite was loaded into 1 ml syringe columns for testing water filtration and biocidal efficiency. A small amount of non-absorbent cotton was used to plug the upper and lower ends of the column to prevent the nanocomposite from leaching.

Testing of the Nanocomposite's Biocidal Activity

1 ml of the microbial culture (*E. coli, B. subtilis, P. aeruginosa* and *S. aureus*) at 0.6 OD_{600} was passed through the cartridge under slight positive pressure in order to maintain a constant flow rate of 1 drop per second. The filtrate was collected in sterilized vials, serially diluted and plated on sterile agar plates. The plates were incubated in an upright position at 37°C for 24 hours. The number of colonies formed were counted and recorded.

RESULTS AND DISCUSSION

Green Synthesis

Addition of 5% musk melon fruit extract to a colorless solution of 1 mM silver nitrate resulted in a change of color to bright yellowish brown within 30 minutes confirming the formation of silver nanoparticles (Figure 1, left panel). The speed of the reaction can be attributed to the corresponding abundance of vitamin C in the fruit extract. The color of the solution may be due to nanosize of particles.

Synthesis of Ag NPs SEM image of Ag NPs SEM image of Ag-AC Nanocompsite

Fig. 1: Synthesis of Silver Nanoparticles (Ag NPs), SEM Image of Silver Nanoparticles, and SEM Image of Silver-Activated Charcoal (Ag-AC) Nanocomposite

SEM Analysis

The SEM image of silver nanoparticles (Figure 1, center panel) shows uniformly distributed nanostructures that are spherical in shape and uniformly arranged. The SEM image of the silver-activated carbon nanocomposite (Figure 1, right panel) reveals a porous morphology and the silver nanoparticles are seen adsorbed onto the porous regions of activated charcoal with the particles uniformly dispersed.

UV-Visible Spectroscopy

The UV-Visible absorption spectrum (Figure 2) of the bright yellowish brown silver nanoparticles solution showed a surface plasmon absorption band with λ_{max} of 432 nm indicating the presence of silver nanoparticles. This red shift from 308 nm (plain 1 mM silver nitrate) to 432 nm confirms the reduction of silver nitrate to silver using musk melon extract. The broadening of peak refers to poly dispersed nanoparticles in the solution.[20]

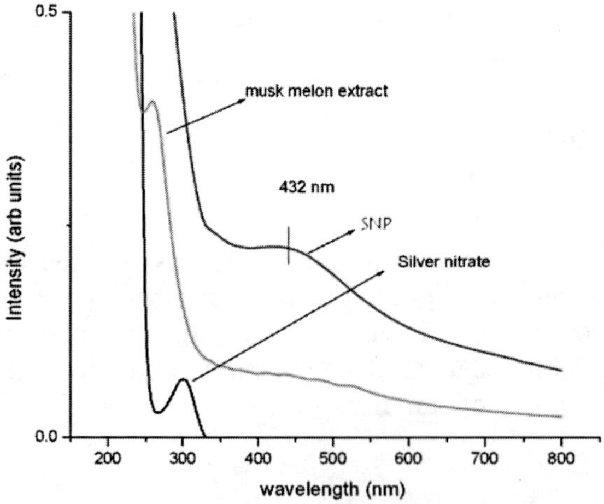

Fig. 2: UV-Vis Spectrum of Silver Nanoparticles Synthesized Using Musk Melon Extract

Anti-Microbial Assay

The silver nanoparticles prepared using musk melon fruit extract as the reducing agent were found to be more effective against *S. aureus* and *B. subtilis* (Figure 3A).[22–24] This might be due to nanometer size of particles formed. Hence this can be used in a composite with activated charcoal for water filtration purposes.

Efficiency Testing of Biocidal Nanocomposite

The efficiency of virgin activated charcoal and prepared nanocomposites were compared in water filtration fixed bed columns. The musk melon silver nanoparticles-activated charcoal nanocomposite filled syringe column was effective against all the four test microbes and the efficiency is shown in Figure 3B. It is seen that the nanocomposite was more effective against *B. subtilis* and *S. aureus* which might be due to the anti-microbial property of silver nanoparticles.

Reuse Efficiency Testing of Nanocomposite

The reusability was checked by passing a known CFU of microbes and testing the efficiency of filtration. The efficiency of the second time usage was comparatively lower than that of the first pass (Figure 3C). However, the nanocomposite showed higher durability which might be due to its anti-microbial capability whereas the activated charcoal's efficiency decreased due to its adsorption property alone. We conclude that the nanocomposite's efficiency is higher compared to that of virgin activated charcoal alone.[8,25,26]

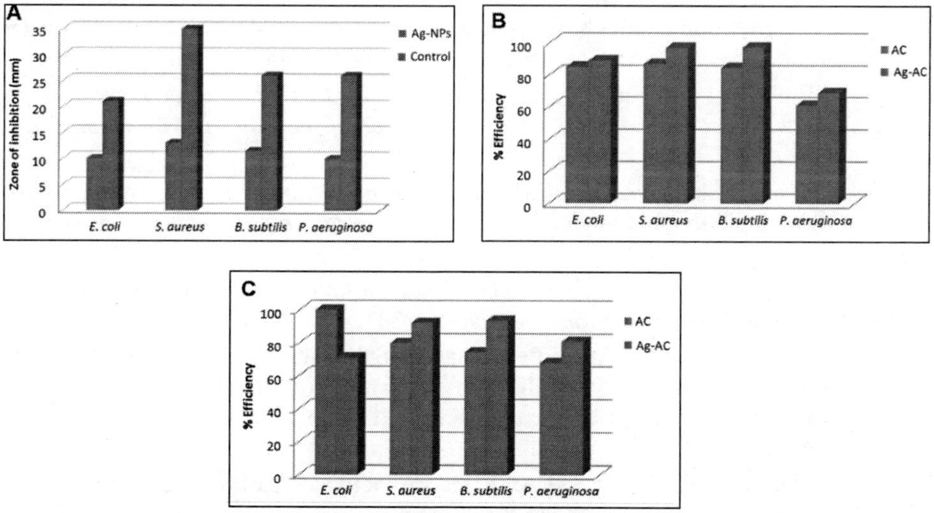

Fig. 3: A – Zone of inhibition caused by silver nanoparticles synthesized using musk melon extract, B – water filtration efficiency of biocidal nanocomposite, AC – activated charcoal, Ag-AC – silver-activated charcoal, C – water filtration reusage efficiency of biocidal nanocomposite.

CONCLUSION

The green synthesis of silver nanoparticles using musk melon fruit extract was studied and it was found that the effectiveness of the nanoparticles was based on the effective reduction reaction happening in the reaction media. It is directly related to the quantity of active reducing agent (L-ascorbate) present in the fruit extract used. The silver nanoparticles exhibited high anti-microbial activity, both in plate culture as well as in fixed bed water filtration column units. With nanocomposite formation, the efficiency of filtration and biocidal activity of activated charcoal is enhanced. This is evident from the reusage efficiency of water filtration fixed bed columns.

ACKNOWLEDGEMENT

Assistance from Karunya University in the form of short term research grants is gratefully acknowledged.

REFERENCES

[1] Jain, D., Daima, H.K., Kachhwahl, S. and Kothari, S.L., Synthesis of plant-mediated silver nanoparticles using papaya fruit extract and evaluation of their anti microbial activities, *Digest Journal of Nanomaterials and Biostructures,* 4, (3), 557–563, 2009.

[2] Leela, A. and Vivekanandan, M., Tapping the unexploited plant resources for the synthesis of silver nanoparticles, *African Journal of Biotechnology*, 7 (17), 3162–3165, 2008.

[3] Faruque, S.M., Naser, I.B., Islam, M.J., Faruque, A.S.G., Ghosh, A.N., Nair, G.B., Sack, D.A. and Mekalanos, J., Seasonal epidemics of cholera inversely correlate with the964 prevalence of environmental cholera phages, Proceedings of the National Academy of Sciences of the United States of America, 102, 1702–1707, 2005.

[4] Young, S., Balluz, L. and Malilay, J., Natural and technologic hazardous material releases during and after natural disasters", a review. *Science of the Total Environment,* 322, 3–20, 2004.

[5] Srinivas, H. and Nakagawa, Y., Environmental implications for disaster preparedness: lessons Learnt from the Indian Ocean Tsunami, *Journal of Environmental Management,* 1350 (89), 4–13, 2008.

[6] Violette, S., Boulicot, G., Gorelick, S.M., Tsunami-induced groundwater salinization in southeastern India, *Comptes Rendus Geosciences* 341, 339–346, 2009.

[7] Ghaedia, M., Sadeghiana, B., Amiri Pebdania, A., Sahraeib, R., Daneshfarb, A., Duranc, C., Kinetics, thermodynamics and equilibrium evaluation of direct yellow 12 removal by adsorption onto silver nanoparticles loaded activated carbon, *Chemical Engineering Journal,* 187, 133– 141, 2012.

[8] Tobin, R.S., Smith, D.K., Lindsay, J.A., Effects of activated carbon and bacteriostatic filters on microbiological quality of drinking water, *Applied and Environmental Microbiology,* 41, 646–651, 1981.

[9] Bell, F.A. Jr, Perry, D.L., Smith, J.K., Lynch, S.C., Studies on home water treatment systems", *Journal of American Water Works Association,* 76, 126–130, 1984.

[10] Demirbas, E., Kobya, M., Sulak, M.T., Adsorption kinetics of a basic dye from aqueous Solutions onto apricot stone activated carbon, *Bioresour. Technol,* 99, 5368–5373, 2008.

[11] Loo, S.-L., Fane, A.G., Krantz, W.B., Lim, T.-T., Emergency water supply: A review of potential technologies and selection criteria, *Water Research* doi: 10.1016/j.watres.2012.03.030. 2012.

[12] Babel, S., Kurniawan, T.A., Low-cost adsorbents for heavy metals uptake from contaminated water: a review, *J Hazardous Mat,* 97, 219–243, 2003.

[13] Jiuhui, Q.U., Research progress of novel adsorption processes in water purification", *A Review Journal of Environmental Sciences,* 20, 1–13, 2008.

[14] Caroli, G., Levre, E., Armani, G., Biffi-Gentili, S., Molinari, G., "Search for acid-fast bacilli in bottled mineral waters", *Journal of Applied Microbiology,* 58, 461–463, 1985.

[15] Wallis, C., Stagg, C.H., Melnick, J.L., The hazards of incorporating charcoal filters into domestic water systems. *Water Research,* 8, 111–113, 1974.

[16] Su, F., Luo, M., Zhang, F., Li, P., Lou, K. and Xing, X., Performance of microbiological control by a point-of-use filter system for drinking water purification", *Journal of Environmental Sciences,* 21, 1237–1246, 2009.

[17] Backer, H., Water disinfection for international travelers, in: Jay S K, Phyllis E K, David O F, Hans D N, Bradley A C, Travel Medicine (Second Edition) Mosby, Edinburgh, 47–58, 2008.

[18] Saier, M.H., Multiple mechanisms controlling carbon metabolism in bacteria, *Biotechnol. Bioeng.* 58 170–174, 1998.

[19] Camper, A.K., Lechevallier, M.W., Broadaway, S.C., Mcfeters, G.A., Bacteria associated with granular activated carbon particles in drinking water, *Appl. Environ. Microbiol,* 52 434–438, 1986

[20] Ankanna, S., Prasada, T.N.V.K.V., Elumalaib, E.K., Savithramma, N., Production of biogenic silver nanoparticles using Boswellia ovalifoliolata stem bark , *Digest Journal of Nanomaterials and Biostructures* 5(2), 369–372, 2010.

[21] Koraa, A.J., Sashidharb, R.B., Arunachalama, J., Gum kondagogu (Cochlospermumgossypium): A template for the green synthesis and stabilization of silver nanoparticles with antibacterial application", *Carbohydrate Polymers* 82, 670–679, 2010.

[22] Balazs, D.J., Triandafillu, K., Wood, P., Chevolot, Y., van Delden, C., Harms, H., Hollenstein, C., Mathieu, H. J. Inhibition of bacterial adhesion on PVC endotracheal tubes by RF-oxygen glow discharge, sodium hydroxide and silver nitrate treatments. *Biomaterials* 25, 2139–2151, 2004.

[23] Melaiye, A. and Youngs, W.J., Silver and its application as an antimicrobial agent, *Expert Opin Ther Pat* 15, 125–130, 2005.

[24] Sondi, I. and Salopek-Sondi, B., Silver nanoparticles as antimicrobial agent: A case study on *E. coli* as a model for gram-negative bacteria, *J Colloid Interface Sci* 275, 177–182, 2004.

[25] Fiore, J.V. and Babineau, R.A., Effect of an activated carbon filter on the microbial quality of water, *Appl Environ Microbiol,* 34 (5), 541–546, 1977.

[26] Stobie, N., Duffy, B., McCormack, D.E., Colreavy, J., Hidalgo, M., McHale, P. and Hinder, S.J., Prevention of staphylococcus epidermidis biofilm formation using a low-temperature processed silver-doped phenyltriethoxysilane sol-gel coating", *Biomaterials* 29, 963–969, 2008.

SCIENTOMETRIC ANALYSIS OF NANO SCIENCE AND NANO TECHNOLOGY RESEARCH OUTPUT IN ASIAN COUNTRIES: A STUDY BASED ON WEB OF SCIENCE DATABASE

S. Kanchana, Suhasini Ernest and R. Balasubramani[1]

Department of Physics, Urumu Dhanalakshmi College, Trichy
[1]Department of Library and Information Science, Bharathidasan University, Trichy
E-mail: lisbala@gmail.com

ABSTRACT

Objective: This study analyses the research output in India in Nano Science and Nano Technology during the period 1999–2011 and the analyses included research growth, rank, citation impact, share of international collaborative papers and major collaborative partner countries, global publications' share, and patterns of research communication in most productive journals. It also analyses the characteristics of most productive institutions, authors and high-cited papers. The publication output and impact of India is also compared with other Asian Countries. Materials and Methods: Web of science Citation database was used for retrieving the publications' output of India and other Asian countries in Nano Science and Nano Technology during 1999–2011. Results: India's global publications' share in neurosciences during the study period was 3.60% (with 240 papers) and it ranked 9th among the top countries in Nano Science and Nano Technology. Conclusion: India is far behind in terms of publication output, citation quality and share of international collaborative papers in Nano Science and Nano Technology when compared to other Asian countries with an emerging economy. There is an urgent need to substantially increase the research activities in the field of Nano Science and Nano Technology in India.

V. Rajendran, R. Yuvakkumar, K. Thyagarajah and K.E. Geckeler (eds.)
Applications of Nano Materials: Electronics, Energy and Environment, pp. 303–310 (2012).

INTRODUCTION

Now a day the scientometrics, research publications are clearly one of the quantitative measures for the basic research activity in a country. It must be added, however, that what excites the common man, as well as the scientific community, are the peaks of scientific and technological achievement, not just the statistics on publications. Many scientometrics studies have appeared in the literature to focus on the performance of science in the field of Nano Technology.

METHODOLOGY

Data was collected from the Science Citation Index (SCI) which is available via the Web of Science (WoS). The WoS is the search platform provided by Thomson Reuters (the former Thomson Scientific emerged from the Institute for Scientific Information (ISI) in Philadelphia). SCI database is one of the very comprehensive databases covering all aspects of science. The study period (1999–2011) is selected as the database is available in machine from since 1982. The search string "Nano Technology" in the "Basic search" field of SCI was used for the year s 1999–2011 to download the records on the subjects 'Nano Technology'. A total of 6669 records were downloaded and analyzed by using the web of science website application as per the objectives of the study.

The global publication shares of the top 25 most productive countries in Nano Technology during 1999–2011 varied between 0.76% and 22.07%. United States of America topped the list, with global publication share of 22.07% during 1999–2011. China ranked second, followed by, Japan, Germany, South Korea, France, England (their global publications share ranging from 5–17%). Italy, India, Taiwan, Australia, Netherlands, Canada and Singapore ranked at 8th to 14th positions (their global publications' share ranging from 2–4%).

India ranked at 9th position among the top 25 most productive countries in Nano Technology with its global Publications' share of 3.60% during 1999–2011. India's global publications share increased from 1.75 to 3.67%.

The Table 2 reveals that the Exponential growth rate of publications in Nano Technology research output. An exponential growth in number of publication was observed during 1999 to 2011, average growth rate is 1.45. The highest growth rate is 7.00 percents found during 2001. It was found very low growth rate (0.38) at the year of 2004 with 9 publications. The total exponential growth rate value is 18.87; it is found the average exponential growth rate is 1.45 percents for sample periods.

India's publications' output in Nano Technology India's during 1999–2011 consisted of 240 papers, with an average of 19 papers per year. Compared to India, the publications' output of USA, China andJapan during the same period consisted of 1472 papers, 1093 and 933 papers, with the average number of papers per year as 114, and 84.72 respectively. In terms of impact and quality, the average citations per paper registered by India's publication output during 2011 was 1.04. Compared to India, USA publications registered higher citation impact of 2.71, compared to India, China and Japan publications are lower citation impact of 0.95 and 0.98 during the same period.

Table 1: Global Publications Output Publications Share and Rank of Top 25 Most Productive Countries in Nano Technology

S.No.	Country	No of Papers			% Share of Papers			Rank		
		1999–2011	1999	2011	1999–2011	1999	2011	1999–2011	1999	2011
1.	Usa	1472	47	183	22.07	16.49	24.02	1	2	1
2.	China	1093	38	152	16.39	13.33	19.95	2	3	2
3.	Japan	933	58	55	13.99	20.35	7.22	3	1	4
4.	Germany	585	30	54	8.77	10.53	7.09	4	4	5
5.	South Korea	424	11	56	6.36	3.86	7.35	5	6	3
6.	France	381	17	36	5.71	5.96	4.72	6	5	7
7.	England	317	17	41	4.75	5.96	5.38	7	5	6
8.	Italy	244	10	36	3.66	3.51	4.72	8	7	8
9.	India	240	5	28	3.60	1.75	3.67	9	11	11
10.	Taiwan	234	7	31	3.51	2.46	4.07	10	9	10
11.	Australia	151	5	22	2.26	1.75	2.89	11	12	13
12.	Netherlands	145	9	15	2.17	3.16	1.97	12	8	19
13.	Canada	137	4	25	2.05	1.40	3.28	13	14	12
14.	Singapore	137	5	18	2.05	1.75	2.36	14	13	14
15.	Switzerland	128	7	15	1.92	2.46	1.97	15	9	18
16.	Russia	119	6	10	1.78	2.11	1.31	16	10	17
17.	Spain	112	4	16	1.68	1.40	2.10	17	15	15
18.	Sweden	105	4	16	1.57	1.40	2.10	18	16	16
19.	Belgium	97	3	8	1.45	1.05	1.05	19	17	21
20.	Poland	84	7	6	1.26	2.46	0.79	20	9	23
21.	Austria	76	7	7	1.14	2.46	0.92	21	9	22
22.	Iran	74	7	34	1.11	2.46	4.46	22	9	9
23.	Israel	63	7	4	0.94	2.46	0.52	23	9	25
24.	Denmark	57	2	10	0.85	0.70	1.31	24	18	20
25.	Scotland	51	1	5	0.76	0.35	0.66	25	19	24
	World	6669	285	762						

Table 2: Exponential Growth Rate of the Nano Technology Research Output in India

S.No.	*Year*	*No. of Records*	*Exponential Growth Rate*
1.	1999	5	–
2.	2000	2	0.40
3.	2001	17	**7.00**
4.	2002	23	1.35
5.	2003	24	1.04
6.	2004	9	0.38
7.	2005	10	1.11
8.	2006	22	2.20
9.	2007	19	0.86
10.	2008	29	1.53
11.	2009	24	0.83
12.	2010	28	1.17
13.	2011	28	1.00
			18.87(1.45)

Table 3: Publications Output and Citations Received by Papers of India, USA, China, Japan

Years	*India*			*USA*			*China*			*Japan*		
	TP	*TC*	*ACPP*	*TP*	*TC*	*ACPP*	*TP*	*TC*	*ACPP*	*TP*	*TC*	*ACPP*
1999	5	51	10.20	47	3287	69.94	38	773	20.34	58	1360	23.45
2000	2	24	12.00	62	2128	34.32	59	1519	25.75	59	1743	29.54
2001	17	390	22.94	99	4708	47.55	87	2216	25.47	114	2594	22.75
2002	23	415	18.04	116	5661	48.80	102	3336	32.71	128	2851	22.27
2003	24	1179	49.12	140	5250	37.50	135	2901	21.49	165	4091	24.79
2004	9	506	56.22	52	2407	46.29	36	309	8.58	40	651	16.28
2005	10	157	15.70	90	2922	32.47	59	626	10.61	44	548	12.45
2006	22	93	4.23	119	2502	21.03	68	988	14.53	67	847	12.64
2007	19	193	10.16	127	3286	25.87	68	1104	16.24	50	582	11.64
2008	29	473	16.31	142	3547	24.98	72	517	7.18	41	300	7.32
2009	24	229	9.54	153	2099	13.72	91	614	6.75	58	290	5.00
2010	28	125	4.46	142	993	6.99	126	552	4.38	54	319	5.91
2011	28	29	1.04	183	496	2.71	152	145	0.95	55	54	0.98
1999–2008	240	3864		1472	39286		1093	15600		933	16230	

TP- Total Papers TC- Total Citations ACPP- Average Citations per Paper.

Research profile of the most productive Indian institutions in Nano Technology. The top 15 most productive Indian institutions involved in research in Nano Technology have each published 4 and more papers during 1999–2011. The list of these Indian institutions along with their research output, collaborative papers, and citations received and h-index values are presented in Table 4. These 14 Indian institutions together have contributed 52.08% share (with 125 papers) of the publications' output of India in Nano Technology, with an average of 8.92 papers per institution. Only four Indian institutions have registered higher publications' share than the group's average. These were Indian Institute of Technology (with 45 papers), followed by Bhabha Atom Research Center (with 11 papers), Indian Institute of Science (with 11 papers), Indira Gandhi Centre for Atomic Research (with 9 papers).

Table 4: Research Output, Citation Impact, H-Index of the Top Most Productive Indian Institutions in Nano Technology 1999–2011

S. No.	*Name of Institution*	*TP*	*TC*	*ACPP*	*h-index*
1.	Indian Institute of Technology	45	1480	32.89	15
2.	Bhabha Atom Research Center	11	93	8.45	5
3.	Indian Institute of Science	11	399	36.27	7
4.	Indira Gandhi Centre for Atomic Research	9	131	14.55	6
5.	Banaras Hindu University	7	42	6.00	4
6.	Indian Association for the Cultivation of Science	6	132	22.00	4
7.	National Chemical Laboratory	6	182	30.33	6
8.	National Physical Laboratory	6	4	0.66	2
9.	Central Glass and Ceramics Research Institute	4	81	20.25	3
10.	Centre for Materials for Electronics Technology	4	24	6.00	4
11.	Nuclear Science Center	4	27	6.75	3
12.	Jamia Millia Islamia	4	3	0.75	1
13.	Tata Institute of Fundamental Research	4	56	14.00	3
14.	University of Rajasthan	4	105	26.25	3

The average citations per paper registered by the total papers of these 14 Indian institutions are 16.08. Among these 14 institutions, Indian Institute of Science has scored the highest impact of 36.27 citations per paper, followed by Indian Institute of Technology (with 32.89 citations per paper), National Chemical Laboratory (with 30.33 citations per paper), University of Rajasthan (with 26.25 citations per paper), Central Glass and Ceramics Research Institute (with

20.25 citations per paper) and Indian Association for the Cultivation of Science (with 22.00 citations per paper).

The average h-index value of these 14 most productive Indian institutions is 4.71. Five Indian institutions have scored higher h-index value than group's average of 4.71.

Table 5: List of Top 10 Productive Indian Journals Publishing in Nano Technology

S. No	*Journal Name*	*Records*	*TLCS*	*TGCS*	*h-index*
1.	Materials Letters	11	1	267	9
2.	Journal of Materials Science	10	1	85	6
3.	Nuclear Instruments and Methods in Physics Research Section B-Beam Interactions with Materials and Atoms	7	3	48	4
4.	Applied Surface Science	6	2	92	5
5.	Sensors and Actuators B-Chemical	6	0	101	6
6.	International Journal of Hydrogen Energy	5	0	128	4
7.	Journal of Nanoscience and Nanotechnology	5	0	1	1
8.	Materials Science and Engineering A-Structural Materials Properties Microstructure and Processing	5	4	66	3
9.	Current Science	4	2	69	3
10.	Digest Journal of Nanomaterials and Biostructures	4	0	21	4

Table 6: Research Output, H-Index of Top 10 Productive Authors of India in Nano Technology

S. No	*Author*	*Records*	*TLCS*	*TGCS*	*h-index*
1.	Agarwala RC	5	1	26	2
2.	Kulkarni VN	5	3	55	3
3.	Kumar S	5	4	21	2
4.	Tyagi AK	5	1	66	3
5.	Ali I	4	0	3	1
6.	Choudhury AR	4	6	33	3
7.	Khanna PK	4	1	20	2
8.	Kumar R	4	2	76	3
9.	Tripathi SK	4	3	16	2
10.	Chakarvarti SK	3	2	12	2

These top 10 most productive Journals have contributed 63 papers, with an average contribution of 6.3 papers per Journal publications output of India during 1999–2011. Among the top 10 most productive journals, only three journals have published higher number of papers than the group's average. These are Materials Letters (with 11 papers), Journal of Materials Science (with 11 papers), Nuclear Instruments and Methods in Physics Research Section B-Beam Interactions with Materials and Atoms (with 7 papers).

Among the prominent authors contributing to Nano Technology research in India, top 10 Indian authors were identified as most productive and they have published more than 3 papers during 1999–2011. Of these 10 most productive authors, top most author Agarwala RC were affiliated to Indian Institute of Technology Roorkee, Uttar Pradesh,Second author Kulkarni VN were Department of Physics, Indian Institute of Technology Kanpur, Kanpur, Third author Kumar S were The Technological Institute of Textile and Sciences, Bhiwani,

These 10 authors together have contributed 43 papers, with an average contribution of 4.3 papers per author publications output of India during 1999–2011. Among the 10 most productive authors, only four authors have published higher number of papers than the group's average. These are Agarwala RC, Kulkarni VN, Kumar S and Tyagi AK (5 papers each)

The average h-index value registered by these 10 authors is 2.3. Four authors registered higher h-index value than the group's average. They were Kulkarni VN, Tyagi AK, Choudhury AR, Kumar R (h-index value of 3 each).

CONCLUSION

In this study the literature on Nano Technology, a promising new material, has been analyzed by scientometric methods. The time evolution of the overall number of citations reveals that the impact increase of the Scientometrics papers is possibly going to outrun the impact increase of the related research fields on Nano Technology. India ranked at 9th position among the top 25 most productive countries in Nano Technology research output. The highest number of publications in India was produced in 2010, 2011. The most productive author is Agarwala RC with 5 papers dealing with Nano Technology. The most productive Journal is Materials Letters with 11 papers.

REFERENCES

[1] Balasubramani, R., Amsaveni, N. and Amsaveni, N., Research Activities in Artificial Cell, 1991–2010: A Scientometric Analysis, *Indian Journal of Applied Research*, Vol. 1, Issue 3, 2011.

[2] Srinivasaragavan, S., Surulinathi, M. and Balasubramani, R., Mapping of Harvard Business Review Publications, Smart Journal of Business Management Studies, Vol. 6, No. 2, 2010.

[3] Balasubramani, R. and Gunasekaran, M., Scientometric Analysis of Artificial Intelligence Research Output: An Indian Perspective, *European Journal of Scientific Research*, Vol. 70, No. 2, pp. 317–322, 2012.

[4] Zitt, M. and Bassecoulard, E., Delineating Complex Scientific Fields by an Hybrid Lexical-Citation Method: An Application to Nanosciences. Information Processing and Management, 42(6), 1513–1531, 2006.

FUNCTIONALIZATION OF CARBON NANOTUBES WITH METAL PHTHALOCYANINE FOR GAS SENSING APPLICATION

PIKA JHA, MOHINI SHARMA, ABHILASHA CHOUKSEY, POORNENDU CHATURVEDI, DEEPAK KUMAR, GARIMA UPADHYAYA, J.S.B.S. RAWAT AND P.K. CHAUDHARY

SSPL, DRDO, Timarpur, Delhi

ABSTRACT

In present research work, various composites which are composed of functionalized Single Walled Carbon Nanotubes (Hydroxyl SWNT, Carboxyl SWNT, or Amided SWNT) and Copper Phthalocyanine (CuPc) has been obtained. The resultant composites have been characterized by Diffuse Reflectance Infrared Fourier Transform (DRIFT) spectroscopy, Raman spectroscopy and FESEM analysis. We observe that as expected, Phthalocyanines get non-covalently adsorbed onto CNTs surface by π–π stacking. From DRIFT spectra it is clear that there is conversion of β- form of Phthalocyanine to α-form, which is a clear evidence of composite formation. The results also reports that only oxidized CNTs i.e. Carboxylated and Hydroxylated Carbon Nanotubes gives reaction with Phthalocyanine because of electron withdrawing nature of –COOH and –OH group which is beneficial for coupling with other π-conjugated material through π–π stacking. No reaction has taken place with Amided Carbon Nanotube because of electron donating nature of long alkyl chain attached to amide group which is formed by functionalization of CNT with octadecyl amine. In the present paper only results of SWNT-OH with CuPc have been reported. It is well known that these CNT-MPc composites are more efficient in improving the relative responses to different analyte gases as compared to the individual CNT or MPc. We have observed high selectivity in the thin film resistors prepared from SWNT-OH/CuPc composites towards nitrogen oxide gases.

V. Rajendran, R. Yuvakkumar, K. Thyagarajah and K.E. Geckeler (eds.)
Applications of Nano Materials: Electronics, Energy and Environment, pp. 311–318 (2012).

INTRODUCTION

Carbon Nanotubes comprise a proximately new class of nanostructures composed totally of carbon atoms. Because of their electronic and mechanical properties, Carbon Nanotubes emerge to be a favorable material for Composites, Electronics Sensing and Biological applications. CNTs-based gas sensors have received considerable attention because of their outstanding properties towards a wide variety of gases that may be detected, such as high sensitivity and especially lower operating temperature, compared with the other types of gas sensors. To date SWNTs have shown sensitivity towards nerve agents and gases such as NH_3, NO_2, H_2, CH_4, CO, SO_2, H_2S, H_2O_2 and O_2.[1–5] To make the CNT based sensors more specific and selective to gases, researchers have focused on a wide variety of functional groups, to be attached either covalently or non-covalently to the CNTs. One such interesting class of functional molecules is metal phthalocyanine (M-Pc).

MPc are a class of coordination complexes composed of a macrocyclic ring with a central metal center. MPcs are used industrially as dyes [6] and catalysts,[7] and as components of recordable media,[8] photovoltaics,[9] and vapor sensors.[10] Phthalocyanine chemical and electronic properties may be tuned through variation of the metal center,[11] or through organic functionalization of the macrocycle.[12] These are chosen mainly because of three reasons: firstly M-Pc are known to exhibit high electron charge transfer due to 18π electron conjugated system of Metal Phthalocyanine.[13] Secondly MPc are selective sensor towards gas vapour.[14] Lastly they have exceptional thermal and chemical stability compared with other materials. Metal Phthalocyanine interacts with CNT by two means: either covalently or non-covalently.

Electron withdrawing group present either on CNT or MPc is beneficial for coupling with other π-conjugated material through π-π stacking. Electron withdrawing substituents enhance binding by withdrawing electron density from the π-cloud of the substituted ring, reducing the repulsive electrostatic interaction with unsubstituted MPc. Conversely, electron donating substituents donate excess electrons into the π-system and diminish the π-stacking interaction.[15] Thus only those SWNTs which are having electron withdrawing groups at the end or surface of a tube give reaction with MPc. No reaction has taken place with amided SWNTs (formed with octadecyl amine) because of electron donating nature of long alkyl chain.

In the present work various composites have been obtained consisting of functionalized SWNT and MPc.Here MPc used were unsubstituted, therefore only those CNTs which have electron withdrawing group at the surface and end of the tube give reaction with MPc.[16] Phthalocyanines (Pcs), both metalated (MPcs) and metal-free (H_2Pc), are metal-organic semiconductors that have been applied as chemiresistive sensors.[17–18] Conductivity in MPc films depends strongly on atmospheric chemical species, particularly oxidants and reductants.[19] As gas molecules interacts with MPc surface-charge transfer interaction occur leading to a large increase or decrease in surface conductivity, thus MPc based gas sensors can be considered as excellent gas sensors.[20–21] These CNT-MPc composites are more efficient in improving the relative responses to different analyte gases as compared to the individual CNT or MPc.[22]

Selectivity is considered as an important factor for the evaluation of hybrid sensor. Results report that the hybrid sensors were shown to be highly sensitive for gas oxidants.

Sensor made from SWNT-OH+CuPc shows sensitivity towards NO_2 and NO. The excellent resistance response of the hybrid sensor is due to the large conjugated π-network of SWNT-OH+MPc composite. As we know NO_2 is a strong electron donor, it gets chemisorbed onto the surface of SWNT-OH+CuPc, thus surface charge transfer interactions occur between CuPc and SWNT-OH. Since SWNT-OH-MPc derivative can form an excellent charge transfer complex, the charge can favorably travel from CuPc to f-SWNT quickly, resulting in a large and fast variation of resistance.

EXPERIMENTAL

Materials

Copper phthalocyanine were purchased from Sigma-Aldrich. Single walled hydroxyl, carboxyl, and amided CNT were purchased from carbon solutions. The samples of composite, CuPc and f-SWNT were diluted in KBr matrix to measure DRIFT spectra on a Varian-680-IR FTIR spectrometer. FESEM (Zeiss-supra-55) was utilized to get detailed information about the surface morphology and structure of prepared composite. Raman spectroscopy was carried out using Horiba Jovin Yvon micro Raman spectrophotometer using excitation wavelength of 514 nm (Ar+).

Preparation of CNT-MPc Composite

The procedure for the preparation of CNT-MPc composite can be described as follows. Requisite amounts of SWNT (hydroxyl, carboxyl or amided SWNT) were dispersed in toluene by using multifrequency ultrasonicator (Blackstone NEY) at optimized frequencies for requisite duration. After dispersion same amount of CuPc were added to the dispersed solution. The above mixture was then kept under stirring for more than 20 hours at temp. below 100°C. The mixture was then removed and again ultrasonicated to obtain the desired composite. The resultant composite was filtered over 0.5 μm PTFE membrane (Millipore) through vacuum filtration. The sample was taken for DRIFT,[23] RAMAN and FESEM analysis.

Fabrication of CNT-MPc Composite Hybridized Sensing Devices

In order to fabricate the SWNT-MPc hybrid devices, a typical procedure was as follows: the as prepared SWNT-MPc derivatives were dispersed in DMF and ultrasonicated at optimized frequencies for requisite durations followed by vacuum filtration. Metallization was done by depositing Cr/Au contact pads through a shadow mask on this CNT film using DC/RF Sputtering system. The gas sensing performance of **CuPc-SWNT** has been investigated at 40 °C with two probe setup Ecopia ESP500 in a chamber. The five gases; CO, SO_2, NH_3, NO and NO_2; of concentration 0.5 ppm were introduced in the chamber. The change in resistance of the sample has been measured as a function of time by using Fluke 289 DMM. Each of the gases was exposed one by one to the sample.

RESULTS AND DISCUSSION

The resultant composites were characterized by Diffuse Reflectance Infrared Fourier Transform (DRIFT) spectroscopy, Raman spectroscopy and FESEM analysis. Figure 1(a) illustrates the DRIFT spectra of SWNT-OH + CuPc. It is clear from the plot that resultant composite shows drastic change in comparison to unreacted CuPc and hydroxylated SWNT over a range from 500–1000 cm^{-1}. A comparision of the IR spectra shows that bands related to in-plane C-N vibration at 1167 cm^{-} becomes broader for CNT-CuPc as compared to pure CuPc. Peak at 1426 cm^{-1} in both CuPc and composite was attributed to C-C isoindole. Peak at 1502 cm^{-1} is observed in both composite and CuPc was assigned to C=N stretching. A broad peak compared to CuPc for C=C macrocycle is observed in composite at 1612 cm^{-1} Figure 2(b).[24]

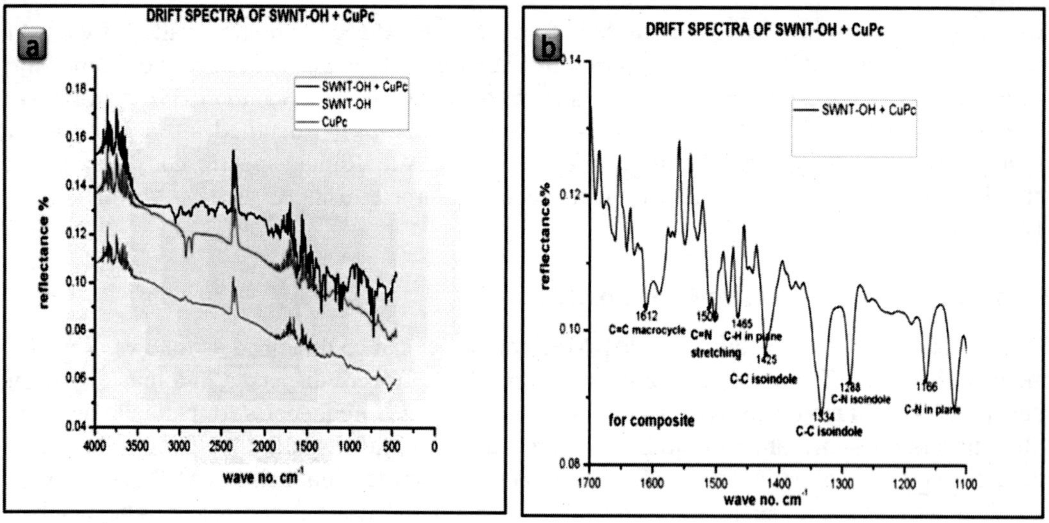

Fig. 1: DRIFT Spectra of (a) SWNT-OH+CuPc and (b) C-C and C-N Stretching Mode in Composite

From DRIFT spectra it is clear that there is conversion of β- form of CuPc to α- form,[25] which is a clear evidence of composite formation shown in Figure 2. The peaks at 769 and 713 cm^{-1} are not present in the composite, while a peak at 769 cm^{-1} with a shoulder at 779 cm^{-1} appears indicating conversion of beta form to alpha form in composite.

As seen from the Figure 3, the G-mode is a tangential stretching mode, so-called graphite mode. The RBM mode is a radial breathing mode; it is characteristic only for SWNT. The D-mode is a double resonance mode and. The ratio of the D-band intensity to the G-band intensity, $R = I_D/I_G$, usually serves as a measurement of the disordered sites on carbon nanotubes walls, serving as the indicator of the level of the covalent functionalization of the CNTs. The higher the ratio, the more disordered carbon is present in a sample. Almost no variation in $I_D \backslash I_G$ ratio of composite confirms non-covalent attachment of MPc. Thus non-covalent attachment can also be confirmed by Raman spectroscopy.

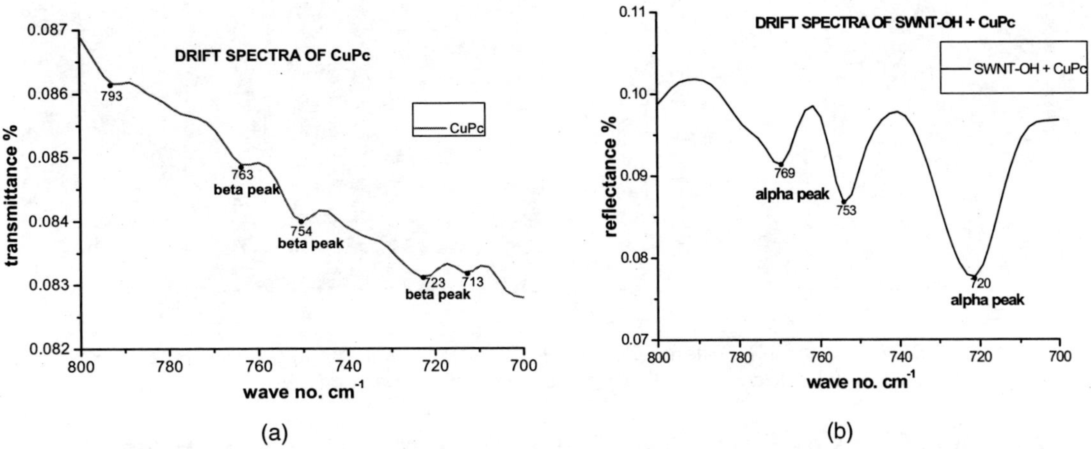

Fig. 2: (a) Conversion of Beta form to Alpha form in SWNT-OH+CuPc and (b) β-form of CuPc

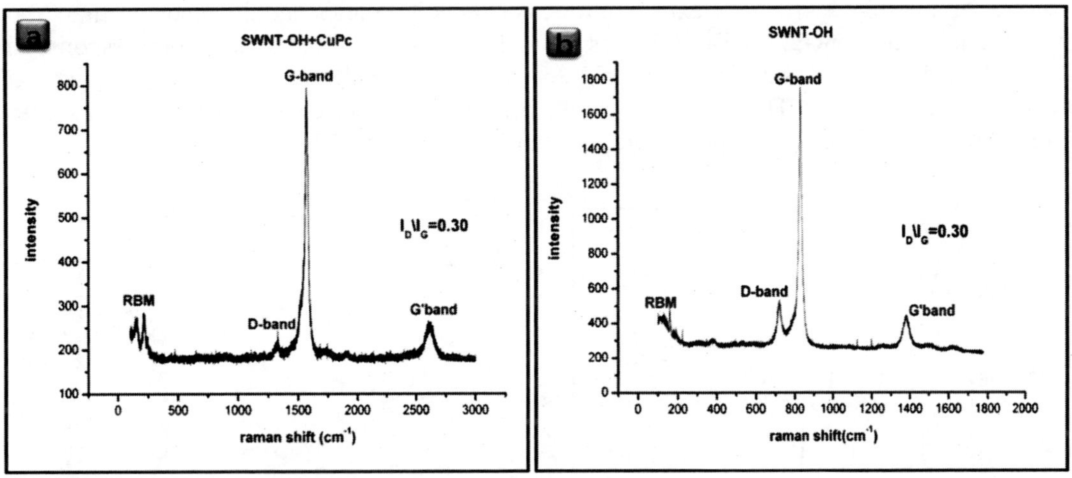

Fig. 3: Raman Spectra of (a) CNT-OH and (b) SWNT-OH- CuPc Composite

FESEM has sufficient resolution that it can be used to obtain direct visualization of the length or diameter. Figure 4a shows the bundles of unreacted SWNT-OH filtered over a polycarbonate membrane, and the diameter of the bundles are around 10–20 nm. The monoclinic crystals of Pc are clear in pure CuPc(Figure 4b). In Figure 4c the FESEM image of the composite formed with hydroxylated CNTs is shown. Spherical particles are clearly seen on the CNT surface which are not present in the unreacted SWNT-OH. However, some unreacted CuPc still exists in the form of monoclinic crystals (Figure 4d).

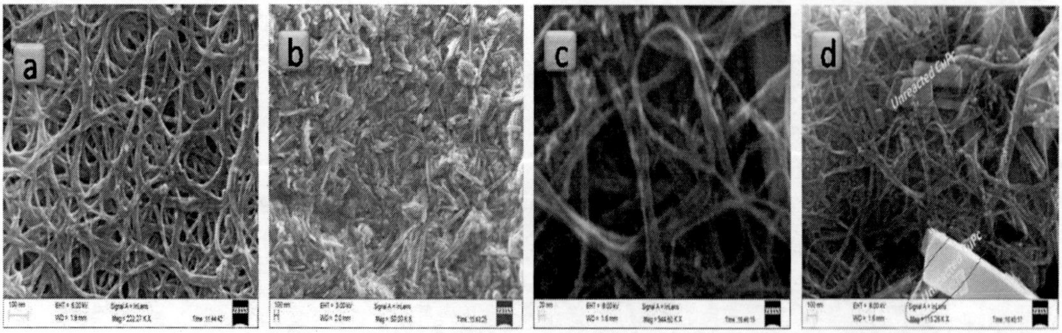

Fig. 4: FESEM Images CNTs Filtered Over Polycarbonate Membrane (a) SWNT-OH (b) CuPc (c) CuPc/SWNT-OH Composite and (d) Unreacted CuPc

Evaluation of Sensing Device Based on f-SWNT-MPc Derivative Hybrids

Figure 4(a) is showing the change in resistance accordingly and comparative response is summarized in Figure 4(b). The exposure time is kept at 3 minutes and the recovery time of 6 minutes at room temperature. Both figures show that the sample is showing small response to CO, SO_2 and NH_3 gases but it is showing 4.86% response for NO and 5.59% for NO_2 even at such a low concentration. This implies that CuPc makes SWNTs selective to NO_x gases. Response time (80%) for NO and NO_2 gas is about 107s and 105s respectively.

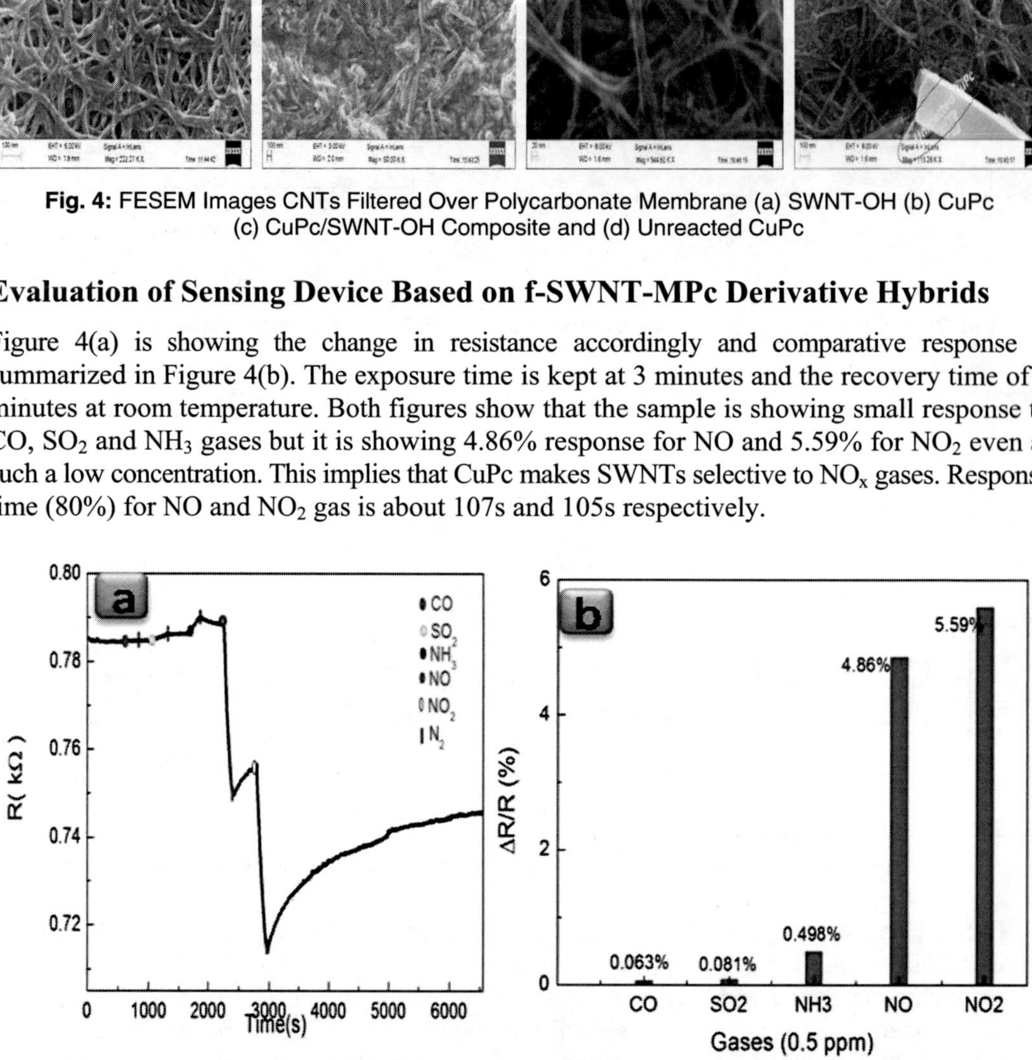

Fig. 4: (a) The Variation of Resistance with Time of Cu-Pc-SWNT TFR for Five Different Gases at 0.5 ppm (b) The Response (%) of Cu-Pc -SWNT TFR for Different Five Gases at 0.5 ppm

CONCLUSION

Various composite were obtained by ultrasonicating and blending CNT with MPc. Only those CNT which have electron withdrawing group at the surface or end of tube gives reaction with Phthalocyanine thus –COOH, and –OH functionalized SWNT gives reaction with MPc. Reactions of CNT with MPc were facilitated by π-π stacking. No reaction has taken place with amided CNT because of electron donating nature of long alkyl chain. These composites were characterized by DRIFT, RAMAN and FESEM analysis. The non-covalent interaction was confirmed with Drfit spectroscopy. No variation in $I_D \backslash I_G$ ratio in composite also confirmed non-covalent attachment. Results also report that hybrid sensor made from SWNT-OH+CuPc was shown to be highly sensitive towards NO_2 and NO.

REFERENCES

[1] Qi, P.F., Vermesh, W., Grecu, M., Jamey, A., Wang, Q., Dai, H.J., Peng, S. and Cho, K.J., *Nano Lett.*, 94, 223122, 2009.

[2] Allen, B.L., Kichambare, P.D. and Star, A., *Adv. Mater.*, 19, 1439, 2007.

[3] Kim, S.N., Rusling, J.F. and Papadimitrakopoulos, F., *Adv. Mater.*, 19, 3214, 2007.

[4] Kauffman, D.R. and Star, A., *Small*, 3, 1324, 2007.

[5] Spaeth, M.L., Sooy, W.R., J. *Chem. Phys.*, 48, 2315–2323, 1968.

[6] Snow, A.W., Barger, W.R., Phthalocyanine Films in Chemical Sensors, Vol. 1, p. 341, 1989.

[7] Liao, M.S., Scheiner, S., *J. Chem. Phys.*, 114, 9780–9791, 2001.

[8] Sakamoto, K. and Okumura, E.O., Syntheses and Functional Properties of Phthalocyanines, *Materials*, 2, 1127–1179, 2009.

[9] Nieuwenhuizen, M.S. and Nederlof, A.J., *Anal. Chem.*, 60, 230–235, 1988.

[10] Bohrer, F.I., Gas Sensing Mechanism in Chemiresistive Cobalt and Metal-Free Phthalocyanine Thin Films, 2007.

[11] Lucas, X., Frontera, A., Quiñonero, D., Deyà, P.M., Substituent effects in ion-pi interactions, *J Phys Chem A.* 2010 Feb. 4; 114(4): 1926–30.

[12] Wang, B.X., Liu, Y.Q., Qiu, W.F. and Zhu, D.B., *J. Mater. Chem.,* 12, 1636, 2002.

[13] Snow, A.W.; Barger, W.R., Phthalocyanine Films in Chemical Sensors. In *Phthalocyanines: Properties and Applications*; Lever, A.B.P., Ed; John Wiley and Sons: New York, 1989; Vol. 1, p. 341.

[14] Eley, D. *Nature* 1948, 162, 819.

[15] Guillaud, G., Simon, J. and Germain, J., *Coord. Chem. Re*V., 1998, 178, 1433–1484.

[16] Schollhorn, B., Germain, J.P., Pauly, A., Maleysson, C. and Blanc, J.P., *Thin Solid Films* 1998, 326, 245–250.

[17] Cao, L., Chen, H.Z., Zhou, H.B., Zhu, L., Sun, J.Z., Zhang, X.B., Xu, J.M. and Wang, M., *Adv. Mater.*, 15, 909, 2003.

[18] Guillaud, G., Simon, J. and Germain, J.P., *Coord. Chem. Rev.*, 178–180, 1433, 1998.

[19] Bouvet, M., Parra, V., Locatelli, C. and Xiong, H., *J. Porphyrins Phthalocyanines*, 13, 84, 2009.

[20] Ye, J.S., Wen, Y., Zhang,W.D., Cui, H.F., Xu, G.Q. and Sheu, F.S., *Electroanalysis*, 17, 89, 2005.

[21] Pillay, J. and Ozoemena, K.I., *Electrochim. Acta,* 54, 5053, 2009.

[22] Chen, X., Yang, Y. and Ding, M.Y., *Anal. Chim. Acta*, 557, 52, 2006.

[23] Guillaud, G., Simon, J. and Germain, J.P., *Coord. Chem. Rev.*,178–180, 1433, 1998.

[24] Teng, L.H. and Tang, T.D., IR study on surface chemical properties of catalytic grown carbon nanotubes and nanofibers, *J. Zhejiang Univ. Sci.* A 9(5):720–726, 2008.

[25] Verma, D. and Dash, R., Role of coordinated metal ions on the orientation of phthalocyanine based coatings, 1180–1186, 2008.

A LAND OF OPPORTUNITIES FOR THE DESIGNED CONSTRUCTION OF FUNCTIONAL INORGANIC AND HYBRID ORGANIC—INORGANIC NANOMATERIALS

A. Kalisdha, M. Subbiah[1], M. Senthilkumar[2] and P. Devapattabiraman[1]

Department of Library and Information Science, Bharathidasan University, Trichy
[1]Department of Performing Arts, Pondicherry University
[2]Meenakshi Ramaswamy Engineering College, Ariyalur
E-mail: kalisdha@gmail.com

ABSTRACT

Through the deep knowledge of materials chemistry and processing, the birth of the molecular engineering of nanomaterials "Chimie douce" based strategies. We describe successively the design and properties of: sol-gel derived hybrids, Nano Building Blocks (NBBs) based hybrid materials, nanostructured porous materials proceeds as thin films and ultra-thin films, aerosol processed mesoporous powders and finally hierarchically structured materials. The importance of the control of the hybrid interfaces via the use of modern tools as DOSY NMR, SAXS, WAXS, Ellipsometry that are very useful to evaluate in situation the hybrid interfaces and the self-assembly processes is emphasized.

INTRODUCTION

In all living organisms nature provides a multiplicity of materials (organic, inorganic, hybrids and composites), architectures, systems and functions. Natural composites, crustacean carapaces and in the human body, bone or tooth tissues, are characterized by hierarchical constructions at scales ranging from nanometers, micrometers, to millimeters that introduce the capacity to answer the physical or chemical demands occurring at these different levels. Indeed, many natural

V. Rajendran, R. Yuvakkumar, K. Thyagarajah and K.E. Geckeler (eds.)
Applications of Nano Materials: Electronics, Energy and Environment, pp. 319–326 (2012).

materials are highly integrated hybrid systems synthesized in mild chemical conditions and have found the best compromise between different properties or functions.

HYBRID MATERIALS AND SOL-GEL CHEMISTRY: GENERAL BACKGROUND

Hybrid inorganic-organic materials can be broadly defined as synthetic materials with organic and inorganic components, intimately mixed. Such materials could be either homogeneous derived from monomers and miscible organic and inorganic components. Hybrid nanocomposites had an explosive development since the 1980s, with the expansion of soft inorganic chemistry processes. The mild synthetic conditions provided by the sol–gel process such as metallo-organic precursors, low processing temperatures and the versatility of the colloidal state allow for the mixing of the organic and inorganic components at the nanometer scale in virtually any ratio.

Today hybrids play an important role in the development of many functional systems as in:

- microelectronics (resistors and molding compounds, spin-on dielectrics in microelectronic interlayer, multilayer dielectric, planarization applications);
- optical systems (luminescent solar concentrators, laser dyes, sensors, photochromic, NLO or photo-voltaic devices);
- micro-optics (free-standing micro-optical elements, such as a lens array on VCSEL elements for fiber coupling, diffractive lenses);
- bioactive hybrids (biosensors, bioreactors, cements in dental applications, controlled
- release of drugs and molecules);
- decorative and protective coatings involving scratch and abrasion resistance applications and barrier systems (solar cells, optics, electronics, food packaging...), electrochemical devices (batteries, photovoltaic and fuel cell application).

SOL DERIVED HYBRIDS VERSUS NBB APPROACHES

Sol-gel processes, and more generally, soft chemistry routes have been used to elaborate a great variety of hybrid organic-inorganic materials, which exhibit many different interesting and useful properties In parallel to these routes based on molecular precursors, the NBB approach has developed with the goal of achieving hybrid materials in which the spatial extension and the architecture of the inorganic domains, as well as the nature of the interface between the organic and inorganic components, would be better defined. NBB approach is based on the assembling of well-defined and preformed objects such as metal oxoclusters (MOCs) and NPs that keep their integrity in the final materials. An important point in this approach is a careful control of the NBBs to provide reactive groups on their surface or anchoring sites that can react with complementary NBBs.

Metal-oxo Clusters

The metallic centers that exhibit coordination numbers between four and six are coordinated to bridging oxo-ligands and terminal or bridging alkoxy OR groups. Among the numerous titanium oxo-clusters, we especially focused our interests on three titanium oxo-clusters, the Ti16 [Ti16O16(OEt)32], the Ti8 [Ti8 O8 (OOCR) 16], and the Ti6 [Ti6O6(OR)6(OOCR')6].

Nanoparticles

Though less well-defined, nanoparticles (NPs) represent a rich and versatile alternative to metallic oxo-clusters in the NBB approach to hybrid organic-inorganic materials.

Mechanical and Optical Properties of the Resulting Hybrids

The regularity of the dispersion of the organic and inorganic phases and the extended and controlled interfaces between the two phases encountered in hybrid materials offer the possibility to control the mechanical properties.

IN SITU EVALUATION OF HYBRID INTERFACES AND THEIR EVOLUTION

Affinity by DOSY NMR in the Colloidal State

The chemical syntheses of nanoparticles (NPs), through a bottom-up approach, very often use organic. The use of NPs also often relies on the functionalization of their surface. This step can be required to make them compatible with the application medium or to create new functionalities at their surface Indeed, many NPs based suspensions can be considered as true hybrid organic-inorganic systems. A fine characterization of the surface chemistry of NPs is therefore highly necessary to understand and improve the synthesis and to rationalize the utilization of NPs. In particular, measuring the affinity of organic molecules for NPs, in suspension, is of prime importance. In 2006, this quantification method was used to report the adsorption isotherm of TOPO on 4.7 nm InP QDs at 295K. The free energy (DG0) for the adsorption was measured and a Fowler isotherm was shown to better describe the adsorption than a simple Langmuir isotherm, pointing out the non-negligible and favor-able role of TOPO–TOPO interactions in the process. In 2007, a complex multicomponent sol-gel derived-system containing TiO_2 NPs (2–6 nm), acetylacetone (acac), p-toluenesulfonic acid (pTs), and butanol was investigated.

Self Assembly: Interface Class

I-SAXS/Ellipsometry

As mentioned previously, one of the major issues leading to a high control of the final nanomaterials lays in the characterization of the hybrid interfaces all along the formation mechanisms. In this way, synchrotron 2D-SAXS techniques, used for the structural characterization of mesophases, are very efficient for three main reasons.

NANOPOROUS MATERIALS

Periodically Organized Mesoporous Thin Films (POMTFs) by EISA

The EISA process implies first the elaboration of an isotropic dilute solution containing mainly inorganic (or hybrid) precursors, solvents, catalysts and molecular templating agents. In a second step during coating processing, the fast evaporation of volatile species (typically ethanol, THF, HCl, H_2O) triggers the self-assembly process by concentrating progressively the system in inorganic precursors and surfactant. In due time, the concentration in surfactant reaches the equivalent of a critical micellar concentration for the system, and then micelles start to form, surrounded by the inorganic precursors. EIMP is particularly useful to create mesoporous layer with difficult systems (RuO2 and multicationic oxides). Depending on the processing conditions, the micelles could self-organize on a large scale leading to a liquid-crystalline phase before the extended condensation of inorganic species will lock the system.

The third step is dedicated to the removal of the surfactant, the stiffening of the inorganic network and eventually the crystallization of the inorganic walls. The EISA approach is not straightforward since it implies competitive processes: inorganic condensation versus organic meso-organization, both of them being influenced by the diffusion of the volatile species. In the following paragraphs, we will emphasize some critical aspects of the synthesis of nanocrystalline mesostructured thin films and hybrid thin films synthesized via EISA approach.

Properties of Mesoporous Thin Films by EISA

These materials exhibit unexpected electrochemical, catalysis and photocatalysis properties because the mesoporous network serves as efficient pathways for molecular transport or/and the connected solid network allow fast transport of electrons, oxygen ions, protons throughout the nanoarchitecture. Because these nanoarchitectures exhibit high surface-to-volume ratios, they also serve as tools to amplify the critical phase of any electrochemical system: the interface.

1. Photocatalyst
2. Photovoltaic
3. Electrochemical properties/Sensing
4. Fuel cells.

Nanopatterned Surfaces by EIMP

TiO_2, Al_2O_3, and ZrO_2 patterns (or masks) composed of ordered nano motifs of various morphologies and typical lateral dimension of less than 60 nm and thickness below 15 nm have been prepared by dip-coating Si-wafer, FTO, glass or even Au-coated substrates in corresponding metal chloride ethanolic solutions in the presence of commercial PB-b-PEO or PS-b-PEO block copolymer micelles. The various nanostructures are obtained by self-assembly during evaporation and are subsequently stabilized at 500 8C. The nano pattern structures and morphologies are usually investigated by combined techniques such as TEM, SEM, AFM, Ellipsometry, Electrochemistry, and GI-SAXS. Here, we chose to illustrate them by AFM images.

Aerosol Route to Mesoporous Submicronic Spheres

Although, in this case, the mother solution containing inorganic precursors, condensation catalyst, structuring agent and volatile solvent is not deposited onto a substrate but is atomized to form an aerosol made of sub-micron size to micron size droplets. This process is nowadays developing very fast, for it presents numerous advantages for industrial production over conventional solution powder generation processes, such as the batch precipitation:

The formation of a mesoporous is thin membrane at the particle surface, exhibiting a repeating distance of about 4 nm. This layer is due to an earlier micellisation of IC-11 surfactant. In addition, the inner part is also mesostructured but presents much bigger pores, similar to those previously reported for aerosol-generated F127–silica particles (inner pore diameter 7 nm).

Properties of Aerosol Generated Mesoporous Submicronic Spheres

The fundamental study of the EISA performed with the spray-drying process allowed us to much simplify the production of mesostructured submicronic particles. This process becoming very attractive for industrial production we developed several families of materials for very specific applications.

HIERARCHICALLY STRUCTURED MATERIALS

Hybrid organic-inorganic materials are increasingly taking their position in the free spaces left between inorganic chemistry, polymer chemistry, organic chem¬istry, and biology. The controlled design of hybrid organic-inorganic interfaces allows the construction of materials presenting complex hierarchical structures: a particularly interesting challenge for the materials chemists. The use of all these various templating strategies combined with "chimie douce" and smart processing methods such as multilayers deposition, aerosols, ink jet printing, electrospinning, foaming, constributes to a very attractive field of research.

Multiple Templating

Monodisperse spherical hollow NPs of mesoporous silica featuring mesopores with a radial orientation in the silica shell can be synthesized via a dual-templating method. Specifically designed polystyrene latexes with anionic or cationic surface charges acted as the core templates, while cetyltrimethylammonium bromide served as a cotemplate to structure the mesopore formation during tetraethoxysilane hydrolysis/conden¬sation. The particles were well-separated and presented homogeneous mesoporous silica shells.

Dynamic Templating

1. Controlled phase separation
2. Breath figures as smart templates

Organogelators Templating

The most important processing methodologies that are available for materials transcription through organogel templating are co-assembly, post-transcription, or self-templating procedures. The main types of intermolecular interactions responsible for the gel-phase formation and some of the principal variables governing the gelation process and controlling the obtained fibrous morphologies (fibers, rods, ribbons, helices, tubes, etc.) have been recently reviewed.

Building Block Assemblies for Photonic Band Gap Materials

1. Latex games
2. Multilayers deposition of POMTF.

CONCLUSION

This new field of research is grounded on the comprehensive study of the material properties from the molecular to the nano-metric or micronic scales. Organized nanostructures and nanostructured materials often exhibit improved or even unique physical, chemical, and other properties. These specific proper-ties are achieved through the design and the control of hybrid organic-inorganic interfaces. A new class of hierarchical structures has appeared after the first phase initiated by the sol-gel community. Biomimetism and bioinspiration are smart tools for the design of innovative materials and systems that are strongly inspiring the materials chemistry community. Indeed, materials found in nature combine many wonderful features such as sophistication, miniaturization, hierarchical organizations, hybridation, resistance and adaptability. Elucidating the basic components and building principles selected by evolution to propose more reliable, efficient and environment respecting materials requires a multidisciplinary approach. Bio-inspired strategies are included in a wider emerging new field coined "bio-inspired or integrative materials chemistry". This field is opening not only new lands for innovative research but should also motivate or even give birth to a passion for research in young scientists in the area.

REFERENCES

[1] Sanchez, C., Arribart, H. and Guille, M.M., Giraud, Nature Materials 4 (2005) 277.

[2] Ozin, G.A., Chemical Communications 6 (2000) 419.

[3] Mann, S., Biomimetic Materials Chemistry, Wiley-VCH, Weinheim, 1997.

[4] Mann, S., Angewandte Chemie-International Edition 39(2000) 3393.

[5] Gomez, R.P. and Sanchez, C., *New Journal of Chemistry* 29 (2005) 57.

[6] Livage, J. and Monde , Le (1977) October 26th;

[7] Livage, J. and Lemerle, J., *Annual Review of Materials Science* 12 (1982) 103;

[8] Livage, J., *Chemistry of Materials* 3 (1991) 578.

[9] Sanchez, C. and Ribot, F., *New Journal of Chemistry* 18 (1994) 1007.

[10] Sanchez, C., Soler-Illia, G.J.A.A., Ribot, F., Lalot, T., Mayer, C.R. and Cabuil, V., *Chemistry of Materials* 13 (2001) 3061.

[11] Soler-Illia, G.J.A.A., Sanchez, C., Lebeau, B. and Patarin, J., *Chemical Reviews* 102 (2002) 4093.

[12] Mann, S., Burkett, S.L., Davis, S.A., Fowler, C.E., Mendelson, N.H., S.D. Sims, D. Walsh and N.T. Whilton, *Chemistry of Materials* 9 (1997) 2300.

[13] Goltner, C.G. and Antonietti, M., *Advanced Materials* 9(1997)431.

[14] Avnir, D., Levy, D. and Reisfeld, R., *Journal of Physical Chemistry* 88 (1984) 5956.

[15] Sanchez, C., Boissiere, C., Grosso, D., Laberty, C. and Nicole, L., *Chemistry of Materials* 20 (2008) 682.

[16] Backov, R., *Soft Matter 2* (2006) 452.

[17] Prouzet, E., Ravaine, S., Sanchez, C. and Backov, R., *New Journal of Chemistry* 32 (2008) 1284.

[18] Sanchez, C., Lebeau, B., Chaput, F. and Boilot, J.P., *Advanced Materials* 15 (2003) 1969.

[19] Sanchez, C., Julian, B., Belleville, P. and Popall, M., *Journal of Materials Chemistry* 15 (2005) 3559.

[20] Mammeri, F., Bourhis, E. Le, Rozes, L. and Sanchez, C., *Journal of Materials Chemistry* 15 (2005) 3787.

[21] Nicole, L., Boissière, C., Grosso, D., Quach, A. and Sanchez, C., *Journal of Materials Chemistry* 15 (2005) 3598.

[22] Rozes, L., Steunou, N., Fornasieri, G. and Sanchez, C., Monatshefte für Chemie 137 (2006) 501.

[23] Brinker, C.J. and Scherer, G.W., Sol-Gel Science. *The Physics and Chemistry of Sol-Gel Processing, Academic Press,* San Diego, 1990;

[24] Klein, L.C., Sol-Gel Technology for Thin Films, Fibers, Pre-forms. Electronics and Specialty Shapes, Noyes Publications, 1988.

[25] Gómez-Romero, P. and Sanchez, C., Functional Hybrid Materials, Wiley-VCH, Weinheim, 2004.

[26] Loy, D.A. and Shea, K.J., Chemical Reviews 95 (1995) 1431; R.J.P. Corriu, D. Leclercq, Angewandte Chemie-International Edition 35 (1996) 1420;

[27] Corriu, R.J.P., Comptes Rendus De L'Academie Des Sciences Serie Ii Fascicule C-Chimie 1 (1998) 83.

[28] While most sol–gel chemistry depends on the use of controlled water quantities to yield metal-oxo polymers, there is an alternative method to generate metal-oxo polymers in non-hydrolytic conditions see e.g. Bourget, L., Corriu, R.J.P., Leclercq, D., Mutin, P.H., Vioux A.J. Non-Cryst. Solids 1998, 242, 81 and references therein.

[29] Judeinstein, P. and Sanchez, C., *Journal of Materials Chemistry* 6 (1996) 511.

[30] Yaghi, O.M., M, O.K., Ockwig, N.W., Chae, H.K. and Eddaoudi, M., *J. Kim, Nature* 423 (2003) 705.

SILVER-ACTIVATED CHARCOAL NANOCOMPOSITE FOR MICROBIAL DECONTAMINATION OF WATER

V. Sherly Arputha Kiruba, Arun Dakshinamurthy and P. Mosae Selvakumar[1]

Department of Nanotechnology, School of Nanosciences and Technology, Karunya University, Coimbatore, Tamil Nadu
[1]Department of Chemistry, School of Science and Humanities, Karunya University, Coimbatore, Tamil Nadu
E-mail: mosae@karunya.edu; arun@karunya.edu

ABSTRACT

Eco friendly chemical synthesis has various advantages over traditional methods and has become the need of the hour. Here we report a greener method for synthesizing silver nanoparticles from musk melon fruit extract and its characterization using XRD, SEM, particle size analyzer and UV-Visible spectroscopy. Its anti-microbial activity was also studied so as to be used along with activated charcoal as nanocomposite for water purification. The nanoparticles formed were effective against S. aureus and B. subtilis which are indicators of water contamination by humans. The nanocomposite formed was found to be more effective compared to virgin activated charcoal and makes it possible to be used for water filtration against microbial contamination. We conclude that the efficiency of virgin activated charcoal can be improved by compositing it with silver nanoparticles.

INTRODUCTION

Nanoparticles are fundamental building blocks of nanotechnology. The most important and distinct property of nanoparticles is that they exhibit a large surface area to volume ratio, the

V. Rajendran, R. Yuvakkumar, K. Thyagarajah and K.E. Geckeler (eds.)
Applications of Nano Materials: Electronics, Energy and Environment, pp. 327–334 (2012).

reason why they have been effectively used in medicine and bio-medical applications such as drug delivery, water filtration and medical diagnostics. Nanoparticles have also been studied for their anti-microbial properties.[1] Among the various methods for nanoparticle synthesis, 'green synthesis' has proven to be the most eco-friendly method.[2]

Natural disasters like flooding can considerably increase microbial contamination of drinking water[3] and can also result in release of hazardous materials into water resources.[4–6] Treatment of water therefore becomes a priority. Silver nanoparticles play an important role in water treatment due to their anti-microbial activity.

The adsorption technique, based on the transfer of pollutants from the solution to the solid phase, is one of the most efficient water treatment methods.[7] Adsorbents can be used to remove toxic substances such as chlorine-disinfectant residues, organics and particulates.[8–10] Due to its high surface area, porous structure, and special surface reactivity, activated charcoal (AC) has been the most popular and widely used adsorbent all over the world.[11,12] AC is highly inert and thermally stable, and can be used over a broad pH range. It has great capacity of adsorbing various organic compounds and can be easily modified by chemical treatment to increase its adsorption.[13] Although virgin AC can adsorb microbes in water, natural organic matter would rapidly occupy the adsorption sites and promote biofilm growth[14] thereby negatively impacting the quality of treated water.[8,15,16]. Activated charcoal can also remove objectionable taste and odorous compounds such as iodine and chlorine by adsorption and chemical reduction.[17] The major drawback of activated charcoal based adsorption that limits its long-term application is that it requires continuous replacement due to saturation, though suitable for short term acute emergencies.[11] Carbon metabolizing bacteria also pose concerns in treatment of water using AC[18,19] In this work we report the development of a nanocomposite of activated charcoal and 'green' synthesized silver nanoparticles for effective water treatment against microbial contamination.

EXPERIMENTAL

Synthesis and Characterization of Silver Nanoparticles

Silver nanoparticles were prepared using musk melon (*Curcumis melo*) extract, as reducing agent since its rich in L-ascorbate. The fruit pulp was mashed with mortar and pestle and centrifuged at 5000 rpm for 15 mins to pellet the debris. The clear supernatant was used for nanoparticle preparation. 1 mM silver nitrate solution was reduced with 5% musk melon fruit extract. The colloidal solution of silver nanoparticles was characterized with UV-Vis spectroscopy, particle size analyzer, SEM and XRD.

Anti-Microbial Assay

Agar well diffusion method was followed for anti-microbial assays. Two gram negative (*Escherichia coli, Pseudomonas aeruginosa*) and two positive (*Staphylococcus aureus, Bacillus subtilis*) microbes were tested. Ampicillin and tetracycline were used as control antibiotics depending on the strain used. The zone of inhibition was measured and recorded.

Preparation of Activated Charcoal (AC) and its Characterization

Washed and dried coconut shells were treated with sulfuric acid for 2 days and then washed over 15 times with distilled water (centrifuged after overnight sedimentation) till the pH reached 7. It was then dehydrated at 110°C overnight and ground to obtain activated charcoal powder, and was then characterized using SEM and XRD before being used along with the silver nanoparticles as a nanocomposite to test its antimicrobial activity during water filtration.

Preparation of Silver-Activated Charcoal (Ag-AC) Nanocomposite and its Characterization

100 mg of the prepared activated charcoal was added to 100 ml of the silver nanoparticles solution. After mixing briefly, the solution was allowed to settle overnight, filtered the next day and air dried to obtain the nanocomposite. The nanocomposite was characterized using SEM and XRD.

Cartridge Fabrication Using Silver-Activated Charcoal (Ag-AC) Nanocomposite

50 mg of the obtained air dried nanocomposite was loaded into 1 ml syringe columns for testing water filtration and biocidal efficiency. A small amount of non-absorbent cotton was used to plug the upper and lower ends of the column to prevent the nanocomposite from leaching.

Testing of the Nanocomposite's Biocidal Activity

1 ml of the microbial culture (*E. coli, B. subtilis, P. aeruginosa* and *S. aureus*) at 0.6 OD_{600} was passed through the cartridge under slight positive pressure in order to maintain a constant flow rate of 1 drop per second. The filtrate was collected in sterilized vials, serially diluted and plated on sterile agar plates. The plates were incubated in an upright position at 37°C for 24 hours. The number of colonies formed were counted and recorded.

RESULTS AND DISCUSSION

Green Synthesis

Addition of 5% musk melon fruit extract to a colorless solution of 1 mM silver nitrate resulted in a change of color to bright yellowish brown within 30 minutes confirming the formation of silver nanoparticles (Figure 1, left panel). The speed of the reaction can be attributed to the corresponding abundance of vitamin C in the fruit extract. The color of the solution may be due to nanosize of particles.

Fig. 1: Synthesis of Silver Nanoparticles (Ag NPs), SEM Image of Silver Nanoparticles, and SEM Image of Silver-Activated Charcoal (Ag-AC) Nanocomposite

SEM Analysis

The SEM image of silver nanoparticles (Figure 1, center panel) shows uniformly distributed nanostructures that are spherical in shape and uniformly arranged. The SEM image of the silver-activated carbon nanocomposite (Figure 1, right panel) reveals a porous morphology and the silver nanoparticles are seen adsorbed onto the porous regions of activated charcoal with the particles uniformly dispersed.

UV-Visible Spectroscopy

The UV-Visible absorption spectrum (Figure 2) of the bright yellowish brown silver nanoparticles solution showed a surface plasmon absorption band with λ_{max} of 432 nm indicating the presence of silver nanoparticles. This red shift from 308 nm (plain 1 mM silver nitrate) to 432 nm confirms the reduction of silver nitrate to silver using musk melon extract. The broadening of peak refers to poly dispersed nanoparticles in the solution.[20]

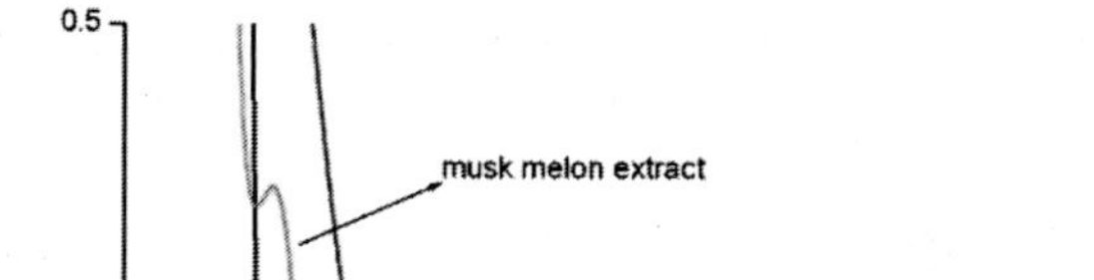

Fig. 2: UV-Vis Spectrum of Silver Nanoparticles Synthesized Using Musk Melon Extract

Anti-Microbial Assay

The silver nanoparticles prepared using musk melon fruit extract as the reducing agent were found to be more effective against *S. aureus* and *B. subtilis* (Figure 3A).[22–24] This might be due to nanometer size of particles formed. Hence this can be used in a composite with activated charcoal for water filtration purposes.

Efficiency Testing of Biocidal Nanocomposite

The efficiency of virgin activated charcoal and prepared nanocomposites were compared in water filtration fixed bed columns. The musk melon silver nanoparticles-activated charcoal nanocomposite filled syringe column was effective against all the four test microbes and the efficiency is shown in Figure 3B. It is seen that the nanocomposite was more effective against *B. subtilis* and *S. aureus* which might be due to the anti-microbial property of silver nanoparticles.

Reuse Efficiency Testing of Nanocomposite

The reusability was checked by passing a known CFU of microbes and testing the efficiency of filtration. The efficiency of the second time usage was comparatively lower than that of the first pass (Figure 3C). However, the nanocomposite showed higher durability which might be due to its anti-microbial capability whereas the activated charcoal's efficiency decreased due to its adsorption property alone. We conclude that the nanocomposite's efficiency is higher compared to that of virgin activated charcoal alone.[8,25,26]

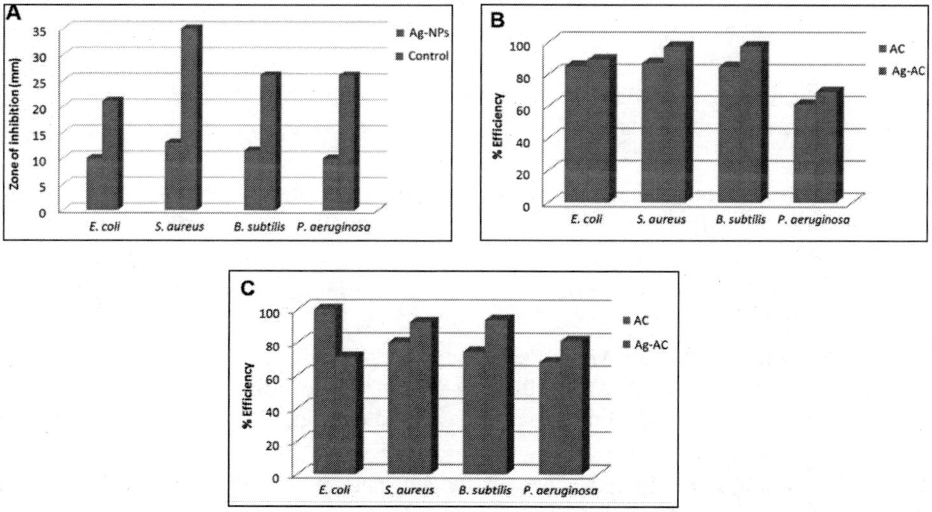

Fig. 3: A – Zone of inhibition caused by silver nanoparticles synthesized using musk melon extract, B – water filtration efficiency of biocidal nanocomposite, AC – activated charcoal, Ag-AC – silver-activated charcoal, C – water filtration reusage efficiency of biocidal nanocomposite.

CONCLUSION

The green synthesis of silver nanoparticles using musk melon fruit extract was studied and it was found that the effectiveness of the nanoparticles was based on the effective reduction reaction happening in the reaction media. It is directly related to the quantity of active reducing agent (L-ascorbate) present in the fruit extract used. The silver nanoparticles exhibited high anti-microbial activity, both in plate culture as well as in fixed bed water filtration column units. With nanocomposite formation, the efficiency of filtration and biocidal activity of activated charcoal is enhanced. This is evident from the reusage efficiency of water filtration fixed bed columns.

ACKNOWLEDGEMENT

Assistance from Karunya University in the form of short term research grants is gratefully acknowledged.

REFERENCES

[1] Jain, D., Daima, H.K., Kachhwahl, S. and Kothari, S.L., Synthesis of plant-mediated silver nanoparticles using papaya fruit extract and evaluation of their anti microbial activities, *Digest Journal of Nanomaterials and Biostructures,* 4, (3), 557–563, 2009.

[2] Leela, A. and Vivekanandan, M., Tapping the unexploited plant resources for the synthesis of silver nanoparticles, *African Journal of Biotechnology*, 7 (17), 3162–3165, 2008.

[3] Faruque, S.M., Naser, I.B., Islam, M.J., Faruque, A.S.G., Ghosh, A.N., Nair, G.B., Sack, D.A. and Mekalanos, J., Seasonal epidemics of cholera inversely correlate with the964 prevalence of environmental cholera phages, Proceedings of the National Academy of Sciences of the United States of America, 102, 1702–1707, 2005.

[4] Young, S., Balluz, L. and Malilay, J., Natural and technologic hazardous material releases during and after natural disasters", a review. *Science of the Total Environment,* 322, 3–20, 2004.

[5] Srinivas, H. and Nakagawa, Y., Environmental implications for disaster preparedness: lessons Learnt from the Indian Ocean Tsunami, *Journal of Environmental Management,* 1350 (89), 4–13, 2008.

[6] Violette, S., Boulicot, G., Gorelick, S.M., Tsunami-induced groundwater salinization in southeastern India, *Comptes Rendus Geosciences* 341, 339–346, 2009.

[7] Ghaedia, M., Sadeghiana, B., Amiri Pebdania, A., Sahraeib, R., Daneshfarb, A., Duranc, C., Kinetics, thermodynamics and equilibrium evaluation of direct yellow 12 removal by adsorption onto silver nanoparticles loaded activated carbon, *Chemical Engineering Journal,* 187, 133– 141, 2012.

[8] Tobin, R.S., Smith, D.K., Lindsay, J.A., Effects of activated carbon and bacteriostatic filters on microbiological quality of drinking water, *Applied and Environmental Microbiology,* 41, 646–651, 1981.

[9] Bell, F.A. Jr, Perry, D.L., Smith, J.K., Lynch, S.C., Studies on home water treatment systems", *Journal of American Water Works Association,* 76, 126–130, 1984.

[10] Demirbas, E., Kobya, M., Sulak, M.T., Adsorption kinetics of a basic dye from aqueous Solutions onto apricot stone activated carbon, *Bioresour. Technol,* 99, 5368–5373, 2008.

[11] Loo, S.-L., Fane, A.G., Krantz, W.B., Lim, T.-T., Emergency water supply: A review of potential technologies and selection criteria, *Water Research* doi: 10.1016/j.watres.2012.03.030. 2012.

[12] Babel, S., Kurniawan, T.A., Low-cost adsorbents for heavy metals uptake from contaminated water: a review, *J Hazardous Mat,* 97, 219–243, 2003.

[13] Jiuhui, Q.U., Research progress of novel adsorption processes in water purification", *A Review Journal of Environmental Sciences,* 20, 1–13, 2008.

[14] Caroli, G., Levre, E., Armani, G., Biffi-Gentili, S., Molinari, G., "Search for acid-fast bacilli in bottled mineral waters", *Journal of Applied Microbiology,* 58, 461–463, 1985.

[15] Wallis, C., Stagg, C.H., Melnick, J.L., The hazards of incorporating charcoal filters into domestic water systems. *Water Research,* 8, 111–113, 1974.

[16] Su, F., Luo, M., Zhang, F., Li, P., Lou, K. and Xing, X., Performance of microbiological control by a point-of-use filter system for drinking water purification", *Journal of Environmental Sciences,* 21, 1237–1246, 2009.

[17] Backer, H., Water disinfection for international travelers, in: Jay S K, Phyllis E K, David O F, Hans D N, Bradley A C, Travel Medicine (Second Edition) Mosby, Edinburgh, 47–58, 2008.

[18] Saier, M.H., Multiple mechanisms controlling carbon metabolism in bacteria, *Biotechnol. Bioeng.* 58 170–174, 1998.

[19] Camper, A.K., Lechevallier, M.W., Broadaway, S.C., Mcfeters, G.A., Bacteria associated with granular activated carbon particles in drinking water, *Appl. Environ. Microbiol,* 52 434–438, 1986

[20] Ankanna, S., Prasada, T.N.V.K.V., Elumalaib, E.K., Savithramma, N., Production of biogenic silver nanoparticles using Boswellia ovalifoliolata stem bark , *Digest Journal of Nanomaterials and Biostructures* 5(2), 369–372, 2010.

[21] Koraa, A.J., Sashidharb, R.B., Arunachalama, J., Gum kondagogu (Cochlospermumgossypium): A template for the green synthesis and stabilization of silver nanoparticles with antibacterial application", *Carbohydrate Polymers* 82, 670–679, 2010.

[22] Balazs, D.J., Triandafillu, K., Wood, P., Chevolot, Y., van Delden, C., Harms, H., Hollenstein, C., Mathieu, H. J. Inhibition of bacterial adhesion on PVC endotracheal tubes by RF-oxygen glow discharge, sodium hydroxide and silver nitrate treatments. *Biomaterials* 25, 2139–2151, 2004.

[23] Melaiye, A. and Youngs, W.J., Silver and its application as an antimicrobial agent, *Expert Opin Ther Pat* 15, 125–130, 2005.

[24] Sondi, I. and Salopek-Sondi, B., Silver nanoparticles as antimicrobial agent: A case study on *E. coli* as a model for gram-negative bacteria, *J Colloid Interface Sci* 275, 177–182, 2004.

[25] Fiore, J.V. and Babineau, R.A., Effect of an activated carbon filter on the microbial quality of water, *Appl Environ Microbiol,* 34 (5), 541–546, 1977.

[26] Stobie, N., Duffy, B., McCormack, D.E., Colreavy, J., Hidalgo, M., McHale, P. and Hinder, S.J., Prevention of staphylococcus epidermidis biofilm formation using a low-temperature processed silver-doped phenyltriethoxysilane sol-gel coating", *Biomaterials* 29, 963–969, 2008.

OPTOELECTRONIC PROPERTIES OF SPRAY DEPOSITED Ni-DOPED ZINC OXIDE THIN FILMS

S.K. PATIL, S.S. SHINDE AND K.Y. RAJPURE

Electrochemical Materials Laboratory, Department of Physics, Shivaji University, Kolhapur
E-mail: rajpure@yahoo.com

ABSTRACT

The transparent thin films of nickel doped zinc oxide are deposited on to the corning glass substrates via chemical spray pyrolysis technique using zinc acetate and nickel nitrate as precursors. Structural, morphological, optical, electrical, dielectrical and impedance properties of the films as a function of nickel doping concentration are investigated. The deposited films are polycrystalline with a hexagonal (wurtzite) crystal structure having (002) preferential orientation. The films are highly transparent in the visible region with a transmittance higher than 86%, and have an optical band-gap ranging from 3.23 to 3.19 eV depending on the nickel doping concentration. High electrical conductivity of doped film is explained on the basis of presence of oxygen vacancies. Decrease of electrical conductivity due to nickel doping is explained on the basis of compensation of oxygen vacancies. The dielectric constant and loss tangent as a function of frequency are reported. We studied inter-particle interactions using complex impedance spectroscopy.

INTRODUCTION

Zinc oxide is an attractive semiconductor for numerous applications and a building block material for optoelectronic devices because of its large band gap (3.3 eV), exciton biding energy (60 meV), transparency, chemical stability etc. It has tremendous applications such as LEDs, transparent FETs, photodetectors,[1] spintronics and piezoelectric devices.[2–3] However, all applications are mostly dependent on intrinsic defects or additional defects.[4] The physical properties of semiconductors are efficiently enhanced by incorporating impurity doping. ZnO

V. Rajendran, R. Yuvakkumar, K. Thyagarajah and K.E. Geckeler (eds.)
Applications of Nano Materials: Electronics, Energy and Environment, pp. 335–344 (2012).

is usually n-type due to electron doping via defects originating from Zn interstitials and O vacancies in the lattice. The understanding of the role of various impurities and defects in ZnO lattice is very crucial. Yan *et al.*[5] reported synthesis and optical properties of Mn and Ni doped ZnO thin films by sol–gel process in methyl alcohol. The effect of Ni concentration on the structural, electrical, optical, and non-linear optical properties of the Ni:ZnO thin films was investigated by Abed *et al.*[6] The effective non-linear quadratic and cubic electronic susceptibilities of thin film samples were also determined by the SHG and THG techniques. Liu *et al.*[7] reported the growth of Ni doped ZnO (Zn1-xNixO) thin films on quartz substrates via magnetron sputtering deposition process with the Ni concentrations of 5, 10 and 20 at%. Pandey *et al.*[8] reported the synthesis of a transparent magnetic semiconductor by incorporating Ni in zinc oxide (ZnO) matrix prepared by fast atom beam (FAB) sputtering. They discussed an interaction between free charges generated by doping and localized d spins of Ni.

These films have been prepared by a variety of techniques like sol–gel,[5] magnetron sputtering,[7] Fast Atom Beam sputtering,[8] Pulsed Laser Deposition,[9] Chemical Vapor Deposition [10] and spray pyrolysis.[11–13] Among many techniques to prepare materials, spray pyrolysis plays an important role on microstructures, optical, electrical and magnetic properties of the magnetic materials. Combining optical properties (better transparency) with the conducting and magnetic behavior of the material is of immense importance in the field of transparent spintronic devices. Therefore, a detailed study on the transport and optical transmission is required for better understanding of this material. In this work, we have studied nickel-doped ZnO thin films in the perspective of their transparent conducting properties. The influence of Ni doping on to the structural, morphological, optical, electrical, dielectrical and impedance properties of zinc oxide thin films in detail has been studied.

EXPERIMENTAL

ZnO thin films were grown by using the chemical spray pyrolysis technique onto corning glass substrates using zinc acetate [$Zn(CH_3COO)_2.2H_2O$] as a precursor. The precursor solution was prepared from a mixture of 0.1 M [$Zn(CH_3COO)_2.2H_2O$] 99.99%, A.R. Grade, Himedia dissolved in double distilled water. To obtain Ni doped ZnO thin films nickel nitrate was used as precursor. The atomic percentages of [Ni/Zn] were 0.5, 1.0, 1.5, 2.0 and 2.5 at%, respectively. It was observed that for higher doping concentration degradation of films occurs probably due to a reaction of the solution with excess nickel. The actual [Ni]/[Zn] ratios in the film are far less than in the solution. Before loading into the system, the substrates were washed with detergent and then completely rinsed in methanol; acetone and double distilled water and finally dried in air. Then the substrates were progressively heated up to the required temperature, before being sprayed on. The resulting solution was sprayed onto the preheated substrates held at optimized substrate temperature of 450°C.[14] Compressed air was used as a carrier gas. Other preparative parameters, such as spray rate (5 cc/min), distance between the atomizer and the substrate (30 cm) were kept constant for all experiments.

The structural characterization of deposited thin films were carried out, by analyzing the X-ray diffraction patterns obtained under CrKα radiation from a Philips X-ray diffractometer model PW-3710 and surface morphology was studied using scanning electron microscope, JEOL JSM-6360, Japan. Room temperature transmission spectra were recorded near to normal incidence using a UV–Vis Shimandzu UV-1800 spectrophotometer, Japan. Resistivity measurements were carried out by two probe method as a function of temperature. The type of semiconductor was examined from thermoelectric power (TEP) measurement. The AC parameters such as capacitance (Cp) and dissipation factor (tan δ) of the films were measured in the frequency range 20 Hz to 1 MHz using LCR meter (HP 4284 A). The dielectric constant (ε') of the Ni doped ZnO thin films was calculated using the relation as reported in.[14]

RESULTS AND DISCUSSION

X-ray Diffraction Studies

X-ray diffraction (XRD) patterns of Ni doped ZnO (NZO) thin films for different Ni doping concentrations from 0.5 to 2.5 at% at optimized substrate temperature 450°C are shown in Figure 1. A matching of observed and standard 'd' values of NZO using JCPDS card No. 05–0664 confirms the deposited films fit well with the hexagonal (wurtzite) crystal structure having preferred orientation along (002) plane. The intensity of (002) plane increases while intensity of (101) plane decreases up to 1.5 at% and vice versa, showing the crystal reorientation effect.[14] The relatively more pronounced ZnO (002) peaks observed may be attributed to the lower levels of Ni content in the films and also attributed to the Ni^{2+} ions which have substituted for the Zn2+ ions to form the primary phase Ni^{2+}:ZnO. Some weak reflections such as (102) and (103) have also been observed but with smaller intensities. No peaks corresponding to metallic Zn or Ni are observed in the XRD patterns. The average crystallite size of ZnO thin films is calculated using the Scherrer's relation.[15] The crystallite size enhances with doping concentration up to 1.5 at% and then decreases as seen in Figure 2. The average crystallite size of the NZO thin films varies from 30–36 nm. It is obvious that the Ni doping modifies the film

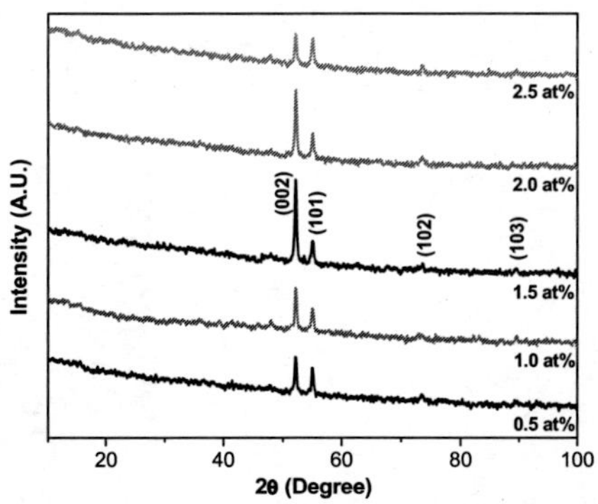

Fig. 1: X-ray Diffraction Patterns of Ni-doped Zinc Oxide Thin Films

growth mechanism. As seen, the growth rate increases linearly with increasing the doping level. It is worth noting that the (002) plane requires lower energy formation. Therefore, growth rate becomes faster with increase in Ni doping concentration; consequently the easiest formation (002) plane is favored.

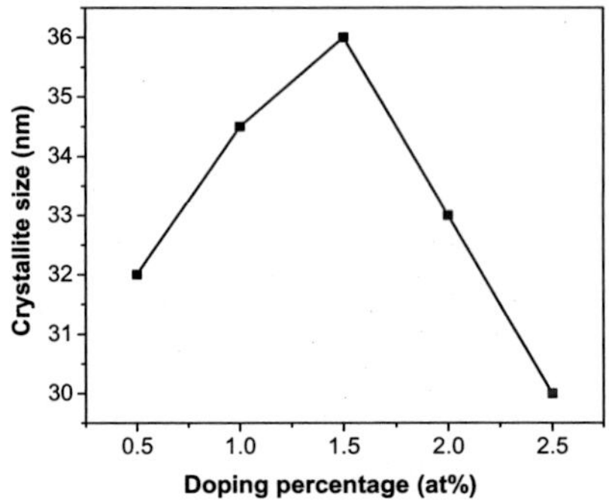

Fig. 2: Variation of Crystallite Size with Ni Doping Concentration

Morphological Analysis

Figure 3 shows the SEM images of Ni-doped ZnO thin films. All the films are homogeneous, dense and adherent and grow with thin columnar grains perpendicular to the substrate. It is

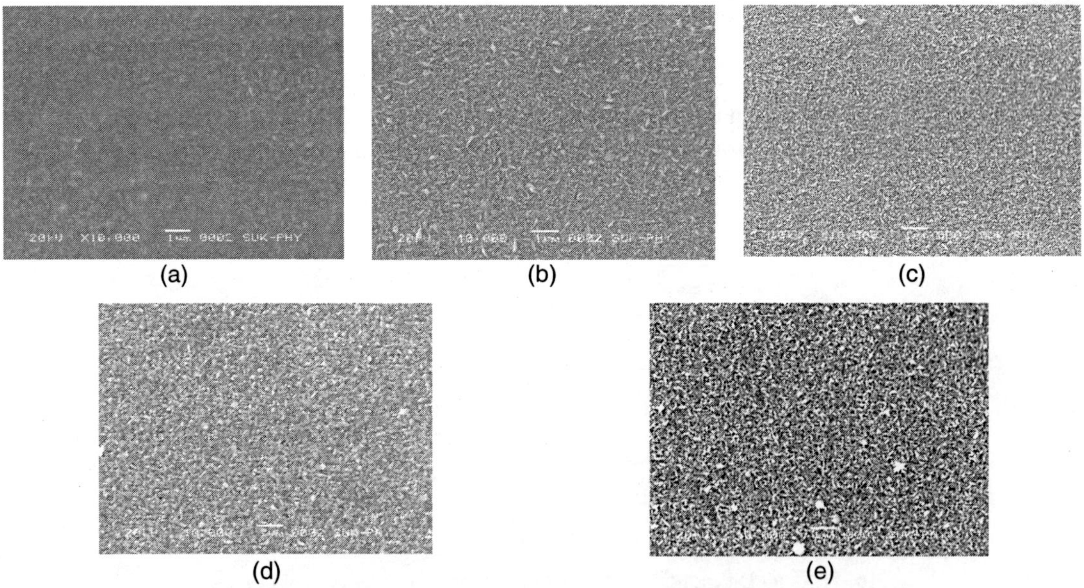

Fig. 3: Scanning Electron Micrographs of Ni-doped Zinc Oxide Thin Films
(a) 0.5 at%, (b) 1.0 at%, (c) 1.5 at%, (d) 2.0 at% and (e) 2.5 at%

apparent from the micrographs that morphology of the films has tightly packed grains and relatively smooth surfaces without any voids and crack. Apparent hexagonal-shaped granules are seen without any noticeable pinholes, which are uniformly distributed, giving a smooth surface morphology. The grain size of the films slightly decreases with increasing the Ni content up to 1.5 at% and then increases. It is attributed to the impurity effect of Ni on the film structure. Namely, the more Ni atoms as impurity can lead to a high nucleation density at the initial stage of the film growth, decreasing the grain size of the film.

OPTICAL PROPERTIES

Figure 4 shows the optical transmittance spectra of NZO thin films deposited at various Ni doping concentrations. They show strong interference fringes due to the multiple reflections at the interfaces; their observation indicates a fairly homogeneous film thickness. All these films show higher transparency in the visible range with an average transmittance of about 85%. The transmittance falls sharply at the band edge which is an indication of the good films crystallinity. For a direct band gap semiconductor, the optical band gap energy (Eg) can be estimated from the absorption spectrum using following relation:[16]

$$\alpha h\nu = C\left(h\nu - E_g\right)^{1/2} \qquad \ldots (1)$$

The optical energy gap (Eg) of the films are obtained by extrapolating the linear part of the spectrum $(\alpha h\nu)^2$ vs hν to $\alpha = 0$. We see that the band gap of NZO films decreases with Ni concentration from 3.23 to 3.19 eV.

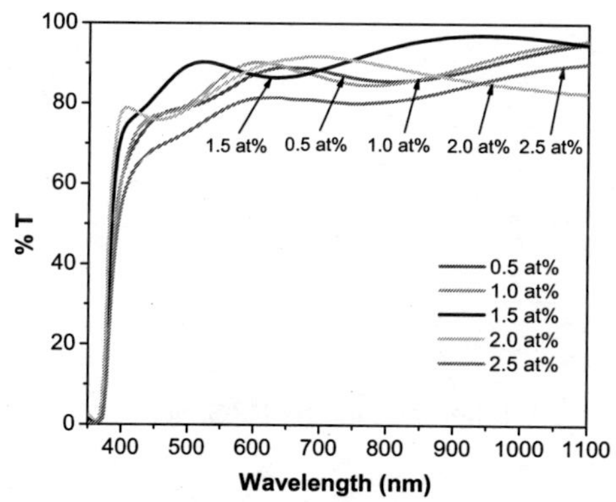

Fig. 4: Optical Transmittance Spectra of Ni-doped Zinc Oxide Thin Films

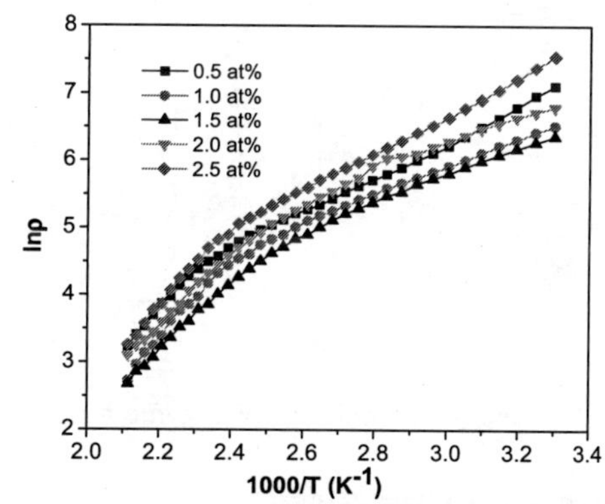

Fig. 5: Variation of Resistivity with Respect to Inverse of Temperature for Ni-doped Zinc Oxide Thin Films

Electrical Properties

Figure 5 shows the variation of lnρ against (1000/T) for Ni-doped ZnO thin films. It is well known that the films microstructure plays an important role in the electrical properties of the semiconductors. The decrease in dc electrical resistivity with increasing temperature confirms the semiconducting behavior of NZO thin films. It is seen that, electrical resistivity decreases with Ni doping concentration up to 1.5 at% and then increases for higher doping concentration. It is the fact that nickel exceeds the limit of maximum solubility in zinc oxide, producing a grain boundary segregation of impurities that causes a dispersion of carriers. The XRD patterns indicate that the increase of crystallite size results in decrease in grain boundaries leading to enhancement in conductivity. However, when the Ni doping ratio exceeds 1.5 at% the conductivity decreases. The higher c-axis preferred orientation results to higher conductivity because of reduction in the scattering of the carries at the grain boundaries.[17]

Thermoelectric Power

The type of conductivity exhibited by NZO thin films is determined from thermoelectric power analysis. The polarity of thermally generated voltage at the cold end is negative which confirms that the films are of n-type. Figure 6 shows the variation of thermo emf against temperature difference for the films deposited for different Ni doping concentrations. It is found that, the relative thermo emf increases with doping concentration up to 1.5 at% and then decreases for higher doping concentrations varies linearly with difference in temperature. Relatively higher thermo-emf is observed for the film deposited at 1.5 at% doping concentration is due to better crystallinity.

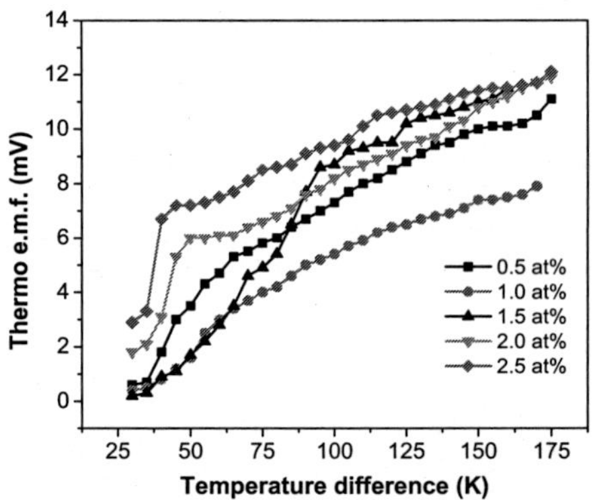

Fig. 6: Variation of of Thermo e.m.f. with Temperature Difference for the NZO Films

Dielectric Properties

The variation of dielectric constant with respect to frequency at room temperature is shown in Figure 7. It is seen that the dielectric constant (ε') decreases suddenly at lower frequencies and remains constant at higher frequencies, showing the dispersion behavior of dielectric constant at lower frequencies owing to the charge transport relaxation time. The dielectric constant increases up to 1.5 at% Ni doping and then decreases. The large value of the dielectric constant is associated with space charge polarization and inhomogeneous dielectric structure viz. impurities, grain structure, defects and pores. The decrease in dielectric constant takes place when jumping frequency of electric charge cannot follow the change in applied electric field beyond a certain critical frequency. Decrease in dielectric constant with frequency is due to decrease in polarization. At lower frequencies, dielectric loss tanδ (Figure 8) is large and it decreases with increasing frequency. The tanδ is the energy dissipation in the dielectric system, which is proportional to the imaginary part of the dielectric constant. The appearance of minimum loss with frequency is a direct consequence of a mathematical analysis of an equivalent capacitance circuit for oxide dielectric films.[18] The higher values of dielectric constant and loss tangent observed at 1.5 at% Ni is due to the large surface charge sorption of the Ni^{2+} ion in ZnO host lattice.

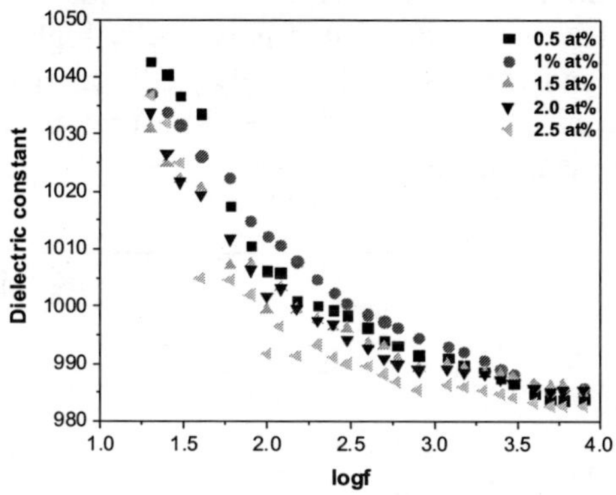

Fig. 7: Dielectric Constant Variation with Frequency for Various Ni-doped ZnO Films

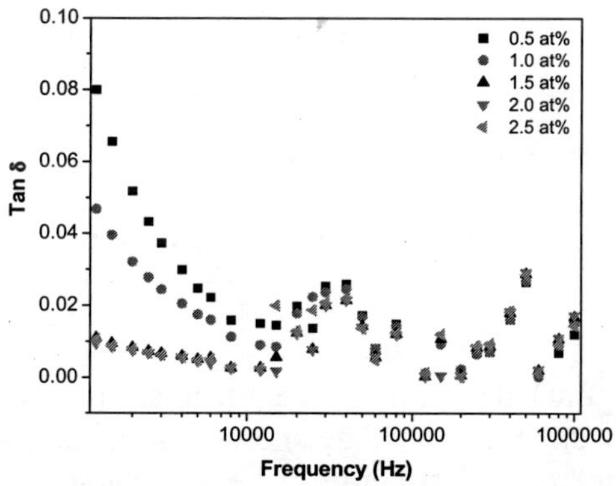

Fig. 8: Variation Loss Tangent with Frequency for Various Ni-doped ZnO Films

Impedance Analysis

Figure 9 represents the real (Zr) and imaginary component (Zi) for Ni-doped ZnO thin films at room temperature. The center of the semicircle is localized below the real axis attributed to a constant phase element. The impedance spectrum is characterized by the appearance of semicircle arcs whose pattern changes, but not its shape, when the doping concentration is increased. Such pattern tells us about the electrical process occurring within the film and their correlation with their microstructures. Intersect of the semicircle with real axis at low frequency is attributed to the total resistance. On the other hand, the impedance response of grain governs at high frequencies and the resistances of grain is deduced from the left intersect of the semicircle with real axis. The intercept of the semicircle with the real axis at low frequency represents the sum of the resistance of grains and grain boundaries, while the intercept at high frequency represents the resistance of the grains only. Grain boundary resistance decreases with doping concentration up to 1.5 at% and then increases, it seems to be due to the fact that the grain boundary effect has assisted in lowering the barrier to the motion of charge carriers. The impedance value is typically higher at lower temperatures in the low-frequency region and decreases gradually with increasing frequency. Also Z' decreases with doping concentration up to 1.5 at% indicating an increase in conductivity as confirmed from electrical measurements.

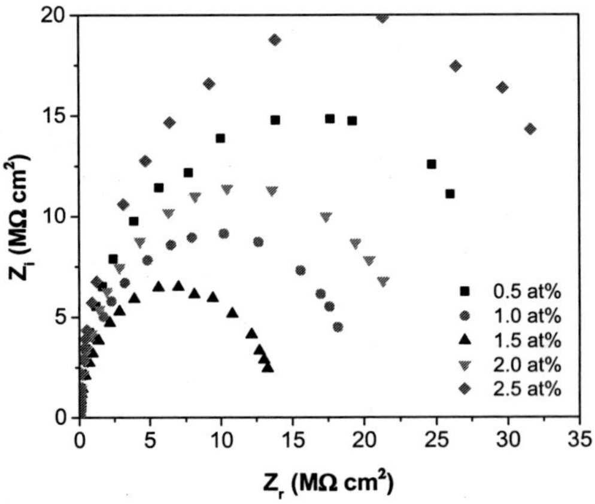

Fig. 9: Complex Impedance Spectra of Ni-doped Zinc OxideThin Films

CONCLUSIONS

The transparent Ni-doped ZnO thin films were successfully deposited onto corning glass substrates in an aqueous medium. X-ray study shows that the doped films are polycrystalline with a hexagonal (wurtzite) crystal structure. The micrographs of the films have tightly packed grains and relatively smooth surfaces without any voids and crack. Deposited films show higher transparency in the visible range with an average transmittance of about 85 %. The dielectric behavior shows the electronic polarizability at higher frequencies due to space charge polarization.

ACKNOWLEDGEMENTS

Authors are very much thankful to UGC and DST for the financial support through UGC-DSA-I, DST-PURSE and DIST-FST-II programmes.

REFERENCES

[1] Liang, S., Sheng, H., Liu, Y., Huo, Z., Lu, Y. and Shen, H., ZnO Schottky ultraviolet photodetectors, *J. Cryst. Growth* 225 (2001) 110–113.

[2] Saleh, S., Elsimary, H., Zaki, A. and Ahmed, S., Design and fabrication of piezoelectric acoustic sensor, *WSEAS Trans. Electron.* 3 (2006) 192–196.

[3] Ozgur, U., Alivov, Y.I., Liu, C., Teke, A., Reshchikov, M.A., Dogan, S., Vrutin, V., Cho, S.J., Morkoc, H., A comprehensive review of ZnO materials and devices, *J. Appl. Phys.* 98 (2005) 041301.

[4] Schmidt-Mende, L. and ManManus-Driscoll, J.L., ZnO-nanostructures, defects, and devices, Mater. Today 10 (2007), 40–48.

[5] Xiaolu Yan, Dan Hu, Hangshi Li, Linxiao Li, Xiaoyu Chong, Yude Wang, Nanostructure and optical properties of M doped ZnO (M = Ni, Mn) thin films prepared by sol–gel process, Phys. B 406 (2011) 3956–3962.

[6] Abed, S., Aida, M.S., Bouchouit, K., Arbaoui, A., Iliopoulos, K. and Sahraoui, B., Non-linear optical and electrical properties of ZnO doped Ni thin films obtained using spray ultrasonic technique, *Opt. Mater.* 33 (2011) 968–972.

[7] Liu, E., Xiao, P., Chen, J.S., Lim, B.C. and Li, L., Ni doped ZnO thin films for diluted magnetic semiconductor materials, Curr. Appl. Phys. 8 (2008) 408–411.

[8] Pandey, B., Ghosh, S., Srivastava, P., Avasthi, D.K., Kabiraj, D. and Pivin, J.C., Synthesis and characterization of Ni-doped ZnO:A transparent magnetic semiconductor, J. Magn. Magn. Mater., 320 (2008), 3347–3351.

[9] Pivin, J.C., Socol, G., Mihailescu, I., Berthet, P., Singh, F., Patel, M.K. and Vincent, L., Structure and magnetic properties of ZnO films doped with Co, Ni or Mn synthesized by pulsed laser deposition under low and high oxygen partial pressures, Thin Solid Films 517 (2008) 916–922.

[10] ShaoMin, Z., HongLei, Y., LiSheng, L., XiLiang, C., ShiYun, L., YaoMing, H., RuiJian, Y., Ning, L., Magnetic properties of Ni-doped ZnO nanocombs by CVD approach, Nanoscale Res. Lett. 5 (2010) 1284–1288.

[11] Benzarouk, H., Drici, A., Mekhnache, M., Amara, A., Guerioune, M., Bernède, J.C., Bendjffal, H., Effect of different dopant elements (Al, Mg and Ni) on microstructural, optical and electrochemical properties of ZnO thin films deposited by spray pyrolysis (SP), Superlatt. Microstruc.52 (2012) 594–604.

[12] Prabakar, K., Heeje Kim, Growth control of ZnO nanorod density by sol–gel method, Thin Solid Films 518 (2010) e136–e138.

[13] Maldonado, A., Tirado-Guerra, S., M. dela L. Olvera, Chemically sprayed ZnO:(F, Zr) thin films: Effect of starting solution ageing time and substrate temperature on the physical properties, J. Phys. Chem. Solids 70 (2009) 571–575.

[14] Shinde, S.S., Bhosale, C.H., Rajpure, K.Y., Structural, optical, electrical and thermal properties of zinc oxide thin films by chemical spray pyrolysis, *J. Mol. Struc.*, 1021 (2012), 123–129.

[15] Sawant, V.S., Shinde, S.S., Deokate, R.J., Bhosale, C.H., Chougule, B.K., Rajpure, K.Y., Effect of calcining temperature on electrical and dielectric properties of cadmium stannate, *Appl. Surf. Sci.,* 255 (2009), 6675–6678.

[16] Pankove, J.I., Optical Processes in Semiconductors, Dover, New York, 1976.

[17] Kim, K.T., Kim, G.H., Woo, J.C., Kim, C., Characteristics of Nickel-doped Zinc Oxide thin films prepared by sol–gel method, *Surf. Coat. Technol.,* 202 (2008) 5650.

[18] Pratt, I.H., Firestone, S.J., Fabrication of RF-sputtered barium titanate thin films. *J. Vac. Sci. Technol.,* 8 (1971), 256.

TIN OXIDE DOPED INDIUM OXIDE (SnO_2:In_2O_3) THIN FILMS PREPARED BY ELECTRON BEAM EVAPORATION FOR ETHANOL SENSING APPLICATION

C. RAVI DHAS, RAJU REGMI, B. ANITHA, A. MOSES EZHIL RAJ[1], K. RAVI CHANDRAN[2] AND C. SANJEEVI RAJA[3]

Department of Physics, Bishop Heber College (Autonomous), Trichirapalli, Tamil Nadu
[1]Department of Physics, Scott Christian College (Autonomous), Nagercoil, Tamil Nadu
[2]Department of Physics, AVVM Sri Pushpam College (Autonomous), Thanjavur
[3]Department of Physics, Alagappa University, Karaikudi, Tamil Nadu
E-mail: cravidhas@gmail.com

ABSTRACT

An alcohol sensor having high sensitivity along with optimized selectivity, which can perform at low operating temperature has been always in great demand in biomedical, chemical, drug and food industries. In this work indium oxide thin film (~0.282 μm in thickness) are deposited onto the glass substrate using electron beam evaporation method. Then tin oxide layers are doped onto the optimized indium oxide films using the thermal energy during the deposition. The thickness of the resulting film after tin oxide doping is varied from 0.301 μm to 0.529 μm in order to study the influence of the tin oxide amount used during doping towards the sensing properties of the films. The sensing properties of the prepared films are studied for 100 and 200 ppm of ethanol vapor. Microstructures of the films are analyzed using X-ray diffraction and Atomic force microscopy. The electrical measurements of the films are carried out and the electrical band gap is calculated. Theoretical description has been done to understand the ionosorption of gaseous species on the film surface, responsible for the sensing properties.

V. Rajendran, R. Yuvakkumar, K. Thyagarajah and K.E. Geckeler (eds.)
Applications of Nano Materials: Electronics, Energy and Environment, pp. 345–356 (2012).

INTRODUCTION

Semiconducting oxide sensors have been intensively studied due to their sensitivity to a large number of gases.[1] Among these oxides, pure or tin-dope indium oxide (ITO) has greatly attracted increasing interest not only for its gas sensing properties but also for its high electrical conductivity and transparency which are exploited in a broad range of engineering applications.[2] The popular sensing mechanism is based on the change in electrical resistance of the metal oxide semiconductors due to ionosorption of gaseous species on their surface. The constituent of ethanol dissociates over the film and activates the spillover process resulting remarkable change in the resistance of the film. The material characteristics and size effects of metal oxides have been explored for its potential in sensing applications.[3] The charge transfer process induced by surface reactions determines the resistance of the metal oxide. The magnitude of change in electrical resistance gives a direct measure of the concentration of the target gas present.

Transparent conducting oxides (In_2O_3, SnO_2) have been used as catalysts for the selective oxidation of hydrocarbon. The interaction between SnO_2 and In_2O_3 is unique because of the similarity of ionic radii and the structures in their highest oxidation state. Although In_2O_3 and SnO_2 thin films have already been studied, their sensing properties combined as a binary oxide are not completely known and in particular the potentials are fully exploited for ethanol detection. In_2O_3 shows electro physical and chemical properties essentially different compared to SnO_2 and this different behavior seems to considerably modify the traditional gas detection mechanism based on the mediation of surface chemisorbed oxygen species.

So far, many methods have been used to deposit ITO thin films such as pulsed laser deposition(PLD),[4,5] sputtering,[6] e-beam evaporation(EBE),[7] sol–gel technique,[8] and spray pyrolysis.[9,10] Among all these methods, e-beam evaporation has the advantages of low cost, low temperature, high purity and high deposition rate.[7,11]

In this work, the e-beam evaporation method was used to deposit SnO_2:In_2O_3 thin films with low resistance and high transparency. The effect of oxygen pressure and substrate temperature on the electrical and sensing properties were studied and correlated with each other. The optimized deposition parameters were discussed and concluded. SnO_2:In_2O_3 thin films deposited under the optimized deposition conditions show the good electrical and sensing properties which is comparable with the best results of SnO_2:In_2O_3 films prepared by other different techniques.

The extensive investigation on the modification of the In_2O_3 thin film by SnO_2 doping can be done by applying thermal energy during the deposition, for improving the selectivity and sensitivity towards the ethanol vapor at room temperature is our main aim in this work.

EXPERIMENTAL DETAILS

The indium oxide target was prepared from 99.99% pure indium oxide powder which was mixed with polyvinyl alcohol and was grinded for 30 minutes. After proper grinding, indium oxide pellets were prepared by using dye, which were finally annealed at 400°C for 4 hours. Tin oxide pellets were prepared using 99.99% pure tin oxide powder adopting the same procedure. Entire depositions were carried out at an approximate pressure of 5×10^{-5} mbar inside the chamber. The samples were grown at different experimental conditions. Target Substrate distance (12 cm), Voltage (5.5 KV) and current value (40 m A) were kept constant.

The deposition of the indium oxide layer was to obtain a good crystalline film which could be used to dope by the tin oxide. Table (1) and. Table (2) summarize the details of experimental parameters used for the deposition of the indium oxide and tin oxide respectively.

Table 1: Experimental Parameters used in the EBE for the Growth of In_2O_3 Layer

Substrate Temperature	*Current Value*	*Deposition Time*
300°C	40 mA	5 and 10 min
350°C	40 mA	5 and 10 min

Table 2: Experimental Parameters Used in the EBE for the Doping of SnO_2 over the optimized In_2O_3 Film

Constant Parameters	*Deposition Time*
Substrate temperature: 350°C Current Value: 40 mA Voltage: 5.5 KV	5, 10, 15, 20, 25 min

Sensitivity measurements can be characterized by any of the two methods, dynamic and static system. We have adopted static system, which comprises an airtight chamber with air admittance and gas inlet valves as shown in Figure 1.

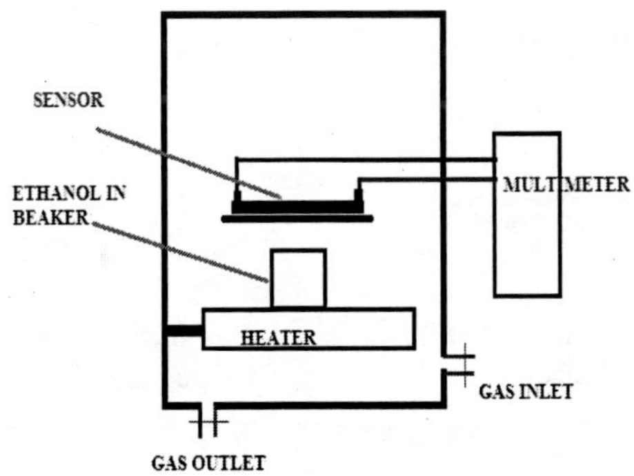

Fig. 1: Gas Sensing Apparatus

The film sensor was placed inside an airtight chamber of known volume 57,000 cm^3. Ohmic contacts were made on the sensor films which were connected with an ohmic meter for the necessary measurement. Initially, the sensitivity measurements were performed at the room temperature in air ambient. This response is considered as a reference response for the calculation of sensitivity. After noting the reference response, the ethanol liquid was placed in a beaker above the heater inside the chamber which is evacuated to a base pressure of 10^{-3} m bar using a rotary pump. All the valves are closed to avoid the leakage of the test gas. The resistance of the sensor is measured for different exposure time in ethanol vapor-air ambient. After completing the measurement, the gas is wiped off completely by operating the rotary pump and by injecting fresh air into the chamber.

RESULTS AND DISCUSSIONS

The prepared indium oxide and tin doped indium oxide thin films were characterized by XRD to obtain the structure parameters as a function of the deposition parameters.

In_2O_3 Film Grown at Different Conditions

The XRD patterns recorded for the deposited In_2O_3 film revealed the (211), (222), (400), (440) and (622) peaks for most of the deposited conditions with (222) as a predominant peak. This is because of the low crystallization energy requirements for the formation of (222) plane.

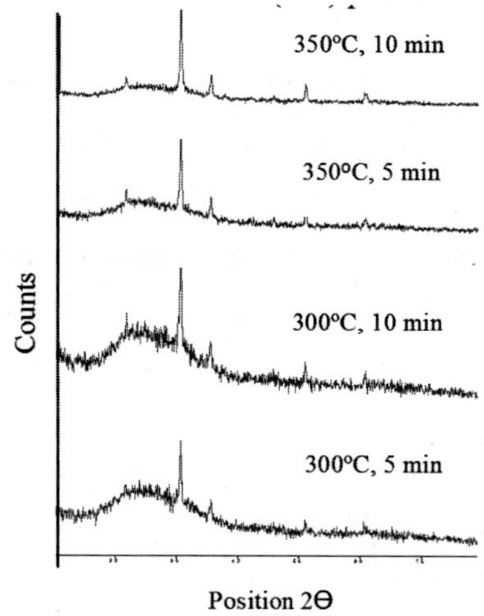

Fig. 2: XRD Spectrum of Indium Oxide Film Grown at Various Conditions

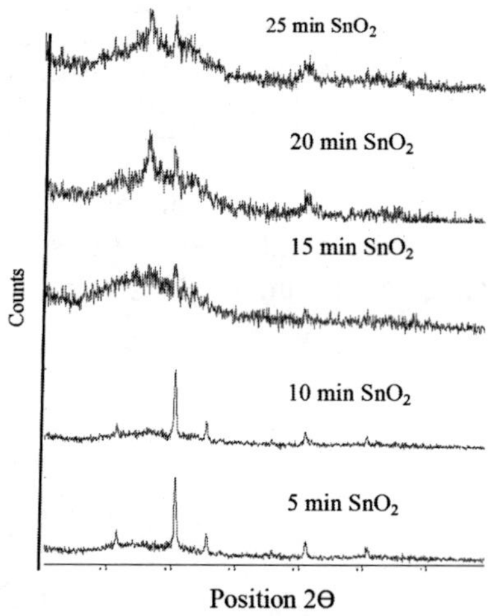

Fig. 3: XRD Result of Tin Oxide Doped over Optimized Indium Oxide Film for Different Deposition Time

The XRD patterns of In_2O_3 thin films prepared for different substrate temperature 300°C and 350°C respectively for 5 min and 10 min time of deposition is shown Figure (2).The other deposition parameter were being kept at the optimum condition throughout the experiments. The XRD spectra clearly indicates that the structural transitions at 350°C is due to the crystallization of the deposited In_2O_3 film and nearly five orientations are observed along (211), (222), (400), (440) and (622) directions and cubical in nature. Further it indicate that In_2O_3 films deposited at 300°C are less crystallization whereas those deposited at 350°C are in crystalline phase with (222) being the dominant orientation. At the substrate temperature 300°C, the deposited films are less crystalline because the available thermal energy may not be sufficient for the formation of In_2O_3 film. At 350°C surface mobility of atomic species is dominant and hence produces the In_2O_3 films are in the well crystallized form.

The grain size was calculated from the XRD pattern using Debye-Scherrer formula.[12] It is observed that the average grain increases as the substrate temperature increases. As the temperature increases the crystalline quality of the In_2O_3 improves as indicated by the fact that the diffraction peak narrows in the (222) orientation. In addition, the intensity ratio of the peak shows the dominant of (222) peak at 350°C.

Table 3: XRD Results for Indium Oxide Thin Films

parameter	*Pos.° 2θ*	*d Space [Å]*	*a Value*	*h k l*	*Grain Size nm*	*Strain mm*	*Dislocation Density Lines/m^2 × 10^{14}*	*N /Unit Area × 10^{16}*
3500°C, 10 min	35.6	2.522	10.088	400	53.2	0.651	3.53	0.188
	37.8	2.3771	10.085	411	79.6	0.435	1.58	0.0561
	60.7	1.5246	10.113	622	54.8	0.633	3.33	0.172
350°C, 5 min	35.6	2.5216	10.086	400	79.8	0.651	1.57	0.061
	51.1	1.7849	10.097	440	29.0	0.435	11.9	1.27
300°C, 10 min	35.5	2.5247	10.098	400	56.7	0.611	3.11	1.11
	51.1	1.7860	10.103	440	43.5	0.797	5.29	2.47
	60.8	1.5232	10.104	622	111.0	0.311	0.806	0.147
300°C, 5 min	35.6	2.5196	10.078	400	31.9	1.086	9.81	2.74
	51.1	1.7846	10.095	440	29.2	1.185	11.7	3.57

An evaluation of crystalline quality is observed with deposition temperature and temperature domain can be distinguished: (i) Less crystalline films deposited at low temperature. (ii) Crystalline with oriented layers at 350°C. Experimentally observed inter planar distance 'd' is an agreement with standard JCPDS values (card no 88:2160). The dislocation density, lattice strain, number of crystallites of In_2O_3 films with temperature is given in Table 3.

From the above XRD results of the film grown at different conditions we found that the film grown at condition 10 minutes, 350°C, shows the good crystalline nature (Figure 2). This optimized film is further used for the tin oxide doping for different deposition time

SnO_2 Doped over Optimized In_2O_3 Film for Different Deposition Time

As it can be seen in Figure (3) at 5 min and 10 min SnO_2 doped over optimized In_2O_3 film, less crystalline phase is appeared at 15 mins, 20 min and 25 min time of deposition SnO_2 doped over optimized In_2O_3 film, good crystalline nature of the multilayer is obtained from the large number of the peak in the XRD pattern Figure 3. However the 20 min time of deposition SnO_2 doped over optimized In_2O_3 film shows the well-defined structure. It can found that at 5 min and 10 min time of deposition SnO_2 doped over optimized In_2O_3 film only less significant peak were obtained for the respective "2θ" values. The low intensity of this appearance suggests that this film consist of less crystalline phase for 5 min and 10 min SnO_2 coating over In_2O_3film. In Figure 3 diffraction pattern for 15 min, 20 min and 25 min time of deposition SnO_2 doped over optimized In_2O_3 film is poly crystalline and strongly dependent on the preparation conditions. Film prepared at these three conditions exhibited peaks for (222), (400), (440) and (622) planes of SnO_2 doped over optimized In_2O_3 film only.

Table 4: XRD Results for SnO_2 Doped over Optimized In_2O_3 Film

Parameter	*Pos.° 2θ*	*d-space [Å]*	*a value*	*h k l*	*Grain Size nm*	*Strain mm*	*Dislocation Lines/l^2 X 10^{14}*	*N /unit area X 10^{16}*
5 min SnO_2	35.5	2.5242	10.096	400	56.2	0.617	3.17	0.16891
	51.0	1.7865	10.106	440	21.6	1.602	0.214	2.95297
	60.5	1.5268	10.128	622	182.0	0.191	30.3	4.99264
10 min SnO_2	35.5	2.5227	10.090	400	44.7	0.775	5	0.38647
	51.1	1.7853	10.099	440	18.2	1.905	0.302	5.73323
	60.7	1.5245	10.112	622	47.0	0.738	4.54	0.33366
15 min SnO_2	30.6	2.9145	10.096	222	14.0	2.43	0.49	4.4888
20 min SnO_2	52	1.7716	10.022	440	12.3	2.821	6.62	24.825
25 min SnO_2	52	1.7605	9.959	440	12.3	2.821	0.662	28.5488
	60.6	1.5266	10.126	622	45.2	0.766	4.89	0.57231

The lattice parameter 'a' evaluated after refinement of all the observed peak is approximately 10.0834 Å units which is nearly equal to standard values (JCPDS card no 89–4596). However the intensity of peaks observed for the film of deposited for 20 min SnO_2 doped over optimized In_2O_3 film are relatively higher than that for the other two conditions namely 15 min and 25 min respectively. This is due to the diffusion of tin species into the respective indium lattice size to form more dense films. In addition the observed d-spacing of the diffraction planes are very close to the standard values. The other structural parameters SnO_2: In_2O_3 film derived from XRD shows in Table 4. The grain size of SnO_2 doped over optimized In_2O_3 film crystallites evaluated with half intensity width of the XRD peaks. The grain size of the film evaluated from the predominant peak (440) is decreased from 21.6 nm to 12.3 nm with increase of time of deposition SnO_2 doped over optimized In_2O_3 film. This information further confirmed good crystalline nature of the films, and reveals that the film deposited at 20 min SnO_2 doped over optimized In_2O_3 film is suitable for gas sensing elements.

Electrical Characteristics

As expected, the deposition conditions varying the electrical resistance of the films. [Figure 4 (b&c)] shows the change in resistance when SnO_2 doping on optimized In_2O_3 film for different deposition rate. With increase in time of deposition of 20 min, SnO_2 over In_2O_3 become less resistive and leads to higher conduction due to the diffusion of tin atoms in to Indium lattice. Observed electrical conductivity from sheet resistance is found to increase with film thickness with respect to time of deposition and their variations can be explained on

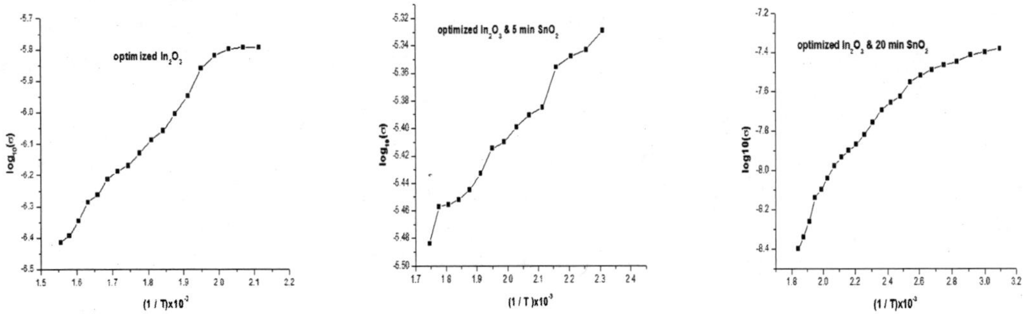

(a) (b) (c)

Fig. 4: Plot for Temperature vs. Conductivity (a) Optimized InO_3 film. (b) 5 min Sn: In_2O_3 Film (c) 20 min Sn:In_2O_3 Film

Table 5: The Thickness and Electrical band Gap Values

	Thickness (μ m)	*Activation Energy (E_a)*
Optimized In_2O_3	0.282	2.878
Optimized In_2O_3 and 5 min SnO_2	0.301	2.569
Optimized In_2O_3 and 10 min SnO_2	0.345	1.727
Optimized In_2O_3 and 15 min SnO_2	0.422	1.439
Optimized In_2O_3 and 20 min SnO_2	0.460	0.720
Optimized In_2O_3 and 25 min SnO_2	0.529	0.491

the basics of the change in mean free path of the conduction electrons. Whenever the specimen becomes thin enough, collision with the surface is a significant fraction in the total number of collisions. Thomsons [13] and Fuchs [14] discovered the possibility that the dimensions of a specimen could influence the conductivity. In particular, Sondheimer [15] extended the work by including mean free path in related theories. All these theories have been reviewed by Campell [16] attained an equation related to film conductivity with thickness.
Ohmic condition, whether ionic or electronic gives exponential temperature dependence, given by $\sigma = \sigma_0 \exp(-E_a/kT)$, where '$E_a$' is the activation energy of the process.

With extrinsic ionic conduction, mobility of the ion is activated and E_a being the energy of ion to h_{0p}. However, the electronic conduction is by excitation into the conduction band, the production of free electrons is activated. Whatever is the ohmic mechanism, an Arrhenius plot ln (ρ) V_s 1000/T usually exhibits increasing linear slopes as temperature is raised. It is noted that the activation energy also depend on the film thickness, which increases as the thickness decreases. This dependence may be due to the combined effect of the following reasons:

The change in barrier height due to the size of the grains in a polycrystalline film, large density of dislocation, quantum size effect, changes in stoichiometry.

The following in the explanation for the conductivity dependence on thickness and grain size.

According to the charge transfer model [16, 17] a certain number of particles are charged by gaining or losing electrons at any temperature above 0K. If the activation energy due to the thermal energy, then at equilibrium the number of charged particles (n) is related to the total number of particles (N) in the film by the expression, $n = N \exp(-E_a/kT) = N \exp(-e^{2/}\varepsilon_r kT)$. Hence, the activation energy $E_a = e^2/\varepsilon_r$ due to thermal agitation is inversely proportional to the average linear dimension (r) of the grains. Since the increase of film thickness by increasing the time of doping of SnO_2 over In_2O_3 is generally accompanied by an increase of 'r', there is lowering of 'E_a'..

Atomic Force Microscopy

Figure 5 shows the 3D Atomic force microscopy image of the SnO_2: In_2O_3 films deposited for 5 min, 10 min, 15 min, 20 min and 25 min. AFM results show spherical grains in all the films varied smoothness. The grain size of SnO_2:In_2O_3 films deposited at 20 min time of deposition is smaller than that of the films deposited at 5 min, 10 min, 15 min and 25 min which was confirmed in XRD result also. However, at 25 min time of deposition, the size distribution is non-uniform. Whereas, in the film deposited at 20 min time of deposition uniform grains are observed. The AFM image gives the r.m.s. roughness value, which will be indicating the film surface smoothness quantitatively. The 2D AFM pictures Figure 6 also show the respective surface profile. The r.m.s roughness ($R_{r.m.s}$) values Table 6 are 0.8324 nm, 1.3109 nm, 3.0549 nm, 24.5325 nm and 26.617 nm respectively.

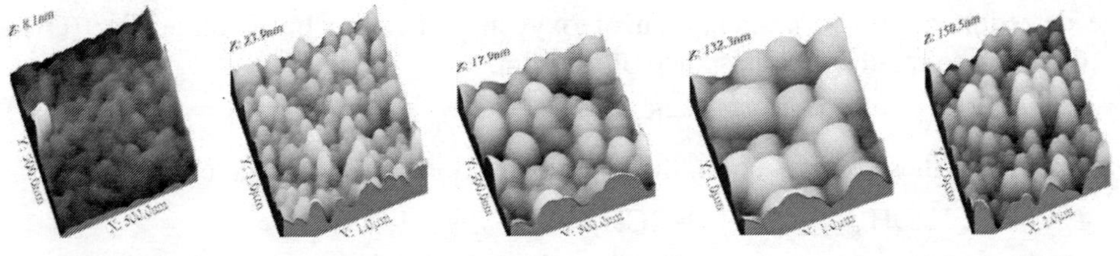

Fig. 5: AFM Image for SnO_2: In_2O_3 Films Deposited for 5, 10, 15, 20 and 25 min in 3D

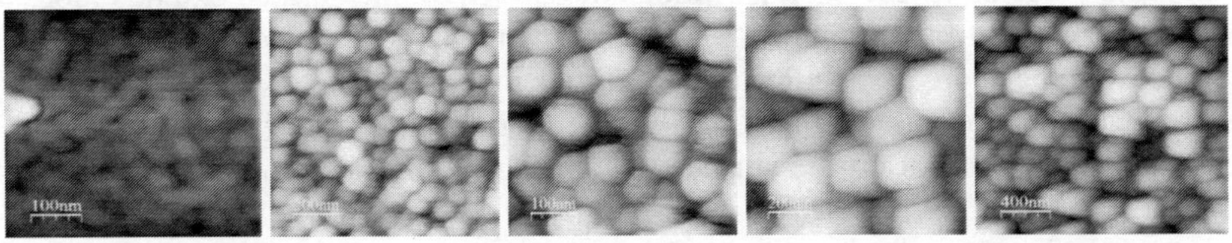

Fig. 6: AFM Image for SnO_2: In_2O_3 Films Deposited for 5, 10, 15, 20 and 25 min in 2D

Table 6: RMS Roughness Values of SnO_2: In_2O_3 for 5, 10, 15, 20, 25 min

Time of Deposition	*RMS Roughness*
5 min	0.8324 nm
10 min	1.3109 nm
15 min	3.0549 nm
20 min	24.532 nm
25 min	26.617 nm

Sensing Properties of SnO_2:In_2O_3 Film

After having optimized the deposition parameters for films the device was fabricated as a resistive type gas sensor.

Figure 7 shows the transient response curve of the optimized In_2O_3 film to ethanol gas. The variation sensitivity was measured for ethanol concentration of 100 ppm and 200 ppm. Upon injection of ethanol gas, the sensitivity reached saturation in about 12 min and when dry air was introduced, the sensitivity is returned to the initial value after 22 min. The response and recovery times are in several min and the obtained value of sensitivity is very low for 100 ppm ethanol when compared to 200 ppm ethanol. The sensitivity and transient response of the In_2O_3 sensor towards ethanol is found poor which may be include by incorporating some suitable dopent like SnO_2over optimized In_2O_3 is one such material having sensitivity in the measurable range, stable microstructure and sensitive to gas response. For sensitizing reaction, reaction gas is observed with contact on the surface of the SnO_2: In_2O_3 film layer to form the schottky contact and the electrical resistance of the sensor is changed:

$$O_2\,(gas) + 2e^- \longrightarrow 2O^- + \Delta H_1$$

The detection reactions, participation of oxygen which can be in molecular (O_2) are in atomic (O^-) form to reduce the resistance of the film.

$$C_2H_5OH_{(gas)} + O^- + \Delta H_2 \longrightarrow e^- + CH_3CHO_{(gas)} + H_2O + \Delta H_3$$

On combining these two reaction, the overall reaction on the surface can be given as,

$$O_2(gas) + 2C_2H_5OH_{(gas)} \longrightarrow 2CH_3CHO_{(gas)} + 2H_2O + \Delta H_4$$

The intermediate product acetaldehyde is also reduced to carbon dioxide and water. The depletion region created by the chemisorptions of oxygen on the surface extends more deeply providing larger scope to more number of gaseous elements to be absorbed there by giving a better response. One more possibility of the partial incorporation of 'Sn' irons in 'In' sites changes the carrier concentration and the work function on the surface. This allows more gaseous elements of ethanol on the surface, leading to an improved output response and the selectivity.[18]

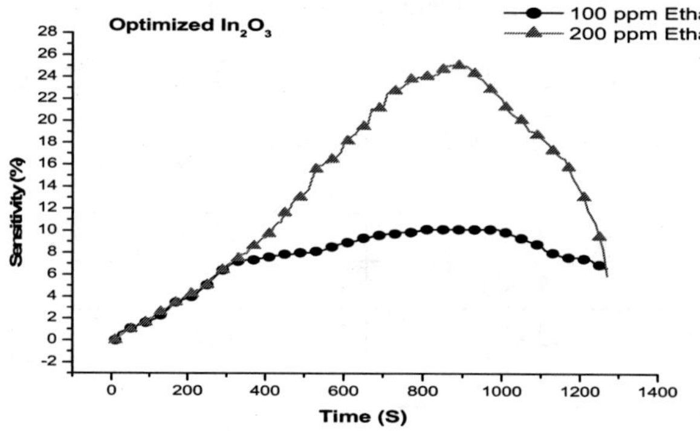

Fig. 7: Optimized In_2O_3 Film

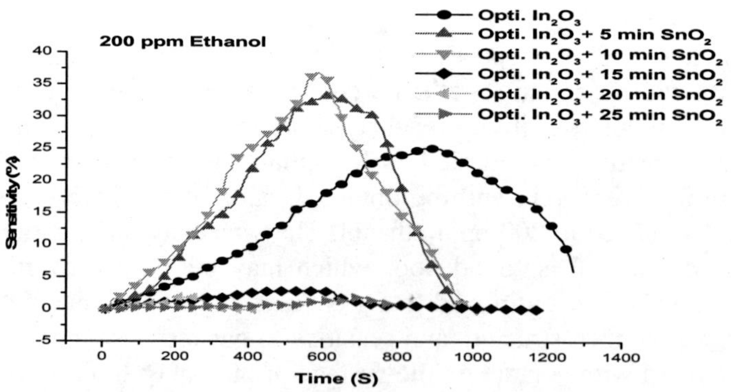

Fig. 8: SnO_2: In $2O_3$ film for 5, 10, 15, 20, 25 min

The transient response curve of the SnO_2:In_2O_3 sensor are show in Figure 8. The variation of the sensitivity are measured for SnO_2 doping over In_2O_3 for 5 min, 10 min, 15 min, 20 min and 25 min and the ethanol concentration 200 ppm. Upon injection on ethanol vapour the sensitivity of SnO_2:In_2O_3 for 20 min deposition sensors reaches saturation is about 8 min which is less than that observed in other sensor. When dry air is introduced the sensitivity reached to the initial values. This fact proves the reversibility of the process. These results shows the SnO_2:In_2O_3 sensor having less doping concentration (5 min, 10 min, 15 min time of deposition of SnO_2 over In_2O_3) has lower response and recovery time than that of the SnO_2:In_2O_3 sensor having high doping of SnO_2 over In_2O_3 (20 min and 25 min). However, when the doping of SnO_2 over In_2O_3 is for 25 min, the resistivity of film to C_2H_2OH decreased. This shows the reverse sensitivity of the film to ethanol vapour 25 min doping of SnO_2 over In_2O_3. The exothermic effect in the surface reaction of the oxide film with ethanol is difficult to proceed with increasing doping concentration of 'Sn' over In_2O_3, thus decreasing their sensitivity.

CONCULSION

The tin oxide doping was carried out onto the optimized indium oxide thin film. The XRD result shows that the resulting film was Tin doped Indium oxide. The morphological results showed that the films are cubical in nature. The thicknesses of the films were in the order of micrometer ranges. The microstructure of the particles was confirmed by the AFM. The optimized films are smooth and the grains are relatively small for the films which showed high sensitivity towards the ethanol vapor. This confirms that the poly crystalline thin films, whose grains have extremely small dimensions, are good for gas sensing applications. The electrical band gap was found to be ~ 2.878 e V for the optimized indium oxide film and the band gap varied from 2.569 e V to 0.491 e V with increase in the tin oxide doping time.The gas sensing experiment shows good response from optimized indium oxide,20min and 25 min tin oxide doping over optimized In_2O_3 film. Those films were more sensitive towards the 200 ppm of ethanol vapor.

REFERENCES

[1] Eranna, G., Joshi, B.C., Runthala, D.P. and Gupta, R.P., *Crit. Rev. Solid State Mater. Sci.,* 29 (2004) 111.

[2] Granqvist, C.G. and Hultaker, A., Thin solid films 411(2002)1.

[3] Nandan Singh *et al.*, Synthesis of In_2O_3- ZnO core shell nanowires and their application in gas sensing; Sens. and Actuators B: Chem. 160 (2011) 1346–1351.

[4] Cristian Viespe, Ionut Nicolae, Cornelia Sima, Constantin Grigoriu, Rares Medianu, Thin Solid Films 515 (2007) 8771–8775.

[5] Craciun, V., Craciun, D., Chen, Z., Hwang, J. and Singh, R.K., Appl. Surf. Sci. 168 (2000) 118–122.

[6] Yanfa Yan, J. Zhou, X.Z. Wu, H.R. Moutinho, M.M. *Al-Jassim, Thin Solid Films, 515* (2007), 6686–6690.

[7] Hamid Reza Fallah, Mohsen Ghasemi, Ali Hassanzadeh, Hadi Steki, *Mater. Res. Bull.,* 42 (2007), 487–496.

[8] Kim, H., Pique, A., Horwitz, J.S., Mattoussi, H., Murata, H., Kafafi, Z.H. and Chrisey, D.B., *Appl. Phys. Lett.*, 74 (1999), 3444.

[9] Ait , M., Aouaj, Diaz, R., Belayachi, A., Rueda, F., Abd-Lefdil, M., *Mater. Res. Bull.*, 44 (2009) 1458–1461.

[10] Moholkar, A.V., Pawar, S.M., Rajpure, K.Y., Ganesan, V. and Bhosale, C.H., *J. Alloys Compd.,* 464 (2008), 387–392.

[11] Hamberg, I. and Graqvist, C.G., *J.Appl. Phys.*, 60 (1986), R123-R159.

[12] Cullity, B.D., Elements of X-ray Diffraction (Addison-Wesley,London), p. 284 (1978).

[13] Thomson, J.J., *Proc. Cambridge Phil. Soc.*, 11(1901), 12.

[14] Fuchs, K., Proc. *Cambridge Phil. Soc.,* 34(1938)100.

[15] Sondheiner, E.H., *Advan. Phys.*, 1(1952)1.

[16] Nengebaur, C.A., Webb, M.B., *J. Appl. Phys.*, 33(1962)74.

[17] Nengebaur, C.A.,"Physics of thin films", *Academic Press Inc.*, New York, Vol. II, 1964.

[18] Vaishnav, B.S., Patel, P.D. and Patel, N.G., Thin Solid Films, 490 (2005) 94.

CONSTRUCTION OF Mg/MnO_2 AND Zn/MnO_2 PRIMARY CELLS USING MANGANESE DIOXIDE NANOPARTICLES

S.R. SRITHER, M. SELVAM, S. ARUNMETHA, R. YUVAKKUMAR, K. SAMINATHAN AND V. RAJENDRAN

Center for Nanoscience and Technology, K.S. Rangasamy College of Technology, Tiruchengode, Tamil Nadu
E-mail: veerajendran@gmail.com

ABSTRACT

In this study, MnO_2 nanoparticles were prepared and their characteristics were studied for Mg/MnO_2 and Zn/MnO_2 primary battery applications. Mg/MnO_2 and Zn/MnO_2 primary cells were constructed and examined using the prepared MnO_2 nanoparticles and micron powders. The storage capacity and energy density of the micron and MnO_2 nanoparticles of Mg/MnO_2 cell are increased, respectively, from 3.1 to 7.31 mAh g^{-1} and 4.1 to 9.8 Wh kg^{-1} whereas those of micron and MnO_2 nanoparticles are 0.4 to 0.5 mAh g^{-1} and 0.42 to 0.52 Wh kg^{-1}. The discharge time of MnO_2 nanoparticles for Mg/MnO_2 increased to approximately 4650 s compared to that of micron powder. But, there was no significant difference in the discharge time for Zn/MnO_2. Thus, this study opens up a new approach to use MnO_2 nanoparticles as a cathode material for primary batteries in future.

INTRODUCTION

Nanotechnology is showing an exponential growth in areas of consumer products, electronics, textiles, cosmetics and sporting goods because of its large surface area to volume ratio properties.[1] Nanomaterials are widely used in energy storage and conversion, such as solar cell, fuel cell, supercapacitors and secondary batteries, due to their higher efficiency, energy and power densities.[2,3] Mostly, manganese dioxide nanoparticles are used in supercapacitors

V. Rajendran, R. Yuvakkumar, K. Thyagarajah and K.E. Geckeler (eds.)
Applications of Nano Materials: Electronics, Energy and Environment, pp. 357–362 (2012).

and rechargeable batteries to provide good capacitive performance for energy storage process when compared with the conventional products.[4] MnO_2 is known for its use in primary batteries to produce current immediately for many portable electrical and electronic devices.[5]

In general, zinc-manganese dioxide (Zn/MnO_2) battery provides an economical power source for devices requiring light-to-moderate drain; and this holds among all in the primary battery market because of the inexpensive materials.[6] However, the output efficiency of Zn/MnO_2 technology is not adequate due to the concurrent growth of electronic technology. In the meantime, magnesium (Mg) has been used as a most attractive, less toxic, and environmental friendly material in the field of electrochemical energy conversion for primary and secondary batteries. It has been successfully used with MnO_2 powders to obtain an Mg/MnO_2 battery system and also delivers doubled capacity when compared to Zn/MnO_2 system.[7] The main drawback is the sudden potential drop after partial discharge. It is expected that a size reduction (<100 nm) will have a major role in increasing power density and stability of the battery.

Thus, MnO_2 nanoparticles are required to provide maximum storage capacity with safe deeper discharge in existing primary batteries. In this investigation, α-MnO_2 nanoparticles are used as a cathode material to construct Mg and Zn cell setup. In addition, micron MnO_2 powders are used and compared with synthesized nanoparticles to explore the difference in discharge curves.

MATERIALS AND METHODS

Two gram polyvinyl alcohol (SD Fine, India) was dissolved in 100 ml double-distilled water and stirred vigorously. In that aqueous polyvinyl alcohol solution, 1.5 ml polyacrylamide (50 wt%; Sigma-Aldrich, Germany) was added and stirred continuously for 2 h. A certain amount of ethanol (99.9%; AR grade, India) was drop-wise added to the mixture. Then, the solution was heated up to 323 K. Afterwards, 1.5 g potassium permanganate (>98.5%; Merck, India) was added slowly followed by 2 h of magnetic stirring. Finally, a brown colloid was formed and then centrifuged eight times with double-distilled water. Then, the colloid solution was dried at 473 K for 1.5 h. The synthesized product was characterized comprehensively using X-ray diffraction spectrometer (XRD; X'Pert PRO; PANalytical, the Netherlands), scanning electron microscope (SEM; JSM-6390LV; JEOL, Japan) coupled with energy-dispersive spectrometer (model JED-2300; JEOL), and transmission electron microscopy (TEM; CM200; Philips, USA). Further, the discharge measurements of constructed Zn/MnO_2 and Mg/MnO_2 cells were obtained using Metrohm Autolab (PGSTAT302N; the Netherlands).

Construction of Mg and Zn/MnO_2 Primary Cells

The Mg and Zn/MnO_2 primary cells were constructed using the following steps:

- Cathode material was prepared by adding MnO_2 nanoparticles and graphite (>99.5%, 50 MM) in a weight ratio of 3:1 with a trace of polyvinylidene fluoride as a binder. The mixture was well ground using a mortar and pestle. The well-ground powder was pelleted using hydraulic pellet press at a pressure of approximately 10 tons for 5 min to obtain a fine-structured pellet.

- 1 M $MgCl_2$ was prepared as an electrolytic solution.
- Magnesium ribbon (0.05 wt%) was used as an anode material.
- The anode material was mounted on a stainless-steel semi-ball, followed by electrolyte-soaked separator (brown envelop sheet). Then, it was covered by a cylindrical Mylar sheet of 3.5 cm. The cathode pellet was kept above the separator that was previously dipped in the electrolytic solution. Further, the stainless-steel semi-ball was placed over the cathode to close the other end. Copper rounds were used to fit the either ends to provide good electrical contacts.

The setup was constructed by arranging the above material in an order as shown in Figure 1. Similar procedure was followed for Zn/MnO_2 cell. Zinc plate was used as an anode whereas $ZnCl_2$ was used as an electrolyte in the place of magnesium ribbon and $MgCl_2$. In addition, MnO_2 micron powders are used and compared with synthesized nanoparticles to evaluate the difference in discharge curves.

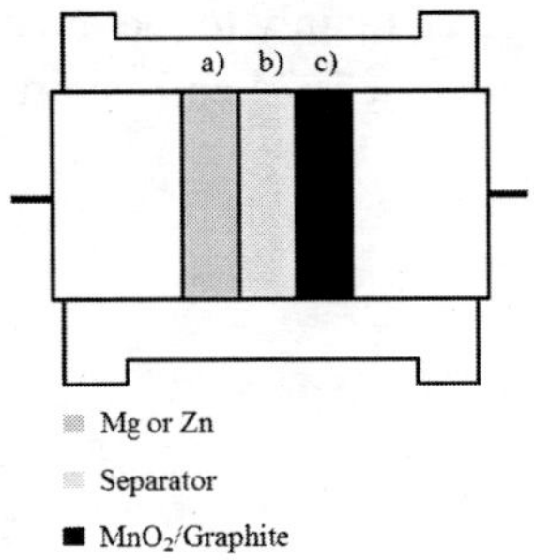

Fig. 1: Schematic Representation of Primary Cell Setup

RESULTS AND DISCUSSION

Figure 2 illustrates the XRD pattern of the synthesized MnO_2 nanoparticles.

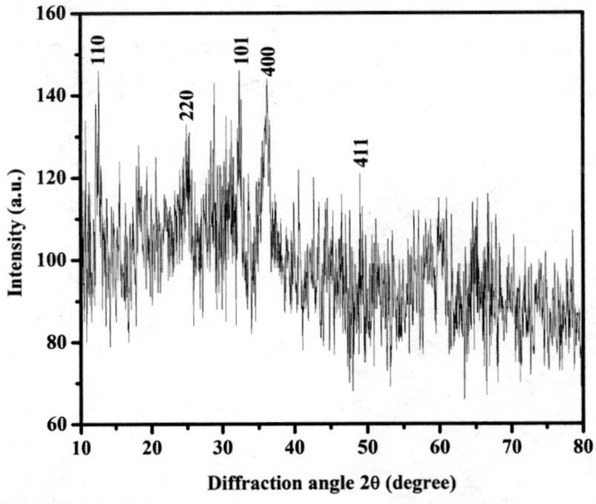

Fig. 2: XRD Pattern of Synthesized MnO_2 Nanoparticles

The observed peaks at diffraction angles (2θ) are well indexed with body-centered tetragonal α-MnO_2 phase (JCPDS file no. 811947). The average crystalline size of MnO_2 nanoparticles is

10.5 nm. Fourier transform infrared spectroscopy (FTIR) of the prepared MnO_2 nanoparticles is shown in Figure 3. The MnO_2 has six vibrating modes, out of which the observed three bands at 693, 580, and 537 cm^{-1} are IR active regions. The observed band at 693 cm^{-1} corresponds to the vibrating mode (Mn–OH).[8] The band observed nearly at 600 cm^{-1} corresponds to the stretching mode, which involves the internal motion of a change in length of the Mn–O–Mn or Mn–O bond.[9] The band recorded around 537 cm^{-1} is due to the bending mode, which is sensitive to the Mn–O–Mn bond angle.[10] The broad peak observed in the region of 400–750 cm^{-1} is attributed mainly to the three modes of 537, 580, and 693 cm^{-1}, respectively. In addition, band recorded at 1624 cm^{-1} shows the presence of Mn trace coordinated with OH bonds.

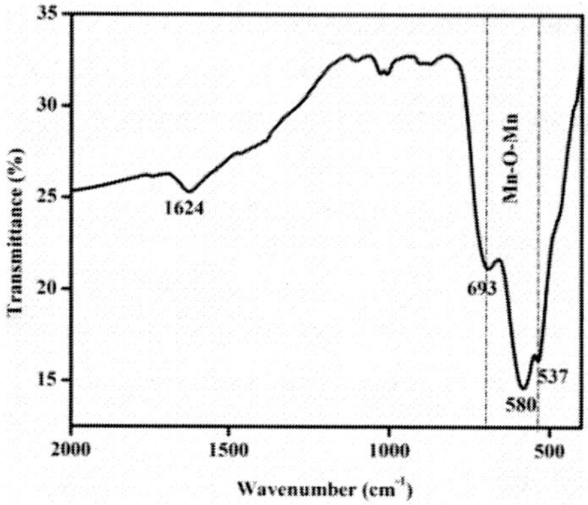

Fig. 3: FTIR Spectra of Synthesized MnO_2 Nanoparticles

The SEM image of the prepared nanoparticles shows spherical structure with little agglomeration (Figure 4a) due to the use of polymers as capping agents. Energy-dispersive X-ray (EDAX) analysis (Figure 4b) confirms the presence of Mn and O with a trace of K.

(a) (b)

Fig. 4: Characteristics of Synthesized MnO_2 Nanoparticles: (a) SEM Image and (b) EDAX Analysis

The prepared MnO_2 particles are in very loose contact and are composed of many small particles as shown in TEM image (Figure 5). The observed selected area electron diffraction pattern shows a broad doughnut shape with a few spots following an obvious ring, confirming the crystalline nature.

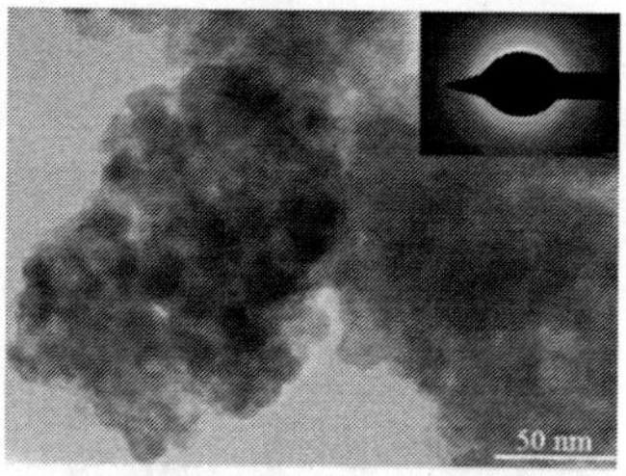

Fig. 5: TEM Image of Synthesized MnO_2 Nanoparticles

Mg/MnO_2 and Zn/MnO_2 Cell Discharge Measurements

The galvanostatic discharge processes were carried out by discharging the Mg/MnO_2 and Zn/MnO_2 cells at a constant current of 1 mA with the cutoff voltage 1 V. Figure 6a shows the discharge curves of Mg/MnO_2 cell prepared from MnO_2 nanoparticles and micron powder. The initial open-circuit voltage of Mg/MnO_2 obtained for the MnO_2 nanoparticles is 1.63 V, which is almost the same as that for the micron powder. During the discharge process, decrease in voltage does not occur initially for both MnO_2 nanoparticles and micron powders. After partial discharge, a sudden decrease in voltage occurs for the micron powder and then the curve quickly reaches to the cutoff voltage. However, MnO_2 nanoparticles deliver sloping discharge and substantial voltage reduction with an increase in the current toward the cutoff voltage. It can be seen that the discharging time of nanoparticles is increased to approximately 4650 s when compared with that of the micron powder. The MnO_2 nanoparticles have a high battery capacity of 7.31 mAh g^{-1} at an energy density of 9.8 Wh kg^{-1}. However, micron powders yield battery capacity of 3.1 mAh g^{-1} at 4.1 Wh kg^{-1} energy density.

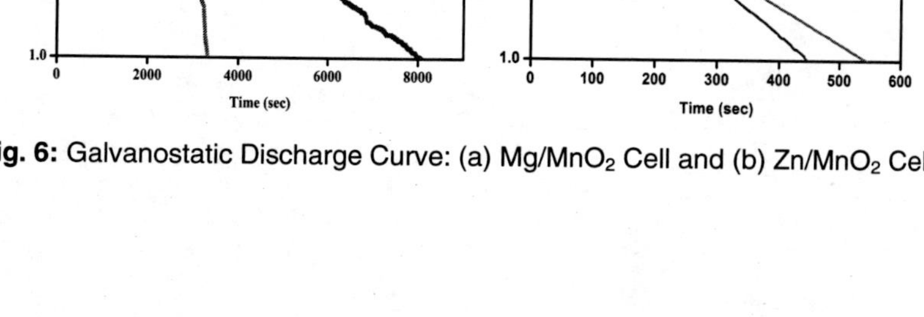

Fig. 6: Galvanostatic Discharge Curve: (a) Mg/MnO_2 Cell and (b) Zn/MnO_2 Cell

Figure 6b shows the discharge curves of Zn/MnO_2 cell prepared from the MnO_2 nanoparticles and micron powder. The initial open-circuit voltage of the micron powder is approximately 1.20 V whereas the synthesized nanoparticles deliver 1.27 V. During the discharge process, no drop in voltage occurs for micron powder. A smooth slope curve is continued up to the cutoff voltage. But, the nanoparticles deliver a slight voltage drop with smooth curve and then continue as a slope to the cutoff voltage. However, the discharge time to cutoff voltage of both curves has not much difference. The MnO_2 nanoparticles show a cell capacity of 0.5 mAh g^{-1} at an energy density of 0.52 Wh kg^{-1} whereas micron powders show it to be 0.4 mAh g^{-1} at 0.42 Wh kg^{-1} energy density.

CONCLUSIONS

In summary, Mg/MnO_2 and Zn/MnO_2 primary cell were constructed using the prepared MnO_2 nanoparticles and micron powders. Electrochemical measurements showed that MnO_2 nanoparticles show high storage capacity and energy density for Mg/MnO_2 cell rather than the Zn/MnO_2 cell. This study confirms that the prepared nano-MnO_2 is a promising cathode material for the Mg/MnO_2 primary batteries in future.

REFERENCES

[1] Kim, S.Y., Jeong, S.H., Lee, E.Y., Park, Y.H., Bae, H.C., Jang, Y.S., Maeng, E.H., Kim, M.K. and Son, S.W., *Toxicol. Environ. Health Sci.*, 2011, 3, 258–261.

[2] Zhao, X., Sanchez, B.M., Dobson, P.J. and Grant, P.S., *Nanoscale,* 2011, 3, 839–855.

[3] Wang, D., Cao, L., Huang, J. and Wu J., *Mater. Lett.*, 2012, 71, 48–50.

[4] Yuan, C., Su, L., Gao, B. and Zhang, X., *Electrochim Acta,* 2008, 53, 7039–7047.

[5] Manickam, M., Singh, P., Touma Issa B., Thurgate S. and Marco R.D., *J. Power Sources,* 2004, 130, 254–259.

[6] Linden, D. and Reddy, T.B., *Handbook of Batteries*, 3rd ed., McGraw-Hill, New York, 2002.

[7] Li, C., Cheng, F., Ji, W. and Tao, Z., *J. Chen. J. Nano Res.,* 2009, 2, 713–721.

[8] Tang, N., Tian, X., Yang, C., Pi, Z. and Han, Q., *J. Phys. Chem. Solids,* 2010, 71, 258–262.

[9] Yuan, C., Gao, B., Su, L. and Zhang, X., *J. Colloid. Interf. Sci.,* 322, 545–550, 2008.

[10] Ai, Z., Zhang, L., Kong, F., Liu, H., Xing, W. and Qui, J., *J. Mater. Chem. Phys.*, 2008, 111, 162–167.

OPTIMIZATION OF CONDUCTIVITY OF GRAPHENE USING ELECTROCHEMICAL AND CHEMICAL REDUCTION METHODS

M. SELVAM, S.R. SRITHER, R. SURIYAPRABHA,
K. SAMINATHAN AND V. RAJENDRAN

Center for Nano science and Technology, K.S. Rangasamy College of Technology
Tiruchengode, Tamilnadu, India
E-mail: ksaminath@gmail.com

ABSTRACT

Electrochemically and chemically reduced graphene sheets were obtained separately and characterized. X-ray diffraction pattern of graphene samples shows a sharp and intensive peak centered at a diffraction angle (2θ) of 26.35° and confirms the structure of graphene was not affected by synthesis methods. Transmission electron microscopy results indicate that the defect density and number of layers of graphene sheets varied with the synthesis method. The hexagonal structure of the sheet and purity of the samples were confirmed as superior in electrochemical process by scanning electron microscopy coupled with an energy-dispersive X-ray analysis and Raman spectroscopy studies. The conductivity of chemically reduced graphene was decreased when compared with that of the electrochemically synthesized graphene. This study explains that graphene with enhanced functional properties can be prepared from effective synthesis methods.

INTRODUCTION

Graphene is a monolayer form of carbon with two-dimensional honeycomb lattice structures and has excellent electronic properties [1,2]. Graphene has superior electrical conductivity

V. Rajendran, R. Yuvakkumar, K. Thyagarajah and K.E. Geckeler (eds.)
Applications of Nano Materials: Electronics, Energy and Environment, pp. 363–368 (2012).

than graphite and other the allotropes of carbon, which is due to its higher surface area and chemical tolerance. The electron mobility of graphene is higher than that of conventional bulk forms of carbon. Synthesis of graphene has been widely studied due to its unique properties such as high conductivity, mechanical strength, thermal conductivity, high energy, and power densities [3]. It is also used in applications such as composite materials [4], batteries [5], solar cells [6], transistors, and biosensors [7]. Nowadays, demand for newer materials is increased for energy storage/conversion synthesis. To understand and create improved materials, devices, and systems, the conventional bulk materials (micro-sized) are converted into nanoscale materials. Therefore, graphene has been widely synthesized using different methods mainly to enhance their size-dependent properties [8].

The various methods used to synthesize graphene sheets are epitaxial growth by ultra-high vacuum graphitization, chemical oxidation of graphite and further reduction, chemical vapor deposition [9], solvothermal synthesis [10], and sol–gel [11]. The electrochemical and chemical reduction methods are widely used due to their low cost and environment friendly nature. Therefore, in this investigation, graphene sheets are synthesized using electrochemical and chemical reduction methods are characterized employing comprehensive techniques. The obtained results are discussed in the light of its electrochemical performance.

MATERIALS AND METHODS

Graphene samples were prepared by two methods namely electrochemical and chemical reduction methods. The procedure discussed in our earlier studies [12] was used to prepare the electrochemically synthesized graphene sample (hereafter termed as EG sample). Hummer's modified method [13] was used to prepare the chemically reduced graphene sample (hereafter termed as CG sample). X-ray diffraction (XRD; X'Pert Pro; PANalytical, The Netherlands) and field emission scanning electron microscope (FE-SEM, JEOL, Japan) studies were used for characterization of the prepared samples. The chemical composition of prepared sample was examined using energy-dispersive X-ray analysis (EDX; S-4200; Hitachi, Japan). Transmission electron microscopy (TEM; HR-TEM 2011; JEOL, Japan) and Raman spectra of EG and CG graphene samples were obtained using Raman spectrum (Japan) with a WITech alpha300 confocal system equipped with an Nd:YAG laser (λ=517 nm). The conductivity measurements of the prepared graphene samples were carried out using four probe conductivity techniques (models 2182A and 6221; Keithley, Germany).

RESULTS AND DISCUSSION

Graphene formation is easily identified during its synthesis by electrochemical rather than chemical method. XRD patterns of the EG and CG samples of graphene are shown in Figure 1. The peaks in XRD pattern of samples confirm the crystalline nature of the graphene. A sharp intensive peak observed around the diffraction angle (2θ) of 26° is highly specific for the crystalline nature of graphene. However, the observed three peaks confirm that the structure of EG and CG samples is hexagonal (JCPDS file no. 411487). The 2θ values for EG and CG samples are 26.35° and 26.52°, respectively. It is noted that the 2θ value of EG sample is less than that of CG sample. This indicates that the EG sample has lower particle size when compared with that of the CG sample, due to the additional stress induced while preparing EG sample. These results further confirm that the structural stability of graphene is not affected by the synthesis procedure.

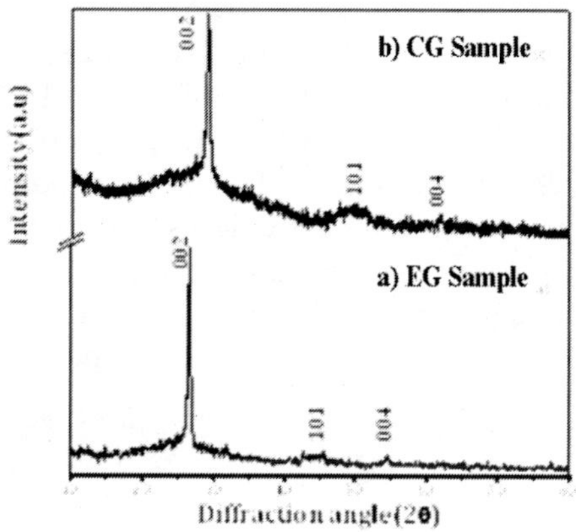

Fig. 1: XRD pattern of prepared sample

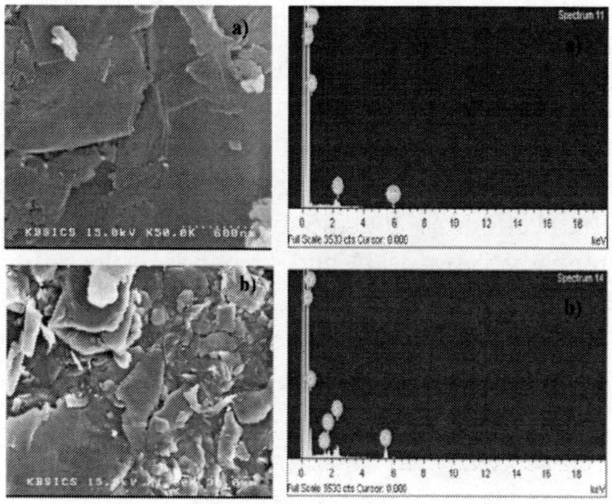

Fig. 2: SEM-EDAX image of a) EG and b) CG sample

SEM and EDX patterns (Figure 2) of prepared samples show a worm-like structure with ultrathin sheet morphology. Well-ordered hexagonal graphite lattices of multilayered graphene flakes are clearly seen in EG and CG samples. However, the purity of the samples is significantly different due to the different synthesis procedure. The purity of the samples is found to be high in EG samples (93.63%) than in that in CG method (79.49%). The EDX pattern of EG sample shows the presence of elements such as C, O, and S whereas elements such as Al, Si, S, and Cr are present in CG samples.

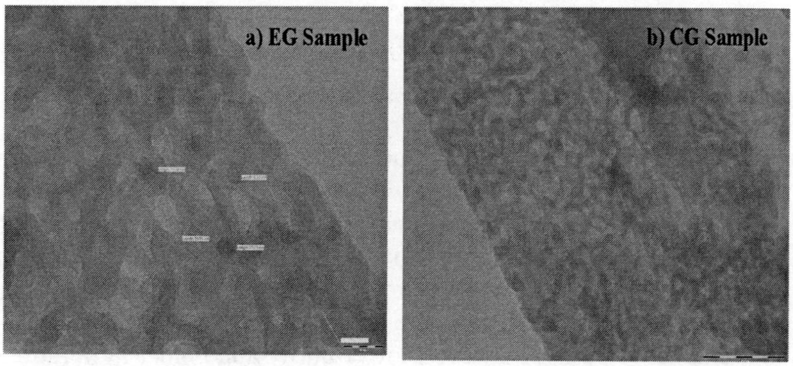

Fig. 3: TEM Images of Prepared Samples

Substructural observations from TEM images (Figure 3) explain that graphene flakes are stacked together and show a multilayered structure. It confirms that the particles are aggregated in a honeycomb structure and have a small particle size in the order of few tens of nanometers. However, the approximate particle size is measured to be 20–40 nm by considering the maximum and minimum particle sizes of the graphene.

The effect of synthesis methods is also explained by Raman spectroscopy, as shown in Figure 4. The D, G, and 2D bands are observed, respectively, around 1350, 1590 and 2700 cm^{-1} for EG sample, whereas for CG samples, the peaks appeared, respectively, around 1420, 1870 and 2732 cm^{-1}. It is noted from the above observation that the bands of CG sample are shifted toward higher wave number than those of EG sample. The increase in Raman shift is due to in-plane band structuring motion of pairs of carbon Sp2 hybridization. This is an important observation as for monolayer with Bernal stacking, computational studies show that the intensity ratio of D and G bands is less than 1, i.e., $I_D/I_G< 1$, whereas for multilayer with Bernal stacking it is $I_D/I_G>1$. The intensity ratio of D and G bands of EG and CG samples are 0.7856 and 0.8490, respectively. This indicates that EG and CG samples are composed by monolayer Bernal stacking of carbon.

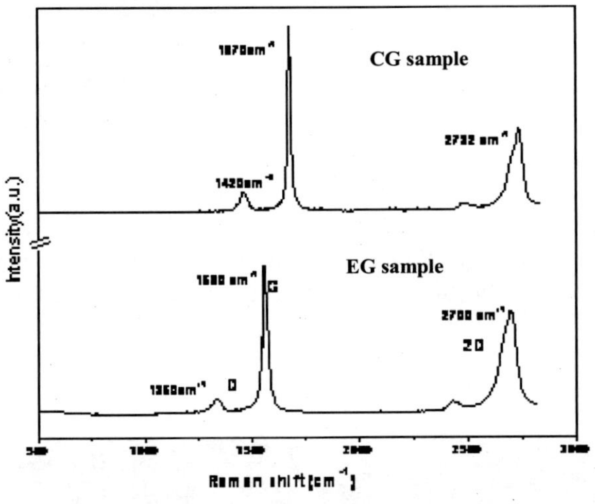

Fig. 4: Raman spectrum of prepared samples

Resistance-dependent conductance of graphene samples is shown in Figure 5. The conductivity measurements are correlated with synthesis method. A sharp change in the conductivity of the CG is observed when compared with that of the EG sample. When compared with CG samples, EG has higher conductance [14]. Further, it is noted that there is a 1000-fold variation in conductance of EG samples than CG samples. Further, measurement of the conductance is compared with that of the Raman spectra and EDX. The EG sample has lower I_D/I_G ratio and higher purity. However, lower purity and higher I_D/I_G ratio are found in CG sample. The conductivity of graphene depends on the I_D/I_G ratio and purity.

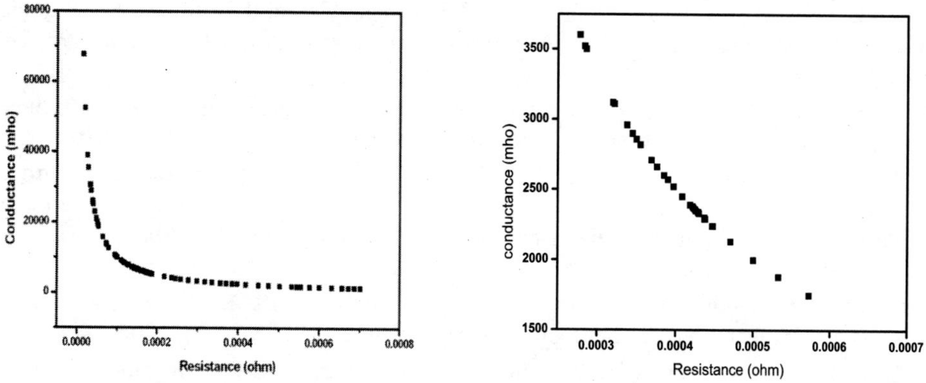

a) EG sample b) CG sample

Fig. 5: Conductance vs. Resistance of prepared samples

4 CONCLUSIONS

A multilayered graphene sheet was produce through electrochemical and chemical reduction methods. Characterization studies reveal that EG sample has high purity with small particle size. The surface morphology of graphene sheet confirms a hexagonal structure. It is concluded that the conductivity of hexagonal morphology with multilayered graphene sheets is attained by appropriate synthesis method suitable for energy storage applications in Li-ion batteries.

ACKNOWLEDGEMENT

One of authors (M. Selvam and K. Sakthipandi) is thankful to Department of Science and Technology (DST) and Council of Scientific and Industrial Research (CSIR), New Delhi for providing the Inspire fellowship to carry out the research.

REFERENCES

[1] Dong, L.X. and Chen, *Q.*, Properties, synthesis, and characterization of graphene, *Chem. Mater. Sci.,* 4, 45-54, 2010.
[2] Park, S. and Ruoff, R.S., Chemical methods for the production of graphenes, *Nat. Nanotechnol.*, 4, 217-224, 2009.
[3] Tang, L., Wang, Y., Li, Y., Feng, H., Lu, J. and Li, J., Preparation, Structure, and Electrochemical Properties of Reduced Graphene Sheet Films, *Adv. Funct. Mater.,* 19, 2782, 2009.
[4] Luo, Z., Lu, Y., Somers, L.A. and Johnson, A.T., High yield preparation of macroscopic graphene oxide membranes, *J. Am. Chem. Soc.,* 131, 898-899, 2009.
[5] Wang, F., Graetz, J., *et al.* Chemical Distribution and Bonding of Lithium in Intercalated Graphite: Identification with Optimized Electron Energy Loss Spectroscopy, *ACS Nano.*, 5, 1190, 2011.

[6] Wu, J.B., Becerril, H.A., Bao, Z.N., Liu, Z.F., Chen, Y.S. and Peumans, P., Organic solar cells with solution-processed graphene transparent electrodes, *Appl Phys Lett.,* 92, 263302–263304, 2008.

[7] Kaniyoor, A., baby, T. and Ramaprabhu, S., Graphene synthesis via hydrogen induced low temperature exfoliation of graphite oxide, *J. Mater. Chem.*, 20, 8467-8469, 2010.

[8] Li, Q., Li, Y., Zhang, L., Li, D., Fan, C. and Long, Y.T., Comparative studies on electrocatalytic activities of chemically reduced graphene oxide and electrochemically reduced graphene oxide noncovalently functionalized with poly (methylene blue). *Electroanalysis,* 22, 2862–2870, 2010.

[9] Stankovich, S., Dikin, D.A., Dommett, G.H.B., Rohlhaas, K.M., Zimney, E.J. and Stach, E.A., et al. Graphene-based composite material, *Nature*, 442 282–286, 2006.

[10] Liu, N., Luo, F., Wu, H., Liu, Y., Zhang, C. and Chen, J., One-step ionic- liquid-assisted electrochemical synthesis of ionic-liquid- functionalized graphene sheets directly from graphite, *Adv. Funct. Mater.,* 18, 1518–1525, 2008.

[11] Yoo, E.J., Kim, J., Hosono, E., Zhou, H.S., Kudo, T. and Honma, I., Large reversible Li storage of graphene nanosheet families for use in rechargeable lithium ion batteries, *Nano Latter.,* 8, 2277-82, 2008.

[12] Selvam, M. and Saminathan, K., *et al.* Synthesis and characterization of electrochemically-reduced graphene, *Bulle. Mat. Sci.,* 35, 10, 2012.

[13] Wu, J., Shen, X., Jiang, L., Wang, K. and Chen, K., Solvothermal synthesis and characterization of sandwich-like graphene/ZnO nanocomposites, *App. Surf. Sci.*, 256, 2826, 2010.

[14] Schwamb, T., Burg, B.R., Schirmar, N.C. and Poulikakos, D., An electrical method for the measurement of the thermal and electrical conductivity of reduced graphene oxide nanostructures, *Nanotechnology*, 20, 405704, 2009.

HIGH-SURFACE SILICA NANOPARTICLES ON THE PHYTOCHEMICAL RESPONSES OF MAIZE (*ZEA MAYS* L.)

R. SURIYAPRABHA, G. KARUNAKARAN, R. YUVAKKUMAR, P. PRABU, V. RAJENDRAN AND N. KANNAN[1]

Centre for Nanoscience and Technology, K.S. Rangasamy College of Technology, Tiruchengode, Tamil Nadu
[1]*Department of Biotechnology, K.S. Rangasamy College of Arts and Science, Tiruchengode, Tamil Nadu*
E-mail: veerajendran@gmail.com

ABSTRACT

The aim of this study was to explore the effect of high-surface area (360.85 m^2g^{-1}) silica nanoparticles (20–40 nm) on the phytochemical changes during maize growth in comparison with bulk silica at 15 kg ha^{-1}. The plant responses to nanosilica and bulk silica treatments were analyzed in terms of biochemical components such as total protein, proline, chlorophyll and organic compounds (gas chromatography–mass spectroscopy), and silica accumulation. The observed metabolic changes show that the expression of organic compounds such as proteins, chlorophyll and phenols favoured to maize treated with nanosilica at 15 kg ha^{-1} compared with bulk silica and control. The expression of nanosilica offers a better defense response in maize than bulk silica, which helps to improve the sustainable farming of maize crop as an alternative source of silica fertiliser.

INTRODUCTION

The properties of nanomaterials and their potential applications are ever-increasing in all fields, particularly in agricultural biotechnology. Transport of certain nanoparticles is involved in a

V. Rajendran, R. Yuvakkumar, K. Thyagarajah and K.E. Geckeler (eds.)
Applications of Nano Materials: Electronics, Energy and Environment, pp. 369–374 (2012).

potential pathway of plants for beneficial outcome. However, investigations regarding the behaviour of nanometal oxides, such as nanosilica, in plants and the mechanism of interaction, effects and agricultural applications are still in rudimentary stage.[1,2] Maize (*Zea mays* L.) ranks third in global cereal production and is considered as an essential food crop.[3] Previous studies on the effect of silica nanoparticles (SNPs) in maize reveal improved growth parameters,[4] but the effects of SNPs on maize physiological components and silica distribution in roots and shoots are essential to understand the functional properties of mineral fertilizers in detail. Determination of essential regulatory and defense compounds in maize, such as proteins and phenols, is necessary to ascertain the biotic and abiotic tolerance mechanisms adapted through silica fertilisation. For application of silica for plant growth enhancement, the availability of an economical silica source with increased functional properties is currently a demanding task. Moreover, the degree of Si solubility from siliceous materials depends on particle size and chemical compositions.

To understand the function of SNPs, it is preferable to study the phytochemical changes of maize cultivars differing in nanosilica absorption. Generally, the ameliorative effect of bulk SiO_2 sources varies genotypically among plant species [5]; hence, we focus to examine nanosilica accumulation and its effect on maize metabolic defense compounds.

EXPERIMENTAL DETAILS

The specific surface area of the prepared SNPs was determined using BET surface area analyzer (Autosorb AS-1MP; Quantachrome, USA). The physisorption analysis was carried out with N_2 adsorption–desorption measurements at liquid nitrogen temperature (−196°C). The total pore volume and average pore diameter were calculated according to the Barrett–Joyner–Halenda method.[6]

Soil Amendment with Nanosilica and Maize Culture

Experimental plots containing sandy loam soil (pH 7.0 ± 0.5) were prepared by adding nanosilica (20–40 nm) as prepared by Suriyaprabha *et al.*[7] and bulk silica (40–150 mesh) at a concentration of 15 kg ha^{-1} (hereafter termed respectively, as N15 and B15), along with control. Maize seeds (*Z. mays* L., TIP TOP) collected from Rasi Seeds Pvt Ltd (Attur, India) were sown in each plot by maintaining an appropriate temperature and humidity.

Determination of Protein, Chlorophyll and Phenol

The total protein content of 20-day-old maize leaves was determined by Bradford method, using bovine serum albumin as a standard.[8] Chlorophyll was extracted from the leaves with 80% acetone and the ratio of chlorophyll *a* to *b* was determined spectrophotometrically (U–2900; Hitachi, Japan). [9] Total phenolic compounds were extracted and estimated spectrophotometrically.[10]

Analysis of Elemental Compositions

Leaf samples were carefully collected after 20 days of maize growth and dried in a hot-air oven at 100°C for 48 h. Then, the samples were burnt in a high-temperature muffle furnace at 800°C for 4 h. The obtained ash was used for elemental analysis using X-ray fluorescence spectrometer (EDX-720; Shimadzu, Japan).

Gas Chromatography–Mass Spectroscopy

Shade-dried maize leaves (20 days old) were extracted using methanol and analysed by gas chromatography–mass spectroscopy (GC–MS; Thermo GC-Trace Ultra Ver. 5.0; Thermo Scientific MS DSQ II). The volatile compounds in the control, nanosilica-treated and bulk silica-treated samples were identified by comparing with the standards or the mass spectrum matched with the inbuilt library (Wiley 9).

Statistical Analysis

Significant differences in chlorophyll and protein contents were analysed by one-way analysis of variance. The means were compared by Duncan's and Tukey's tests at 5% level ($p < 0.05$) with the statistical software SPSS.

RESULTS AND DISCUSSION

The obtained surface area for the prepared silica particles is found to be 360.85 m^2g^{-1} (Figure 1). The total pore volume and average pore diameter are observed to be 0.8479 cc g^{-1} and 9.3 nm respectively. Maize plots amended with nano and bulk silica are assessed for the difference in phytochemical components.

Leaf samples collected from the experimental plots exhibits an increase in chlorophyll *a* and *b* content in nanosilica (Table 1) in contrast to bulk (0.012 and 0.03 μg ml^{-1}) of 20 day-old samples of maize leaves. Moreover, higher chlorophyll content is achieved in N15 (0.045) than in B15. However, the actual effect of nanometal/metal oxides on chloroplasts and phyto compounds in plants is not well ascertained.[3] The acceleration may be due to increased leaf area that renders better light absorption and photosynthetic activity of chlorophyll *a* and *b*.

Table 1: Variations in Total Chlorophyll (*a* and *b*), Protein, Phenol and Proline Contents of Maize Samples Amended with Silica Sources

Sample	*Chlorophyll a (μg ml^{-1})*	*Chlorophyll b (μg ml^{-1})*	*Chlorophyll a to b Ratio*	*Protein (mg g^{-1})*	*Phenols (μg ml^{-1})*
Control	0.017	0.010	1.647	25.28	0.364
N15	0.045	0.043	1.049	29.08	0.060
B15	0.012	0.007	1.598	26.77	0.323
$LSD*_{0.05}$	0.006	0.003	0.000	–	–

*Least significant difference at 5% level.

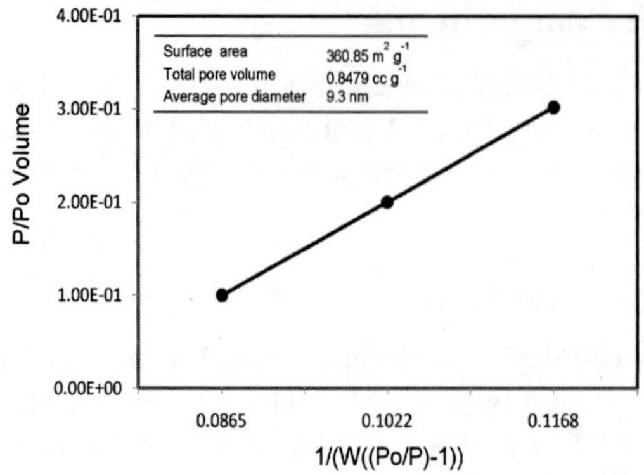

Fig. 1: BET Plot of Prepared Nano Silica

Changes in total maize biochemical components such as content of protein and total phenols of 20 days-old maize roots and leaves are presented in Table 1. Protein content is linearly accelerated in N15 (29.08 mg g^{-1}) with negligible differences in total quantity. The increase in protein content with nanosilica and bulk silica may be attributed to the presence of metabolic balance between inductions of proteins, such as cell wall transporters, and damping off stress-responsive enzyme activities as a function of nanoscale fertilization in maize. Total phenol content of leaves is reduced in N15 (0.06 μg ml^{-1}) even though it is higher in B15. Thus, the observed silicification may be substituted by production of structural phenolic compounds.[11] In addition, the observed low phenolic content in N15 (0.06 μg ml^{-1}) may reflect induction of the stress tolerance mechanism in maize.

Elemental studies on dry leaf ash of maize show a decrease in the nutritional content exerted by nano and bulk silica (Table 2). In N15, SiO_2 content is 19.18% (Table 2). Reduction of K and P contents due to increased SiO_2 deposition has been reported earlier.[12] The increase in K and P content is found to be more in nanosilica than that in bulk silica. Hence, the accumulation

Table 2: Elemental Compositions of Leaf Ash

Analyte (%)	*Control*	*N15*	*B15*
K_2O	90.35	77.39	91.35
SiO_2	4.93	19.18	4.39
Fe_2O_3	0.28	0.31	0.16
SO_3	0.22	0.31	0.13
MnO	0.13	0.22	0.10
CuO	0.05	0.06	0.03
P_2O_5	4.01	2.49	3.34
ZnO	0.03	0.04	0.02

of silica and other elements in maize shoots varies greatly. However, the extent of SiO_2 deposition in maize leaves is maximum in N15. In bulk silica, the content of silica is the same as that of the control (4.4%). This may be due to increased particle size (micron size) which is unable to use by plant. Therefore, Si uptake of maize seems to be promoted by nanosilica treatment in the soil.

The modulations in total organic compounds according to the treatments of nano and bulk silica are explained from GC–MS results. A relative abundance of volatile compounds in leaf extract with respect to the retention time is shown in chromatograph (Figure 2). Area percentage of aldehydes, ketones, alkanes, alcohols, acid-esters and ethers of the bulk and nanosilica-treated leave samples is evaluated using Wiley library 9 and the abundance is plotted in Figure 2. Stress-tolerant mechanisms mechanisms are reflected by the abundance of the phenolic compounds in the leaf extract. Total organic compounds such as phenols, aldehydes and ketones are found to be reduced in maize leaves grown in nanosilica-amended soil. The expression of such compounds is responded more in control as well as in bulk. It is noticed that other organic compound contents are not varied considerably. These results are correlated with the study on the protective mechanism of silicon in rice through phenols.[13] In this study, the responses of biochemical components to silica treatments, SNPs show an augmented accumulation of silica and regulated expression of defense compounds.

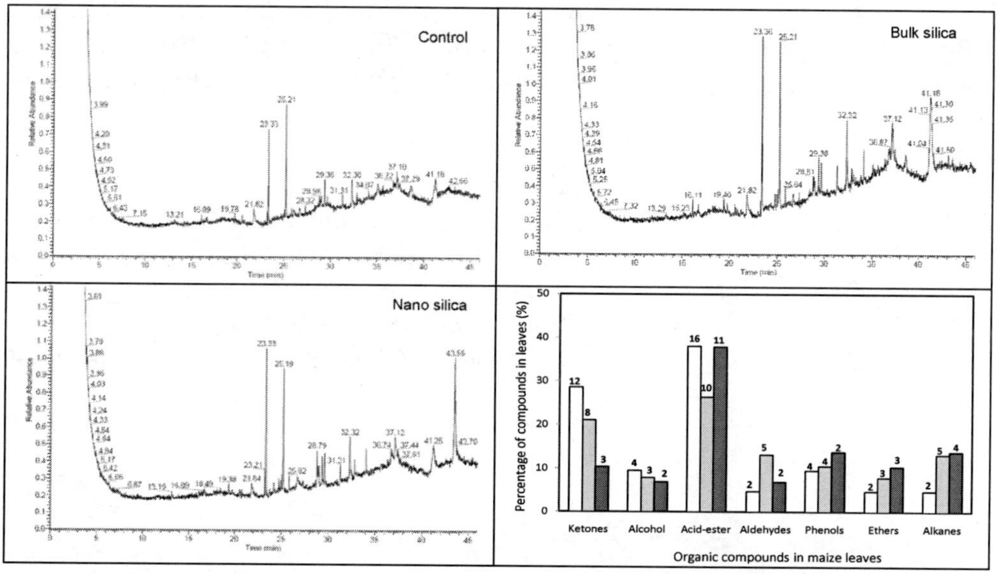

Fig. 2: GC–MS Chromatogram and Respective Area (%) of Organic Compounds in Leaf Extract

CONCLUSIONS

High-surface-area SNPs synthesised from economic method are used for maize growth. Biochemical transformations due to nanosilica fertilisation considerably enhance the expression of essential components than those due to bulk. Moreover, the regulation of phytocompounds such as proteins and phenols also favour maize growth. Hence, this study helps to choose the better silica source to enhance yield.

ACKNOWLEDGEMENTS

The authors acknowledge the financial assistance provided by Defence Research and Development Organisation, New Delhi, India to carry out this research project (ERIP/ER/0905113/M/01/1216).

REFERENCES

[1] Debnath, N., Das, S., Seth, D., Chandra, R., Ch. Bhattacharya, S. and Goswami, A., *J. Pest. Sci.* 84, 99–105, 2011.

[2] Ghormade, V., Deshpande, M.V. and Paknikar, K.M., *Biotechnol. Adv.* 29, 792–803, 2011.

[3] Mahmood, S., Hussain, A., Saeed, Z. and Athar, M., *Int. J. Environ. Sci. Technol.* 2(3), 269–274, 2005.

[4] Yuvakkumar, R., Elango, V., Rajendran, V., Kannan, N. and Prabu, P., Int. *J. Green Nanotechnol.* 3, 180–190, 2011.

[5] Kulikova, Z.L. and Lux, A., *Bull. Environ. Contam. Toxicol.* 85, 243–250, 2010.

[6] Barrett, E.P., Joyner, L.G. and Halenda, P.P., *J. Am. Chem. Soc.* 73, 373–380, 1951.

[7] Suriyaprabha, R., Karunakaran, G., Yuvakkumar, R., Prabu, P., Rajendran, V. and Kannan, N., *Curr. Nanosci.* 8, 6, 2012.

[8] Bradford, M.M., *Anal Biochem.* 72, 248–254, 1976.

[9] Arnon, D.I., *Plant Physiol.* 24, 1–15, 1949.

[10] Wakabayashi, K., Hoson, T. and Kamisaka, S., *Plant Physiol.* 113, 967–73, 1997a.

[11] Carver, T.L.W., Robbins, M.P., Thomas, B.J., Troth, K., Raistick, N. and Zeyen, R.J., *Physiol. Mol. Plant Pathol.* 52, 245–257, 1998.

[12] Miao, B.H., Han, X.G. and Zhang, W.H., *Ann. Bot.* 105, 967–973, 2010.

[13] Goto M., Ehara H., Karita S., Takabe K., Ogawa N., Yamada Y. *et al. Plant Sci.*, 164, 349–356, 2003.

TOXICOLOGICAL EFFECT OF NANO- AND BULK Al_2O_3 PARTICLES ON SOIL AND ITS NUTRIENTS

G. KARUNAKARAN, R. SURIYAPRABHA, P. MANIVASAKAN,
R. YUVAKKUMAR, V. RAJENDRAN AND N. KANNAN[1]

Centre for Nanoscience and Technology, K.S. Rangasamy College of Technology, Tiruchengode, Tamil Nadu
[1]Department of Biotechnology, K.S. Rangasamy College of Arts and Science, Tiruchengode, Tamil Nadu
E-mail: veerajendran@gmail.com

ABSTRACT

Nanometal oxides are extensively used in various industries such as medicine, cosmetics, paints, UV protection and food. Because of the increased use of nanoparticles, it is necessary to understand their environmental effect. In this study, we focused on the ecotoxicity of nano- and bulk Al_2O_3 particles in soil bacteria. In addition, the effects of nanoparticles and bulk particles on soil and its nutrients (nitrogen, phosphorus and potassium) were also analysed using soil analysis techniques. This study shows that Al_2O_3 nanoparticles are toxic to soil. When the microbial population decreases in the soil, the available form of nutrients also decreased. Thus, precaution should be taken for the use and disposal of toxic nanomaterials to prevent their harmful effects on the soil.

INTRODUCTION

The benefits of nanotechnology are being seen in all fields of science and technology, including defence, electronics and biomedical applications. Alumina (Al_2O_3) is used in many applications such as medical devices, cell labelling, biosensors, cancer therapy and drug delivery systems [1]. Owing to its high mechanical strength and less toxicity, Al_2O_3 is used as a biomaterial, for prostheses and medical implant applications. The knowledge about the toxicity of nanoparticles is essential for their effective application in the biomedical field.

V. Rajendran, R. Yuvakkumar, K. Thyagarajah and K.E. Geckeler (eds.)
Applications of Nano Materials: Electronics, Energy and Environment, pp. 375–378 (2012).

The ecotoxicity of metal oxide nanoparticles has been investigated in *Daphnia similis* [2], soil earthworm *Eisenia fetida* [3] and bacteria [4]. The purpose of this study was to examine the effect of nano- and bulk Al_2O_3 particles on soil bacteria. In addition, the effects of nano- and bulk particles on soil properties and their influence on nutrient are revealed with a view to understand the nature of toxicity.

EXPERIMENTAL DETAILS

Analytic reagent (AR) alumina (Al_2O_3; 99%) was procured from Loba Chemie and nanoalumina was produced from raw bauxite by mineral acid precipitants (HNO_3) [5]. The nano-Al_2O_3 was produced from raw bauxite by adding sodium hydroxide (5 N) for aluminium hydroxide (Bayer liquor) formation followed by mineral acid precipitant (6 M HNO_3) at pH 7 to obtain the precipitate. The white precipitate of aluminium hydroxide was obtained and kept at 353 K for 12 h followed by incubation at 393 K for 1 h. The obtained white precipitate was further calcined at 773 K for 6 h. The white powder obtained after calcination was characterised by employing different techniques and used without any further purification. Stock solutions of the above-mentioned metal oxide particles at a concentration of 1 mg ml^{-1} were dispersed in ultrapure water by sonication to obtain a uniform colloidal suspension. Soil samples were collected from agricultural land at Tiruchengode, India. The upper layer of soil (depth: 0–15 cm) was air-dried and sieved using a square-hole sieve of 2 mm mesh. Air-dried soil (100 g) was taken in each of the 20 plastic bags, followed by addition of 100 ml sterile distilled water. Nano- and bulk metal oxides of Al_2O_3 (concentrations 0.1, 10, 100 and 1000 mg kg^{-1}) were added to all bags except the control bags and were then incubated for 30 days at 298 K. 10^{-5} dilution was taken and spread on nutrient agar plates in triplicates and the plates were incubated at 300 K for 24 h, followed by counting the colonies using colony counter. The results were compared with those of the control. In addition, the treated and untreated soil samples on the first day and after 30th day were analysed further for the soil nutrients such as total nitrogen (N) [6], phosphorus (P) [7] and potassium (K) [8].

RESULTS AND DISCUSSION

The cubic nature of the synthesised nano-Al_2O_3 powder is confirmed by X-ray diffraction pattern. The Fourier transform infrared band observed at 1100 cm^{-1} corresponds to the Al–OH vibration. The peak observed at 1633 cm^{-1} corresponds to the stretching mode of –OH and H_2O groups. At 1410 cm^{-1}, the bending vibration of C–C group is observed. The band observed at 502 cm^{-1} corresponds to the bending and stretching vibrations of Al–O bond. Transmission electron microscopic image of the prepared powder reveals the particles to be spherical. Particle size distribution confirms the average size of the powder to be 58 nm. Al_2O_3 nanoparticles show average particle size (d50) of 68 nm in saline. The hydrophobic potential in terms of contact angle of the particles shows 35.67 Å (nano-Al_2O_3) and 17.37 Å (bulk Al_2O_3). The characteristics of Al_2O_3 nanoparticles are shown in Table 1.

Table 1 Characteristics of Nano- and Bulk Metal Oxide Particles

Sample	*Crystalline Phase*	*Purity (%)*	*Particle Size Distribution (mean size)*		*BET*	*Reference*
			Water (nm)	*Saline (nm)*	*(m^2g^{-1})*	
Nano-Al_2O_3	Cubic	≥99	58	68	190	Manivasakan *et al.*[5]
Bulk Al_2O_3	–	≥99	–		–	Merck (1344-28-1)

Nanoalumina (Figure 1) incorporated soil shows a decrease in soil bacterial population from 2.8 × 105 to 1 × 105 CFU at a concentration of 0.1 and 1000 mg kg^{-1}. Similarly, bulk Al2O3-incorporated soil shows a sharp decrease in soil bacterial population from 3.6 × 105 to 3.1 × 105 CFU $g^{-1}$of soil at a concentration of 0.1 and 1000 mg kg^{-1}. The incorporation of nano-Al2O3 on the soil reveals that the total organic carbon content is increased from 1.89 to 1.9 g kg^{-1}, nitrogen from 0.13 to 0.12 g kg^{-1}, phosphorus from 7.4 to 6.5 mg kg^{-1} and potassium from 506 to 498 kg ha^{-1} when compared with control. Bulk Al_2O_3 shows that the total organic carbon content was increased from 2.1 to 2.45 g kg^{-1}, nitrogen from 0.12 to 0.18 g kg^{-1}, phosphorus from 7.5 to 9.8 mg kg^{-1} and potassium from 503 to 597 kg ha^{-1} when compared with control.

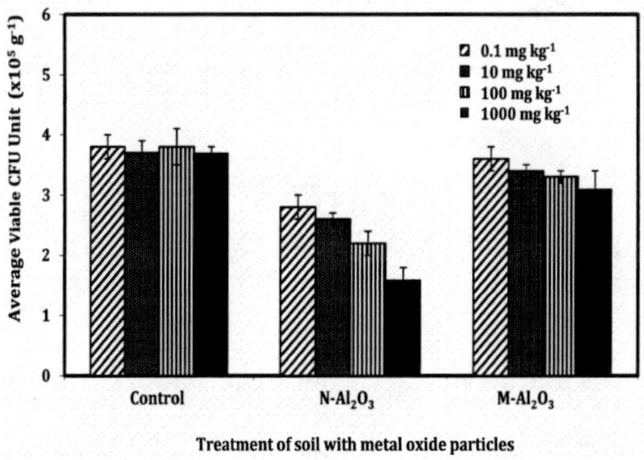

Fig. 1: Average Viable Soil Bacterial CFU with Respect to Metal Oxide Particles at Different Concentrations

Al_2O_3 nanoparticles show highest toxicity when compared with other treated particles. It is evident from the earlier study [9] that nano-Al_2O_3 is less toxic and is not toxic to soil directly. However, it requires other metals such as arsenic (As) to stimulate toxicity different from our observations. The proposed toxicity of metal nanoparticles in soil population is investigated, and it has been concluded that the impact of metal nanoparticles on bacteria is measurable [10].

The result of soil nutrient analysis (Table 2) reveals the effect of Al_2O_3 nanoparticles and bulk particles on available nitrogen, phosphorus and potassium (NPK) levels in soil. The variation in available nutrient after the addition of nanoparticles and bulk particles may be due to their effect on total soil microbial population. It is already proven that microbes such as Azotobacter, Bacillus megaterium and B. mucilaginous are more effective in enhancing the soil properties and its nutrient by improving the soil NPK values [11]. When nanoparticles having antimicrobial activity are released into the soil, nutrient value of the soil decreases. The results of this investigation revealed the importance of particle size and the correlation between soil microbial populations and soil nutrient values. Therefore, it is important to consider the soil and its microbial population health before using the metal oxide nanoparticles into the environment.

Table 2 Nano- and Bulk Al_2O_3-Incorporated Soil Analysis (1000 mg kg^{-1})

Available nutrients	*Control*		*Nano-Al_2O_3*		*Bulk Al_2O_3*	
	0th day	*30th day*	*0th day*	*30th day*	*0th day*	*30th day*
N (g kg^{-1})	0.13 ± 0.03	0.17 ± 0.05	0.13 ± 0.05	0.12 ± 0.03	0.12 ± 0.02	0.18 ± 0.05
P (mg kg^{-1})	7.5 ± 0.04	9.5 ± 0.02	7.4 ± 0.03	6.5 ± 0.05	7.5 ± 0.03	9.8 ± 0.02
K (kg ha^{-1})	502 ± 0.02	598 ± 0.03	506 ± 0.04	498 ± 0.06	503 ± 0.03	597 ± 0.02

CONCLUSIONS

This study confirms the higher toxicity of Al_2O_3 nanoparticles as compared to their bulk counterparts. Sensitivity of bacteria depends on the nature and size of the materials. Also, nano- and bulk Al_2O_3 particles affect the soil nutrient value by influencing bacterial growth. The results of this study signify the effect of nano- and bulk-sized metal oxides on the soil.

ACKNOWLEDGEMENTS

The authors acknowledge the financial support (ERIP/ER/0905113/M/01/1216) provided by the Defence Research and Development Organisation, New Delhi, India to carry out this research.

REFERENCES

[1] Lanone S., Rogerieux F., Geys I., Dupont A., Maillot-Marecha E., Boczkowski J., et al., Part. Fibre Toxicol. 6, 14, 2009.
[2] Marcone G.P.S., Oliveira A.C., Almeida G., Umbuzeiro G.A., Jardim, W.F., J. Hazard. Mater. 211-212, 436–442, 2012.
[3] Hu C.W.M., Li, Cui Y.B., Lia D.S., Chen J, Yang L.Y., Soil Biol. Biochem. 42, 586–591, 2010.
[4] Jiang W., Mashayekhi H., Xing B., Environ. Pollut. 157, 1619–1625, 2009.
[5] Manivasakan P., Rajendran V., Rauta P.R., Sahu B.B., Panda B.K., Powder Technol. 211, 77–84, 2011.
[6] Bremner J.M., Nitrogen-total. In: Sparks, D.L. (Ed.), SSA Book Series No. 5, ASA and SSSA, Madison, WI, 1085–1121, 1996.
[7] Frank K., Beegle D., Denning J., 1998. Phosphorus. In: Brown, J.R. (Ed.), Recommended Chemical Soil Test Procedures for the North Central Region, North Central Regional Research Publication No. 221 (revised), Missouri Agric. Exp. Stn., Columbia, MO, 21–26.
[8] Stanford P., English W.L., Use of frame photometer in rapid soil tests for potassium and calcium. Agron. J. 41, 446–447, 1949.
[9] Doshi R., Braida W., Christodoulatos C., Waznea M., O'Connorb G., Environ. Res. 106, 296–303, 2008.
[10] Shah V., Belozerova I., Water Air Soil Pollut. 197, 143–148, 2009.
[11] Wu S.C., Cao Z.H., Li Z.G., Cheung K.C., Wong M.H., Geoderma, 125, 155–166. 2005.

NANO MAGNESIUM ALUMINATE ($MgAl_2O_4$) AS A THICK FILM LAYER FOR GAS SENSOR APPLICATIONS

N. NITHYAVATHY, K. MEGALA, S. ARUNMETHA[1] AND V. RAJENDRAN[1]

Department of Mechatronics, Kongu Engineering College, Erode, Tamil Nadu
[1]Centre for Nano Science and Technology, K.S.R. College of Technology, Tiruchengode, Tamil Nadu
E-mail: nithyavathy@kongu.ac.in; meguk89@gmail.com, sarunmetha@gmail.com; veerajendran@gmail.com

ABSTRACT

Nanotechnology has created many new nanocomposite materials for gas sensor systems. Initially, the preparation of $MgAl_2O_4$ spinel is made with combination of MgO and Al_2O_4 with a polymer polyethylene glycol for gas sensing application. This was done by sol gel technique and then to a sintering of 673 K is made. $MgAl_2O_4$ (2:1) of 0.1N is added with PEG to from magnesium aluminate fluid and then coated on a glass substrate to form a thick film coating. The coating of nanosensor layer is obtained from magnesium aluminate using spin coating technique. This semiconducting metal oxide nanomaterial has a high surface to volume ratio, offers a new method of gas sensing in sensor technology. The material property analyses are done through SEM/EDX analysis and UV spectroscopic variations. The variation in the conductivity of the coated nanocomposite material is measured through four probe tests and the results are characterized.

INTRODUCTION

The promise driving nanocomposites over more traditional composites lies in taking advantage of the new and often exciting properties afforded by nanostructures. Thus, a significant amount of research is being targeted in understanding the properties for both the individual and composite nanomaterials. Multi metaloxide nanocomposite materials give response to a number of gases. A detail study is made for analysing the sensitivity, selectivity and stability of the composite material used for gas sensing. In general nanocomposites can be broadly defined as

V. Rajendran, R. Yuvakkumar, K. Thyagarajah and K.E. Geckeler (eds.)
Applications of Nano Materials: Electronics, Energy and Environment, pp. 379–384 (2012).

a mix of dissimilar components that can be zero-dimensional, one dimensional, two dimensional or three dimensional that are mixed at the nanoscale. $MgAl_2O_4$ spinel is normally used for humidity sensing, refractory application and microwave and millimetre wave applications. It has a wide use of gas sensing also due to its absorption properties because of rich surface reactivities. The sensitivity of $MgAl_2O_4$ to O_2 and H_2 is increased by the addition of polymer polyethylene glycol. The sensor mechanism is developed to sense range of gases and it is characterised by spectral analysis.

The magnesium aluminate spinel powder reveals exotic properties like high melting point (2,135° C), good mechanical strength, high chemical inertness, low dielectric constant and good optical properties. Among the many inorganic solids, magnesium aluminate spinel ($MgAl_2O_4$) system has long been considered as an important ceramic material due to its superior mechanical, chemical and thermal properties. The synthesis of $MgAl_2O_4$ spinel powder requires higher temperature and then longer reaction time. However, the synthesised powder suffers from agglomeration. The proper distribution of various metals ions in co-precipitation method results into stoichiometric and smaller particle sizes. Nanocomposites material involving organic or inorganic materials is a rapidly rising area of research and application development.

EXPERIMENT

Materials

Analytical reagent of magnesium oxide [MgO] from Himedia, Polyethylene glycol [PEG 6000] from RFCL, Ethanol from Jiangsu Huaxi were obtained and used in the present study. All reagents were used for the preparing the nanocomposites without any further purification. The stirring process was carried out by a magnetic stirrer. The proportionate mixture of MgO, Al_2O_4 (1:1 and 2:1) and polyethylene glycol (0.1 N) was prepared by solgel process in double deionised water by applying an ultrasonicator. The mixture was sintered for 673 K. The obtained powder was then ball milled to obtain the particle size of approximately 50 nm. The nanocomposite $MgAl_2O_4$ powder was dispersed in dispersal medium and thus, spin-coated on a glass plate of 6 mm × 9 mm × 0.5 mm thickness.

Nano Powder

The nanocomposite $MgAl_2O_4$ powder is mixed with potassium bromide binding material and pellets were prepared. The pellets were sintered at 673 K for 6 h. The starting precursor mixture of MgO and Al_2O_4 were mixed with the aqueous solution undergoes constant stirring using an ultrasonicator. It is then ride to a powder mixture for 12 h at a temperature of 394 K for 20 mins. The powder was sintered at 473 K and then calcinated at 673 K. The obtained powder is ball milled to attain a uniform material structure.

The precursor of sol gel techniques was calcinated at low temperature compared with other form of powder preparation. Characterisation results show that an increase of calcination temperature will increase the sintering ability and hardness of $MgAl_2O_4$ spinel. The anhydrous magnesium oxide and aluminium oxide with molar ratio 2:1 was thus prepared. The magnesium aluminate spinel powders without calcination and with calcination were formed.

The chemical reactions take place for synthesis of magnesium aluminate nanopowders as:

$$MgO + Al_2O_3 \longrightarrow MgAl_2O_4 \qquad \text{(i)}$$

The process of magnesium aluminate spinel powders by using anhydrous magnesium oxide MgO and anhydrous aluminium oxide Al_2O_4 at Mg:Al mole ratio 2:1 was dissolved in de-ionized water at room temperature using hot plate magnetic stirrer with this PEG is added with 0.1 N. The thick film structure of the compounds were fabricated over a glass substrate (6 mm × 9 mm × 0.5 mm in thickness), followed by drying at 353 K for 6 h. Then spin-coatings precipitate was evaporated and then coated on a glass substrate. The thick film was dried at 673 K for 10 min each time before proceeding for next coatings.

Glass Substrate

The nanocomposite powder of $MgAl_2O_4$ was mixed with ethylene as a dispersal medium and then coated on a glass plate in a designed format as given below. The coating was done through techniques like spin coating and dip-coating techniques. The coated samples of $MgAl_2O_4$ were placed in the chamber, where gas is to be sensed and is allowed to stay through 3 mins. The content of H_2 and O_2 are detected based on the conductivity results before and after passing the gases using four probe tests.

RESULTS AND DISCUSSION

The distribution of grain size depends on the range of the specimen pressed. The measured stoichiometry ratio of $MgAl_2O_4$ spinels behaves in the same manner as calculated values of 17.09 and 37.83 weight% as shown in the Figures 4 and 5. The results of SEM/EDX patterns prove that the detected peaks of $MgAl_2O_4$ matches with the standard pattern. The coated $MgAl_2O_4$ on a glass substrate using a desired pattern shows the nature of sensing the gases like H_2 and O_2 based on the acceptor or donor electron gas, the conductivity either decrease or increase. The conductivity also changes for the change in gas concentration and surface gas reaction.

The coated $MgAl_2O_4$ metal oxide based gas sensor has a very high performance on gas sensing. The electrodes are connected to the glass substrate and operated at a temperature of 573 K by a temperature controller and made to a stabilisation of 1 h. H_2 and O_2 gases are maintained at a 1 bar pressure and passed to the chamber with flow rate of 10 m^3/min. The output electrode is connected to a measurement device to measure the variation in the conductivity of the gas sensor. SEM image is taken to absorb the absorption rate of each of the gas passed. Figures 4 and 5 reveals that as the temperature increases the breaking of complex material takes place.

Conductivity Test by Four Probe Method

Conductivity test is carried to determine the conducting property of the nanocomposites materials, by passing electric voltage of 0–12 V across the four probes attached to both ends of the substrate materials. The difference in input voltage to change in output voltage is measured. The conductivity test made for different samples namely aluminium oxide, magnesium oxide, composite of aluminium oxide and magnesium oxide and the graphs are shown in Figures 1, 2 and 3 as a plot of voltage versus conductivity. The values of conductance increase at a constant phase and remain constant after a certain input voltage.

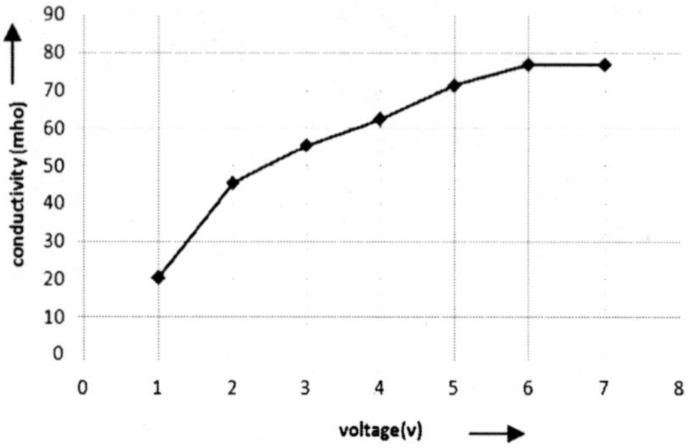

Fig. 1: Conductivity of Nanocomposite $MgAl_2O_4$ Obtained from Precursors Mg $(NO_3)_2$ and Al $(NO_3)_2$

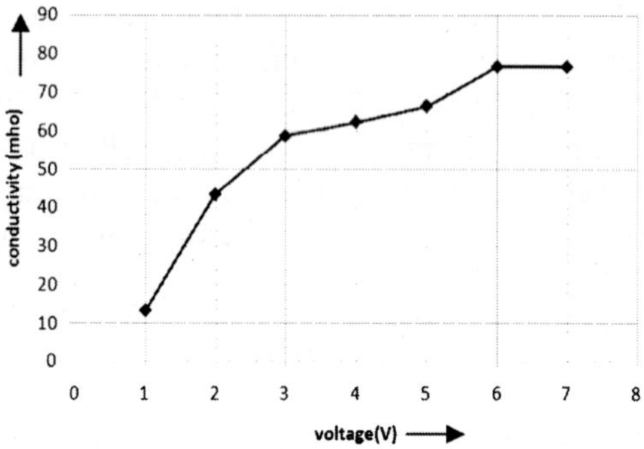

Fig. 2: Conductivity of Nanocomposite $MgAl_2O_4$ from MgO, Al_2O_3 (2:1) and PEG before Calcinations

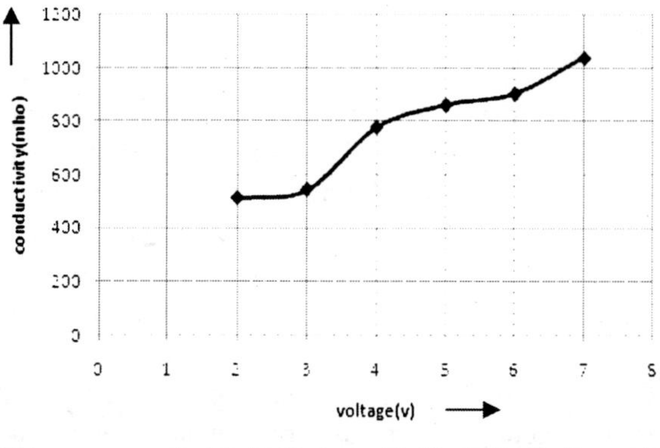

Fig. 3: Conductivity of Nanocomposite $MgAl_2O_4$ from MgO, Al_2O_3 (2:1) and PEG after Calcinations

It is necessary to place these compounds in a desiccated environment when these analyses are performed. The above presentation of the compounds helps from absorbing atmospheric air which may result in different physical and chemical properties of the compounds coated over the glass substrate. Similarly, Figure 2 shows the conductivity nature of magnesium oxide combined with aluminium oxide and Figure 3 indicates the conductivity of magnesium oxide combined with aluminium oxide. Figures 2 and 3 show an increase in conductivity value with respect to the input changes. It is also evident that both these compounds exhibit similar values after certain increase in input values and both these process of composite mixture reveal similar conductivity phenomena. Thus, it seems that $MgAl_2O_4$ nanocomposites elements show increased range of conductivity in Figure 3.

SEM image shows the surface morphology of magnesium oxide (MgO) and aluminium oxide (Al_2O_4) powders. The ratio of 2:1 of magnesium aluminate ($MgAl_2O_4$) SEM analysis shows the data before calcination and after calcinations. Figure 4 shows that the $MgAl_2O_4$ samples before calcination shows fewer rods and peeks when compared to the SEM of $MgAl_2O_4$ samples after calcination. Generally, an increase in temperature treatment due to the sintering process increases the crystallite sizes of the powders. Therefore, it can be concluded that the synthesized spinel powders are exposed to higher temperatures than conventional heating prepared powders. The SEM image (Figure 6) of the calcinated and synthesised sample reveals small rod like shape with round discs scattered over to prove that Mg powders were dispersed in aluminate powders with a homogenous morphology.

Fig. 4: SEM Micrographs of the $MgAl_2O_4$ before Calcination

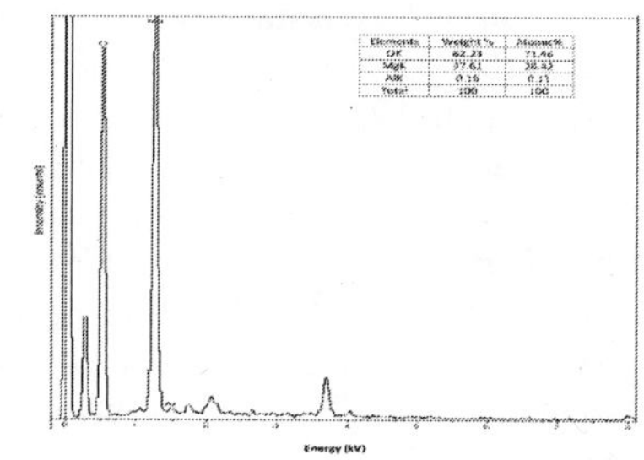

Fig. 5: SEM/EDX before Calcination of $MgAl_2O_4$

Fig. 6: SEM Micrographs of the $MgAl_2O_4$ after Calcination at 573 k

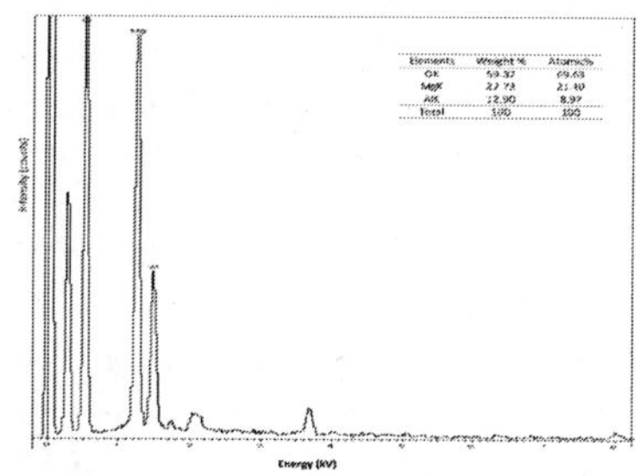

Fig. 7: SEM/EDX after Calcination of $MgAl_2O_4$

CONCLUSION

The metal oxide nanocomposite $MgAl_2O_4$ gas sensors have a high gas response and surface reaction. The additions of polyethylene glycol to $MgAl_2O_4$ precursor form a polymer chain by increasing the sensitivity. The high sensitivity gas sensor paves way for the detection of highly toxic gases monitoring in real-time using conductivity level output. Thus, the nanosize particles of the $MgAl_2O_4$ spinel powder are used to perform gas sensing applications. The characterization results shows that increasing of calcination temperature increases the sintering ability and hardness of $MgAl_2O_4$

REFERENCES

[1] Rashad, M.M. El-Shall, H., 'A Novel approach for synthesis of nanocrystalline $MgAl_2O_4$ powders by co-precipitation method. *J Mater Sci* 2009; Vol. 44, pp. 2992–2998.

[2] Zhen-Xing, Tang, 'Preparation of nano-MgO using ultrasonic method and its characteristics'. ECLETICA 2008, Vol. 33, pp. 15–20.

[3] Wu, Chen Chou, Cao, Xiaoan, Wen, Qiang, Wang, Zehua, Gao, Qianqian and Zhu, Huichang (2009). 'A vinyl acetate sensor based on cataluminescence on MgO nanoparticles', *Talanta,* Vol. 79, pp. 1223–122.

[4] Graf, M., Gurlo, A., Barsan, N., Weimar, U. and Hierlemann, A., 'Microfabricated gas sensor systems with sensitive nanocrystalline metal-oxide films', *Journal of Nanoparticle Research* 2005, Vol. 8, pp. 823–839.

[5] Esmaeili, Elaheh, Khodadadi, Abasali and Mortazavi, Yadollah (2009). 'Microwave-induced combustion process variables for MgO nanoparticle synthesis using polyethylene glycol and sorbital', *Journal of the European Ceramic Society,* Vol. 29, pp. 1061–1068.

[6] Lee, Jong-Heun (2009). 'Gas sensors using hierarchical and hollow oxide nanostructures: Overview', *Sensors and Actuators B*, Vol. 140, pp. 319–336.

[7] Ganesh, Ibram (2008). 'Fabrication of Magnesium aluminate ($MgAl_2O_4$) spinel foams' Elsevier, Vol. 37, pp. 2237–2245.

[8] Feng, Tong Jian and Ming, Chen Da (2012). 'A Novel way to synthesis Magnesium aluminate spinel($MgAl_2O_4$) powders by polymer gel reaction', *Materials Science/Engineering,* Vol. 512, pp. 600–603.

[9] Rooi Ping, Lim, Azad, Abdul-Majeed and Dung, Teng Wan (2001). 'Magnesium aluminate ($MgAl_2O_4$) spinel produced via self-heat-sustained technique', *Pergamon,* Vol. 36, pp. 1417–1430.

[10] Rashad, M.M., Zaki, Z.I. and H. El, Shall (2009). 'A novel approach for synthesis of nanocrystalline $mgal_2O_4$ powders by co-precipitation method', Springer Science + Business Media, LLC.

[11] Shah, A. Tokeer, 'Principles of Nanoscience and Nanotechnology', *Naroosa Publishing House,* New Delhi, 2010.

[12] Fu, Xiaorong, Song, Zhitang, Wu, Guangming, Huang, Jipo, Duo, Xinzhong and Lin, Chenglu (1999). 'Preparation and Characterization of MgO Thin Films by a Novel Sol-Gel Method', *Journal of Sol-Gel Science and Technology*, Vol. 16, pp. 277–281.

[13] Tao, Ying, Cao, Xiaoan, Peng, Yan and Liu, Yonghui (2010). 'A novel catluminescence gas sensor based on MgO thin film', *Sensors and Actuators,* Vol. 148, pp. 292–297.

[14] Hyoun-Eekim, Zinkle, Steven J. and Allen, William R. (1991). 'Preparation of Magnesium aluminate spinel ($MgAl_2O_4$) containing controlled amount of ^{17}O isotope', *J. American Sci*, Vol. 74, pp. 1442–44.

[15] Ali Saberi, Farhad Golestani-Fard, Hosein Sarpoolaky, MonikaWillert-porada Thorsten Gerdes and Simon, REinhard (2008). 'Chemical synthesis of nanocrystalline Magnesium aluminate spinel ($MgAl_2O_4$)', Vol. 74, pp. 142–146.

[16] Borhania, Bahareh and Mohsen-Niaa, Mohsen (2012). 'Preparation and characterization of polypropylene/$MgAl_2O_4$.MgO nanocomposite'. TNT 2012, Madrid, Spain.

AUTHOR INDEX